FUTURE INFORMATION ENGINEERING AND MANUFACTURING SCIENCE

International Research Association of Information and Computer Science

IRAICS Proceedings Series

ISSN: 2334-0495

VOLUME 2

PROCEEDINGS OF THE 2014 INTERNATIONAL CONFERENCE ON FUTURE INFORMATION ENGINEERING AND MANUFACTURING SCIENCE (FIEMS 2014), BEIJING, CHINA, 26–27 JUNE 2014

Future Information Engineering and Manufacturing Science

Editor

Dawei Zheng

International Research Association of Information and Computer Science, Beijing, China

CRC Press
Taylor & Francis Group
Boca Raton   London   New York   Leiden

CRC Press is an imprint of the
Taylor & Francis Group, an **informa** business

A BALKEMA BOOK

CRC Press/Balkema is an imprint of the Taylor & Francis Group, an informa business

© 2015 Taylor & Francis Group, London, UK

Typeset by V Publishing Solutions Pvt Ltd., Chennai, India
Printed and bound in Great Britain by CPI Group (UK) Ltd, Croydon, CR0 4YY

Published by: CRC Press/Balkema
　　　　　　 P.O. Box 11320, 2301 EH Leiden, The Netherlands
　　　　　　 e-mail: Pub.NL@taylorandfrancis.com
　　　　　　 www.crcpress.com – www.taylorandfrancis.com

ISBN: 978-1-138-02644-5 (Hbk)
ISBN: 978-1-315-76075-9 (eBook PDF)

Table of contents

Preface

We cordially invite you to attend the conference in Beijing, China during June 26–27, 2014. The main objective of the conference was to provide a platform for researchers, engineers, academicians as well as industry professionals from all over the world to present their research results and development activities in Information Engineering and Manufacturing Science. This conference provided opportunities for the delegates to exchange new ideas and experiences face to face, to establish business or research relations and to find global partners for future collaboration.

The conference received over 240 submissions which were all reviewed by at least two reviewers. As a result of our highly selective review process about 87 papers have been retained for inclusion in the proceedings, less than 40% of the submitted papers. The program of the conference consisted of invited sessions, and technical workshops and discussions covering a wide range of topics. This rich program provided all attendees with the opportunity to meet and interact with one another. We hope your experience was a fruitful and long lasting one. With your support and participation, the conference will continue to be successful for a long time.

The conference is supported by many universities and research institutes. Many professors play an important role in the successful holding of the conference, so we would like to take this opportunity to express our sincere gratitude and highest respects to them. They have worked very hard in reviewing papers and making valuable suggestions for the authors to improve their work. We also would like to express our gratitude to the external reviewers, for providing extra help in the review process, and to the authors for contributing their research results to the conference. Special thanks go to our publisher. At the same time, we also express our sincere thanks for the understanding and support of every author. Owing to time constraints, imperfection is inevitable, and any constructive criticism is welcome.

We hope you had a technically rewarding experience, and used this occasion to meet old friends and make many new ones.

We wish all attendees an enjoyable scientific gathering in Beijing, China. We look forward to seeing all of you at next year's conference.

The Conference Organizing Committee
June 26–27, 2014
Beijing, China

Organizing committees

GENERAL CHAIRS

Prof. M.N.B. Mansor, *University Malaysia Perlis, Malaysia*
Prof. E. Ariwa, *London Metropolitan University, UK*

TECHNICAL COMMITTEE

Prof. M.N.B. Mansor, *University Malaysia Perlis, Malaysia*
Prof. A.M. Leman, *Universiti Tun Hussein Onn Malaysia, Malaysia*
Dr. F.J. Shang, *Chongqing University of Posts and Telecommunications, China*
Prof. A. Srinivasulu, *Vignan University, India*
Prof. M.D.H, *Mangalore University, India*
Prof. C. Kumar, *Siddhant College of Engineering, India*
Prof. R. Latif, *Ibn Zohr University, Morocco*
Dr. L.T. JayPrakash, *International Institute of Information Technology-Bangalore (IIIT-B), India*
Prof. S.H. Ali, *R.M.K. Engineering College, India*
Dr. K. Arshad, *University of Greenwich, School of Engineering, UK*
Prof. M.F.S. Ferreira, *University of Aveiro, Portugal*
Dr. B.S. Ahmed, *Salahaddin University, Kurdistan*
M.M. Siddiqui, *Integral University, India*
P.C.C. Anyachukwu, *University of Nigeria, Nigeria*

Future Information Engineering and Manufacturing Science – Zheng (Ed)
© 2015 Taylor & Francis Group, London, ISBN 978-1-138-02644-5

Effect evaluation of interactive teaching method in pediatric dentistry

L.M. Wei, Y.H. Pan, X.M. Xu & J.F. Wang
School and Hospital of Stomatology, Wenzhou Medical University, Wenzhou, China

ABSTRACT: To evaluate the effect of interactive teaching method in pediatric dentistry. Forty-six students were divided into two groups by completely randomized way. Control group were taught by traditional teaching method and experimental group were taught by interactive teaching method. Knowledge lectures and cases discussion in interactive teaching method were distributed as ratio of 3:1 in class time, and network interactive teaching was carried out after class. Exam content included one-choice question (1 point × 100 questions) and case-analysis question (10 points × 10 questions). Scores were statistically analyzed. Average scores of experimental group were 80.26 ± 6.61 and 75.67 ± 7.04, which were higher than that of control group (72.88 ± 5.73 and 69.89 ± 5.26). There was a significant difference in case-analysis between two groups of learning difficulty ($P < 0.05$). Interactive teaching method is more efficient than traditional teaching method in learning motivation, ability of independent thinking and test score.

Keywords: Pediatric dentistry; Interactive teaching method; Case discussion

1   INTRODUCTION

During pediatric dentistry teaching, the key is that the memory and exercise of basic knowledge are strengthen. At present, domestic pediatric dentistry course mainly adopts Traditional Teaching Method (TTM), in which the teachers are the subject who simply focus on imparting knowledge, and the students are the object who only accept knowledge passively. TTM often results in that student lack the ability of independent thinking and using knowledge to design treatment plan flexibly. The teaching process should include two mutual aspects, that is, teaching and learning (Zhou et al. 2011). The advantage of Interactive Teaching Method (ITM) is that teachers and the students can participate in the learning process collectively. The aim of this study is to evaluate the effect difference between of two kinds of method, and to improve pediatric dentistry teaching level.

2   MATERIALS AND METHODS

2.1   *The research object*

Forty-six dental students of grade 2010 studying in Wenzhou Medical University were included in this study. Using completely randomized way, all of the students were divided into two groups equally, that is, experimental group (ITM group) and control group (TTM group). Both teachers and students did not be told the anticipated effect of two methods. The students from two groups participated in pediatric dentistry course simultaneously, and were taught by a same teacher, who has passed through the training of case-analysis teaching before the start of the semester.

2.2   *Teaching method*

Knowledge of every unit was taught using PPT and blackboard in control group, the relevant homework and learning task was arranged after class too. Each teaching unit is divided into two parts in experimental group.

The first part was completed in the classroom and focused on the content of textbooks. The specific allocated-hours proportion of basic-knowledge teaching and interactive teaching was 3:1, that is, in a 4-hours lesson, basic-knowledge teaching was allocated to 3 hours, interactive teaching 1 hour. After completed all above, all students of the experimental group was divided into the 4 teams. Two cases, which were closely related to the main knowledge points of the course, were illustrated through the PPT. Free discussion about two cases was organized by teacher among members of each team, the final conclusion was obtained and recorded within a limited-time of 20 min. Conclusions were reported and discussed by each team, every student can ask questions related to the case discussion and based on the knowledge earned from this lesson, within limited-time of 3–5 min. Finally, simply analysis, answer, review and summary of the case were done by the teacher.

The second part was completed extracurricularly. First, the network learning platform was established. Second, teachers and students log in and participate in learning activities. The network learning platform included three sections: (1) Homework section, curriculum-related questions or cases were presented by the teacher and student could answer and discuss freely. (2) Explanation section, student could ask any question encountered in the process of learning, then, teacher and other students could explain freely and in time. (3) Instructional videos section, curriculum-related videos were uploaded to the network by teacher. Videos could help student to consolidate the knowledge learned in the classroom, and expand knowledge to clinical practice. Students were required to write what they learned, discuss collectively after they watched videos.

2.3 Teaching effect evaluation method

At the end of the course, to compare the teaching effect of two groups, a question-examination was designed that included: (1) one-choice question. There were 100 questions and 1 point per question. It focused on inspecting the student's memory ability of basic knowledge. (2) case-analysis question. There were 10 questions, and 10 point per question. It focused on inspecting the student's ability of combining theoretical knowledge and clinical practice. Full scores of examination paper was 200 point, within a completion-time of 120 min. The paper was graded by a same teacher with blinding to ensure reliability of the examination results.

After the test, the questionnaires of learning difficulty about the course were distributed to all the students. Give three kinds of different degree of choice, that is, negative, positive and neutral. All of 46 questionnaires were effective after the questionnaire results were checked.

2.4 Statistical analysis

Two sample mean comparison of hypothesis test (t-test) was used to analyze the scores of exam with SPSS 13.0, significance level $\alpha = 0.05$.

3 RESULTS

The exam results and statistical analysis of the two groups are shown in Table 1. The results in two parts of the experimental group were significantly higher than the control group. There was a statistically significant difference ($P < 0.05$).

χ^2 test (Pearson no-correction method) of two groups of learning difficulty are shown in Table 2. There was a statistically significant difference in case-analysis question ($P < 0.05$).

4 DISCUSSION

Pediatric dentistry is higher requirements to clinical practice. For resident doctors who have never

Table 1. Two groups of exam results and t-test ($n = 23$, $\bar{x} \pm s$).

Group	Scores	t	P
One-choice (control group)	72.88 ± 5.73	4.0459	0.0002
One-choice (experimental group)	80.26 ± 6.61		
Case-analysis (control group)	69.89 ± 5.26	3.1543	0.0029
Case-analysis (experimental group)	75.67 ± 7.04		

Table 2. χ^2 test of learning difficulty (n%).

Group	Negative	Positive and neutral		
One-choice (control group)	9(39.1)	14(60.9)	1.6429	0.1999
One-choice (experimental group)	5(21.7)	18(78.3)		
Case-analysis (control group)	8(34.8)	15(65.2)	4.6000	0.0320
Case-analysis (experimental group)	2(8.7)	21(91.3)		

exposed to specialized clinical practice, knowledge of pediatric dentistry was very abstract, and not easy to understand. In the clinic, the symptoms of the children's dental disease were complicated, and each patient's growth situation is also different, so that, its treatment plan, process and prognosis were accordingly different. As a result, students felt more difficult to flexibly apply what they have learned theoretical knowledge. The results of this study showed that the scores of experimental group was significantly better than that of control group, which prompted ITM could get better teaching results. Due to the influence long-term traditional teaching, passive acceptance of knowledge has become the main way of student when they learned pediatric dentistry (Sun et al. 2012). If teacher simply imparted knowledge into students, the difficult of learning in the classroom through the traditional method would increased markedly. In the meantime, the ability of combining theoretical knowledge and clinical practice would reduced. The ability of using it to analyze cases would reduced too. During the process of interactive teaching, both teacher and students could participate in the same time (Staton et al. 2013). They also could help and inspire each other among students, which effectively reduced study requirement. Plentiful opportunities of discussion and thinking in the classroom and improved student ability to analyze cases (He et al. 2010). In addition, mutual discussion about cases in the classroom could not only create a positive learning environment, but also laid the foundation for an online interactive learning after school. However, there were some disadvantages in cases discussion, such as lower efficiency, take more time, which could indirectly affect the explanation of theoretical knowledge (Cahill et al. 2013). Therefore, further research was needed to find out a poised point about how to allocate rational time between elaborating theoretical knowledge and discussing curriculum-related cases.

Memory of knowledge in the classroom was affected not only by instant memory ability, but also the review efficiency after class. On consolidating the knowledge, the effect of network interactive way was more effective than the arrangement of the passive one-way teaching task (Kalwitzki et al. 2010). Students had to focus on how to improve their scores underneath the traditional exam-oriented education. Some students were worried that they might not gain a good score in the examination, even though they spend a lot of time and effort on CBL mode. Network interactive part of interactive teaching method provided a good communication platform between teachers and students, which changes the passive learning into active learning. Across time-space and interactive features of network teaching could make up for the lack of TTM (Hu et al. 2010). Observed in the classroom and network interaction, this study found that, silence negative student in the other major course often behave more active in pediatric dentistry. Flexible use of case discussion and network teaching method in the classroom could break through the limitation of the classroom teaching, improve the learning interest (Engel et al. 1994). In this study, the survey result of two teaching methods showed that learning difficulty was relative to the learning interest and cognition, which prompts that higher learning interest of experimental group resulted in a better grades. The degree of mastering the knowledge (scores and examination pass rate), degree of using the knowledge (the diagnostic accuracy and rationality of the treatment plan) were significantly higher than that of control group, and the difference was statistically significant between two groups. However, the network interactive teaching also had its disadvantages (Adolfi et al. 2005). First, the network interactive process was done in student's spare time, the teacher could not ensure that every student participated in the same degree exactly. Second, some students may coped with it negatively, which caused that their ability of thinking independently did not exercise properly, even cultivated the habit of laziness and plagiarism.

It is important to note that TTM has been a main position in the long-term. If major changes of teaching method in a short period could bring a bad influence on normal teaching. It can be suggested that (Chen et al. 2014): (1) based on TTM as the foundation, appropriately combining ITM with TTM is a better way. (2) Using ITM into the more difficult units, such as plentiful cases discussion, relative videos, could improve the teaching effect significantly. It is the better that cases discussion time will be controlled within half an hour each lesson. (3) During the process of curriculum development, gradually increasing ITM proportion can make the reform of teaching method smooth and reliable, and achieve the goal of improving the teaching effect.

ACKNOWLEDGEMENTS

This study was supported by the Outstanding Young Teacher Project of Wenzhou Medical University, Educational research fund of Wenzhou Medical College, Higher research education project of Zhejiang College Education Association.

REFERENCES

[1] Zhou, Y.L. et al. 2011. The evaluate of teaching efficiency for physiology PBL teaching. *Southern Medical Education* (3): 34–35.

[2] Sun, X. & Chen, J.H. 2012. Exploration of oral prosthodontics teaching mode. *Health Vocational Education* 30 (8): 78.

[3] Staton, L.J. et al. 2013. A multimethod approach for crosscultural training in an internal medicine residency program. *Med Educ Online* 18: 20352.

[4] He, B. et al. 2010. Application of case teaching in oral pro sthodontics teaching. *Medical Information* 5 (2): 335–336.

[5] Cahill, M. et al. 2013. Enhancing interprofes sional student practice through a casebased model. *J Interprof Care* 27 (4): 333–335.

[6] Kalwitzki, M. et al. 2010. Differences in the perception of seven behaviour modifying techniques in paediatric dentistry by undergraduate students using lecturing and video sequences for teaching. *Eur J Dent Educ* 14 (4): 247–253.

[7] Hu, J. et al. 2010. Construction and exploration of webteaching platform for prosthodontics. *Acta Universitatis Medi cinalis Nanjing (Social Science)* 10 (4): 354–356.

[8] Engel, F.E. & Hendricson, W.D. 1994. A case based learning model in or thodontics. *J Dent Educ* 58 (10): 762–767.

[9] Adolfi, D. et al. 2005. Achieving aesthetic success with an immediate function implant and customized abutment and coping. *Pract Proced Aesthet Dent* 17 (9): 649–654, 656.

[10] Chen, L.J. et al. 2014. Bilingual teaching efficiency of prosthodontics in different teaching methods. Frontier and Future Development of Information Technology in Medicine and Education. *Lecture Notes in Electrical Engineering* 269: 1487–1491.

Future Information Engineering and Manufacturing Science – Zheng (Ed)
© 2015 Taylor & Francis Group, London, ISBN 978-1-138-02644-5

The application of Markov method for upper limb rehabilitation robot's dynamic evaluation

Jiao Li
College of Information Science and Engineering, Northeastern University, Shenyang, China
Northern Institute of Technology, Shenyang University of Aeronautics and Astronautics, Shenyang, China

Jianhui Wang
College of Information Science and Engineering, Northeastern University, Shenyang, China

Shuang Li
Xingcheng Sanatorium of Shenyang Military Region, Xingcheng, China

ABSTRACT: Based on the characteristics of upper limb rehabilitation robot and the principle of stochastic process, Markov chain model is established. According to the analysis of the trend of muscle force, balance function, range of motion, and coordination function, the transfer matrix is built and the feasibility is tested. A new idea for upper limb rehabilitation robot dynamic evaluation is provided.

Keywords: upper limb rehabilitation robot; Markov chain; state transfer matrix; dynamic evaluation

1 INTRODUCTION

At present, people's health has been paid more and more attention, and then the accurate evaluation for patient recovery has become an important part in medical rehabilitation. In the past, usually with the static rehabilitation evaluation, the result can only reflect the current situation of the patients, but the situation of patient themselves are dynamic. Therefore, it is necessary to make dynamic rehabilitation evaluation at different times to reflect the recovered situation of patients.

A Markov chain is a finite process, which models a sequence of events that has a fixed number of situations and their specified probabilities, and given the present situation of the future and past that are independent. So far, it has been used for a variety of applications, especially in control applications. Patient's rehabilitation is affected by many uncertain factors, and has a strong randomness. Therefore, the Markov method of upper limb rehabilitation robot's dynamic evaluation is proposed, by simulating patient rehabilitation process of the change and constructing the transfer matrix method. And then the rehabilitation of patients can be evaluated dynamically.

2 THE MARKOV METHOD OF UPPER LIMB REHABILITATION ROBOT

2.1 The definition of Markov chain

A stochastic process is a collection of random variables

$$\{X_n, n \in T\}$$

Typically, T is the discrete time $= \{1, 2, 3 \ldots\}$. Given a finite discrete state space $I = \{i_1, i_2, i_3 \ldots\}$, each X_n belongs to the same subset of I, for all $n \in T$ and all $i_1, i_2, i_3 \ldots i_{n+1} \in I$,

$$P\{X_{n+1} = i_{n+1}/X_0 = i_0, X_1 = i_1, \ldots X_n = i_n\}$$
$$= P\{X_{n+1} = i_{n+1}/X_n = i_n\} \tag{1}$$

$\{X_n, n \in T\}$ is called the Markov chain. In other words, the probability of being in a particular state at the $(n+1)$th step only depends on the state at the (n)th step. We only consider Markov chains for which this dependence is independent of n (that is, time-homogeneous markov chains).

2.2 Classification of state

According to the characteristics of the patients and the actual needs, it is assumed that there are

k indicators in the upper limb rehabilitation robot's evaluation, and each is divided into N levels, the state space of the upper limb rehabilitation robot $I_k = \{1, 2, 3 \ldots N\}$. For example, if muscle force is divided into three levels of small, middle, and large, then $I_k = \{1, 2, 3\}$. If muscle force is divided into five levels of smaller, small, middle, large, and larger, then $I_k = \{1, 2, 3, 4, 5\}$.

2.3 Transition probability

For all states $i_1, i_2, i_3 \ldots$ and all $n \geq 0$

If $P_{ij}(n) = P\{X_{n+1} = j / X_n = i\}$

$P_{ij}(n)$ is called the one-step transition probabilities. It means the probability of transitioning from state i to state j. If there are k indicators in the upper limb rehabilitation robot's evaluation, there are l_k evaluation of regional in the (k)th indicators. A patient is observed m_k times in a certain period of time, n_{ijk} is the total number of from state i to state j between two adjacent. Then:

$$P_{ijk} = \frac{n_{ijk}}{\sum_{j \in I} n_{ijk}} \qquad (2)$$

If P_k is composed of the one-step transition probabilities P_{ijk}, and the state space $I = \{1, 2, 3 \ldots N\}$

$$P_k = \begin{bmatrix} P_{11k} & P_{12k} & \cdots & P_{1nk} \\ P_{21k} & P_{22k} & \cdots & P_{2nk} \\ \vdots & \vdots & \vdots & \vdots \end{bmatrix}$$

P_k is the matrix of one-step transition probabilities. It has the following properties:

1. $P_{ijk} \geq 0, i, j \in I$
2. $\sum_{j \in I} P_{ijk} = 1, i \in I$

2.4 Stationary distribution

The rehabilitation of patients is a limited state. According to the properties of the Markov, the Markov chain must have a unique distribution. The (k)th indicator should meet as follows:

$$\begin{cases} \pi_{jk} = \sum_{i \in I} \pi_{ik} p_{ijk} \\ \sum_{j \in I} \pi_{jk} = 1, \pi_{jk} \geq 0 \end{cases} \qquad (3)$$

The rehabilitation of patients will be evaluated according to the stationary distribution probability.

3 EXAMPLE

3.1 Determination of evaluation indicator

After a large number of experiments, a brain stroke hemiplegic patient who is male and 60 years old is selected to the research. The rehabilitation stage of hemiplegia patient is divided into flaccid paralysis period, cramp period and improved period. Passive motion is performed in flaccid paralysis period and cramp period. Active movement is performed in improved period. Based on 5-DOF upper limb rehabilitation robot that is designed and developed by Harbin Institute of Technology, flaccid paralysis period and cramp period are selected for research. It transforms motor function evaluation contents into indicator system of motor function for 5-DOF upper limb rehabilitation robot, and makes assessment criteria for each indicator, including muscle force, balance function, range of motion, and coordination function.

Muscle force is the maximum force that is generated by muscle contraction. It is mainly used for evaluation of muscle function. Muscle force is divided into five levels of smaller, small, middle, large, and larger. $I_1 = \{1, 2, 3, 4, 5\}$.

Balance function is to understand whether the object has balance disorders, it is divided into five levels of poorer, poor, middle, good, and better. $I_2 = \{1, 2, 3, 4, 5\}$.

Range of motion refers to the arc of Movement of the joints, it is divided into five levels of very non-standard, non-standard, middle, standard, and very standard. $I_3 = \{1, 2, 3, 4, 5\}$.

Coordination function mainly refers to an expression of the whole limb patterns of movement, it is obtained when performing EMG signal acquisition, and it is divided into five levels of poorer, poor, middle, good, and better. $I_4 = \{1, 2, 3, 4, 5\}$.

3.2 The structure of the transfer matrix

The rehabilitation of patients with stroke hemiplegia are the best during 1 to 3 months. Therefore, the patient is trained and observed for 15 times in the first three months, the changes are as follows:

From the Table 1 to Table 4, we can get the transfer matrix of rehabilitation of patients. They are as follows:

$$P_1 = \begin{bmatrix} 2/7 & 5/7 & 0 & 0 & 0 \\ 1/7 & 10/21 & 1/3 & 1/21 & 0 \\ 0 & 3/10 & 3/10 & 2/5 & 0 \\ 0 & 0 & 1/11 & 6/11 & 4/11 \\ 0 & 0 & 0 & 2/5 & 3/5 \end{bmatrix}$$

Table 1. The chang table of muscle force.

The number of observations	Deltoid	Biceps	Brachioradial muscle	Pronator teres
1st	2	1	2	1
2nd	2	1	1	1
3rd	2	2	2	2
4th	2	1	2	2
5th	3	2	2	1
6th	2	2	3	2
7th	2	3	2	2
8th	3	2	2	3
9th	3	3	3	2
10th	4	3	3	4
11th	5	4	4	3
12th	5	4	4	4
13th	5	5	4	4
14th	5	4	5	4
15th	5	5	4	4

Table 2. The chang table of force balance function.

The number of observations	Shoulder 0° elbow flexion 30°	Shoulder 0° elbow flexion 60°	Elbow extension shoulder flexion 30°	Elbow extension shoulder flexion 60°
1st	2	1	1	2
2nd	2	2	1	2
3rd	2	1	2	2
4th	3	2	1	2
5th	3	2	2	3
6th	3	3	2	2
7th	3	2	2	2
8th	3	2	3	2
9th	3	3	2	2
10th	4	4	3	3
11th	4	3	3	3
12th	4	4	4	3
13th	5	5	4	4
14th	5	4	5	4
15th	5	5	4	4

$$P_2 = \begin{bmatrix} 1/5 & 4/5 & 0 & 0 & 0 \\ 2/21 & 4/7 & 1/3 & 0 & 0 \\ 0 & 3/16 & 1/2 & 5/16 & 0 \\ 0 & 0 & 1/10 & 1/2 & 2/5 \\ 0 & 0 & 0 & 1/2 & 1/2 \end{bmatrix}$$

$$P_4 = \begin{bmatrix} 1/2 & 1/2 & 0 & 0 & 0 \\ 1/16 & 3/8 & 9/16 & 0 & 0 \\ 0 & 2/9 & 7/16 & 7/16 & 0 \\ 0 & 0 & 1/4 & 7/12 & 1/6 \\ 0 & 0 & 0 & 1/3 & 2/3 \end{bmatrix}$$

$$P_3 = \begin{bmatrix} 2/5 & 3/5 & 0 & 0 & 0 \\ 1/12 & 1/2 & 5/12 & 0 & 0 \\ 0 & 1/6 & 5/12 & 5/12 & 0 \\ 0 & 0 & 1/5 & 3/5 & 1/5 \\ 0 & 0 & 0 & 1/3 & 2/3 \end{bmatrix}$$

It is irritable and aperiodic. According to the properties of the Markov, the Markov chain must have a unique distribution. The solutions are as follows:

$$\pi_1 = (\pi_{11}, \pi_{21}, \pi_{31}, \pi_{41}, \pi_{51})$$
$$= (0.0141, 0.0707, 0.0898, 0.4323, 0.3930)$$

$$\pi_2 = (\pi_{12}, \pi_{22}, \pi_{32}, \pi_{42}, \pi_{52})$$
$$= (0.0092, 0.0775, 0.1378, 0.4308, 0.3446)$$

Table 3. The chang table of range of motion.

The number of observations	Outreach/adduction of shoulder	Flexion/extension of elbow	Flexion/extension of wrist
1st	1	2	1
2nd	1	2	2
3rd	2	2	1
4th	2	3	1
5th	3	3	2
6th	2	3	2
7th	2	3	3
8th	3	4	2
9th	3	3	2
10th	4	4	3
11th	4	4	4
12th	4	4	3
13th	5	4	3
14th	5	5	4
15th	5	4	4

Table 4. The chang table of coordination function.

The number of observations	Deltoid	Biceps	Brachioradial muscle	Pronator teres
1st	2	1	2	1
2nd	2	1	1	1
3rd	2	1	2	1
4th	3	2	2	2
5th	3	2	3	2
6th	3	2	2	3
7th	3	3	3	2
8th	3	3	2	3
9th	4	3	3	3
10th	4	4	3	3
11th	4	3	4	3
12th	4	4	3	4
13th	5	4	4	3
14th	5	5	4	4
15th	5	4	4	4

$$\pi_3 = \left(\pi_{13}, \pi_{23}, \pi_{33}, \pi_{43}, \pi_{53}\right)$$
$$= \left(0.0116, 0.0835, 0.2088, 0.4350, 0.2610\right)$$

$$\pi_4 = \left(\pi_{14}, \pi_{24}, \pi_{34}, \pi_{44}, \pi_{54}\right)$$
$$= \left(0.004, 0.0878, 0.2712, 0.4362, 0.2043\right)$$

From this result we can conclude that muscle force is large, balance function is good, range of motion is standard, and coordination function is good. The results of dynamic evaluation are compared with the clinical results of the evaluation, they match closely.

4 CONCLUSION

From the above-mentioned analysis, we can know that the Markov method that is used in upper limb rehabilitation robot's evaluation can overcome the shortcoming of the traditional static evaluation, and reflects the patient's recovery and trends in a more intuitive. This model is relatively simple and requires less basic data on the patient, with a strong practical. It provides a new idea for upper limb rehabilitation robot dynamic evaluation.

REFERENCES

[1] Liu Cihua, Stochastic Processes and Applications Higher Education Press, Beijing, 2004.7.
[2] Liu Jinfu, Sun Hongxiang. Applied Stochastic Processes, Tsinghua University, Beijing, 2006.9.
[3] Wang Jianhui, Zhang cheng, Ji wen, Gu Deying. Rehabilitation Evaluation Methodology of upper Limb Rehabilitation Robot based on AHP-fuzzy Comprehensive Evaluation. Journal of Shenyang University, 24(3), 47~51, 2012.
[4] Wang Yong, Han Xueshan, Ding Ying, Shen Jie. Markov Chain-Based Rapid Assessment on Operational Reliability of Power Grid. Power System Technology, 37(2), 405~410, 2013.

Future Information Engineering and Manufacturing Science – Zheng (Ed)
© 2015 Taylor & Francis Group, London, ISBN 978-1-138-02644-5

Steam mop detection system design based on STM32 and μC/OS-II

Jieyong Huang

Zhongshan Institute, University of Electronic Science and Technology, Zhongshan City, China

ABSTRACT: Steam mop is the best equipment for cleaning in average families or companies. This topic is about the performance testing for these products before ex-factory. It introduces a design of steam mop detection system which is based on STM32F103ZET hardware development platform and embedded μC/OS-II operating system, with both hardware and software system of the design to be introduced.

Keywords: steam mop; STM32; μC/OS-II; UCGUI

1 INTRODUCTION

Steam mop, based on a high-temperature and high pressure working principle, will have a temperature of 130°C steam injection in the vent at the bottom of the mop when used. Currently, there are two methods for the detection of steam mop in China, one is manual inspection, which requires testers to manually record the data, implement all operations by hand step by step and observe if every step is completed successfully. Sometimes, testers need repeated operations for the certain performance parameter and determine whether the product is qualified compared with the testing standard data. As the steam mop is multi-functional, it takes much more time for manual inspection of indicators. Manual testing is more used for small household electrical appliance enterprises in nowadays. The other method is to build up the detection environment with PLC. To detect with PLC, users need to write detection program for testing. It is more difficult for ordinary testers, because it requires professional PLC programmers to involve in the detection work and make summaries and analysis about the testing data. The shipment volume of the household electrical appliance with steam mop is comparatively large, which requires a larger number of testers. Therefore, the cost of using PLC for detection is relatively high.

From the perspective of product application, this topic proposes to design a detection system that is able to react in real-time and transmit data through the network. The system will detect many performance parameters of steam mop, including galvanic current, voltage power, consumption of steam and so on.

2 SYSTEM HARDWARE DESIGN

2.1 *Overall design of the system hardware*

This system hardware design system block diagram is shown in Figure 1. Hardware system uses STM32F103ZET as the master CPU, adopts switching power supply for system's electricity supply, and has 8 inputs and 8 outputs detection interface that can detect four devices simultaneously. Also, it uses modular design for input and output design, increases or decreases modules depending on the increasing or decreasing of the input and output volume, and configures DM9000 Ethernet Chip that is capable of synchronous data transmission. The display interface is with 7-inch TFT LCD screen and touch screen, number keys and function keys as well, which are more convenient to set up parameters and operational data in the backstage. Also, it provides SD card interface for storing test data and achieving program updating.

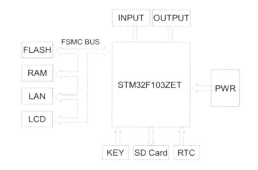

Figure 1.   The block diagram of the system hardware design.

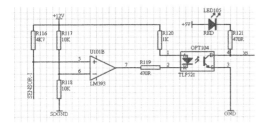

Figure 2.   Steam input detection circuit.

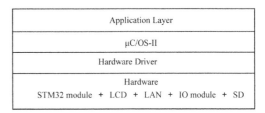

Figure 3.   The block diagram of the system software design.

2.2   *Detection module design*

The detection system is mainly used for testing the amount of steam in the steam mop. The detection input principle is shown in Figure 2 below. Amplify the signal of detecting the amount of steam SENSOR-I in the operational amplifier by flow sensor, and then input it to the input port of the system by the photon coupled isolator TLP521.

3   SYSTEM SOFTWARE DESIGN

System software design is as shown in Figure 3 below. The system needs to process the input of the external steam detection signal, display the results to the LCD and show the detection status through output module (Y0–Y7), record the detection data to the SD card, make a summary and send it to the server through the network module. During the detection process, the system also needs to respond to the operation of the external keys or set up GUI parameters. Therefore, the system transplanted μC/OS-II and conducts transition management for these tasks, which improves the reliability and stability of the system. The application layer of the system software is mainly for achieving the logic processing of detection's input and output as well as the operation of UCGUI.

3.1   *μC/OS-II task manage*

The μC/OS-II is a real-time multi-tasks operation system kernel, which is portable, preemptive, and able to implant ROM and cut out. It is widely applied in micro-controllers, micro-processors and digital signal processors. The system embeds μC/OS-II, which makes the debugging program easier. The task status scheduling of μC/OS-II is shown in Figure 4.

The entire detection system is divided into several tasks: LCD reading and writing, LAN transmission, key response, input and output control and so on. Each task is relatively independent and then set with timeout function. When the time is

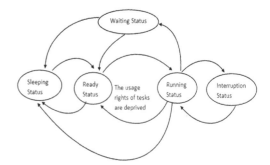

Figure 4.   μC/OS-II task scheduling.

over, the task must surrender the usage rights of CPU. Even one task exception occurs, it will not affect the operation of other tasks. This will not only improve the reliability of the detection system but also makes the debugging easier.

3.2   *UCGUI graphical interface design*

The detection system display modules consist of 7-inch TFT LCD screen and touch screen. Display interface realizes the interface drawing by transplanting UCGUI. UCGUI is an embedded graphics supporting system under application, which is designed to provide efficient and independent processor as well as LCD controller's graphical user interface for any application that is displayed via LCD graphics. UCGUI is not only suitable for single-task or multi-tasks system environment but also for any LCD controller and the real display or virtual display of any size under CPU. Its design construction is modularized and consists of different layers of different modules. The LCD driver layer is including all the detailed graphic operation for LCD. The display interface of the detection system is shown in Figure 5.

3.3   *SD card file system*

The system configures SD card interface, which is mainly for achieving programs' BootLoader

Detected Voltage 0V	Detected Current 0.0A		Power 0W
(105V 135V)	(0.1A 15.0A)		(10W 2025W)
Work flow	Current Status		
Step 1	Degree One	Degree Two	Degree Three	Neutral
Step 1	Degree One	Degree Two	Degree Three	Neutral
Step 1	Degree One	Degree Two	Degree Three	Neutral
Step 1	Degree One	Degree Two	Degree Three	Neutral
Detected Qty 8	Steam Volume of Degree One	Steam Volume of Degree Two	Steam Volume of Degree Three
Qualified Qty 1	0r/s	0r s	0r s
Qualification Rate 12%	Current Steam Volume 0r/s		

Figure 5. Display interface of detection system.

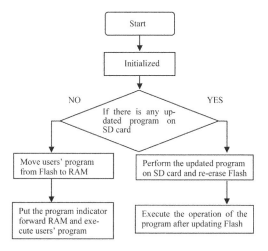

Figure 6. BootLoader flowchart.

and recording detection data. This function actually realizes FatFS. FatFS is a small SD card file system. To transplant FatFS on STM32F103ZET, it requires modifying several key documents, namely: diskio.c, integer.h and ffconf.h.

- diskio.c is the implementation code of the middle interface between the bottom layer of file system and the SD card driver. During transplanting, the several functions in the diskio.h declaration are required to be rewritten and the code in ff.c will be called.
- integer.h is the data type definition that FatFS uses, which will be modified based on the transplant platform.
- ffconf.h is a part of the configurations of FatFS system module. After FatFS is successfully transplanted, the detection data can be recorded on the SD card.

SD card interface of the detection system has the function of system updating at the same time. According to the users' needs, it modifies BootLoader program and stores it into the SD card so that the system can be automatically updated. The flowchart is shown in Figure 6 below. The principle of BootLoader is actually positioning the updated program at the starting and ending point of SD card. Each time when programming is under execution, the BootLoader will detect the SD card and perform if there is any updated program.

4 CONCLUSION

Steam mop uses the high-temperature steam for cleaning, decontamination, disinfection and sterilization. After that, there is very little residual water on the floor and will evaporate soon. Nowadays, many families or companies are using it for cleaning. This detection system design in this topic will strengthen the reliability and safety of the household electric appliance enterprises products with steam mop. Currently, there is not yet any product to detect the steam mop in the market. If the enterprises detect the steam mop by using the detection system, not only the production capacity, or namely the management capacity for the steam mop technique will be greatly improved, but also the development and research costs of the enterprises will be saved largely.

ACKNOWLEDGEMENT

The Quality and Teaching Reform Project of University of Electronic Science and Technology of Zhongshan Institute Code number JY201410.

REFERENCES

[1] CHEN Qijun, *Embedded System and Its Applications—System Design and Development Based on Cortex-M3 Kernel and STM32F103 series of Microcontroller*, Tongji University Press, 2011.5.
[2] Labrosse (USA), *Application and Development of Embedded Real-time Operating System μC/OS-III—Based on STM32 microcontroller*, Beihang University Press, 2012.12.
[3] *FatFs—Generic FAT File System Module*. http://elm-chan.org/fsw/ff/00 index_e.html.
[4] *UCGUI Chinese Manual*. http://www.ucgui.com.

Future Information Engineering and Manufacturing Science – Zheng (Ed)
© 2015 Taylor & Francis Group, London, ISBN 978-1-138-02644-5

Research on information technology application major practical teaching system based on ITAT contest

Fahai Li & Li Chen

School of Electrical and Information Engineering, Hubei University of Automotive Technology, Shiyan, China

ABSTRACT: It's an important method to cultivate and select high-level applied talents by ITAT contest in universities. As an example of teaching reform of IT application major in the university, the construction ideas and implement ways of practical teaching system on the ground of ITAT contest are introduced, and the "four in one" practice teaching mode is discussed. The practical result has proved that ITAT contest could improve the quality of practice teaching, and provide an important reference for building professional courses.

Keywords: ITAT; contest; practical teaching; reform

1 INTRODUCTION

In recent years, with the continuous development of information technology and its application, and with the employing increasing demands of enterprises and institutions, mostly IT related occupation requires candidates received professional skills training, and have the appropriate project experience. This requirement exists with the current educational content provided by colleges and universities between the obvious faults, college graduates with the skills and level of experience difficult to match the requirements of the enterprise, which led to the "Graduates Employment" has become a wide spread social problem. For this reason, based on existing computer programs, we focus on the development trend of information technology and the application direction of information technology, combining the ITAT competition's content. We build up a system that is based on working process, systematic program led by action around computer application, software engineering and network engineering. We develop applied project, set up teaching content carrying the project and the "Teaching, Learning and Action" theory. We adopt project-driven and mission-led teaching method. We create a teaching team which has higher teaching level, strong practical capability, reasonable major structure and we also establish a multiple-featured, advanced equipment computer training base [1]. Though ITAT course's reform in teaching content and method, we can make the trained talent more in line with the demand of employing in enterprises, furthermore, we can strengthen student's competition ability in job hunting, really achieve "zero distance" contact.

2 THE CONSTRUCTION IDEAS AND IMPLEMENTATION METHODS OF PRACTICE TEACHING SYSTEM

2.1 *We promote the construction of information technology application major, taking competition as the driving force*

We are based on the specialized construction of information technology application at the provincial level, with the combination of "information technology application level" competition content: computer application, the key technology of software engineering and network engineering, and so on. We bring up an applied talents training pattern that takes professional working task and employing ability as the guidance, project driven, work-integrated learning, and we have improved core curriculum standard. To establish a modular curriculum and training system, to complete four supporting teaching modules which include taking training the students' employment ability as the center, grasping the basic knowledge, cultivating professional quality, raising their core post ability, and expanding synthetic professional ability. We cooperate with companies such as Microsoft, Cisco to set up experimental training base, and introduce the enterprise standard, project, occupational situation, and enterprise engineers into schools directly for the cultivation of the talent in software development, and network engineering, and so on. Create a new mode of teaching to the project as a carrier, task-driven, take 'teaching, learning and doing' as one [2]. In teaching take competency training as the goal, and take students as main body, teaching while learning, learning while doing, to make "teaching" organically integrated, the content of the course was done in the training room, which greatly improved the students' learning enthusiasm, initiative, and improved the students' practical ability, the teaching effect and teaching quality.

2.2 *We promote practical teaching system reform, taking contest as the pusher*

Propose the professional curriculum practice teaching system reforming program, and to develop professional curriculum for practice teaching projects in view of the talents training object for information technology application, combing with the enterprise related requirements, through the enterprise research and teachers going to the factory training, in order to adapt to the rapid development of the computer industry and a highly concerned IT trend [3]. We cooperate with enterprises positively and introduce real projects as teaching case, adopt a "project-oriented" approach, to spin-off integrated modules and reconstruct the "project", making it accord with teaching and training process. Students with further study, complete "project" of the various "sub-task" and finally grasp the main knowledge and technology gradually. During the implementation of teaching curriculum, the curriculum team has to pay attention to student's learning status, designing a number of small projects completed by student groups. Each project group equips with a leader and several members, everyone has to manage his own business. Teachers and instructors guide the whole process and supervise the whole process which students do.

2.3 *We promote the practice base construction, taking contest as the starting point*

When to build training base on campus for computes information major and internet major on the basis of existing computers and network laboratories, through new construction, alteration, extension, building computer networks and professional. This includes software laboratory, embedded laboratory, networking equipment laboratory, networking cabling and information security laboratory. There are 6 high configuration servers, 400 Lenovo computers, which is mainly aimed at experimental training teaching of computer software modules in software laboratory. There are 200 hosts, 100 and 100 sets of single-chip microcomputer experiment box, 50 boxes in the embedded development lab, which is mainly aimed at experimental teaching training for embedded modules. There are

24 Cisco routers, 8 Cisco switches, 2 Cisco secure firewall devices, 2 Cisco integrated wired/wireless Wlan devices such as switches, 3 Cisco wireless AP in Network interconnection device Labs, which is mainly aimed at the building blocks of computer network experiment teaching training. There are many years PDS training platform, simulation wall and multiple PDS supporting to establish in Network cabling laboratories, which is mainly aimed at the experimental training teaching of network system integration module. There are SimplelSES safety equipment and practical training platform SimplelSES in Information Security Lab, which is mainly aimed at the experimental training teaching of network information security module. The campus practice base offers a good hardware condition for talent training.

2.4 *We improve teachers' practical ability for application of information technology, taking contests as pusher*

The contest put teachers to continually learn through competition, and get them quickly improved. We send teachers to participate in a variety of practice ability of training before the contest. There have been teachers participated in Microsoft Engineer, Cisco Network Engineer, Red Hat Linux Certified Engineer and South Shield Information Security Engineer training, they got appropriate engineer certificate respectively. Teachers who are in this process have gained considerable advanced knowledge from the enterprise. They applied project-oriented teaching method to the course construction, and they had a flexible teaching method. The resources of cooperation between schools and enterprises have been greatly used, and team-building is effective. Teacher's reputation is greatly improved, and they can serve for enterprise services, which forming a virtuous circle [4].

2.5 *We perfect the incentive mechanism, taking contests as pusher*

Academic competitions are the effective means of testing students' comprehensive quality. Through participation in various academic competitions, we can not only exercise and study student's practical ability, thinking skills and creativity, but also promote the improvement of undergraduate teaching level, increase the school's visibility as well. Our college developed and implemented a number of administrative measures which committed to the discipline that contest is of normalized and institutionalized, organized based on the "high starting point, high standards and strict and effective requirements" formed to an effective mechanism that contest-led the reform, competition-driven teaching and research, contest-testing the scientific research. We developed an object award system in order to make organized management standardized scientific and institutionalized for student's participating in academic competitions, effectively to participate in the purpose of the competition. At the same time, we establish and improve a system of contest awards discipline, giving the excellent students some preferential policies such as certificates, awards, honors, employment incentives. It has two important influence on the implementation, on the one hand, it can inspired student's learning interest, exercise students of innovation thinking, improve student's capacity of settlement actual problem, which cultivate high quality innovation talent, encourage students participate in various students discipline race; on the other hand it will be a condition for teachers who wants to be promoted, which may motivate general teachers positive in participating in guiding and training. This will makes teacher's contribution get some reward.

3 NEW PRACTICE TEACHING MODE

3.1 *Implementation of "four in one" mode of practice teaching*

In curriculum system of "work- learning, project-oriented" curriculum, implements "four in one" mode of practice teaching of item practice training, comprehensive training, projects training, practical training, and top-post internship. Adopts "teaching, learning and doing" teaching method to highlight students' practical training and establishment of engineering thought though introducing practical project and real work environment of enterprise inside or outside school. And leads college, teacher, enterprise to promote interaction between teaching and production, cooperation between college and enterprises, upgrading IT major student's ability quality hence, we can make talents cultivation and enterprise needs seamlessly.

3.2 *Bridging the gap between school teaching and practical work, effectively improving students' professional abilities and employment*

Strengthen practical teaching reform in practical teaching content and methods. First, we deal with it from the source of major construction, developing training programs, doing job analysis, establishing professional employment goals and core competencies project, setting conducive to improving students' ability of integrated curriculum which can guarantee the number of hours of practice teaching. Second start with practical teaching mode, constitutes a variety of practical teaching methods, finishing the process of practice teaching step by step.

To arrange experiment teaching and courses teaching synchronization, strengthen students understanding on theory knowledge; arrange concentrated real training week for special practice training, achieving courses' small project designing and making, which is beneficial for improving students design capacity; set specifically and creative courses, these courses mainly focus on practice and innovation, training students on products of innovation capacity; arrange graduated integrated internship and graduated design, which is an integrated item and complement project practice, turning student's theory integrated into practical ability through this items practice teaching. Meanwhile, to narrow the gap between school teaching and practice work through the internship outside real training base of, makes students directly participate practice project before graduation, rising students professional capacity and employment effectively. Arrange students for participating in professional qualification certification exam, to strengthened students professional capacity and courses knowledge of integrated capacity through training, help students set employment consciousness, from the point of employment capacity to strengthen student's hands-on capacity. Organize students to participate in various levels of competition, training students' comprehensive qualities of innovative consciousness and innovative abilities of science and technology, upgrading student's professional skill and teamwork spirit in pre-match training [5].

4 PRACTICAL TEACHING EFFECT

4.1 *To meet the development needs of the information technology application major construction in our department*

Form the deeply degree cooperation with government information department, IT associations, network equipment vendors, system integrators, IT companies, Science and Technology Industrial park, around computer, and software engineering and network engineering this three big professional, through the implementation of "2.5 + 0.5 + 1" talent training mode. Carry out the training mode that combines working and learning positively. Push the cooperation between universities. Introduce enterprise culture, increased outside real training intensity. Explore the talents training mode for project-driven, competency-oriented, top-post practice. Perfect four supporting teaching mode for cultivating job ability, commanding basic reputation knowledge, developing professional quality, upgrading, integrate capacity. All these can make IT talents be in line with community needs.

4.2 To construct practice teaching system of application of information technology

Information technology application skills project focuses on assessing students' comprehensive capability at computer application, software engineering and network engineering, as well as management skills such as teamwork and planning organizations, whether they are familiar with the work process in designing-adjusting-maintenance. Integrate with the talent market's demand and the requirement of professional skill identity closely according to the competition rules and the real working environment in enterprises to set up cooperative project developed by universities and enterprises, also, based on the "three combinations, three suit" design training content, namely practical training and close connection with the national education reform. Let training adjust to progress of the implementation of the reform programs to meet the changing requirements of the market, constantly adjusting to high skilled talents construction in the new situation in the country. The indoor training base focus on cultivating and training students' necessary professional quality and practical operation ability according to the national standards and its "skill ability" in professional skill test as well as the produce process, quality standard and working method.

4.3 Form a good role model by preparation project textbook

The IT application textbook and training guide book reflects the application of information technology in professional features since they introduced the industry standards and the project implementation process. Materials and instructions not only provide the teaching guide for our college's teachers and students, but also form a good demonstration and promotion role, showing the education achievement of undergraduate education in Hubei province.

5 CONCLUSIONS

It is an important method to select talents by holding university IT contest and establishing a competition mechanism in which there are contests in every college, there are selection in every level, and there national contest in our country. The development of IT contest has a great difference on deepening the reform of higher education, enhancing the attractiveness of higher education, promoting the status and role of higher education, as well as the social fashion-every occupation has its top one, which play an important role in forming good atmosphere of care, attention and support of higher education of the whole society. The IT contest promotes the training of information technology professionals, which transferred creative and practical excellent talent to the software and information technology industry, and realized "zero distance" between Universities and enterprises, and enhanced students' employ ability.

REFERENCES

[1] Liu, T. & Zhou, M.Z. 2013. Research on practical teaching system for the training of IT applied talents. *Journal of Anqing Teachers College (Natural Science Edition)* 19(4):113–116.
[2] Fang, H. & Lei, L. 2013. Practice of training mode for electronic information innovative project application-oriented personnel. *Research and Exploration in Laboratory* 32(11):308–310.
[3] Lu, B.Y. 2011. Construction of teaching aids network mode of talents training for information management and information system specialty. *Journal of Fuyang Teachers College (Natural Science)* 28(1):76–78.
[4] Liu, Y.F. & Zhang, L.J. 2013. Computer network experimental teaching for cultivating innovative ability. *Experimental Technology and Management* 31(4):28–31.
[5] Mou, P. & Ding, J.X. 2014. Improving research learning ability of students based on practical teaching. *Experimental Technology and Management* 31(4):157–159.

Future Information Engineering and Manufacturing Science – Zheng (Ed)
© 2015 Taylor & Francis Group, London, ISBN 978-1-138-02644-5

The studies for transforming information in control system based on TBS model

Yi Gan, Qiang Gao & Fujia Sun
College of Mechanical Engineering, University of Shanghai for Science and Technology, Shanghai, China

ABSTRACT: It was proposed that the TBS model was applied to the studies for building model of control system. TBS model is based on the ideas that transmit from top to bottom and feedbacks from bottom to top. The plug assembly line was taken for example to analyze and build TBS model of its control system. In TBS model of its control system, Modular and hierarchical configuration was introduced, and specific applications were analyzed. It provides the reference for control system to adapt to the changing industrial control needs.

Keywords: control system; TBS; transforming information; assembly line of production

1 INTRODUCTION

With the development of automatic control technology, more functions and greater performance are required gradually for automation control system in industrial enterprise. Therefore, optimization mechanism that includes adaptive control, fuzzy system and neural network was gradually introduced into the automation control system to implement the modular and hierarchical flexible configuration. It shortened the development cycle of product and enhanced the stability of the control system and production efficiency [1,2]. All manufacturing activities were conducted under the domination of manufacturing information [3]. The gradually implementation of control process in control system were accompanied with the flow, configuration, and operation for manufacturing information. Control system is a process of decision-making, configuration and optimization for manufacturing information. Control system must be able to process a large number of complex manufacturing information. It is basic issue for application of manufacturing information in automatic control system to choose suitable analysis method to build a multifunctional, stable and efficient control system in processing a large number of complex manufacturing information to optimize control process [3,4].

Compared with foreign countries, due to long-term "extensive" mode of production, material performance gap, lack of modern design and analysis methods, etc factors, an adverse impact was brought to the development of control system [5,6]. The researchers had conducted extensive analysis research on methods of modern design. For example, Yuliang Li [7,8] proposed integration collaborative software system framework of product according to the top-down design process characteristics; Bottom-up method was applied to Traditional Chinese Medicine (TCM) mechanical platform design by Weiqin Ma [9]; Shannon [10] proposed the formulas of probability measure; Zhengxin Liu [11,12] adopted model designing idea of the TBS assembly model to analyze concurrent design of Pro/E software platform mold. But it is still rare to apply modern design analysis method to manufacturing activities of the control system in the existing research. In this paper, TBS model was applied to studies for transforming information in control system from the perspective of manufacturing informatics [3,4], which optimized the process of processing manufacturing information in control system to meet the diversified needs of control field.

2 TBS MODEL OF CONTROL SYSTEM

TBS (Top basic skeleton) model [11–13] is based on top-down design ideas. It takes the overall framework of structure and function as information exchanging platform, which reflects the design concepts from macro to micro, abstraction to concreteness.

According to the top-down ideas, this paper applied TBS model to control system modeling, namely the CTBS (Control system TBS) model. CTBS is to build a basic framework on the top floor in the initial stages according to the basic

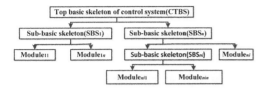

Figure 1. TBS model of control system.

function and requirements of the control system. CTBS model is shown in Figure 1.

In Figure 1, CTBS model consists of parallel and tandem communication structures. Every sub-basic skeleton inherits critical information from associated upper framework. Three states are presented in CTBS model, namely CTBS, SBS and Module. CTBS includes the overall design of parameters and concepts of control system. Relevant information of control system gradually expands in the transmitting process of top-down and forms sub-basic skeleton and modules.

Compared with TBS model that was used in designing and assembling mechanical products, some characteristics of control system is added in CTBS model.

As control system consists of hardware and software, content of CTBS model is no longer limited to hardware information such as mechanical parts etc. Content of CTBS model can also be composed of software information such as program modules etc. In the process of building CTBS model, manufacturing information is divided layer upon layer according to the function and demands by top-down method.

There exists coupling modules in CTBS model. They are limited and controlled to each other. CTBS model can implement the changing building for the overall control system quickly. When the information of modules is changed, the changed information can be transmitted to the top CTBS from bottom to top. Then top CTBS transmits changed information to relevant coupling modules from top to bottom. Thus the overall information sharing and update in CTBS model can be completed conveniently.

3 APPLICATION EXAMPLE OF CTBS

3.1 *The building of CTBS model for production line*

Control system of certain assembly production line for car lock stores and monitors the real-time production information and data that is generated in the process of assembling. And there are 41 working positions in the production line. Every working position has its different assembling

tasks that are done by its multiple corresponding working procedures. The working procedures may be changed to meet different production requirements. The manufacturing information data of production line is changing all the time during the working process of production line. Scene data needs to be scanned and collected circularly to implement the real-time monitoring for different working positions, and be answered respectively. Thus the control system may be unstable.

Working process of control system for the production line is a process that data source of hardware exchanges with data source of software real-timely. In order to facilitate exchanging data for data source of hardware and software, in recent years Microsoft cooperated with some process control company and created the communication interface protocol for object linking and embedding of process control, namely OPC [14]. It enabled hardware and software to exchange data in Microsoft operation system. As a result, OPC was applied to communication system of assembly production line in this paper. Meanwhile PLC was adopted to control execution of tool (baffle, fixture, etc) action (block, clamp, etc) in working position of production line.

CTBS model for the production line can be divided into two parts, namely PLC control (SBS-PLC) and communication system (SBS-COM), as is shown in Figure 2. Both SBS-PLC and SBS-COM include its corresponding hardware configuration and software design. The hardware configuration and software design include the corresponding configuration of the overall framework and the building of every module in framework.

3.2 *Transforming information of CTBS model for production line*

The working process of CTBS model for the production line is completed with the cooperation of SBS-PLC and SBS-COM, as is shown in Figure 3. PLC is connected with sensors in SBS-PLC. When one production pieces reach working position,

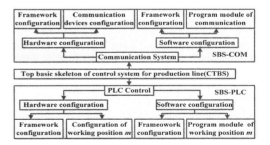

Figure 2. CTBS model for production line.

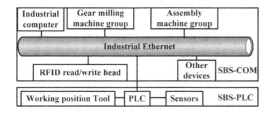

Figure 3. Transforming information of CTBS model.

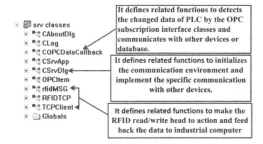

Figure 4. Program modules of application program.

production pieces touch the sensor. Then the sensor is triggered to make the data of the storage address of PLC changed; Application program of industrial computer in SBS-COM detects the changed data of PLC, and communicated with other devices and database. And the communication results would be answered to PLC again; According to the communication results, PLC controlled the execution of tool (fixture, baffle etc) action (block, clamp, etc) in working position of production line.

Application program in industrial computer is shown in Figure 4. In Figure 4, program modules are defined in the form of class. And specific functions are defined in related classes to communicate with different devices or database.

In the process of practical work, SBS-PLC feedback relevant information of PLC control from CTBS to communication system. SBS-COM got relevant information of communication system from CTBS, read and acquired data from SBS-PLC, then feedback corresponding data to PLC. At the same time, SBS-COM operated the communication with other devices and database. Thus the working process of control system for the production line was completed in this way.

3.3 *Application an analysis CTBS model*

As working procedures of working positions may be changed in assembly production line, the partial change will influence the stability of control system structure. SBS-PLC in CTBS model controls

the execution for actions of tools in all working positions.

Control program of PLC in CTBS model is stored in the memory card of PLC in the form of program modules, as is shown in Figure 5. Amount of working procedures in each working position will not exceed 10. In order to facilitate the administration, in Figure 5, the last number of each program modules is set as the sequence of working procedures in certain working position, the rest numbers of each program modules are set as the sequence of working positions. For example, FC22 represents the second program modules in the second working position. Child program modules of working procedures in the same program modules is in series with each other. And each independent program module is parallel with other program modules. For example, F22 can control the execution of multiple working procedures in the second working position in sequence. When the execution of one working procedures that FC22 controls fails, subsequent working procedures in FC22 can't be executed; FC22 is parallel with FC102. When the execution of FC22 fails, the result does not influence the execution of FC102, and vice versa.

In CTBS model of production line, as is shown in Figure 2, design of all program modules in working positions m can be taken as top basic-skeleton of program modules, and SBS_n can be established by program modules of working procedures.

Two Program modules that control blocking and freeing production pieces in working position m control and implement the tasks of assembling lock. The TBS model of program modules is shown in Figure 6.

According to production needs, a new clamping working procedure needs to be added to the original two working procedures. The clamping working procedure is required to execute in series on the basis of the original working procedures. The TBS model of program modules for working positions

Figure 5. Inner program modules of PLC.

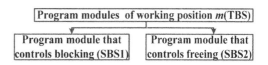

Figure 6. TBS model of program modules.

19

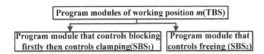

Figure 7. Tandem TBS model of program modules.

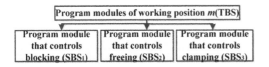

Figure 8. Parallel TBS model of program modules.

is shown in Figure 7. In Figure 7, program module SBS₁ controls the execution of two working procedures that control blocking production piece firstly and clamping it later.

When the production needs are changed again to require the new clamping working procedure to execute in parallel with the blocking working procedure. In order to meet the production needs, the TBS model of program modules are to be adjusted. The program content of SBS₁ needs to be modified. Moreover, new program modules SBS₃ is created to control the execution of the clamping working procedure, as is shown in Figure 8. The program module that control blocking, freeing and clamping the production piece are parallel with each other.

In working position m, when the new working procedure is added, the amount of program modules in parallel increases while the amount of program modules in series stays the same; the program module that controls the execution of new added working procedure in series is combined with the original program modules into one program modules to execute. On the basis of not changing the overall framework structure, it only needs to modify and adjust the program modules and hardware configuration that are related with the execution of working procedures to meet the production needs. Specific course needs to be considered from the function of control system, the relationship of current working procedures and other working procedures, control complexity, devices, and technology, etc. Then relative optimal solution is determined.

4 CONCLUSIONS

The constraint relationship of whole framework and each module in CTBS model of control system in the initial stage of design has been considered. After completing whole CTBS model, each bottom module is designed in detail under whole framework. In practical application of production line, the modular and hierarchical configuration makes the structure of control system convenient and flexible to be adjusted. Thus it can meet the changing industrial control needs. Meanwhile, information exchanging and coordination control in CTBS model are done by combining methods of top-down and bottom-up. Therefore, it promotes the efficiency of building control system.

ACKNOWLEDGMENT

Project was supported by the National Natural Science Foundation of China (No. 51375314).

REFERENCES

[1] Yongxiang Lu. 2010. Stepping into Green and Intelligent Manufacturing-Chinese Developing Road of Manufacturing. *China Mechanical Engineering* 21 (4): 379–386, 399.
[2] Xiaoli Cui et al. 2004. Vehicle Semi-active Suspension Adaptive Fuzzy Control Based on Neural Network Control Theory. *China Mechanical Engineering* 15 (2): 178–181.
[3] Bopeng Zhang. 2003. *Manufacturing informatics.* Beijing: Tsinghua university press.
[4] Congqian Qi. 2003. *An introduction to manufacturing industry informatization.* Beijing: China aerospace press.
[5] [e]Guoning Qi et al. 2004. Summary on Mass Customization Production Mode. *China Mechanical Engineering* 15 (14): 1240–1245.
[6] Shiguang Hu. 2008. *Research on Technology of Building Open CNC System with Intelligent Characteristic.* Tian jin: Tianjin University.
[7] Yuliang Li. 2007. Integrated Joint Policy Decision for Top-down Design Process. *Chinese Journal of Mechanical Engineering* 43 (6): 154–163.
[8] Shuting Zhang et al. 2009. Variation propagation approach for collaborative top-down assembly design. *Computer Integrated Manufacturing Systems* 15 (8): 1468–1477, 1492.
[9] Weiqin Ma. 2004. Platform Design of a Chinese Traditional Medicine Machine Based on Bottom-up Process. *China Mechanical Engineering* 15 (16): 5–8.
[10] Zhengxin Liu. 2005. *Parallel Design and manufacturing of Mold Based on TBS Model Assembly.* Chongqing: Annual Meeting for Institution of China Mechanical Engineering.
[11] Yi Gan et al. 2006. Parallel Design of Product Based on TBS Model Assembly. *Mechanical Design* 23 (5): 5–6, 32.
[12] Yi Gan et al. 2013. Studies on Information States Measurement for Modeling Design. *Applied Mathematics Information Sciences* 7 (2): 627–632.
[13] Gan Yi et al. 2013. Information State and Information Measurement for Modeling Design Based on TBS. *China Mechanical Engineering* 24 (16): 2131–2135.
[14] Sicheng Ren. 2002. The New Generation Specification of Software Interface for Industry Process and Control-OPC Technology. *Chinese Journal of Science Instrument* 23 (S3): 265–267.

Future Information Engineering and Manufacturing Science – Zheng (Ed)
© 2015 Taylor & Francis Group, London, ISBN 978-1-138-02644-5

Experimental research on arc discharge parameters of high-current ion source for EAST-NBI

Z.J. Tian
Zaozhuang University, Zaozhuang of Shandong Province, China

C.D. Hu & S. Liu
Institute of Plasma Physics, CAS, Hefei of Anhui Province, China

ABSTRACT: In order to achieve 2–4MW neutral beam injection in EAST, a high-current ion source was independently developed in the Institute of Plasma Physics of Chinese Academy of Sciences. At present, the first set of neutral beam injector engineering has been debugged successfully. Measurements indicate that the ion sources can discharge steadily; meanwhile the uniformity of the plasma is also better than those of other similar system. In this study, partial result of ion source discharge during debugging was reported and theoretically analyzed.

Keywords: neutral beam injection; high-current ion source; arc discharge parameter; discharge characteristic

1   INTRODUCTION

The national great science engineering EAST has been built with high quality and put into operation comprehensively. However, this in only first step of the whole science program. For achieving the all aims of EAST, high power auxiliary heating and current drive system has to be equipped [1–3]. As the highest heating efficiency and the most clear physical mechanism, Neutral Beam Injection (NBI) has been a best auxiliary heating method. Megawatt high-current ion source is the key to the neutral beam injector parts, which performance determines the neutral beam injection parameters. In order to achieve 2–4MW neutral beam injection in EAST, a high-current ion source was independently developed in the Institute of Plasma Physics of Chinese Academy of Sciences is shown in Figure 1 [4–5]. Discharge chamber size is 650 mm × 260 mm. The depth of the multi pole magnetic high-power plasma generator is 300 mm. Ion source is consisted of 16 credit card type filaments which diameter is 1.5 mm and length is 15 cm, arranged alone two lines. All of the filaments were parallelly fed, in order to reduce the nonuniformity of plasma density caused by uneven heating power. Four electrode structures is employed in the extraction system, which owns many advantages, such as simpler structure, good uniformity of plasma, large area extraction, larger flow intensity, high discharge efficiency and high

proton ratio [6–7]. Debugging experiments were also carried out using hydrogen. Measurements indicate that EAST-NBI-1 has the ability of neutral beam injection for EAST. A extraction beam with energy of 80 kev, current of 40A, the pulse width of 1 second has been achieved with Filament voltage of 6.5–8V and current of 2600A after running with 10 s, and arc voltage of 80–130V, arc discharge power of 60–115KW after running with 1–2 s, which performance is better than the initial design objectives. In order to further increase heating efficiency and improve its performance, the study on the source discharge characteristic is quite needed.

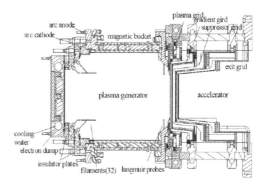

Figure 1.   Ion source of EAST-BI.

2 MEASUREMENT SYSTEM AND MEASUREMENT METHODS FOR DISCHARGE PARAMETERS

Currents of filament (I_f) and arc (I_{arc}) were measured using the magnetic balanced hall current sensor. Voltages of filament (V_f) and arc (V_{arc}) were obtained by a divider resistance. The pressure in discharge chamber was measured by a diaphragm vacuum gauge. Saturation ion current density of plasma (J_i) was measured using the electrostatic probe [8].

When NBI works, ion source is suspended in high pressure with plasma electrode as the reference potential. Therefore, measured voltage value on the ion source is relative to the plasma electrode potential. For the safety of measurement and data collection system, this pressure is ground connected as shown in Figure 2.

3 RESULTS AND DISCUSSION

During the debugging of ion resource, direct-current main with 0–5500A and 0–20V was used as filament supply, and direct-current main with 0–3000A and 0–200V was used as power supply of arc. Nitrogen with 0.5 Pa was used as the working gas, direct-current main with 0–100KV and 0–100A was used as the plasma grid supply.

3.1 Relationship between filament current and arc current

Since hot cathode arc discharge was used to form plasma, pressure was limited to 0.5 Pa. When increasing the filament current, increase in arc current was observed is shown in Figure 3, which can be attributed to the fact that the electrons needed for ionizing hydrogen gas was provided by filament

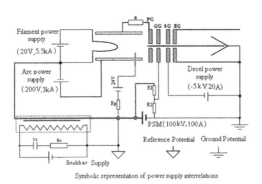

Figure 2. Measurement circuit.

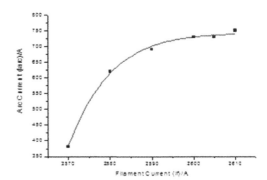

Figure 3. Relationship between filament current and arc current.

current. According to theory calculation, saturated thermionic emission current density and Emitter absolute temperature have a relationship:

$$J_e = AT^2 e^{-\frac{e\phi}{kT}} \qquad (1)$$

3.2 Relationship between filament current and arc current

Since hot cathode arc discharge was used to form plasma, pressure was limited to 0.5 Pa.

From the above equation, it can be observed that saturated thermionic emission current density will increase with increasing filament temperature. It is obvious that increasing filament current will increase filament temperature. Therefore, increasing filament current could lead to the increase in saturated thermionic emission current.

According to the theory of plasma discharge, the stability criterion of double cathode sheath layer is:

$$J_e \le \left(\frac{M}{m}\right)^{1/2} J_i \qquad (2)$$

where J_i is the positive ion current density reaching the cathode surface, and M and m are the masses of ion and electron. Because $M \gg m$, so $J_e \gg J_i$. Therefore, $Ia = Ii + Ie \approx Ie$. This means discharge current is approximately equal to current from cathode. Therefore, the arc current also increases with filament current. When filaments current increase to 2600A, the arc current does not longer increase with filament current. This is because the filament emission has been space charge limit state, and the number of electron is only determined by arc voltage.

3.3 Relationship between emission current density of filament and arc voltage

The effect of arc voltage on the filament emission electron density was also investigated. According to the equation:

$$I_e \approx n_i C_S e s_L$$

where C_S is the sound speed of ion, S_L are the overall electron loss area, n_i is the plasma density. The change in arc voltage with emission current density of filament can be obtained by using emission current density of filament–temperature curve, arc current-saturation ion current curve and arc voltage-current curve. The pressure and filament current remain as 0.5 Pa and 2600A respectively. The \ characteristic curve in Figure 4 shows that current density of filament increases with increasing arc voltage. This implies that on the precondition of keeping emission current density unchanged, an increase in lifetime of filament can be achieved by decreasing filament current and increasing arc voltage.

3.4 The relationship between arc current and plasma density

From Figure 5, it can be observed that during a course of discharge, the arc current and saturation ion current density has a similar tendency. Many discharge data show that arc current is proportional to saturation ion current density (see Fig. 6). Meanwhile, it is very known that beam current is also proportional to saturation ion current density. Therefore, saturation ion current density can be used to estimate the beam current and arc current. The measured relationship better arc current and saturation ion current density is meaningful.

The stability of electric is in quiet important for high-current ion resource, especially for the beam.

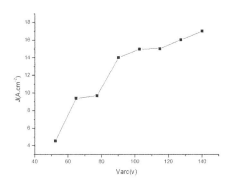

Figure 4. Relationship between emission current density of filament and arc voltage.

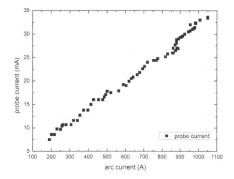

Figure 5. Arc current vs probe current.

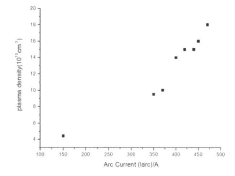

Figure 6. Arc current plasma density.

Ion current of the beam I_{acc} has a relationship with the voltage of the beam V_{acc}

$$I_{acc} = P V_{acc}^{3/2}$$

where P is the extraction perveance. Since the proportionality between ion current and plasma density, ion current also has a proportionality relationship with arc emission current. From volt-ampere characteristics of ion source, it can be observed that no negative resistance region, indicating better stability of arc discharging. The improvement in the discharge power supply is needed, as good constant current output characteristics can increase the stability of ion discharging, which can be observed from the efficiency equation of arc.

$$\eta_{arc} = \frac{I_{acc}}{P_{arc}} = \frac{I_{acc}}{V_{arc} \times I_{arc}} \tag{3}$$

4 CONCLUSION

From this study, it can be observed that changes in arc discharging parameters of ion resource have a influence on the discharge characteristic. The systematic study will help to enhance stability of running of resource, increase arc efficiency and improve power of beam. An initial investigation has been performed. More works need to be performed to give further explanation in the future.

REFERENCES

[1] Hu C.D. The Development of a Megawatt-Level High Current Ion Source. Plasma Science & Technology, 2012, 14(1):75–77.

[2] Wan Y.X., Li J.G. and Weng P.D. 2006, Plasma Science and Technology, 8: 253.

[3] Hu L.Q., Wan B.N., Hu C.D. et al. Rev. Sic. Instrum., 75: 3496, 2004.

[4] Hu L.Q., Zhang X.D., Yao R.H. 2006, Nuclear Techniques, Vol. 49, No. 2 149:152.

[5] Xu Y.J., Hu C.D., Liu S. et al. 2012, Chinese Physits Letters, 29: 035201.

[6] String W.L., Ryan P.M. Tsai C.C. et al. 1979, Rev. Sci. Tnstrum. 50: 102.

[7] Okumura Y., Horiike H., Mizuhashi K. 1984, Rev. Sci. Tnstrum. 55:1.

[8] Hu C.D. NBI Team. 2012, Plasma Science and Technology 14:871.

Future Information Engineering and Manufacturing Science – Zheng (Ed)
© 2015 Taylor & Francis Group, London, ISBN 978-1-138-02644-5

On the security of recharging social security cards

Jianchun Yang

School of Business Administration, Guizhou University of Finance and Economics, Guiyang, China

ABSTRACT: This paper, based on the program of embedding financial functions into social security cards, finds out a solution to security problems on recharging social security cards through the security technology of PSAM cards. This method could guarantee the security and reliability of transactions through the keys saved in PSAM cards. Particularly, it introduces the real-time security authentication of sophisticated data stream and the combined verification of cards and servers through network to ensure the each transaction not forgeable so that the security of recharging social security cards could be guaranteed.

Keywords: social IC cards; recharging; security technology of PSAM cards

1   INTRODUCTION

With the constant promotion of social security cards, a growing number of financial functions are incorporated into these cards, which could promote the cooperation among multiple fields. It could avoid wasting resources to develop the system of social security cards based on the established network platform of banks. Such a move could better achieve information sharing and build a resource-saving society. Social security cards with financial functions could support not only the non-financial applications, like identity credential, information query and settlement of medical costs, but also the financial applications, such as payment for social security fees and obtainment of wages and welfare. What's more, it could also support the financial functions of debit cards. For the cards with financial functions, the security is their most important property. And with the development science and technology, there emerge different methods of committing financial crimes, thus put higher requirements for the security of the applications with financial functions. How to ensure the security of recharging social security cards through the existing IC cards technology is a big issue to be solved for the program of injecting financial functions into social security cards. Only such security is guaranteed, could the IC cards with financial functions be widely promoted.

2   PRINCIPLES OF THE SECURITY DESIGN

As the core of security management of terminal equipment, PSAM cards are applied into the terminals of bank outlets, merchants' POS machines and all kinds of self-service terminal equipment. In order to achieve the security of recharging social security cards, the recharging terminals should utilize specific PSAM cards. Being a special type of IC cards, PSAM cards are mainly used to save authentication keys. Therefore, as long as the authentication keys are saved safely, the security of the overall application could be enhanced.

2.1   *Basic principles of IC cards*

The internal structure of IC cards is composed of CUP, ROM, RAM, EEPROM and I/O, and constructs a complete computer security system. Users' data is placed in the EEPROM protected by the logic encryption; the COS mask is in the ROM and the RAM stores the intermediate variable when COS is running. For instance, users' process keys are placed in the RAM when they are formed, and if the system loses power, it would lose the keys automatically, thus ensuring the security of users' keys.

The file system within the IC cards is organized according to the *Chinese Specifications on Financial Integrated Circuit Card*, and it includes Master File, Dedicated File and Elementary File. The specific file structure could be seen in the Chart 1.

2.2   *Security principles of PSAM cards*

PASM cards have a specific key for authentication at terminal equipment. This key could be utilized in the recharging process for authentication so that the central system would authorize the use of terminals for recharging. The security of PSAM

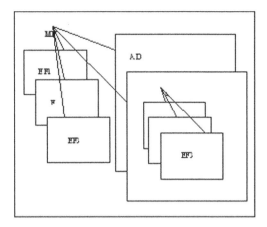

Chart 1. Hierarchical structure of the file system of IC cards.

cards is realized through control the generation of keys within the internal structure, and the security of the central system, users' cards and terminals of recharging is also guarantee through the dual authentication: external authentication and internal authentication.

One of the features of PSAM cards is their universality. After going through the special treatment of initialization, they could be used in different types of machines. Since they support the mechanism of issuance by multiple parties, at the initial stage of initialization, different issuers could create file structures and load different keys at both the cards and terminals based on their specific demands so as to achieve different security control.

3 SECURITY DESIGN OF RECHARGING PROCESS

3.1 Methods of achieving security authentication

The security mechanism of PSAM cards is to use the cards' property of saving keys in the whole authentication process. This could effectively reduce human's management on keys so as to achieve the security of the system, users' cards and terminal equipment though different authentication methods.

Before issuing cards to users, the master key is saved in the PSAM cards, and generates a sub key after the encryption on the unique bytes (such as the applied sequence number) of users' cards, and then the sub key is incorporated into the users' cards. Since the uniqueness of each applied sequence number, each card has a unique sub key.

Once the key is injected into to the card, it would not emerge outside of the card. The sub key generated by the master key of the PSAM card is stored in the RAM for encrypt and decrypt data. And during each authentication, in order to avoid the risk of data being cracked because of the direct participation of the sub key into computation, the unique bytes of the sub key are diversified so that the generated process key would encrypt the data that need authentication. After authentication, the process key would disappear.

The authentication process involving PSAM cards is stated as follows:

1. *The Process of Authenticating the Validity of Application System (External Authentication)*
   The application would ask the user's card to send a random number X to it, and the sub key produced at the internal authentication within PSAM cards would encrypt the random number (to replace the application to achieve the encryption through the key and computation) and produce a cipher text Y and also send Y to the application. And the application would send the cipher text to the user's card for external authentication. The user's card would decrypt the cipher text Y and produce the result Z through the external authentication, and also compare X and Z. If they are the same, it means the application is valid.

2. *The Process of Authenticating the Validity of Users' Cards (Internal Authentication)*
   The application would ask the PSAM card to generate a random number X and send the number to the user's card. The internal authentication within the user's card would encrypt the random number and send back the produced cipher text Y to the application (to replace the application to produce the cipher text). And the application would send the cipher text Y to the PSAM card and ask it to authenticate the text. The PSAM card would decrypt the cipher text Y through its external authentication and produce the result Z, and then compare X and Z. If they are the same, it means the user's card is valid.

3. *The Process of Authenticating the Validity of Recharging Terminals (External Authentication)*
   The recharging terminal would ask the PSAM card to generate a random number N and send the number back to it. It would encrypt the random number and produce a cipher text M (the key is stored in the PSAM card and recharging terminal). The terminal would send the cipher text to PSAM for its external authentication. The PSAM card would decrypt the cipher text M and produce a result N1, and also compare N and N1. If they are the same, it means the recharging terminal is valid.

3.2 The encryption and decryption computation of data

3.2.1 The encryption computation of data

The procedures of data encryption are stated as follows:

The first step: LD stands for the length of clear data, and the new data block produced by LD is added in front of the clear data.

The second step: the data block in the above step would be dissembled to 8-bytes data blocks, namely D1, D2, D3 and D4. The length of the last data block might be less than 8 bytes.

The third step: if the length of the last (or only) data block is just 8 bytes, the fourth step begins. If it is less than 8 bytes, add the number of "80" in the hexadecimal number system on its right; if the length of the latter data block reaches 8 bytes, the fourth step begins. Otherwise, the 1-byte number of "0" in the hexadecimal number system would be added on its right till the length reaches 8 bytes.

The fourth step: The corresponding key of each data block is encrypted by SingleDES or TripleDES based on the length of the key.

The fifth step: after the computation completes, all encrypted data blocks are connected in the original order (encrypted D1, encrypted D2, encrypted D3 and encrypted D4). And then, the produced data block is transmitted to the data field of command.

3.2.2 The decryption computation of data

The procedures of data decryption are stated as follows:

The first step: the data block in the data field of command would be disassembled to 8-bytes data blocks, namely: D1, D2, D3 and D4. The decryption processes of each data block are as follows.

The second step: after the computation completes, all decrypted data blocks are connected in the original order (decrypted D1, decrypted D2, decrypted D3 and decrypted D4). The produced data block is composed of LD, clear data and padding characters.

The third step: since LD stands for the length of clear data, it is used to recover the clear data.

The Realization Process of Recharging Module.

Before recharging terminals are used, they to be registered in the settlement center. After registration, the recharging terminals could handle transactions such as recharging. During each recharging in the terminals, they have to exchange data with the settlement center three times.

4 CONCLUSION

This paper, based on the program of injecting financial functions into social security cards, finds out a solution to security problems on recharging social security cards through the security technology of PSAM cards. This method could guarantee the security and reliability of transactions through the keys saved in PSAM cards. Particularly, it introduces the real-time security authentication of sophisticated data stream and the combined verification of cards and servers through network to ensure the unforgeability of each transaction so that the security of recharging social security cards could be guaranteed.

REFERENCES

[1] Cai Songyuan. 2013. On the Social Security Cards with Financial Functions: Exampled by the Promotion of Social Security Cards in Heilongjiang Province. *Finance Theory and Teaching*. (in Chinese).
[2] Zheng Bin & Zhang Renbin & Song Wei. 2010. On the Security System of Social Security Cards. *Computer Knowledge and Technology*. (in Chinese).
[3] Wang Aiying. 2009. Smart Card Technology: IC Cards and RFID Tag. Tsinghua University Press. (in Chinese).
[4] EMV integrated circuit cards Certification for payment system, Version 4.3, Book3: Application specification, Nov 2011.

Future Information Engineering and Manufacturing Science – Zheng (Ed)
© 2015 Taylor & Francis Group, London, ISBN 978-1-138-02644-5

On the risks control of the centralized authorization system in banking business

Yu Han

School of Finance, Guizhou University of Finance and Economics, Guiyang, China

ABSTRACT: The centralized authorization system in banking business is a separation system for front office business and back office business, which needs to be real time and could control risks and improve efficiency. This paper explores the design of the centralized authorization system in banking business and the relevant risk control issues in its realization process, from the aspects of the security control of authorization data, security control of authorization business, supervision on authorization business, the differentiation and control of true and false information, the stable guarantee of business after centralized authorization and concurrent risks control of the system, so as to find out effective solutions to problems in banking business authorization.

Keywords: banking business; centralized authorization; risks control

1 INTRODUCTION

Banking business develops rapidly with the fast development of the world's current science and technology, and clients are raising increasingly high requirements for the quality of banking service. Facing a growingly fierce market competition in the industry, banks find it imperative to increase their service efficiency and quality. Meanwhile, with the rapid expanding of banking business, the contradictions between the on-site authorization and business development within banks have become more visible, and they also become the bottlenecks which limit the business efficiency of bank tellers, leaving banks some certain security risks. The existing mode on business authorization has fallen behind the business development, so banks need to change their current authorization mode and find an alternative one.

2 INTRODUCTION TO THE CENTRALIZED AUTHORIZATION SYSTEM

The centralized authorization system in banking business is a new and web-based authorization mode, in which bank tellers transmit the video screen images of clients, images of slips and certificates and voice messages that need authorization to the terminals of authorization officers at the designated Centralized Authorization Center and those officers examine and complete authorization at their terminals. This system is a real-time separation system for front office business and back office business, and it could control risks, improve efficiency and optimize personnel allocation, thus serving as an effective solution to the business authorization in banks.

Such a system is specifically designed for bank tellers at bank outlets when conducting remote centralized authorizations, and it needs no altering on the back office and front office procedures of the business system. The software of applying for business authorization operates at the bank tellers' devices within the business system in a weak coupling manner, but the software of agreeing to the authorization and the server software operates on

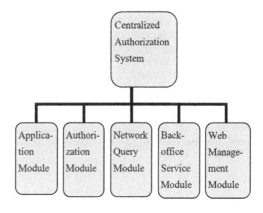

Chart 1. Function modules of centralized authorization system.

devices outside of the business system. The centralized authorization system is composed of the application module, authorization module, network query module, back-office service module and Web management module (see Chart 1).

3 SYSTEM RISKS CONTROL

3.1 *Security control of authorization data*

In business authorization, the security of authorization data is the premise for successful authorization. Within the system, the authorization data includes business data and image data.

Business data achieves its synchronization in a point-to-point manner. Therefore, once the business that needs authorization has produced business links in the synchronization process, any altering on business data requires a real-time synchronization of the data on both application module and authorization module first. The changes on business data from either module is realized transparently and no covert operations at one end would occur so that the altering on business data is transparent. Besides, the data communication on the web is achieved through simulated synchronization. If the data is manipulated on the web, the data format would not meet the requirement for correct format of simulation display. Therefore, there would be no correct business images on both the authorization module and application module, and business operations could not be realized.

Through the control on the data analysis format after transparent data synchronization, the correctness and security of authorization data thus could be ensured in business operations.

The images of authorization business certificate are needed in business authorization, and the imaging devices in the system are also specifically designed to achieve security control. After the images are transmitted, the system would also ensure that the images are captured on the corresponding imaging instruments. The authorization would not pass in the system without scanned pictures or with other images.

As for the security control on imaging devices, an encryption module is designed in the devices. Therefore, a verification of encryption is needed when images are captured; otherwise, the devices would not be allowed to be connected to the authorization system.

3.2 *Security control of authorization business*

The security control of authorization business is achieved through the identity authentication of authorization officers and bank tellers. Only when both the authorization officers and bank tellers pass the identity authentication at the remote authorization system, could they complete the authorization business with a logging status at the system.

What's more, the system classifies authorization business of each authorization officer, which ensures that each authorization operation is reasonable. Such classification is made by the system automatically and authorization officers are prohibited to conduct operations that are incompatible with their authority. At the same time, authorization officers are also designated by the system automatically, so bank tellers do not know the authorization officers when applying for authorization, which could avoid the purposeful application of bank tellers.

3.3 *Supervision on authorization business*

The system would keep a detailed record on the information of bank tellers, authorization officers and authorization business. Based on the detailed record, the system also has an examination and query module for the business, so that each authorization business would be queried and examined, which ensures that once there are some problems or disputes on the business, such a detailed and accurate record could provide the foundation for in-depth analysis and judgment.

3.4 *Differentiation and control of true and false information*

Once the central authorization system is implemented, the original authorization operations are changed in their forms. In the authorization process of banks, authorization directors have to not only examine the correctness of the business information, but also help bank tellers to identify whether the certificates and information of clients are true or false. Since authorization directors are not on site after the implementation of the system, it increases the difficulty for them to differentiate just through the images of those certificates. Thus, in response to the above situation, the following advice is proposed:

Firstly, banks should explicitly stipulate that bank tellers all responsible for distinguish between the true and false receipts, and authorization directors just play a supporting role in the process.

Secondly, the imaging instruments have to guarantee that their shooting effect is good enough to ensure the high similarity between the images of certificates and receipts and the original ones.

Thirdly, the system should recognize the source channels of the images of certificates and receipts. Through the technological and procedural controls on the source channels, the system could stop bank tellers from replacing the original images with other ones.

Fourthly, as for the verification of the information of clients, the system could achieve it through the check on clients' 2nd-generation ID Cards and their head portraits taken by auxiliary cameras.

3.5 *Stable guarantee of business after centralized authorization*

Since the authorization operations are transacted in a centralized approach after the implementation of the centralized authorization system, the system could ensure the stable operations of banking business after centralized authorization through the following aspect.

Firstly, the centralized authorization is part of the business system, and it uses the same network as banks' host computers do. Therefore, there are no higher risks for the centralized authorization system in the same network. If the network malfunctions, both the banks' host computers and the centralized authorization system would be suspended.

Secondly, the system also has risk controls on the network flow. Since the system is set on the main network, it would only function well when the speed of broadband network reaches 2 megabits per second.

Thirdly, the centralized authorization system is deployed in hot-standby manner so as to ensure that the system is under constant operation to the largest extent.

Fourthly, based on the current situation of banking management, all outlets would still retain outlet directors or business managers. Each outlet could retain one authorization director. Under normal circumstances, the director does not need to get involved in the authorization affairs; but in events of emergency, he could come to the rescue.

3.6 *Concurrent risks control of the system*

The implementation of the system would inevitably lead to reduced efficiency of business transaction during the business peak period. Therefore, authorization centers have to be deployed in a flexible manner so that the business requirements could be fulfilled reasonably.

Based on the actual situation of banks, the systems are proposed to be deployed at the level of cities, which could both decrease the banks' business pressure and reduce the network risks. In each authorization center, the ratio between authorization directors and bank tellers is one to twenty, but during business peak periods (such as the Spring Festival and other holidays), the ratio could increase to one to fifteen so as to fulfill the bank's business demands and maintain the business efficiency.

4 CONCLUSION

This paper explores the design of the centralized authorization system in banking business and the relevant risk control issues in its realization process, from the aspects of the security control of authorization data, security control of authorization business, supervision on authorization business, the differentiation and control of true and false information, the stable guarantee of business after centralized authorization and concurrent risks control of the system, so as to find out effective solutions to problems in banking business authorization.

REFERENCES

[1] Shari Lawrence Pfleeger (American), Joanne M. Atlee (Canadian). 2010. Soft Engineering: Theory and Practice. (4th Edition) Beijing: Posts & Telecom Press. (in Chinese).
[2] Ashok Iyengar (American). 2009. Websphere Business Integration Primer. Beijing: China Machine Press. (in Chinese).
[3] Xu Chengxian. Financial Information Security. 2013. Beijing: Tsinghua University Press.
[4] Li Gaicheng. Financial Information Security Project. 2010. Beijing: China Machine Press.
[5] Julia H. Allen & Sean Barnum & Robert J. Ellison. 2009. Software Security Project. Beijing: China Machine Press.

Future Information Engineering and Manufacturing Science – Zheng (Ed)
© 2015 Taylor & Francis Group, London, ISBN 978-1-138-02644-5

Optimal CVRP depot location determination using two-tier PSO

R.-M. Chen & Y.-A. Chen

Department of Computer Science and Information Engineering, National Chin-Yi University of Technology, Taichung, Taiwan, R.O.C.

ABSTRACT: Vehicle Routing Problems (VRPs) currently encountered may vary as a result of different restrictions. Since the location of a depot has a very significant effect on the transportation cost, this study therefore intends to propose a two-tier PSO, inner and outer layers, to obtain the best location to establish a depot and the optimal vehicle routes using the determined depot as the center. The inner-layer PSO is applied to obtain vehicle routes while the outer-layer PSO is to establish the most ideal depot location. A new particle encoding is proposed for the inner-layer PSO to reduce the customer assignment efforts. Decrease of inertia weight and addition of local searches are adopted to improve the quality of solutions. In the end, capacitated vehicle routing problems from the OR Library were applied to run tests and compare the cost difference between having planned the optimal depot location and not having done so, while the experiment results are presented to show the importance of depot locations.

Keywords: Meta-heuristic; Optimization; Particle Swarm Optimization (PSO); Logistics Management; Scheduling problem; Capacitated Vehicle Routing Problem (CVRP)

1 INTRODUCTION

The Vehicle Routing Problem (VRP) is a scheduling problem encountered in logistic arrangement, an extension from the traveling salesman problem. There are dissimilar types of VRPs; this work focuses on Capacitated Vehicle Routing Problems (CVRP). In Operation Research (OR), VRPs have been proven to be NP-hard. For this reason, many researchers have come up with a variety of meta-heuristic methods in recent years to cope with vehicle routing problems, including the Genetic Algorithm (GA) (Munawar et al. 2009), Particle Swarm Optimization (PSO) (Chen et al. 2006) (Wang et al. 2009), and Ant Colony Optimization (ACO) (Yu et al. 2009). Among them, PSO has the advantage of requiring less parameters and faster convergence rates, and are therefore adopted by many scholars to solve various problems.

The objectives are mostly to plan the least costly vehicle routes when locations of depots and customers are already known. However, the transportation cost is an extremely important consideration for logistic businesses. Besides fine vehicle route planning, good choice of depot locations is also an important way to reduce business costs. Therefore, two PSO algorithms, inner-layer and outer-layer PSO, are adopted in this work respectively for vehicle routing and depot location planning. The outer-layer PSO is first applied to establish

the optimal depot location before the inner PSO is used to produce the optimal routing. In the meantime, five different decreasing approaches are adopted to adjust the inertia weight of the PSO, and six local search strategies are also employed to improve solution quality. The results then serve as the basis to observe the effect of the depot location on cost reduction.

2 PROBLEM DESCRIPTION

The VRP was first proposed by G.B. Dantzig & J.H. Ramser (1959). The problem involved the demand of each of many customers scattered in different places. The depot had to assign vehicles to visit all the customers and satisfy their needs by planning the shortest total travel distance without violating any restrictions.

In a CVRP, there are a fixed number of customers and a depot. The locations of each customer and the depot are known. Set $C = \{c_0, c_1, c_2, ..., c_n\}$ stands for the set composed of the depot and customers. The c_0 is the depot and n is the number of customers. The depot will send out a fleet comprising several vehicles. The vehicle fleet $V = \{v_1, v_2, ..., v_k\}$, in which k is the number of vehicles. Q_a stands for the capacity of No. a vehicle. Each vehicle must leave from the depot and return to the depot in the end. Each customer has to be visited once and once only. The Objective and restrictions of the CVRP are defined as follows:

$$Fitness = \min \sum_{i=0}^{n} \sum_{j=0}^{n} \sum_{v=1}^{k} d_{ij} X_{ij}^{v} \quad i \neq j \qquad (1)$$

In Eq. (1), the objective function of the VRP is defined as to obtain the shortest total travel distance. d_{ij} is the distance from customer i to customer j and X_{ij}^{v} stands for whether vehicle v will go from customer i to customer j. When $X_{ij}^{v} = 1$, it means vehicle v travels from customer i to j. On the other hand, when $X_{ij}^{v} = 0$, vehicle v does not travel from customer i customer j.

3 A COMBINATION OF TWO-LAYER PARTICLE SWARM OPTIMIZATION WITH HEURISTICS

3.1 Particle Swarm Optimization (PSO)

Particle swarm optimization is a type of collective intelligence. It was put forward first in 1995 by Kennedy & Eberhart (1995). Suppose that an N number of particles are scattered in an L-dimension space. The position of the ith particle ($i = 1, ..., N$) is composed of L vector components. $X_i^V = \{X_{i1}, ..., X_{iL}\}$ indicates the position vector of particle i, in which X_{ij} stands for the jth vector component of the ith particle. The velocity vector of the ith particle is also composed of L components – $V_i = \{V_{i1}, ..., V_{iL}\}$. The optimal individual experience of the ith particle is thus represented as $P_i = \{P_{i1}, ..., P_{iL}\}$ whereas the optimal swarm experience (Gbest) is $G = \{G_1, ..., G_L\}$. These velocity and position update rules are as shown in Eq. (2):

$$\begin{cases} V_{ij}^{new} = w \times V_{ij} + c_1 \times r_1 \times (P_{ij} - X_{ij}) + c_2 \times r_2 \times (G_j - X_{ij}) \\ X_{ij}^{new} = X_{ij} + V_{ij}^{new} \end{cases}$$
$$(2)$$

In Eq. (2), w is the inertia weight used to determine the level of effect of the previous velocity on the new velocity. Meanwhile, c_1 and c_2 are learning factors which have an effect on particles' learning of global experience and individual experience, whereas r_1 and r_2 represent random numbers between [0, 1].

3.2 Solution representation

In PSO, the components of position vector are real numbers; therefore, before the inner-layer PSO performs visit order decision and fitness calculation, the (X_i^V) position vector has to be converted into the visit sequence of a vehicle while each customer the vehicle is assigned to have to be

determined before assessment can be conducted. Hence, this work suggests a new particle encoding scheme to reduce the customer assignment efforts for the inner-layer PSO which includes customers and vehicles assigned, as shown in Figure 1.

The visit sequence of each vehicle and each customer a vehicle is assigned to are determined by using a random key scheme. Take six customers and three vehicles for example. Figure 2 shows a solution (X_i^V) obtained. The key values are associated to the components of X_i^V to generate a set of visit sequence and this sequence is then defined as the vehicle assignment with Veh_k as the dividing points.

The particle for the outer-layer PSO is conducted by using two components respectively representing the X and Y coordinates of the depot, i.e., $X_i^D = \{X_{i1}^D, X_{i2}^D\}$. The fitness calculation is then performed by transferring the depot coordinates (X_i^D) to the inner-layer PSO for optimal routing calculation and the resulted optimal total travel distance is adopted as the fitness of the outer-layer PSO.

3.3 Inertia weight adjustment

In PSO algorithms, the inertia weight w has an effect on the size of search paces of particles. It is usually used to control the search paces. Many kinds of linear and non-linear curves are applied to control the decrease of w value to adjust search paces.

Linear decrease (Bratton et al. 2007) is stable reduction of search paces with increasing numbers of iterations, as shown in Eq. (3).

$$w_{iter} = (iter_{max} - iter)(w_{max} - w_{min})/iter_{max} + w_{min} \qquad (3)$$

In Eq. (3), w_{iter} is the inertia weight of the current iteration, w_{max} the maximum inertia weight in the beginning, w_{min} the final inertia weight, $iter$ the

1	1	2	⋯	n-1	n	n+1	⋯	n+k-2	n+k-1
X_i^V	X_1^V	X_2^V	⋯	X_{n-1}^V	X_n^V	X_{n+1}^V	⋯	X_{n+k-2}^V	X_{n+k-1}^V
Key	Cus_1	Cus_2	⋯	Cus_{n-1}	Cus_n	Veh_1	⋯	Veh_{k-1}	Veh_k

Figure 1. Solution representation (inner-layer PSO).

Key	Cus_1	Cus_2	Cus_3	Cus_4	Cus_5	Cus_6	Veh_1	Veh_2
X_i^V	2	1.3	0.8	2.4	1.9	0.2	1.2	2.1

Sorting in ascent order ⬇

X_i^V	0.2	0.8	1.2	1.3	1.9	2	2.1	2.4
Key	Cus_6	Cus_3	Veh_1	Cus_2	Cus_5	Cus_1	Veh_2	Cus_4

Assign customer to vehicle ⬇

Vehicle assignment	Cus_6 Cus_3	Cus_2 Cus_5 Cus_1	Cus_4
	Veh_1	Veh_2	Veh_3

Figure 2. The solution decoding process (inner-layer PSO).

34

current iteration number, and $iter_{max}$ the maximum iterations.

A non-linear decline curve (Lei et al. 2006) is a search strategy applied as shown in Eq. (4). When s is set to 0.7 (lower crescent), global exploration is intended; local exploitation is intended as s is set to 10 (upper-crescent).

$$w_{iter} = (1 - iter/iter_{max})/[1 - s \times iter/iter_{max}] \qquad (4)$$

A Sigmoid decreasing curve (Malik et al. 2007) has the advantages of both the non-linear low-crescent decline curve and the non-linear upper crescent decline curve strategic, as shown in Eq. (5).

$$w_{iter} = 1/\left(1 + e^{u \times (iter - n \times iter_{max})}\right) \qquad (5)$$

In that, u and n are the control parameters for the Sigmoid decreasing curve, $u = 0.01$ and $n = 0.5$ are commonly used to adjust the decline position of the curve.

An asymmetric non-linear S decline curve (Chen et al. 2013) was proposed in our previous studies, as shown in Eq. (6).

$$w_{iter} = 1/\left\{1 + [[iter_{max} - (iter_{max} - iter)]/ \right. $$
$$\left. (iter_{max} \times 0.47)]^8\right\} \qquad (6)$$

3.4 Local search

The local search is a search strategic to generate new solutions in neighbor of the current solution to try finding a better solution. A number of local searches are conducted and the best solution is selected to be the starting point of the next iteration.

In this work, the local search is applied only in the inner-layer PSO; Swap, Insertion, Reverse, Section swap, Swap reverse (Szeto et al. 2011), and double swap (Ziarati et al. 2011) local searches were tested.

3.5 Simulated Annealing (SA)

Simulated Annealing (SA) was proposed by Kirkpatrick in 1983 (Kirkpatrick et al. 1983). SA is a scheme commonly applied for escaping from local optimal. Hence, to avoid PSO trapping on local optimal, simulated annealing is adopted in this investigation.

Solutions found via local searches are first assessed to determine whether they are better than the original solution. If yes, SA is not performed and the better solution is adopted as the starting point of the next iteration; if not, Eq. (7) is applied to calculate the acceptability and compare with a random number to evaluate whether the new solution can be accepted as the starting point of the next algorithmic calculation.

$$P_{SA} = \exp(\Delta E / t)$$
$$\Delta E = \left(f(x') - f(x)\right)/f(x) \qquad (7)$$
$$t = \alpha^{iter} \times t$$

In that, P_{SA} is the acceptability of the new solution, ΔE the energy difference, $f(x')$ the fitness value of the local search solution selected, $f(x)$ the fitness of the original solution, t is the temperature, α is the temperature decrease rate, and $iter$ the number of iteration.

4 EXPERIMENTAL RESULTS

To verify the performance of the method proposed in this article, simulations of famous Augerat benchmarks were conducted.

The experiment data were processed in two stages. The first stage is to find out the best dynamic inertia weight curve and the local search scheme of the inner-layer PSO. The second stage is comparison the results of inner-layer PSO which with and without outer-layer PSO. The inner-layer PSO parameters were 20 particles, the learning factors $c_1 = c_2 = 2$, and the number of iterations was $500 \times$ customers. The outer-layer PSO parameters are 8 particles. 20 iterations were conducted. All the data were the best deviations from ten trial times. The deviation is shown in Eq. (8).

$$DEV(\%) = \frac{Obtained\ makespan - BKS}{BKS} \times 100\% \quad (8)$$

Five different inertia weight adjustments as stated above were tested. Linear Decrease (LD), Lower-crescent Decline curve (L_D), Upper crescent Decline curve (U_D), Sigmoid decreasing curve (S_D) and Asymmetric nonlinear S Decline curve (AS_D). The range of decrease of w is 0.9~0.4 and the test outcomes are as shown in Figure 3. The test outcomes indicate that the average deviation with the asymmetric nonlinear S-decreasing curve is the lowest. Hence, the asymmetric nonlinear S-decreasing curve was used to adjust the inertia weight in subsequent local searches.

In local searches, Swap (LS_1), Insertion (LS_2), Reverse (LS_3), Section swap (LS_4), Swap reverse (LS_5) and Double swap (LS_6) were tested and the tests also include with and without SA involved. The results are as shown in Figure 4. They indicate that the swap method with SA produces the best

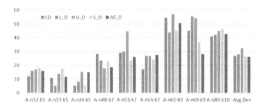

Figure 3. Result of different inertia weight adjustments.

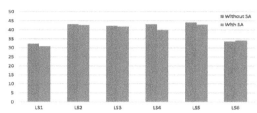

Figure 4. Result of using SA and without using SA.

Table 1. Result of two-layer PSO.

	Without outer PSO			With outer PSO			
	Best	Average	Worst	Best	Average	Worst	Gap
A-n32-k5	919	1022	1190	709	745.6	785	−210
A-n33-k5	769	840.9	873	679	698.1	712	−90
A-n34-k5	871	987.8	1163	698	734.3	776	−173
A-n48-k7	1339	1525	1664	1033	1105.3	1163	−306
A-n53-k7	1318	1536.6	1720	1140	1193.2	1252	−178
A-n54-k7	1501	1716.5	1883	1177	1243.9	1286	−324
A-n65-k9	1895	2076.5	2236	1545	1661.9	1739	−350
A-n69-k9	1792	2008.2	2185	1530	1629.3	1755	−262
A-n80-k10	2551	2870.8	3095	1787	1885.5	1971	−764
Avg							**−295.2**

outcome; therefore, the data are compared with the results which outer-layer PSO is added.

In this section, the results before and after the outer-layer PSO was applied are compared. The results are as shown in Table 1.

5 CONCLUSIONS

This study proposes using two-layer PSO to establish the optimal depot location and corresponding vehicle routing. A new particle encoding scheme is also proposed to reduce the customer assignment efforts. Various adaptive inertia weight schemes and a number of local search methods were tested. The experiment results reveal that involving asymmetric S-decreasing inertia weight with the swap approach produces the best outcome. Meanwhile after including the SA mechanism, the averaged best deviation also reduce 1.43%. Moreover, after applying the outer-layer PSO to adjust the depot location, the fitness of the questions for each scale significantly improves, by 210 for A-n32-k5 and by 764 for the A-80-k10. The average improvement is 295.2, proving that the location of a depot can indeed affect vehicle routing costs and the method proposed in this study can effectively obtain the best location to set up a depot.

ACKNOWLEDGEMENT

This work was partly supported by the National Science Council, Taiwan, under Contract NSC 102-2221-E-167-018.

REFERENCES

[1] Bratton, D., & Kennedy, J. (2007, April). Defining a standard for particle swarm optimization. In Swarm Intelligence Symposium, 2007. SIS 2007. IEEE (pp. 120–127). IEEE.
[2] Chen, A.L., Yang, G.K., & Wu, Z.M. (2006). Hybrid discrete particle swarm optimization algorithm for capacitated vehicle routing problem. Journal of Zhejiang University Science A, 7(4), 607–614.
[3] Chen, R.M., & Wu, D.S. (2013, April). Solving Scheduling Problem Using Particle Swarm Optimization with Novel Curve Based Inertia Weight and Grouped Communication Topology. In International Journal of Digital Content Technology and its Applications, 2013. JDCTA 2013, 7(7), 94–103.
[4] Dantzig, G.B., & Ramser, J.H. (1959). The truck dispatching problem. Management science, 6(1), 80–91.
[5] Kennedy, J., & Eberhart, R. (1995, November). Particle swarm optimization. In Proceedings of IEEE international conference on neural networks (Vol. 4, No. 2, pp. 1942–1948).

[6] Kirkpatrick, S., & Vecchi, M.P. (1983). Optimization by simmulated annealing. *Science, 220* (4598), 671–680.

[7] Lei, K., Qiu, Y., & He, Y. (2006, January). A new adaptive well-chosen inertia weight strategy to automatically harmonize global and local search ability in particle swarm optimization. In *Systems and Control in Aerospace and Astronautics, 2006. ISSCAA 2006. 1st International Symposium on* (pp. 4-pp). IEEE.

[8] Malik, R.F., Rahman, T.A., Hashim, S.Z.M., & Ngah, R. (2007). New particle swarm optimizer with sigmoid increasing inertia weight. *International Journal of Computer Science and Security, 1*(2), 35.

[9] Munawar, A., Wahib, M., Munetomo, M., & Akama, K. (2009). Implementation and Optimization of cGA+ LS to solve Capacitated VRP over Cell/BE. *International Journal of Advancements in Computing Technology, 1*(2), 16–28.

[10] Szeto, W.Y., Wu, Y., & Ho, S.C. (2011). An artificial bee colony algorithm for the capacitated vehicle routing problem. *European Journal of Operational Research, 215*(1), 126–135.

[11] Wang, Z., Li, J., Zhou, M., & Fan, J. (2009, November). Research in capacitated vehicle routing problem based on modified hybrid particle swarm optimization. In *Intelligent Computing and Intelligent Systems, 2009. ICIS 2009. IEEE International Conference on* (Vol. 3, pp. 289–293). IEEE.

[12] Yu, B., Yang, Z.Z., & Yao, B. (2009). An improved ant colony optimization for vehicle routing problem. *European journal of operational research, 196*(1), 171–176.

[13] Ziarati, K., Akbari, R., & Zeighami, V. (2011). On the performance of bee algorithms for resource-constrained project scheduling problem. *Applied Soft Computing, 11*(4), 3720–3733.

Future Information Engineering and Manufacturing Science – Zheng (Ed)
© 2015 Taylor & Francis Group, London, ISBN 978-1-138-02644-5

Deformed target location based on pattern recognition

Y. Chen & F.R. Huo
Changchun University of Science and Technology, Changchun, Jilin, China

L.Q. Zheng
Changchun New Industries Optoelectronics Technology Co. Ltd., Changchun, Jilin, China

ABSTRACT: Optical correlation technology is one of the key technologies for pattern recognition. But when target deformation exists, such as scale deformation and angular deformation, the intensity of correlation peaks will decrease to a great extent, which leads to failure of target recognition directly. In this paper, the principle and development of Maximum Average Correlation Height (MACH) filter algorithm is given. To make MACH filter suitable for real-time joint transform correlator, the improved MACH algorithm is formed by adjusting controlling parameters of the filter. Various targets with different attitudes are synthesized in frequency domain and projected to space domain. Amounts of simulative and optical experiments show the improved MACH filter have high deformation tolerance and suppression of background noise. As an example, taking an airplane as the target, simulative and optical experiments are carried out. The experimental results show the scale deformation tolerance can reach up to ±22% and angular deformation tolerance can reach up to ±13 degrees. The actual effect of the improved MACH filter algorithm has been confirmed very well.

Keywords: pattern recognition; scale deformed target; maximum average correlation height filter; frequency domain; space domain

1 INTRODUCTION

Technology of target recognition in the field of optical information processing has been developed very fast in recent years. In this paper, we adopted the Hybrid Optoelectronic Joint Transform Correlator (HOJTC) controlled by computer as the effective method of target recognition.

Generally speaking, the target is not stationary relative to the location of detector. Therefore, the target will produce spatial deformation (scale and rotation deformation) relative to the current template. Many methods are used to solve the recognition problem of deformed targets. Currently, there are many variants of Synthetic Discriminant Function (SDF) filter, such as Minimum Variance Synthetic Discriminant Function (MVSDF) filter, Minimum Average Correlation Energy (MACE) filter and MACH filter (Yuan, X.Y. et al. 2002).

MACH filter has been proven to be a powerful correlation filter among all the possible correlation recognition methods (Yuan, X.Y. et al. 1999). MACH filter has high deformation tolerance and can suppress the cluttered scene effectively. In this paper, we proposed an improved MACH filter and use it in JTC to sharpen correlation peaks for spatial deformed targets.

2 PRINCIPLE OF OPTICAL CORRELATION AND EXPERIMENTAL SETUP

Hybrid optoelectronic joint transform correlator is an optical recognizer, which is used to detect the similarity of target function and reference function by optical method. HOJTC composes of many key devices such as Spatial Light Modulator (SLM) with high resolution and square law detector CCD. The schematic diagram is shown in Figure 1.

There is a pinhole with the diameter 10 to 20 microns between the collimating system and the

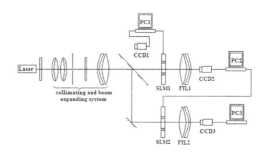

Figure 1. Schematic diagram of HOJTC.

beam expanding system, which has the function of spatial filtering. Collimating beam produced by beam expanding system is divided to two parts by the reflective mirror before SLM1, to illuminate SLM1 and SLM2 respectively. CCD1 is used to record outside cluttered images in real time, in which the target image is possibly involved. Before the target recognition, the target image $R(x, y)$ to be recognized is needed to be inputted to PC1 as the reference template in advance, which is located at the screen corner.

Suppose the detected target function is $T(x, y)$ and the cluttered scene function is $n(x, y)$. The joint image $F(x, y) = T(x, y) + n(x, y)$ recorded by CCD1 is inputted to PC1 for us to observe. The joint image $F(x, y)$ and the reference template $R(x, y)$ are transmitted to SLM1 at the same time, which is illuminated by collimated coherent light beam. After Fourier transform by Fourier Transform Lens 1 (FTL1), the Joint Transform Power Spectrum (JTPS) is recorded by CCD2 and shown in PC2, which is transmitted to SLM2 simultaneously.

The position of a target can be determined through the position of cross correlation signals. Autocorrelation outputs, are noise signals. They locate at the center of the screen, so they are easy to be separated from the useful cross correlation signals (Wang, W.S. et al. 2005).

3 MAXIMUM AVERAGE CORRELATION HEIGHT FILTER

3.1 Principle of maximum average correlation height filter

The Maximum Average Correlation Height (MACH) filter, a deformation-invariant filter, is a variant of well-known Synthetic Discriminant Function (SDF). SDF filter has four conflicting performance measures, namely, Output Noise Variance (ONV), Average Correlation Height (ACH), Average Similarity Measure (ASM) and Average Correlation Energy (ACE). The MACH filer offers good performance for three major criteria simultaneously: easy detection of correlation peak, good deformation tolerance and the ability to suppress clutter noise. The purpose of the designed filter is to make ACH largest, while minimizing ASM. However, ACE and ONV also need tradeoff to meet different applications. In practice, targets always locate in cluttered background. Thus the resulting correlation output will have noise variance which is noise background dependent. To obtain stronger intensity of correlation peaks, ONV should be minimized obviously. Of course, for moving targets, the ONV are always changing with the different positions

of targets in detected images. ONV expression is shown as equation (1).

$$ONV = h^+ P h \qquad (1)$$

where, P is a diagonal matrix whose diagonal entries are obtained from power spectral density and h is the SDF filter. "+" denotes the conjugates transpose of a complex vector (Refregier. 1991).

Meanwhile, to achieve good recognition results, it is necessary to reduce correlation function levels at all points except at the centers of the correlation peaks. Thus, ACE needs to be minimized as the equation (2) shows.

$$ACE = h^+ D_x h \qquad (2)$$

In equation (2), D_x is the diagonal matrix with the average power spectral density values as its diagonal entries. Minimizing ACE usually leads to low sidelobes in the correlation output.

ASM is used to minimize the dissimilarity between the reference template image and target image. In other words, ASM can increase the similarity between the joint images relatively, which is very important to target recognition. ASM, which can be thought of an average dissimilarity measure, is shown as equation (3).

$$ASM = h^+ S_x h \qquad (3)$$

where, S_x is a matrix containing the average Fourier transform of the training images.

ACH present the correlation energy of the reference template image and the target image. To recognize the target successfully, ACH needs to be maximized. ACH is shown as equation (4).

$$ACH = \left| h^T m_x \right| \qquad (4)$$

where, m_x is average value of vectors.

The ideal filter should minimize ONV, ACE and ASM and maximize ACH. The best method in such a situation is the optimal trade-off approach introduced by Refregier (Kumar, E.V.K.V. et al. 1994). Based on this theory, the following energy function (5) need be minimized (Shang, J.Y. et al. 2013):

$$\begin{aligned} E(f) &= \alpha(ONV) + \beta(ACE) + \gamma(ASM) - \delta(ACH) \\ &= \alpha h^+ P h + \beta h^+ D_x h + \gamma h^+ S_x h - \delta \left| h^T m_x \right| \end{aligned} \qquad (5)$$

This equation is a weighted sum of the four conflicting performance measures. Theoretically, $\alpha^2 + \beta^2 + \gamma^2 + \delta^2 = 1$. To minimize the energy function, ACH should be maximized since ACH is positive and all the three remaining criteria, which are ONV, ACE and ASM, should be minimized. The equation (5) can be rewritten as

$$E(h) = h^+ Dh - \delta|h^T m_x| \qquad (6)$$

where, $D = \alpha P + \beta D_x + \gamma S_x \qquad (7)$

In order to get minimum value of equation (8), set the derivative of equation (6) equal to zero. We can obtain

$$f = \frac{\delta}{2} \frac{p_x^*}{D} = \frac{\delta}{2} \frac{m_x^*}{\alpha P + \beta D_x + \gamma S_x} \qquad (8)$$

The equation (10) is the desired MACH filter. The Superscript * denotes conjugate. $\delta/2$ is a constant, which has no influence on the performance of MACH filter. Therefore, set $\delta/2 = 0.1$.

Thus, MACH filter is rewritten as equation (9).

$$f = \frac{m_x^*}{\alpha P + \beta D_x + \gamma S_x} \qquad (9)$$

where, α, β and γ are non-negative. Selecting different values of α, β and γ can control the performance of MACH filter in order to meet the different application requirements.

3.2 Improved MACH filter algorithm

In order to use MACH filter in JTC, the MACH filter need to be synthesized in frequency domain and projected to space domain through inverse Fourier transform. Then for a given deformed target, MACH reference image can be obtained, which contains various forms of scale and angular deformation. MACH reference image and target are input into JTC simultaneously, the recognition of deformed target can be carried out in JTC based on the MACH algorithm.

When synthesizing the MACH filter in frequency domain, the values of control parameters α, β and γ need to be optimized. By choosing different values of α, β and γ, we can control the optimal MACH filter's performance to suit different application requirements. For example, if $\alpha = \gamma = 0$, the filter is more like a MACE filter, which generally gives sharp correlation peaks and good background noise suppression but is sensitive to deformation. If $\alpha = \beta = 0$, the filter is a MACH filter which is designed with high tolerance for spatial deformation. Through amounts of repeated experiments, when $\gamma = 1$, $\beta = \alpha = 0.3$, the MACH reference image projected to space domain

has high deformation tolerance, the resulting correlation peaks are the brightest.

First of all, the edge features of detected target need to be extracted. In this paper, the edge extraction algorithm based on lifting wavelet is adopted to process the reference image of the detected target (Zhang, S. et al. 2013). To describe the effectiveness of the improved MACH filter, a group of experiments for an airplane target in the sky are implemented in this paper. The experiments aim at recognizing the spatial deformed targets.

Twenty five training image are adopted. They are divided into 5 angular groups changing from −16 degrees to 16 degrees, in which the angular step is 8 degrees. In each group, there are 5 images form 80% to 120% of the processed reference template images whose scaling step is 5% for adjacent training images. According to the equation (9) and above improvement, MACH filters can be synthesized in frequency domain using different training images, and the inverse Fourier transform is projected to space domain as MACH reference image.

4 SIMULATION AND OPTICAL EXPERIMENTAL RESULTS

MATLAB is applied to produce MACH reference image for the detected target images and conduct simulation experiments in this paper. In order to prove the feasibility of MACH reference in JTC and determine the recognition scope, an airplane above the clouds in the sky is tested. Original image of the airplane is shown in Figure 2. Figure 3 is the result of edge extraction based on lifting wavelet. Figure 4 shows the target rotated 10 degrees and scaled 80% relative to original airplane template. Figure 5 is the simulation result of correlation peaks with MATLAB software. Bright spots in Figure 6 are the optical experiment result. It can be seen from optical experiment result that the intensity of correlation peaks has decreased obviously, which shows the recognition ability of JTC for spatial deformed targets is inverse proportional with the

Figure 2.   Original airplane image.

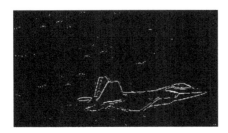

Figure 3.   Edge extraction based on lifting wavelet.

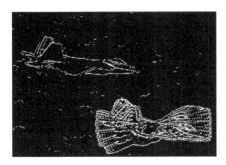

Figure 4.   Target rotation 10 degrees and scaled 80%.

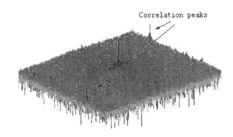

Figure 5.   Simulation result of correlation peaks.

Figure 6.   Correlation peaks.

deformed extent. The intensity of correlation peaks will decrease with the increase of deformed extent. The deformation tolerance of angle and scale can reach up to ±13 degrees and ±22% for the recognition of the airplane in this experiment. There is no certain quantitative relation of target deformation tolerance for the complexity of different kinds of targets.

5   CONCLUSIONS

In this paper, an improved MACH filter is designed and the control parameters of the algorithm are optimized. Improved MACH filter is projected to space domain through inverse Fourier transform. Because traditional JTC is very sensitive to deformed targets, the designed filter algorithm has improved this defect. Experimental results show that the performance of MACH reference image can sharpen correlation peaks and suppress background noise. The deformation tolerance of angle and scale can reach up to ±13 degrees and ±22% for the recognition of the airplane in this experiment. But due to different outlines of different targets, the recognition scope for deformed targets will also be different. Amounts of experiments show the improved algorithm can extend recognition scope of deformed targets to a great extent. The intensity of correlation peaks will decrease with the increase of deformed extent. There is no certain quantitative relation of target deformation tolerance for the complexity of different kinds of targets.

REFERENCES

[1] Yuan, X.Y., Yu, D.Y., Tao, C.K. 2002. Deformation-invariant joint transform correlator based on phase-encoding technique, *Acta Photonica Sinica*, 31(10): 1239–1243.
[2] Zhou, H.Y., Chao, F.H. 1999. MACH filter synthesizing for detecting targets in cluttered environment for gray-scale optical correlator, *Proc. SPIE* 3715: 394–398.
[3] Wang, W.S., Chen, Y., Liang, C.P. 2005. Hybrid opto-electronic joint transform correlator for the recognition of target in cluttered scenes, *Proc. SPIE* 5642: 204–212.
[4] Refregier. 1991. Optimal trade-off filters for noise robustness sharpness of the correlation and Horner efficiency, *Opt. Lett*, 16: 829–831.
[5] Kumar, E.V.K.V., Carlson, D.W. 1994. Optimal trade-off synthetic discriminant function filters for arbitrary devices, *Optical Society of America*, 19(10): 1556–1558.
[6] Shang, J.Y., Zhang, Y., Zhang, Q.B. 2013. Recognition technology of hybrid-deformation target in cluttered background, *Acta Photonica Sinica*, 42(3): 342–347.
[7] Zhang, S., Wang, W.S., Xu, C.Y. 2013. Optical correlation detection technology of moving target under low contrast environment, *Chinese Journal of Scientific Instrument*, 34(2): 319–325.

Future Information Engineering and Manufacturing Science – Zheng (Ed)
© 2015 Taylor & Francis Group, London, ISBN 978-1-138-02644-5

Agriculture robot and applications

Yonghui Wang, Kevin Lee, Suxia Cui, Eric Risch & Jian-ao Lian
Prairie View A&M University, Prairie View, TX, USA

Yubin Lan
Southern Plains Agricultural Research Center, College Station, TX, USA

ABSTRACT: An Unmanned Agricultural Vehicle (UAV) was designed, constructed, and operated in outdoor farm fields. With various kinds of sensors mounted, different signals can be automatically collected and processed for farmers to make intelligent decisions. This system is simple and easy to construct. It meets the initial requirements and is very useful for small farms. Such a ground-based automatic crop condition measuring system will help farmers maximize the economic and environmental benefits of crop pest management through precision agriculture.

Keywords: agricultural robot; unmanned agricultural vehicle; precision agriculture

1   INTRODUCTION

Based on the most recent census data, the world population has reached seven billion in the year of 2012 and it is still growing (United States Census Bureau 2012). With the burden of producing food for the whole world, agricultural industry is facing a great challenge since agricultural land is not increasing. One approach is to utilize available technologies in the form of more intelligent machines to reduce energy inputs in more effective ways than in the past. As a result, precision agriculture is proposed and it is defined as "satellite farming" or "Site Specific Crop Management (SSCM)," which is a farming management concept based on observing, measuring and responding to inter and intra-field variability in crops (Singh, et al, 2008). The goal of precision agriculture is to increase field productivity and at the same time conserve resources. Here resources refer to nature resources, like water; man-made resources, like fertilizer; and manpower. It is also reported in 2011, that the fatality rate for agricultural workers from work-related injuries was seven times higher than the fatality rate for all workers as average. In order to improve human safety in workforce, a good solution is to develop autonomous agriculture robots.

The idea of robotics in agriculture field existing for decades. Many engineers have developed driverless tractors in the past but they have not been successful as they did not have the ability to embrace the complexity of the real world. Most of them assumed an industrial style of farming where everything was known before hand and the machines could work entirely in predefined ways—much like a production line (Blackmore, et al, 2005). While in the reality, each farm field is unique. The irrigation system, crop type, and pest condition vary from land to land. Thus requires the robot not only can "walk" in field, but also can "sense" while it is walking. The approach is now to develop automatic machines that are capable to collect data automatically from natural environment. Then the data can be analyzed and help farmers to make intelligent decisions.

Some modern agricultural robotics systems are introduced on the web page at: http://www.unibots.com/Agricultural_Robotics_Portal.htm.

To answer this demand, an Unmanned Agricultural Vehicle (UAV) was designed, constructed, and operated in fields. The UAV is a precision agriculture vehicle platform for mounting multiple sensors, including crop height sensor, crop canopy analyzer, Normalized Difference Vegetative Index (NDVI) sensor, multispectral camera, and hyperspectral radiometer. Such a ground-based automatic crop condition measuring system will help farmers maximize the economic and environmental benefits of crop pest management through precision agriculture.

The rest of the paper is organized as follows. Section 2 describes in detail about how the UAV is designed and constructed. Applications are introduced in Section 3. Some concluding remarks and discussions are found in Section 4.

2 UAV DEVELOPMENT

2.1 *Overall design*

In fact, accurate and reliable information technology is the basis of precision agriculture. Remote sensing has been widely used to obtain and map the temporal and spatial variability of crops in fields. Information on crop condition can be used to assess and monitor crop growth status, predict crop yield, or develop program for optimizing application of nitrogen fertilizer, fungicide, and growth regulator in precision agriculture. Successful information acquisition relies on the ability of sensors and instrumentation in detecting these crop canopy variables, which are indicative of crop growth (Goel, et al, 2003).

To automatically collect the abovementioned data, an Unmanned Agricultural Robotics System (UARS) is desired, especially for small farms. The common requirement of such a UARS is to acquire accurate and reliable information about the environment and the objects. This UARS needs to be an integrated system which includes a UAV and the sensors. Appropriate sensors are essential in the information acquisition process. The UARS will integrate sensors, which are critical for collecting related data, and GIS/GPS to provide accurate and comprehensive information for decision making.

To fulfill this requirement, the UAV must be a vehicle with the following capabilities. The structure must be sturdy enough for outdoor farm field environment. The vehicle must be wide enough for covering one or two ridges without destroying crop plants. There must be an open space underneath the vehicle for crop plants to pass trough when the vehicle is traveling. A height adjustable platform is need for measuring various plants with different heights. The vehicle must be able to turn direction for easy manipulation. Based on the requirements, the design diagram is shown in Figure 1.

2.2 *Construction*

Based on the requirements and the design diagram shown in Section 2.1, the following materials are chosen. First, to ensure constructing a sturdy vehicle, grooved T-slotted aluminum frames (Fig. 2) are used for building the vehicle structure. The frame is 1.5 inches wide by 1.5 inches height, which makes sure that the vehicle is sturdy enough but not too heavy.

A linear actuator (Fig. 3) is used for turning direction. The actuator is mounted at the front side of the vehicle.

Two actuators (Fig. 4) are used for the height adjustable platform. These track actuators are ideal for applications where a small closed length is needed. The actuators have 200 pounds force which can make sure that the platform is capable for holding several sensors.

The motor chosen for the vehicle is shown in Figure 5. This motor is a 24VDC 127 RPM gear motor. The motor's peak output is about 250 Watts

Figure 2.   T-slotted aluminum frame.

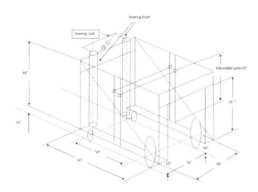

Figure 1.   Design diagram of the UAV.

Figure 3.   Direction actuator.

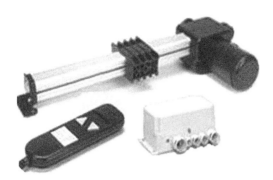

Figure 4.   Adjustable platform actuator.

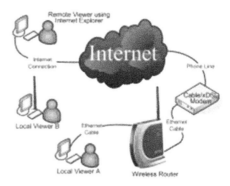

Figure 5.   Motor.

Figure 7.   Wireless communication for remote control.

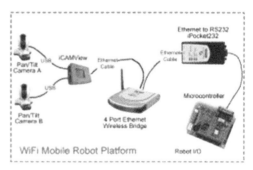

Figure 6.   24V 10000 mAHr NiMH Battery Pack.

Figure 8.   The constructed UAV.

of power or about 1/3 HP. These powerful motors ensure that the vehicle can travel with full load at 5 miles per hour.

To ensure enough power for outdoor farm field applications, a 20 cell (24VDC) 10000 mAhr Rechargeable NiMH battery pack (Fig. 6) is used for the vehicle. The batteries are NiMH, so they do not have the memory issues common with NiCad battery packs. The cells are D size inter-connected via soldered strips and covered with a PVC wrapping (color may be different than photo). Two 14 AWG ~24" wire leads extend out of the pack. These packs also have a built in thermal overload set for 25 Amps that will reset upon cooling.

The vehicle is remotely controlled through WIFI, as shown in Figure 7. The control signal is trans-mitted from the computer, through a router, to the

bridge which is mounted on the vehicle. Then with the help of the Ethernet to RS232 iPocket232, the control signal is sent to the main control board for manipulating the vehicle.

Figure 8 shows the constructed UAV. The UAV meets all of the requirements. It is sturdy, can move fast, has adjustable platform, and has enough space for crop plants to pass through.

3 APPLICATIONS

Various sensors can be mounted on this UAV to measure crop conditions (as illustrated in Fig. 9).

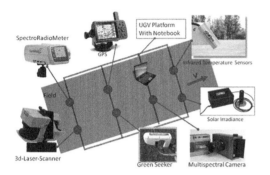

Figure 9. UARS-based crop conditions measuring system.

Figure 10. Rice field test.

To measure crop height and volume, laser or radio detection can be used; to measure biomass of plants, LAI or NDVI can be used; for canopy temperature measurement, infrared temperature sensors can used.

The UARS system was tested in rice field, as shown in Figure 10. In the test, various data was collected with the sensors mounted on the UAV.

4 CONCLUSIONS

A UAV is designed and constructed for precision agriculture. This UAV can be used as an automated vehicle for a UARS system. With various kinds of sensors mounted, different signals can be collected and processed for farmers to make intelligent decisions. This system is simple and easy to construct. It meets the initial requirements and is very useful for small farms. Future plans include upgrading the vehicle to a fully automated vehicle and integrating all the sensors into a smart grid system.

REFERENCES

[1] United States Census Bureau www.census.gov.
[2] M. Singh, P. Sigh, and S.B. Sigh, 2008, Decision Support System for Farm Management, *World Academy of Science, Engineering, and Technology,* vol. 39, pp. 346–349.
[3] S. Blackmore, B. Stout, M. Wang, and B. Runov, 2005, Robotic Agriculture—The Future of Agricultural Mechanisation, *5th European Conference on Precision Agriculture,* ed. J. Stafford, V. The Netherlands, Wageningen Academic Publishers. pp. 621–628.
[4] P.K. Goel, S.O. Prasher, J.A. Landry, R.M. Patel and A.A. Viau, 2003. Estimation of crop biophysical parameters through airborne and field hyperspectral remote sensing. *Transactions of the ASAE* 46(4): 1235–1246.

Future Information Engineering and Manufacturing Science – Zheng (Ed)
© 2015 Taylor & Francis Group, London, ISBN 978-1-138-02644-5

Experimental study of laser quenching based on resonant scanner

G. Zhang

Precision Engineering Research Center, Guangzhou Institute of Advanced Technology,
Chinese Academy of Sciences, Guangzhou, China

S.Z. Zhang

Precision Engineering Research Center, Guangzhou Institute of Advanced Technology,
Chinese Academy of Sciences, Guangzhou, China
School of Mechanical and Automotive Engineering, Fujian University of Technology, Fuzhou, China

C.L. He

Precision Engineering Research Center, Guangzhou Institute of Advanced Technology,
Chinese Academy of Sciences, Guangzhou, China
Machinery and Electronics Engineering, Shantou University, Shantou, China

Y.P. Wang & X.S. Chen

Precision Engineering Research Center, Guangzhou Institute of Advanced Technology,
Chinese Academy of Sciences, Guangzhou, China

ABSTRACT: Using of a resonant scanner and a fiber laser is very convenient for local quenching of some large heavy parts. This laser quenching method is simple and available, which could quench accurately on the specify sites of the parts and be provided with the characteristics of the high precision. An experimental study of laser quenching based on resonant scanner for 41Cr4 steel and C45E4 steel used as workpiece materials respectively through putting up experimental platform of laser quenching is conducted in this study. The optimal machining parameters with the scanning speed of 16 mm/s, the focal distance of 40 mm and the overlapping rate of 40%–50% are obtained. Using these process parameters, the surface hardness is increased to 49.1% maximum for 41Cr4 steel, 66.9% maximum for C45E4 steel respectively. Results show that this laser quenching method is feasible and provided with certain practical application significance.

Keywords: laser quenching; resonant scanner; hardness; experimental study

1 INTRODUCTION

Laser surface modification is a very effective way to improve the surface properties of the metal material, and laser quenching used as one of the laser surface modification methods had been most widely used (see, e.g., Xi et al., 2004; Liang et al., 2001; Chao et al., 2003; Putatunda et al., 1997). Laser quenching is provided with a series of advantages, such as fast heating and cooling, small heated affected zone, laser quenching is a kind of very effective surface strengthening method.

Many scholars have performed the research and made a lot of achievements. Yang Huixiang got a gear that the tooth has ideal hardening layer uniformly distributed alone it, the tests showed that the crystallite size of laser hardening gear hardening layer was significantly less than the crystallite size of gear after carburizing and quenching, after laser hardening, the hardening layer contained lots of residual, the gear lifetime after laser quenching could be greatly improved (Yang et al., 2007). Zhu (1990) measured the Cr12 steel's grain size, the original group was 12 levels, while it was 15 levels after laser treatment, which meant to refine the grain size obviously. The length of the original martensite structure in Cr12 steel was 6 µm, it was smaller than 2 µm after laser quenching, and the martensite was refined obviously.

At present, the study of laser quenching is concentrated by the using various types of laser such as CO_2, Nd:YAG, diode, and excimer laser, using those lasers to quench on the flat surface of parts is very appropriate. However, as for some parts that have the irregular appearance, quenching becomes complex.

In this study, an experimental investigation of laser quenching for 41Cr4 steel and C45E4 steel used as workpiece materials respectively through putting up experiment platform of laser quenching is conducted based on resonant scanner. The resonant scanner is the equipment that uses flying optical path technology, which can rapidly and accurately control the laser light path. In the resonant scanner system, we can use a variety of vector graphics, such as the fold line and the helical line. This method adopts the computer graphics software to transfer to the graphics processing mode, which is provided with the advantage of high efficiency, very good graphic precision, without distortion, it can be a very good solution of laser quenching on the surface of the complex parts. In application of laser quenching, laser quenching process parameters have a great influence on the quenching effect, constantly seeking the relationship between the laser quenching process parameters and the effect of laser quenching, which could improve the quality of laser quenching, had always been the goal of many researchers (see, e.g., Zhang et al., 2004; He et al., 2012; Heitkemper et al., 2003).

Getting the optimum process parameters of laser quenching acts as the key role of laser quenching. Because the laser quenching temperature can not be directly measured, we usually got the appropriate laser quenching process parameters by simulations and experiments (see, e.g., Xu et al., 2013; Kuang et al., 2011). The hardness of material surface after laser quenching was the most important performance assessment and standard for the laser quenching process (see, e.g., Hua et al., 2002; Yu et al., 1990). In this paper, three key parameters of laser quenching process, such as the laser scanning speed, the focal distance and the overlapping rate, which have great effect on the hardness of the hardening layer and the quality of the surface are investigated.

2 THEORY

For transient conduction heat transfer in the workpiece subjected to a given heat source, the governing heat conduction equation may be expressed as follows:

$$\rho(T)C_p(T)\frac{\partial T}{\partial t} = \frac{\partial}{\partial x}\left(K(T)\frac{\partial T}{\partial x}\right) + \frac{\partial}{\partial y}\left(K(T)\frac{\partial T}{\partial y}\right)$$
$$+ \frac{\partial}{\partial z}\left(K(T)\frac{\partial T}{\partial z}\right) + g$$

(1)

where K is the thermal conductivity, ρ is the density, C_p is specific heat and g is the internal heat source.

Here the energy generation term is neglected (i.e., $g = 0$) as there is no appreciable source of heat generation within the system/domain. The laser input heat source is imposed on the boundary. We further assume that convection heat transfer from the workpiece to the ambient air can be neglected, and all the thermal properties are constants. Equation 1 could be simplified to:

$$\rho C_p \frac{\partial T}{\partial t} = K\left(\frac{\partial^2 T}{\partial x^2} + \frac{\partial^2 T}{\partial y^2} + \frac{\partial^2 T}{\partial z^2}\right)$$

(2)

The Gaussian laser power input can be expressed as:

$$I = \frac{2P}{\pi r_s^2}\exp\left(-\frac{2r^2}{r_s^2}\right)$$

(3)

where I is the power density of light source, P is the light source power, r_s is the spot radius of light source and r is the distance from reference point to the spot center, that is defocusing amount.

3 EXPERIMENTAL PROCEDURE

Experimental analyses have been performed at a laser machining facility under actual machining conditions. Figure 1 depicts a image of the experimental setup comprised a fiber laser (continuous wave) with the laser intensity of 200 W (SPI Lasers Ltd. UK), a resonant scanner (GSI Group Inc. USA) with the model of 61714PS1XY2-YP, the maximum scanning speed of 2 m/s, the focal length of 163 mm and the smallest laser spot size of 30 μm, and a hardness tester (SMART Ltd. Hong Kong) with the model of AR936.

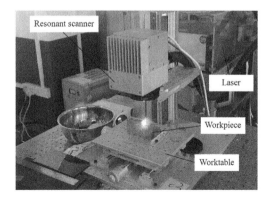

Figure 1.   Image of the experimental setup for the laser quenching process.

In order to reduce experimental error, each hardness value needs to be measured for three times, and then be averaged over it. Experiments are used by negative from coke to determine how close the coke deposition, good light spot size is determined, in accordance with the corresponding spot size and the overlapping rate to calculate the laser spot of the offset.

About the present experimental analysis, the workpiece is fixed on the worktable. Two different materials such as 41Cr4 steel and C45E4 steel are used as workpiece materials respectively, in order to get a higher cooling speed, we choose a big size of the workpiece with 50 mm × 20 mm × 100 mm.

Before the experiment, the workpieces need to be polished to remove microcracks, oxide layer and scratches on the surface, and then to be degreased decontamination with alcohol and sandpaper in order to avoid interference with the experimental results. Before laser quenching, the hardness of 41Cr4 steel and C45E4 steel is 175HB and 160HB respectively measured by the hardness tester above.

4 RESULTS AND DISCUSSION

Many factors could affect the laser quenching, and various factors are correlated with each other, thus the optimum quenching process parameters must use the orthogonal tests. However, the general orthogonal tests are not only cumbersome, but also difficult to deal with experimental data.

This paper makes exploratory experiment through the 41Cr4 steel and C45E4 steel, the optimum process parameters of an interval are obtained, then we use this process parameters for the orthogonal tests, the best process parameters from the orthogonal tests are obtained, and next we carry out the extended experiment through the various process parameters. From the extended experiments, the main technological parameters on the influence of quenching hardness and surface quality are verified and analyzed. In addition, this experiment compares two kinds of steel to prove the validity of the optimum process parameters.

Figure 2 presents the surface of 41Cr4 steel after laser quenching using the laser scanning path of helical line and Figure 3 represents the surface

Figure 2. The surface of 41Cr4 steel after laser quenching using the laser scanning path of helical line.

Figure 3. The surface of C45E4 steel after laser quenching using the laser scanning path of fold line.

Table 1. The hardness before and after laser quenching with different scanning speed.

Scanning speed (mm/s)	8	12	16	20	24
C45E4 steel					
Before (HB)	160	160	160	160	160
After (HB)	226	237	265	238	243
Increment (%)	41.3	48.1	65.6	48.8	51.9
41Cr4 steel					
Before (HB)	175	175	175	175	175
After (HB)	240	249	261	237	226
Increment (%)	37.1	42.3	49.1	35.4	29.1

of C45E4 steel after laser quenching using the laser scanning path of fold line under the certain processing parameters. From the figure, we can observe some surface quality is poor. Because the defocusing amount is short, the laser energy density is very high, melting has been appeared on some surface where the light spot center go through.

Table 1 details the hardness of using different scanning speed under the overlapping rate of 50% and the focal distance of 40 mm. It can be observed from the picture when the scanning speed is 16 mm/s, two kinds of steel have the highest hardness, this is because when the scanning speed is quick, the surface of material does not get enough laser energy, when hardening layer is cooled, it can not get enough of supercooling degree, hardening layer of dislocation, martensite content, the carbon content of martensite will be reduced, and the hardness is reduced accordingly. However, when the scanning speed is too low, somewhere in surface will be melted, so that the hardness will reduce accordingly. The material surface's melting phenomenon will make the surface rough, the surface quality becomes poor.

Table 2 reveals the hardness value of using different focal distances under the overlapping rate of 50% and the scanning speed of 16 mm/s. We can observe from the picture when focal distances is too long, the hardness on the surface of the material is quite low, this is mainly because the focal distances is too long, thus the laser spot becomes very big and greatly reduces the energy density. On the other hand, due to the overlapping rate unchanged,

Table 2. The hardness before and after laser quenching with different focal distances.

Focal distance (mm)	20	40	60	80
C45E4 steel				
Before (HB)	160	160	160	160
After (HB)	249	267	221	209
Increment (%)	55.6	66.9	38.1	30.6
41Cr4 steel				
Before (HB)	175	175	175	175
After (HB)	226	259	215	197
Increment (%)	29.1	48.0	22.9	12.6

Table 3. The hardness before and after laser quenching with different overlapping rate.

Scanning speed (mm/s)	20	30	40	50	60
C45E4 steel					
Before (HB)	160	160	160	160	160
After (HB)	235	241	266	267	227
Increment (%)	46.9	50.6	66.3	66.9	41.9
41Cr4 steel					
Before (HB)	175	175	175	175	175
After (HB)	228	225	261	260	237
Increment (%)	30.3	28.6	49.1	48.6	35.4

the scan spacing increased, material absorption of energy per unit area will be less, all these lead to the hardening layer's hardness decreased. At the same time, when focal distances is too long, it can make the quality of laser beam become very poor, and the energy density unevenly distributed. After quenching, material surface quality becomes poor using technical parameters with the focal distance of 60 mm, the scanning speed of 16 mm/s and the overlapping rate of 50%. It can be observed from the table that the focal distance of 40 mm is more appropriate. Because the focal spot size is only 30 μm, the energy density is very big, so when the focal distances is small, light energy density is high, it can make the material surface be melted.

Table 3 exhibits the hardness value of using different overlapping rate under the focal distance of 40 mm and the scanning speed of 16 mm/s. As we can observe from figure, when the overlapping is from 40% to 50%, two briquette's hardness after quenching is the highest. When the overlapping rate is higher, processing time will become longer, and somewhere on the surface of the material will become be melted. On the other hand, the intensity of a laser beam is Gaussian distribution, away from the spot center, the laser energy density was exponentially reduced by Zhang et al. (2008), and the energy density of the place that far from the spot center of the laser beam will be less. When the overlapping rate is low, the laser energy density will be greatly decreased in somewhere of material surface, thus the hardness is reduced. Taken together, the best overlapping rate is from 40% to 50%.

Using the optimal laser quenching parameters, that is the scanning speed of 16 mm/s, the focal distance of 40 mm and the overlapping ratio of 40%–50%, a comparison chart of the mean and maximum hardness value before and after laser quenching for 41Cr4 steel and C45E4 steel based on 6 repetitive experiments is represented in Figure 4. As evident in Figure 4, after the laser quenching, the hardness value of 2 materials are higher than those before laser quenching.

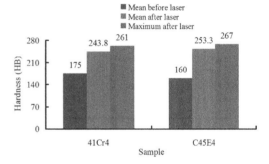

Figure 4. Comparison chart of mean and maximum hardness value before and after laser quenching with different materials.

As for 41Cr4 steel, the mean and maximum hardness are 243.8HB and 261HB respectively. Compared with laser quenching before, the hardness is increased about 39.3% and 49.1% respectively. Then with regard to C45E4 steel, the mean and maximum hardness are 253.3HB and 267HB respectively. Compared with laser quenching before, hardness value is improved about 58.3% and 66.9% respectively.

It can be observed from the Figure 4 that laser quenching enhances surface hardness, the effect on the C45E4 steel is more evident.

From the above detailed experimental results, the laser quenching process based on resonant scanner plays a great supporting role for surface, and the improving effect on the C45E4 steel is very obvious.

5 CONCLUSION

This paper proposes an experimental investigation of the laser quenching process based on resonant scanner for 41Cr4 steel, C45E4 steel used as workpiece materials respectively, conclusions are as follows.

The optimal process parameters are that the scanning speed is 16 mm/s, the focal distance is 40 mm and the overlapping rate is 40%–50%.

Compared with laser quenching before, surface hardness is improved about 49.1% maximum for 41Cr4 steel, 66.9% maximum for C45E4 steel respectively in the laser quenching process.

The paper is only focused on the hardness and surface quality, furthermore research should be investigated the effect of laser processing parameters on the depth of the hardening layer, the wear resistance and tissue.

ACKNOWLEDGMENTS

This study is supported by the National Natural Science Foundation of the People's Republic of China (grant number 51307170) and Innovation Project of Nansha District, Guangzhou, People's Republic of China (grant number 201201020). The author gratefully acknowledges the help of Precision Engineering Research Center, Guangzhou Institute of Advanced Technology, Chinese Academy of Sciences.

REFERENCES

[1] S.M. Xi & J.G. Zhang & X.Y. Sun. 2004. 38CrMoAlA steel's Laser Quenching and Nitriding Compound Treatment. *China Laser* 6(31): 761–766.

[2] E.J. Liang & L.P. Du & J.T. Gu. 2001. Wide-band Laser Quenching of 45# Steel. *Laser Journal* 2(22): 52–58.

[3] M.J. Chao & Y.D. Du & E.J. Liang. 2003. Study on Laser Surface Hardening for Mould Cutting Roller. *Laser Technology* 6(27): 500–506.

[4] Putatunda, S.K & Nambiar, M & Clark, N. 1997. Influence of laser hardening on mechanical properties of low alloy steel. *Surface Engineering* 5(13): 407–415.

[5] H.X. Yang & H. Wang & H.Y. Dong. 2007. Analysis of Laser Hardened Gear and Carburized Gear X Ray Diffraction. *Journal of Changchun University of Technology* 2(28): 187–190.

[6] J. Zhu. 1990. Microstructure and Performance of Steel Materials after Laser Treatment. *Ph.D dissertation Tsinghua University*: Beijing: 3–6.

[7] L.B. Zhang & Q.W. Fan & Y.S. Zuo. 2004. Application of Hybrid Intelligent Technology in Laser Quenching Process Optimization. *Materials Science and Technology* 6(12): 654–668.

[8] C.L. He & S.K. Chen & Z.H. Zhou. 2012. Study on the Process Parameters 718 Laser Quenching of Die Steel. *Hot Working Technology* 20(41): 169–175.

[9] Heitkemper, M & Bohne, C & Pyzalla, A. 2003. Fatigue and fracture behavior of a laser surface heat treated martensitic high-nitrogen tool steel. *International Journal of Fatigue* 25: 101–105.

[10] H.W. Xu & Y. Liu & D.G. Wen. 2013. Study on the Optimization of Technical Parameters of Laser Quenching Experiments. *Chinese Printing and Packaging study* 5(1): 44.

[11] M. Kuang & Y.C. Su & C.M. Deng. 2011. Study on the Performance of the Layer of Die Steel Laser Surface Transformation Hardening of 3Cr2W8V. *Materials Research and Application* 20(5): 105–109.

[12] Y.Q. Hua & J.C. Yang & T. Hu. 2002. Study on the Application of ANSYS517 Software to Simulate the Temperature Field of Laser Quenching. *Metal Heat Treatment* 8(27): 37–40.

[13] P. Yu & Kik, A.A & Medvedev, A. 1990. Effect of pulsed laser quenching on the mechanical properties of structural steels in high-rate deformation. *Physics and Chemistry of Materials Treatment* 24(6): 582–583.

[14] D.Y. Zhang & Q.M. Zhang. 2008. Laser manufacturing technology. *Tsinghua University Press*: Beijing: 126–130.

Future Information Engineering and Manufacturing Science – Zheng (Ed)
© 2015 Taylor & Francis Group, London, ISBN 978-1-138-02644-5

A RFID path tracking information system based on data warehousing

Lei Chen & Jian Guo

School of Information, Beijing Wuzi University, Beijing, China

ABSTRACT: The RFID path data management is always a technical problem for application of RFID technology in logistics management. By analyzing the characteristics of RFID data generated in RFID-enabled warehouses, a RFID path tracking information System is established to manage RFID path data. After detailed analysis of system architecture and functional requirements, a tracking and management of the logistics path management prototype system is established. At last, a combination of online and offline data warehousing model is set up to realize the RFID data storage and management.

Keywords: RFID; path data; data mining; data warehouse

1 INTRODUCTION

RFID technology is an important way of information acquisition in warehouse management, all the data produced by RFID technology will be saved into the data warehouse finally. Fusion of RFID technology and data mining in warehouse management is to construct RFID data warehouse. The research of RFID data warehouse includes moving commodities in groups, data generalization, RFID path data fusion and expansion. Different from the traditional OLAP, the RFID data warehouse architecture deals with data compression, query and processing efficiency of a specific kind of data—path data.

As a result, a RFID path tracking information system is designed. The system realizes automatic information acquisition, tracking moving objects in the warehouse, forecasting moving trend from path data quickly and accurately, and finding out the abnormal movement of commodities.

2 RFID PATH TRACKING INFORMATION SYSTEM ARCHITECTURE

The basic principle of tracking path information using radio RFID technology is to set RFID readers in each storage location in the warehouse. When a commodity passing through the coverage area of a RFID reader, the reader will automatically capture TagID and record the timestamp. The system then produces an information node including <TagID, ReaderID, Time>, and a series of information nodes arranged in time sequence make up a complete path that a commodity pass through in the warehouse. The RFID path data can be used for tracking, querying historical status and position of commodities.

The function of a complete path information tracking system includes radio frequency acquisition of tag data, pretreatment, update, query and storage of path information. The system architecture is shown in Figure 1.

The path information tracking management system includes two modules: one is the path data information management module, the other is a data processing module.

2.1 Path data management module

Information management module includes three submodules.

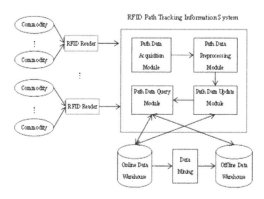

Figure 1. Architecture of RFID path tracking information system.

2.1.1 *Path data acquisition*

When commodities entering, leaving or staying in the warehouse, RFID readers read the tags and pass the information into path data acquisition system, which will generate path data according to multiple physical nodes information and RFID reader location.

2.1.2 *Path data preprocessing*

With the circulation of commodities in the warehouse, massive RFID data are produced by tag readers. All these data may be redundant because of the repeat scanning of the same position or the reason that the tags need to be processed in batches. The redundant data must be preprocessed by data cleaning and aggregating. After being preprocessed, the path data for further knowledge discovery will be saved into data warehouse.

2.1.3 *Path data update*

When a commodity passing through a tag reader coverage area, a new record <TagID, ReaderID, Time> is produced and will be added into path data automatically.

2.1.4 *Path data query*

The system administrator can execute the query operations through real-time access to the Internet to achieve all the path information of commodities and all the relevant information about products arriving and leaving the warehouse.

2.2 *Path data mining module*

The path data in warehouse management will be stored in a data warehouse finally. Different from the traditional online transaction processing system, data warehouse has the characteristics of subject oriented, integrated, relatively stable and reflecting historical changes of data. So it is usually used for management decision support. By means of Extraction Transformation and Loading (ETL), the path data are stored and arranged according to different subject domain. And after finishing establishing the data warehouse, data mining will carry out based on this data warehouse platform.

3 CONSTRUCT RFID DATASET

3.1 *RFID coding standard*

RFID data set is based on the RFID coding system. The mainstream RFID coding system includes EPC Global of America and Europe, Ubiquitous ID of Japan, and ISO international standard system. The ISO international standard is a kind of general standard including four parts: RFID technical standard, data contents and coding standard, performance and consistent standard, application standard. The UID and EPC standards mainly deal with three parts, including electronic product code, radio frequency identification system and information network system, which are similar in most cases. Since EPC Global is designed mainly for the application of logistics and supply chain management, the path tracking system uses EPC as coding standard.

3.2 *RFID original data set*

When commodities with RFID tags come into the signal coverage area of RFID readers, RFID readers will read tag data and generate a RFID original data set, which can be recorded as RawData(CID, Location, Time). CID represents a unique identification code of a commodity, Location and Time stand for location and time information obtained when a RFID tag is read by RFID readers.

3.3 *RFID data cleaning*

RFID readers read the tags regularly and repeatedly, when commodities in the coverage area stay for a period of time, many RFID data records with the same value of CID and location but different time information will be produced. Such redundant data will be cleaned by merging records with the same attributes of CID and location. Cleaned data can be defined as CleanData(CID, Location, Time_in, Time_out), Time_in and Time_out stand for the time when commodities enter and leave the reader coverage area separately. Obviously database size is significantly reduced after data cleaning.

3.4 *RFID data aggregation*

During supply chain management, commodities are transported in batches. A lot of RFID data records in clean dataset with the same time and location values but different CID information can be processed as a batch, so aggregation operation can be used to reduce the scale of data set furtherly. After aggregation, the data set for a batch of commodities can be denoted as Aggregate Data (SID, Location, Time_in, Time_out) and a Map table. SID is a set of other SIDs or CIDs, MAP table is a mapping ensemble between SID and SID, or SID and CID.

3.5 *RFID path data*

As a batch of commodities passing through nodes of supply chain, path data are generated according to the Aggregate data set. RFID path data of a batch of commodities can be expressed as permutations of $(l_1, t_1), (l_2, t_2), ..., (l_n, t_n), i = 1, 2, ..., n$. (l_i, t_i) is a path segment of the path, li stands for the location of path segment *i*, *ti* is duration time the commodities stay there. So the length of path is equal to the number of path segments.

4 COMBINATION OF ONLINE AND OFFLINE RFID DATA WAREHOUSE MODELING

The combination of online and offline data warehousing technology is used to store and process RFID data. Since the main memory database has the characteristic of higher access speed, it is designed as an online data warehouse to save the latest update of RFID data. According to the rule of saving and updating data as soon as possible, the online data warehouse supports information frequent updates and real-time query, improve the efficiency of real-time processing of RFID data.

The disk database has the characteristic of high capacity; it is designed as an offline data warehouse to save a large amount of historical RFID data. According to the storage mode of historical data, the offline data warehouse supports storage and query of the historical RFID data.

4.1 Online RFID data warehouse model

The purpose of building online data warehouse is to deliver a more immediate response for storage and update of the latest state of RFID data. The path data is obtained mainly based on position information of commodities. When a new RFID data record <TagID, ReaderID, Time> is read into the data warehouse, the first thing is to search the original record for TagID in the online data warehouse, and then update the record. In order to improve the efficiency of query and update, the location of commodities is denoted by the position of RFID tag reader. Because every commodity in the warehouse has a fixed position to be placed, the fixed position is named as a home location of a commodity. Assumed that every RFID reader monitors a certain area, so the location information of a RFID data record can be denoted by the number of RFID readers, that is ReaderID.

The online RFID data warehouse model includes several tables, they are ReaderInfo table, ActiveTagList table, and several Tag tables named according to different RFID reader coverage area as TagList_i. ReaderInfo table contains the correspondence relation between tag readers and tag list. TagList_i table stores the latest detailed information of each RFID tag. And ActiveTagList table is in charge of recording the information of RFID tags which leave the home locations. All the attributes and data type of tables are shown in Tables 1–3.

The off-line data warehouse model not only greatly reduces the search number of tags when

Table 1. ReaderInfo table.

Field name	Field type	Remarks
ReaderID	Varchar (20)	ID of tag readers, primary key
TagList_Name	Varchar (20)	The name of TagList table corresponding to the reader on the home location.

Table 2. TagList_i table.

Field name	Field type	Remarks
TagID	Varchar (26)	ID of tags, primary key
Flag	boolean	Indicate whether tag i is in the home location or not
PathID	Int	The ID of path when a tag passing through
Path_length	Int	The length of the path
Time_in	Data time	The time when tag i arrive at the last node of path
Time_out	Data time	The time when tag i leave the last node of path

Table 3. ActiveTagList table.

Field name	Field type	Remarks
TagID	Varchar (26)	ID of tags, primary key
ReaderID	Varchar (20)	Indicate current location of tag
TagList_Name	Varchar (20)	The name of TagList that the current tag belongs to

updating commodities data, but is very easy to do real-time query as well. For example, in order to query the position of a commodity in the warehouse, the only thing to do is to look up ReaderID information in the ActiveTagList table.

4.2 Offline RFID data warehouse model

Offline data warehouse was designed on disk database, mainly used for massive historical RFID data query and storage. The total quantity of historical data is very huge and need to be cleaned and aggregated before they are used to produce path data for tracing commodities. Offline data warehouse model not only greatly reduces the data volume for storage, but also embodies the concept of path, which is easier for high semantic level queries. Three tables are used to describe the model. They are Path table, TagPath table and PathTime table.

The table named TagPath records the correspondence relationship between each tag and path. Path table contains path information. And the time information of nodes in the path is recorded in the PathTime table. All the attributes and data type of tables are shown in Tables 4–6.

This model separates the path information from the time information of commodities. All the common paths are saved in one record, only when a new path data is produced, a new record will be inserted into the data warehouse. When a commodity is at a certain path node, the Time_out information varies with time. Since the latest data of the commodity is already saved in the main memory, a new record will be added into the PathTime table when the location information changes. Due to the real-time requirements for historical data is not high, correspondence relationship between tag and path does not require immediate update. Some delay strategy such as updating data after a certain time interval can be adopted.

Table 4. TagPath table.

Field name	Field type	Remarks
TagID	Varchar (26)	ID of tags, primary key
PathID	Int	The ID of path when a tag passing through

Table 5. Path table.

Field name	Field type	Remarks
PathID	Int	The ID of path when a tag passing through, primary key
Path	Varchar (100)	Path $(r_1, r_2, ...)$, r_i is ReaderID of tag reader i, $i = 1, 2, ...$
Pre_PathID	Int	The sub path after removing the last node
Curr_ReaderID	Varchar (20)	The last node of the path
Path_length	Int	The length of the path

Table 6. PathTime table.

Field name	Field type	Remarks
TagID	Varchar (26)	ID of tags
Order	Int	The position of node in the path, r_i
Time_in	Data time	The time when tag arrive at the node r_i
Time_out	Data time	The time when tag leave the node r_i

This model describes path data of commodities, and improves data storage efficiency by reducing update time and times.

5 CONCLUSIONS

The combination of RFID technology and data mining applied in warehouse management can effectively improve the performance of the supply chain, reduce enterprise operating costs, and has a huge market prospects in the future. By analyzing the characteristics of RFID data in warehouse management, a RFID path data tracking information system realizes the acquisition, processing, query and storage of RFID real time data and historical data. The combination of online and offline data warehousing is put forward to save real time path data and historical path data.

A single commodity's path data is the research object in the path tracking information system, and the further work can be expanded to research on path data of a batch of commodities. Because warehouse management is a very important part of supply chain, the research of path data mining can also extend to the whole supply chain management.

ACKNOWLEDGMENT

Many people have made contributions to my research. I would like to express my warmest gratitude to them. This paper is supported by such projects: "Scientific Research Bases—Technology Innovation Platform—Research of Information and Control Technology on Modern Logistics" (Project Code: PXM2014_014214_000086); and "Key Laboratory of Beijing Municipal Science & Technology Commission—Beijing Key Laboratory of Intelligent Logistics System (Project Code: BZ0211)" (Project Code: PXM2014_014214_ 000092).

REFERENCES

[1] Gonzalez H., Han Jiawei, Li Xiaolei. "Warehousing and Analyzing Massive RFID Data Sets", Proc of the 22nd International Conference on Data Engineering, 2006:83–83.
[2] Lee C'hun-Hee, Chung Chin Wan. "Efficient Storage Scheme and Query Processing for Supply Chain Management Using RFID", Proc of the 2008 ACM SIUMOD International Conference on Management of Data, 2008: 291–302.
[3] Wang F., Liu P. "Temporal Management of RFID Data", Proc of VLDB'05, 2005:1128–1139.
[4] Chawathe S.S., Krishnamurthyy V., Ramachandrany S. "Managing RFID Data", Proc of VLDB'04, 2004: 1189–1195.

Future Information Engineering and Manufacturing Science – Zheng (Ed)
© 2015 Taylor & Francis Group, London, ISBN 978-1-138-02644-5

Enterprise vitality evaluation method: A perspective of enterprise business age model

Bijiang Ai
Xinjiang University of Finance and Economics, Xinjiang, China
College of Management, CAS University, Beijing, China

Yueyue Ma
Xinjiang Engineering College, Xinjiang, China

Yanmei Xu
College of Management, CAS University, Beijing, China

ABSTRACT: Enterprise vitality has immediate bearing on the rise and fall of enterprises, leaving the evaluation of it a focus of business management. The paper analyzes the connotation and influencing factors of enterprise vitality, defines the relation between enterprise vitality and enterprise business age to expound as a focus the business age models of enterprises at home and abroad and offer enlightment for reference.

Keywords: enterprise vitality; enterprise business age; business age model of enterprises

1 INTRODUCTION

An enterprise is a "living organism" with vital signs that survives, grows and regenerates in modern economic and social environment. In reality, some enterprises are vibrant and vigorous while some are lifeless and on the verge of collapse. The root cause of longevity or decay of the enterprise is directly related to enterprise vitality. The existing economic theory views profit as the fundamental purpose of enterprises; people are accustomed to regard the enterprise as a machine to rake in profit, with few in-depth study on the survival, growth and regeneration of enterprises from ecological perspective—assuming the enterprise is "life organism". To examine the life course of the enterprise from ecological perspective and to explore the mechanism of business survival and longevity—enterprise vitality, is an effective way to explain the ups and downs of the enterprise as one of the main topics of enterprise research. Currently, few are bionic research-based outcomes on enterprise vitality from enterprise business age perspective. Among these studies, the names the scholars have used are various (eg: the vitality of enterprises, business longevity, sustainability development, business vitality, etc.), while the objects of study are consistent in nature. This paper draws on the existing research outcome to discuss business

vitality evaluation method basing on enterprise business age model.

2 ENTERPRISE VITALITY

Enterprise vitality is the ability to show its life intensity, the manifestation of which is its internal metabolic status and its ability to adapt to environmental change; enterprise vitality has a direct impact on sustainable development of the enterprise. To reveal the characteristics of the enterprise from the perspective of bionics, concentrating on value creation, longevous enterprise, continuous and sustainable development, is not only a new way of thinking on enterprise management, but also an enrichment and supplement to functional enterprise management theory.

Scholars in China and Japan working on business vitality study are: Sun Minggui (1995), Hu Yang, Hu Pei and Hou Liwen (1997), Han Furong (2001), Li Weian (2001), Fang Zhou (2002), Diao Zhaofeng (2003), Zhao Hong (2004), Yang Kaiyuan (2007), Cui Zhenzhen (2009), Xu Yanmei (2012).

Akimoto Toshio (2002), Tanaka Sujun et al. (2004), Takeda Momo (2004), Kato Takehiko (2004), Tanaka Sujun (2010). Apparently researchers attach their attention to different points of enterprise vitality, their goal being to reveal the current situation of enterprise operation and the possible trend of development.

3 ENTERPRISE VITALITY AND ENTERPRISE BUSINESS AGE

"Age", the length of existing time from birth, is often represented by "years". It is a natural sign with biological basis; since the birth of beings, their age grows with the passage of time, which is in line with irresistible natural law. Physiological age is calculated according to the normal anatomic and physiologic development status of human body. It shows the degree to which the actual structure and physiological function is aging. Thus it can be used to predict the future physical condition of a living body and estimate its lifetime. For example, a person of 60 years old may be 60, 50 or 70 years old physically. Enterprises can use "age" to describe their life course from birth to recession. The same as people's physiological age and mental age, the age we are discussing in exploring enterprise age theory bears two aspects of meaning, namely: natural age and business age. Natural age of the enterprise is the time span from registration to demise; business age is one that evaluating enterprise vitality through hard and soft power revealed by the enterprise's ability of survival, growth and regeneration. Business age uncovers the operation state of the enterprise and its potential, having nothing to do with the time span of its existence. Metering method for natural age and business age are different, the results of which mean different for operation. Natural age reflects the duration of the enterprise's existence and is irreversible. However, business age has no direct relation with its duration. Being a function of business performance and operational status, business age does not necessarily increase with the passage of time and thus it is reversible and volatile. With quantitative analysis method we can accurately work out the business age to reveal the enterprise's "state of life", determine the strength of enterprise vitality, and provide basis for enterprises to develop appropriate strategies.

Enterprise business age is the quantification of enterprise vitality. Its magnitude is restricted by factors influencing enterprise vitality. The relation between enterprise vitality and enterprise business age can be described as Figure 1.

Some scholars said: "The comprehensive analysis integrating enterprise life cycle theory shows that its vitality growing stronger when the enterprise is young. However, due to operational and managerial experience, capital accumulation and other factors, enterprise business age has not reached the

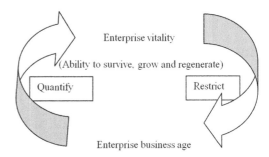

Figure 1. Concept map of the relation between enterprise vitality and enterprise business age.

peak. While on the contrary, when it passes its heyday, enterprise vitality goes downward due to the bureaucracy and other problems. The relationship between enterprise business age and enterprise vitality is an inverted 'U' type." This paper draws an analogy between the enterprise and the living organism to show the strength of business vitality by dividing enterprise business age range into 0–30 years, 30–40 years and 40–70 years, or into 20–40 years and 40–80 years.

4 ENTERPRISE BUSINESS AGE

4.1 *Foreign enterprise business age model*

In 1993, NIKKEI published a book called *Enterprise Age—Hint on Secrets of Organization Life Cycle,* one of the representative works to introduce enterprise age which analyzes in detail of enterprise age in Japan. The author pointed out: "enterprise age, a mirror of business operational condition, is the basis deciding enterprise strategy; the analysis of listed company age distribution, from both macro and micro perspective, will reveal to us industrial structure and transition in Japan." Gu Ligang and some other domestic scholars (2000) have analyzed in detail the model concerned in the work.

The enterprise age model proposed by Japanese scholars takes average sales growth rate (X_1), average staff age (X_2) and the service time of equipments (X_3) as indicators (X_2 and X_3 are based on the data of the most recent year, and the depreciation rate of the equipment is drawn by straight-line depreciation method). It divides the listed companies into two groups in terms of their established time, from which sample companies are drawn for empirical analysis with the above mentioned indicators in the last five years. Once the index is three times higher than the plus or minus standard deviation of the sample companies' indicators, the value will be three times of the plus or minus standardized deviation of the average value.

Composite value of sample companies:

$$Z_i = 0.3640 + 0.10413X_{i1} + 0.14263X_{i2} + 0.14201X_{i3}$$

Average composite value of sample companies:

$$\bar{Z} = 0.30486 + 0.10413 \cdot \bar{x}_1 + 0.14263 \cdot \bar{x}_2 + 0.14201 \cdot \bar{x}_3$$

Based on average composite value of sample companies $\bar{Z}$, max value Z_{max} and minimal value Z_{min}, composite value Z_i transited into enterprise business age SL_i through linear interpolation, which limits the enterprise business age within the scope of 20 to 80 years.

Ie: when $Z_i > \bar{Z}$,

$$SL_i = 40 + \frac{\bar{Z} - Z_i}{\bar{Z} - Z_{min}} \times 40$$

With Z_{max} as the max composite value of the sample companies;

When $Z_i < \bar{Z}$,

$$SL_i = 40 - \frac{Z_i - \bar{Z}}{Z_{max} - \bar{Z}} \times 20,$$

With Z_{min} as the minimal composite value of the sample companies;

Japanese scholars have calculated the business age of 2,352 listed companies and non-listed ones and carried out empirical research. The result shows that all excellent enterprises have competitive business age, and that the business age of excellent enterprises of all walks of life is far younger than its natural age, which manifests enterprise business age is a mirror of enterprise vitality.

Chart 1. Natural age and business age of excellent companies in Japan (excerpt).

Company name	Date of establishment	Natural age	Business age
Toyota	Aug. 1937	56	36.1
Honda	Sept. 1948	45	38.6
Sony	May 1946	47	33.6
Panasonic	Mar. 1918	75	36.5
NEC	July 1899	94	35.2
Fujitsu	June 1935	58	33.9
Canon	Aug. 1937	56	33.1
Ricoh	Feb. 1936	57	36.6
Fuji	Jan. 1934	59	42.6

From: Gu Ligang, Han Furong, Xu Yanmei. On Company Age, Foreign Economy and Management. 2000:9.

Enterprise business model of Japan is advantageous in its innovation and operability, which analyze the case from a macro perspective; on the other hand, it is inferior in its simple design of indicators, unreasonable division of business age intervals and short of quantitive and qualitative-combined evaluation.

4.2 Research of enterprise business age in China

1. Sun Minggui (1996) has an introduction for analysis method in operation safety rate and comprehensive analysis system of enterprise vitality, or Z value analysis method. He points out that Z value analysis method, different from the former practice of measuring enterprise vitality with individual index, adopts a compounded index. If Z value is positive, it means the enterprise is under "safety", if negative, it means it is under "alert". The model is as follows:

$$Z = -5.705 - 0.797X_1 + 0.0817X_2 + 0.00292X_3 + 0.00542X_4 + 0.335X_5 + 0.0580X_6 - 0.0784X_7$$

where: X_1 stands for burden rate for financial cost, X_2 for current account rate, X_3 for capital flow per capita, X_4 for sales profit growth rate per capita, X_5 for fixed asset growth rate per capita, X_6 for pure capital rate accounting for residual amount and X_7 for marginal change of rate for current capital to non-current capital.

From the perspective of bionics, the writer notes that these indicators can be respectively used to analyze different aspects of enterprise vitality. X_1 reflects enterprise's cardiac cost; X_2 enterprise's digestion and absorption capacity; X_3, X_4 and X_5 enterprise's vitality and competitiveness; X_6 enterprise's energy reserves and X_7 enterprise's latent symptoms. Japan has analyzed the Z value of 42 bankrupt enterprises in the last 22 years, finding that the Z value of these bankrupt enterprises were all negative, none being positive. The time series analysis on the Z value of different phases will draw up clearly the locus of the enterprise from its start of downturn to its bankruptcy.

2. HuYang et al. (1997) indicates the relation and difference between enterprise vitality and quality to structure a evaluation index system. Its relational expression is: ①enterprise quality = 0.3 × basic quality + 0.7 × behavior quality, with basic quality = 0.4 × personnel quality + 0.5 × technique quality + 0.1 × scale quality, and behavior quality consisting of capacity of production and operation, capacity

of scientific and technological progress, capacity of funds application and economic benefits. ②enterprise vitality (index) = 0.68 × growing ability in current situation + 0.32 × growing ability in developing trend. He points out that quality is the internal basis to decide the vitality while vitality is energetic manifestation of quality, both of which are identical in orientation of goal and complementary in changes, with vitality paying more attention to behavior and variation trend.

3. Xi Zhaofeng, Li Zhicheng (2003) put forward comprehensive quotient of enterprise business age and its determination method by defining enterprise natural age, element age, mechanism age, psychological age and their determining quotient. Under this context, enterprise business age refers to enterprise vitality and life condition reflected synthetically by perceived age, mechanism age and psychological age. ①Enterprise perceived age indicates its "IQ"; enterprise IQ = 0.4 × standardized value of average staff age+0.4 × standardized value of average equipment age + 0.2 × standardized value of newness degree of fixed assets. ②Enterprise mechanism quotient indicates its "EQ"; enterprise EQ = 0.4 × standardized value of average sales growth rate in the last five years + 0.3 × standardized value of total assets turnover + 0.3 × standardized value of ratio of cost to profit. ③Enterprise psychology age quotient indicated its "WQ"; enterprise WQ = 0.2 + 0.3 × standardized value of new product research and develop rate + 0.3 × standardized value of average years of education of the staff + 0.2 × standardized value of social contribution rate. Comprehensive quotient of enterprise business age = 0.2 + 0.3 × enterprise IQ + 0.3 × enterprise EQ + 0.2 × enterprise WQ, among which, 0.2 is correction parameter decided by unpredictable factors and other coefficients are index weights. Comprehensive quotient is transited into enterprise business age through linear interpolation, which limits the enterprise business age within the scope of 20 to 80 years.

4. Xu Yanmei et al. (2003) come up with the improved mathematic model of business age based on the enterprise business age model proposed by Japan scholars. Enterprise business age is limited between 1 and 30 years.

$$Y_{it} = \frac{X_{it} - X_{imin}}{X_{imax} - X_{imin}} \times 29 + 1$$

X_{it}—the score of enterprise No.i;
X_{imin}—the minimal score of the sample enterprises;

X_{imax}—the max score of the sample enterprises;
Y_{it}—enterprise business age of enterprise No.i.

Design formulas for total age of sample enterprises:
$SL_t = \sum_{i=1}^{p} Wi * Yit$ (t: 1, 2 … n)
SL_t—business age of enterprise No.t
W_i—weight of index No.i
Y_{it}—score of enterprise No.t on index No.i

5. Sun Jianwei (2010) carrys out empirical analysis on enterprise business age of listed textile enterprises in China on the basis of revised business age transfer approach of Japan scholars; composite value of enterprise vitality is transited into enterprise business age through linear interpolation, which limits the enterprise business age within the scope of 20 to 80 years.

Firstly, composite value of enterprise vitality.

$$(Z) = 0.1918*X_1 + 0.4742*X_2 + 0.1611*X_3 + 0.1728*X_4$$

X1—average total asset turnover in the last three years. X2—average return on equity in the last three years. X3—ratio of staff with college degree or above to the total. X4—average ratio of intangible assets to main business income. Secondly, linear transformation.

① When the synthetical value of the enterprise is smaller than the average synthetical value of sample enterprises,

$$SL_i = 40 + \frac{\bar{Z} - Z_i}{Z - Z_{min}} \times 40$$

② When the synthetical value of the enterprise is larger than the average synthetical value of sample enterprises,

$$SL_i = 40 - \frac{Z_i - \bar{Z}}{Z_{max} - Z} \times 20$$

6. Xu Yanmei, Liu Long et al. (2011) select 13 indicators to describe business viability, capacity of growth and regeneration, and examine the discrimination of these indicators. Original extent of enterprise business range of age is modified, resulting in the business age defined within ranges of 0–30 years, 30–40 years and 40–70 years.

Firstly, synthetic value of business vitality (Z_i):
$Z_i = \sum_{i=1}^{13} w_i \times X_i$, with w_i standing for index weight and X_i for the index.

Secondly, transformation of synthetic value into business age.

When $Z_i > Z_{1/4}$, conduct linear interpolation in the range of 30–40 years to get the business age:

$$SLi = 30 + \frac{Z_i - Z_{1/4}}{Z_{max} - Z_{1/4}} \times 10,$$

with $Z_{1/4}$ standing for upper quartile of the synthetic value of sample enterprises.

① When $Z_i < Z_{1/4}$, and enterprise natural age smaller than industry average natural age, conduct linear interpolation in the range of 0–30 years to get the business age:

$$SLi = 30 - \frac{Z_{1/4} - Z_i}{Z_{1/4} - Z\min} \times 30$$

② When $Z_i < Z_{1/4}$, and enterprise natural age larger than industry average natural age, conduct linear interpolation in the range of 40–70 years to get the business age:

$$SLi = 70 - \frac{Z_{1/4} - Z_i}{Z_{1/4} - Z\min} \times 30$$

7. Xu Yanmei et al. (2012) select "consumer goods and services", "information technology" and "medicine" as the industry sample, 578 enterprises as sample enterprises, borrow hierarchical clustering analysis in ecology to classify the factors affecting the business vitality; use principal component analysis to identify common factors so as to construct a system of business vitality indicators; and calculate the weight of each dimension to the entirety through the explanatory contribution ratio of the total variance. The business vitality assessment model thus constructed is as follows:

$$CV = -0.43\,(X_1 + X_2 + X_3 + X_4) + 0.34\,(X_5 + X_6 + X_7) + 0.23\,(X_8 + X_9 + X_{10} + X_{11})$$

where: X_1-flow ratio, X_2-quick ratio, X_3-cash current liabilities ratio, X_4-main business revenue growth rate, X_5-inventory turnover rate, X_6-current asset turnover rate, X_7-total asset turnover rate, X_8-intangible asset, X_9-R&D investment, X_{10}-total assets, X_{11}-total number of staff.

By statistical processing, enterprise business age model is defined as:

$$CA = 50 + 3\,CV.$$

where: CA represents corporate business age, CV enterprise vitality.

5 ENLIGHTMENT

Firstly, by analyzing the current situation of the research on enterprise vitality, we can conclude: "Z value method" and "enterprise business age" are the two basic methods measuring the condition of enterprise vitality; "Enterprise business age" can, more comprehensively and scientifically, composite financial indicators and non-financial indicators to intuitively measure the condition of the enterprise vitality. Domestic scholars have made continuous improvement and innovation for comprehensive value calculation of enterprise vitality, range definition and indicator selection in the application of comprehensive business age model, and use the modern mathematical statistics method to scientifically and reasonably analyze project and index data, making the enterprise business age model more scientific and rational. At the same time, we will use the enterprise business age model to make empirical analysis on life status of listed companies in our country, which provides scientific basis to accurately grasp the present situation and the future development trend of enterprises and industry. However, considering the present situation of the research, the literature directly measuring the strength of enterprise vitality through "enterprise business age" being scarce, the research is still in the stage of promotion and development.

Secondly, from the view of influence factor of enterprise vitality, three dimensional structure (ability to survive, grow and regenerate) are widely recognized. The specific factors in the three dimensional structure vary according to different industries and enterprise features.

Thirdly, non-financial indicators play an important role in improving enterprise performance management and promoting the construction of enterprise culture and management system. Analysis of the sustainability of the enterprise from the combination of financial indicators and non-financial indicators is of realistic significance to enterprise's operation and management; however, due to the difficult obtain of non-financial indicators (qualitative data), qualitative research literature which reflect the enterprise culture, organization structure, management system and social contribution are relatively deficient in the research literature of enterprise vitality, especially in the construction of index system.

REFERENCES

[1] Han Furong, Xu Yanmei. Enterprise Bionics. Enterprise Management Publishing House, 2002.
[2] Collins, Poles. Translated by Yu Lijun. Built to Last. China Citic Press, 2009.

[3] Zhao Hong, Zhao Hong. Research on Growth of Eco-Intelligent Enterprises in the View of Bionics. Huazhong University of Science and Technology, 2004.

[4] Gu Ligang, Han Furong and Xu Yanmei. Research on Enterprise Business Age Model: an Ecological Perspective. Natural Science Funds, 2012.

[5] Tang Wencan, Li Panfeng. Enterprise Growth Theory Basing on Three Dimensions. Soft Science, 2005:17–33.

[6] Sun Jianwei. Business Age Model and Evaluation of Textile Industry Vitality. Institute of Management, Graduate School of Chinese Academy of Sciences, 2010. 7. Research Group on "Calculating Methods and Empirical Research on Enterprise Life in China". Theory and Practice of Enterprise Life Measurement. Statistical Research, 2008:20–32.

[7] Diao Zhaofeng. On Enterprise Business Age and Its Determination. Scientific and Technological Progress And Countermeasures, 2003(9):137–138.

[8] Cao Lijun. Motivation, Mechanism and Implementation of Enterprise Growth. Technology and Management, 1996:6–9.

[9] Sun Minggui. Two Methods For Evaluating Enterprise Vitality. Enterprise Vitality, 1996:6–9.

[10] Hu Yang, Hu Pei and Hou Liwen. Research on Indicator System for Evaluating Enterprise Quality and Vitality. Soft Science, 1997(4):10–15.

Future Information Engineering and Manufacturing Science – Zheng (Ed)
© 2015 Taylor & Francis Group, London, ISBN 978-1-138-02644-5

Combined economic effects of mass customization and operation chain

Xiang Jun He

School of Safety and Environment Engineering, Capital University of Economics and Business, Beijing, China

ABSTRACT: Mass customization is an operation chain mode that is different from the traditional low-cost mass operation mode and the low-cost and differentiated operation mode. And according to the differences in the points of customization, this paper classifies mass operation into three levels, i.e., general, medium and high mass customization. Thanks to the advantages of the operation chain alliance in specialization, low transaction cost and resource sharing among member enterprises, this innovative path of mass customization is able to combine the traditional low-cost mass operation mode with the differentiated operation mode, and accordingly, to achieve composite economic effects benefiting from the integration of internal economies of scale with economies of scope.

Keywords: mass customization; operation chain; combined economy

1 INTRODUCTION

With the development of information technology and the Internet, the operation mode transformed from mega-marketing to one-to-one marketing to meet the increasing and changing individual requirements. Such change has posed a challenge to the traditional mass production mode, which characterizes in standardized products, unified markets and enough long product life cycles providing security for the operation of the enterprises with competitive factors like operation scale, cost and quality. In traditional enterprises, weak economies of scale means certain low-cost economies realized by mass conditions depending on single enterprise resources, while economies of scale for the industry refers to mass and low-cost economies based on single products. However, this mass production mode has a defect, i.e., the inevitable loss of customers. Under such circumstances, many traditional enterprises begin to adopt very limited customization so as to satisfy the individualized requirements of customers. In this way, they can obtain narrow economies of scope with certain flexibility, but they also have to deal with high cost, low efficiency and reduction of customer service quality, and eventually lose customers inevitably. While, for operation chain, it means a value-added chain which maximizes the overall interests of all affiliated enterprises on the chain due to the seamless joint and collaborative operation of strategic management activities such as "market management", "technology management", "production management" and "sales management". The enterprises realize economies of scale under mass and low-cost conditions relying on resource alliances based on the industry chain, and accordingly avoiding price wars with competitors [1]. Relying on the collaborative coordination and overall flexibility of all operation processes, the operation chain can realize low-cost economies of scale based on a high flexibility with certain scale, but the advantage of low cost and differentiation is limited [2]. Along with ever-changing customer requirements in the new environment, customers need more individualized products and services. Besides, the time of delivery becomes shorter and shorter. Thus, the competition becomes more intense. It is not difficult to find that the competitive advantages of both the traditional low-cost mode and the general differentiated mode have become increasingly ineffective while fluctuations in demand and disturbances in market have disturbed the front lines of mass production and limited differentiation. According to the natural law—Survival of the fittest, the operation chain has to choose mass customization as the operation mode.

2 CLASSIFICATIONS OF MASS CUSTOMIZATION AND THE DIFFERENCES WITH OTHER OPERATION MODES

In the long term, it is hard for people to determine whether to provide mass-produced, standardized and low-cost products or services or to provide customized or highly differentiated products or services generated from small scale production; of course, the latter strategy has a relatively higher cost. In other words, people have to make a choice between highly efficient mass production and special and innovative products.

According to the degree of customization, customization is classified as shown in Figure 1 from pure customization to pure standardization [3] and the degree of customization gradually reduces till it is nearly identical to mass production. According to the difference in the flexibility of customized products, ranked from the smallest to the biggest in flexibility, mass customization operation chain mode can be classified into general mass customization, medium mass customization and high mass customization.

High mass customization mode is a mode that provides the mass customization for the whole operation chain. All operation processes of the whole operation chain, from market development, R&D and design to delivery, adopt mass customization, which provides any product or service that a customer needs. That is to say, each process of the operation chain from upstream to downstream is customized according to customer requirements. The mode is featured by high differentiation in products. And based on the substantial differences in structures and functions of such products, there will be very distinctive differences between the products generated in this mode and products produced by competitors. Besides, this mode is of great advantage in meeting customer requirements and catering to customer's needs. Obviously, this is

an optimal mass customization approach, but, it is a mode that is rather difficult to achieve innovation. In order to meet individualized customer requirements, each operation process of the operation is required to have a high flexibility and requires a huge investment.

Medium mass customization is a mass customization operation chain mode based on medium flexibility. The entry point of customization occurs in the stage of R&D and design, so that we can implement changes from the designs of products. In this way, we can realize the customization to the greatest extent through combining, producing and assembling components of products in different ways. However, this realization process is extremely complex. Thus, this mode is only suitable to operation chains with strong design capability and flexible manufacturing capacity. This mode is featured by distinctive products that have novel appearances and innovative properties, so that it holds certain appeal for customers.

General mass customization is a typical mass customization operation chain mode based on general flexibility. First, the entry point of customization occurs in the stage of manufacturing. In the context of standard designs, this mode meets special requirements of customers by changing different components. This change is limited to the change in dimensions and materials of components, so that this mode is suitable for enterprises with strong flexible manufacturing capacity.

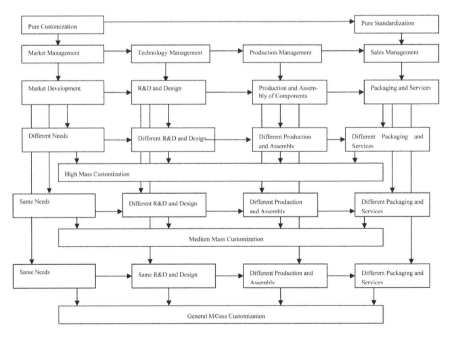

Figure 1.   Classification of mass customization.

Table 1. Comparison between mass customization mode and other operation chain modes.

Type/ content	Traditional mass and low-cost mode (focused on single products)	Traditional low-cost and differentiated mode (focused on diversified products, non-customized)	Mass, low-cost and differentiated mode (mass customization)
Objective	Develop, produce, sell, deliver products and services at a cost lower than competitors	Provide diversified products more distinctive than that of competitors and affordable for customers	Provide diversified customized products that are affordable for customers and can meet customer requirements
Production driving mode	Basically organize production in accordance with market research and customer requirement forecasts	Carry out processing and production mainly based on market forecasts and secondly based on customer requirements	Carry out processing and production entirely based on customer orders, but the overall market scale needs to be forecasted and predetermined
Strategy	One-dimensional operation chain strategy Low-cost economy with a cost lower than traditional single corporate economies of scale	Two-dimensional operation chain strategy Low-cost product diversification economy focused on certain differentiation	Three-dimensional operation chain strategy Mass, low-cost and differentiated combined economy
Main features	Basic and single products, stable mass demands, specific equipment, long development cycle	General-sized and differentiated or small-sized low-cost diversified products, short development cycle	Low cost and differentiation combined strategy, passing combined economy based on diversified products by virtue of the flexibility and fast reaction of the operation chain
Scope of application	New products with stable requirements and rapidly listed mass markets	Certain-scale markets with occasionally changing requirements and certain competitive influence	Products with changeable customer requirements and diversified mass markets

Second, the entry point customization occurs in the stage of assembling. This mode meets individualized requirements of customers by assembling standard components in a special product structure, so that it is suitable for enterprises specialized in the production of prefabricated products. Third, the entry point of customization occurs in the sales stage. This mode generally meets customer requirements by matching finished products and services and customizing the time, place and method of delivery in the sales stage. The new trend is to postpone the assembly phase and assemble products according to customer requirements in the sales stage. This mode has advantages in many aspects such as mass production and low cost, but the flexibility of customization successively declines according to the form of customization. Thus, it is suitable for enterprises with different flexibility and can meet the requirements of general customers in terms of products.

In conclusion, the traditional operation mode can be classified as mass and low-cost operation mode and low-cost and differentiated operation mode. By virtue of modern information techniques and flexible manufacturing techniques, mass customization can deal with diversification properly, produce diversified products efficiently, expand product lines, and enlarge market space. Mass customization becomes an important innovation for enterprises to expand market and win customers in the condition of surplus economy and excessive competition. See Table 1 for the detailed condition of the comparison between mass customization and various operation modes.

3 INNOVATION PATH OF MASS CUSTOMIZATION AND ITS INTERNAL EFFECT ON OPERATION CHAIN

3.1 *Innovation path of mass customization*

For the reverse business mode taking customer as a starting point, or the operation chain mode which subdivides market to limitation to produce only for one customer, the concepts contained are interlinked: namely to meet specific need of each customer to maximum extent and to reflect concerns of each customer. So, fundamentally, the operation chain may be possible to obtain customers, create market, win composite economic and make innovation by means as following.

Professional advantages of operation chain alliance. The mass customization is usually to form a network chain enterprise structure through establishing operation chain relationship so as to promote rapid response ability and competition ability. Owing to each member enterprise on the chain concentrating resources on its core capability, the member enterprise is to be formed with internal specialization and scale mode so as to realize advantage of scale economy for the member enterprises.

Advantages of profession in member enterprise: As the operation chain having realized resource sharing among the member enterprise by integrating external resources, and breaking through limitation of resources of the traditional enterprise, the operation chain is enabled to reach the requirements of economy of scope in short term in order to reduce the expenses of information search of the operation chain in the market and cost of scale operation, the alliance entirety is enabled to form advantages of scale economics in the aspects of production capacity, capital, technology development in complementary advantages.

The advantage of saving transaction cost. The rise of the network technology, improvement of management information system, development of flexible manufacturing technology, popularity of electronic commerce, the collaboration of the traditional enterprise to cross time and space barrier becomes possible relying on the operation chain; and also the circulation, configuration and application of resources are possible in global range, the transaction mode of market becomes more and more convenient so as to greatly reduce the costs of enterprises for external transaction. At the same time, the open-type and flat organizational structure brought by informatization makes market strain capacity of enterprises strengthened, the information communication is sped up, which is beneficial to break through limitation of management level and management span of the traditional enterprises in order to avoid potential problems of bulky organization, overstaffing in organization, poor communication and information distortion so as to make transaction cost to be internalized and greatly save transaction costs.

Under operation chain operation mode of mass customization, the economy of scale and personalization are complementary relation, and the economy scale here fully reflects characteristics of personalization and diversification. However, the characteristics of individuation and diversification make its scale economy and scope economy different from the traditional enterprise, to form internal economic effect advantage [4] of Trinitarian operation chain. Up to now, they have appeared on new market competition stage taking the form of mass customization, such as customized marketing. According to different needs of different customers, different

balance of payments and different requirements of consumer level as well as market competition, the operation chain can take part in the competition through different ways of mass customization respectively, which becomes a sharp instrument of carrying out skills and realizing ambitions of different operation chain enterprises, so as to acquire competitive advantages based on composite economics of low cost and different integration.

3.2 *Composite economic competitive advantage based on general mass customization*

Mass production is a standardized scale production taking production as the principal, which pursues a kind of batch scale economy so as to lower average cost. However, due to customers having more and more initiative, the individual needs restrict to a certain extent realization of traditional economy of scale. The general mass customization refers to less peculiarities of product, paying attention to perfection and accuracy, [5], which are generally the improved products, such as new style, new packaging, or innovation with a certain physical shape etc., and is mass production that gives priority to scale and the personalized customization product as supplement.

Therefore, as compared with mass production, it is a mass production giving priority to general demands to meet the needs of most customers, it not only ensures customer to be able to express a certain custom demand, but also does not cause high complexity and high cost by providing proper customization options to customers, pursuing for less varieties, multiple batches and response of economy of scope giving priority to economy of scale, which is actually a fusion of strong economy of scale and weak economy of scale effect, so as to obtain effect of lower average cost as the whole. Taking Dell Company as an example [6], the company produces about 4 million computers, laptops, servers and workstations each year, but its production mode still follows a set of concept inherited from "Dormitory Enterprise"—just producing computers according to customers' orders. Dell does not have many core technologies, for example, major parts of the computer, chips and operating systems were all pursued by the principle of copinism, and the customization options offered are only limited varieties. It obtains customers through providing convenient and considerate services to response customer demands in time, which helps them won customers.

Generally, customers just dial telephone number 800 or login website of the company, put forward his own configuration of computer, and wait for quotation appearing from screen of the company, input number of credit card, press enter key, that's all. Dell company receives the order information, and passes it to related department through its advanced logistics management software immediately; The supplier will deliver the spare part to assembly workshop with in one hour; the microcomputers being assembled are flowed quickly in each zero-inventory and zero-backlog of the plant; the microcomputers produced according to order will be assembled after 36 hours, goods can be delivered; generally, the customers can receive the goods within 3 days. By means of this kind of general mode of mass customization, Dell occupies numerous shares in computer market, and becomes powerful competitor of IBM and HP.

3.3 *Competitive advantages of composite economics based on medium mass customization*

Moderate large mass customization mode generally aims at more complex customer requirements, which meets the needs of various kinds of requirements of customers through special customized design and special customized production and assemblies. Under the medium mass customization mode, it is not only to improve appearance such as style and packaging etc. to make it novel and obviously distinct it from the products of competitors, and a certain innovation has been made for local performance though adding functions which the competitive products don't have, such products with differences have no doubt a certain attraction to the customers. A product portfolio of operation chain is integration of multiple different individual batches. But if operation chain has met individual requirements on the premise of having production capacity or resources sharing effect, or has carried out high correlation diversification management, effective range saving can be obtained, and then the average cost of each kind of product may be reduced.

The first is that the degree of association among multiple products based on related technology of operation chain is higher than that of competitors, because the higher the association is, the stronger the related economic effect of technological products and ability responding to customers than the competitors. The second is that the operation chain based on market related business scope is more than that of the competitors for pursuing economy of scope or compatibility of economy of scope and economy of scale. The third is to reasonably arrange batches of each product and successive processing order, to make multiple products of the operation chain to lie in economy of scale or to make its average production scale to be greater than that of the competitors so as to obtain lower average cost.

Scope economy is realized under mass customization mode; while the relevance between products and technology or market is only an issue of quality, there should be an issue of quantity. If one kind

of technological resource can play a role in different kinds of products (reuse), then in a case that the resource is shared, the bigger the relevant range ($N = 1 \dots n$, number of types of related products), the better. As the value input one time ($C0$) can be shared by a bigger denominator. When $O < Ne < N$, the average cost of operation chain F is lower than the average cost of operation chain E:

$$\frac{AC_1 + \dots + AC_n - (N_e - 1)C_0}{N_e} \rangle$$
$$\frac{AC_1 + \dots + AC_n - (N_f - 1)C_0}{N_f}$$

According to this inequality, we are sure that pursuing low cost under mass customization mode is actually to pursue the economic efficiency of scale and scope (of course, in term of enterprise development, after N is increased to a certain degree, the opportunity loss of operation chain will be increased.).

The competitive advantage will be finally represented in the corresponding market of a number of competitors and same products managed. The scale economy of scope is represented by reduction of average cost of operation chain; this saving effect also is represented by reduction of average cost of each kind of product. And as a principal part of competition, operation chain can also be able to convert the low cost advantage of this business cost into low cost advantage of individual product.

3.4 *Competitive advantage of composite economics based on high-degree mass customization*

High-degree mass customization mode is based on upstream and downstream products of entire operation chain or mass customization services, which is suitable for those more scattered and cluttered market segmentations where the customers require to be provided with very special requirements of products, which puts forward very high requirement for marketing, research and design, production and sales, it not only needs a wider product line, but also a variety of flexible designs etc. What are different from the two modes mentioned above, the lower relevance of product or market of high-degree mass customization mode are its prominent characteristics due to orientating to high-end customers.

To make high-degree mass customization to be close or reach the cost of mass production, the customers must be subdivided, combined and classified according to different requirements; and then the each kind of product complying with related diversification should be designed for standardization and modularization, and then they are delivered according to categorization to factory with corresponding production scale for manufacturing. The cost of this mass customization mode is mainly reduced through the following ways so as to realize large-scale low-cost and differential integration.

The first is that its development and production mode must be changed from traditional single product to product cluster orientation [7], namely developing a series of derivative products based on product platform; at the same time, only standardization and modular design have to be carried out, can the scope economy based on product development and related design technology be realized.

The second is to select a number of factories of different scales to organize production respectively according to related principles of products. At first, mass production for the products complying with economic scale shall be carried out to realize scale economy, and then, classified production is carried out according to flexibilities in order to achieve scope economy; substantially, production according to different classifications will decompose strong scale economy and wide scope economy into superposition and synthesis of a series of weak scale economy and narrow scope economy.

For example, in the form of industrial organization "Enterprise series system" [8] (Basic form of development of large enterprises of Japanese, such as Toyota series, Hitachi series, Panasonic series and NTT series etc.), through mass customization of enterprise, under the condition of lower correlation degrees between varieties, the core enterprise assign the production tasks without sharing effect to specialized subcontractor in order to save wage cost, and under the production system of "Subcontract system", the long-term industrial cooperative relationship formed among the core enterprises, subcontractors and further subcontractors against long-term contract becomes "Community of survive" (small enterprises "eating backed against mountain", large enterprises get saving and risk reduction). So, improvement of division of labor based on specialization and working efficiency on the collaboration chain is certain to reduce production costs.

In addition, the core enterprise is responsible for unified sales or final sales of products; it can also take use of correlation of market to obtain saving effect brought along by resources sharing of market. Therefore, operation effect of "Enterprise system series" also prompts to competitors that using external resources win-win form is an effective way to seeking for low cost advantage. In the period of "World factory" in Japan, the self-operation rate of Japanese enterprises is greatly lower than the enterprises in U.S. For example, the self-operation rate of automobile industry enterprises in Japan is generally below 30%, while the self-operation rate of automobile industry enterprises in U.S. is more than 50%. A reason of enterprise series system in Japan is that the scale economy effect of enterprise

group is larger than the scale economy effect of enterprise group in Europe and U.S.; it is also the reason for series production of Japanese companies to have a greater competitive advantage in the international market [8].

4 CONCLUSION

Mass customization is put forward under the condition that simple mass production mode can't meet diversified and personalized requirements of customers; its most prominent characteristic is product customization according to special requirements of customers.

As compared with the traditional model of large-scale low-cost mode, the mass customization not only has competitive advantage of low-cost strong scale economy, but also at the same time has the competitive advantage of scope economy with different size to meet the diversified requirements of customers;

As compared with the traditional model of low cost and differences mode, the mass customization have been extended and integrated in both aspects of scale and flexibility so as to make the scale of operation can be larger on the basis of requirements of customers and the cost is must lower. At the same time, relying on flexible manufacturing technology makes overall flexibility of operation chain greater and meets diversified requirements of customers to realize greater difference advantages and to achieve operation chain competitive advantage of large-scale low-cost and difference integration.

REFERENCES

[1] He Xiangjun Wang Lijie, Brief discussion on (business mode) of enterprise competitive advantage based on innovation chain. *Coal Economic Research*, 2004 (12).

[2] He Xiangjun Wang Lijie, Brief discussion on competitive advantage based on rapid response of flexible chain speed economy. *Coal Economy Research*, 2004 (11).

[3] Xu Liangli, discussion on mass customization production mode to be carried out by the enterprises in our country. *Economic Management, New Management*, 2003 (18).

[4] Shi Liping, Change of mass customization production and scale economy. *Logistics Technology*, 2003 (1).

[5] Zhou Xiaodong, Xiang Baohua, Discussion on mass customization and implementation assessment. Technological Economy, 2003 (2).

[6] Adrian J. Slywotzky (U.S.), Value Transfer. Beijing: China Foreign Publication and Distribution Company, 1998.

[7] Dan Bin, Functional assessment and selection method of product cluster orienting to mass customization. *Journal of Management Engineering*, 2004 (1).

[8] Xu Minli, discussion on mass customization production mode to be carried out by the enterprises in our country. *Economic Management, New Management*, (18).

Future Information Engineering and Manufacturing Science – Zheng (Ed)
© 2015 Taylor & Francis Group, London, ISBN 978-1-138-02644-5

The study of evolution design for Chinese traditional medicine granulators by simplified evolution pattern

Suxin Wang
Northeastern University at Qinhuangdao, Qinhuangdao, Hebei, China

Leizhen Wang & Silei Wu
Northeastern University, Shenyang, China

Yongqing Li
Tangshan Normal University, Tangshan, Hebei, China

Li Zhao
Cangzhou Tecnimetal Ltd., Cangzhou, China

ABSTRACT: Granulating structure is a core of granulator system for Chinese traditional medicine granulator and need innovation. Granulator structure concepts are designed by the application evolution line of the TRIZ theory. The new structures have technical conflict, and conflict is described by standard engineering parameter, then invention principle can be found to solve conflict. According to structure convolution, rub-spheronization structure is formed. According to time convolution and invention principle number, divided structure of auto-granulator is formed by division principle, rack rubbing structure and internal gear rolling structure are formed by reverse principle.

Keywords: evolution pattern; conflict; TRIZ; Chinese traditional medicine granulator; design theory

1 INTRODUCTION

Pill is a kind of good form of Chinese traditional medicine, but it cannot be produced by Western medicine facility for its process and physics characteristic. The present conditions are long process and high energy consumes and great labor intension, so it is important to develop one-off and continue forming and shorten process to meet the need of modern times.

2 TRIZ THEORIES

TRIZ is an acronym for Russian words for theory of inventive problem solving, which is created by Genrich Altshuller, who begun work on the theory in 1946 and attempted to formalize methods of creative thinking which led to new inventions. He observed that patents could be split into two major categories: some that were an incremental enhancement of an existing idea, but others were truly innovative. Altshuller realized that there were parallels between different technologies, if the principles behind the majority of inventions could be identified, the creation of new ideas would become much easier. He developed his ideas into TRIZ theory, which is a creative frame work based on the study of 200000 patents and observed trends in the evolution of technical systems.

TRIZ, is a powerful methodology for producing systematic innovation and improving the designer's thinking process for technical advantages. The fundamental idea in this framework is the principle of abstraction where it involves establishing a system of classification in a problem area, finding a system of operators that map the problem categories into corresponding categories of solutions, Altshuller discovered this principle and this idea is shown in Figure 1.

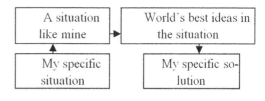

Figure 1. TRIZ generic problem-solving framework.

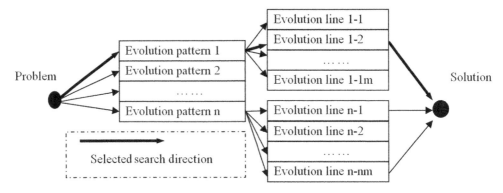

Figure 2.   Search strategy based on evolution.

Figure 3.   Major phases of problem solution.

Altshuller discovered that the emergence and implementation of innovation were not random but follows specific evolutionary patterns driven by this Ideality Law. The importance of the laws of evolution is that they can be used to estimate which phases the stage is going to be. So it is possible to predict what changes will be in future and to develop a strategic plan for new stage.

The analysis of historical patterns of evolution has revealed a number of trends and sub trends that drive system evolution. The search strategy based on evolution is shown in Figure 2.

Knowledge evolution trends can be used as a conceptual method to generate solutions to problems; it can suggest how to improve an interaction or system by considering evolution pattern and line as an inventive principle and as a future evolution stage. Major phases of problem solution are shown in Figure 3.

3   SIMPLIFIED EVOLUTION PATTERN

The system structure evolution is from simple to complex, and from the simple to the complex. In the new project system designers attempt to merge the system unit or element to a minimum number of components, the number included in engineering system decrease. Less critical components can be removed from the system, remaining element to realize their functions. In this pattern the evolution route of formation is trimming and convolution.

3.1   Structure convolution

In Figure 4, traditional Chinese medicine groove shaped mold cavity pill in the automatic pill making machine, pill cycle is 0.7, the pills roundness and brightness are poor, and needs to be rolled. In this paper, structure convolution is used, and evolution line is form a plurality of single system to cross function, and to the system features the same merge into a single system, as shown in Figure 5. The structure of the tooth edge need edge rounding, hobbing process to reduce the drop pill scratches on the pill.

3.2   Time convolution

In Figure 4, the pill and rub with the pill are incorporated in structure and time. Time convolution evolution line is from a plurality of continuous operation to combine similar or connected operation in time. This design structure can improve work efficiency, and streamline patterns of evolutionary process.

The pill and rub with the pill combined in structure and time, the pill is formed in the molding process, and the two rubbing knives are not good molded. One side of the conflict is to simplify the structure, and have good manufacturability, on the other side is the pill round and bright degree is not good, namely the pill round shape is not good. The conflict description of standard engineering parameters is as follows:

Characteristic parameters improved: simplified structure and good manufacturability.

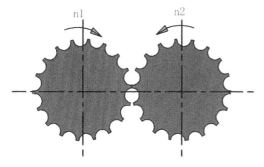

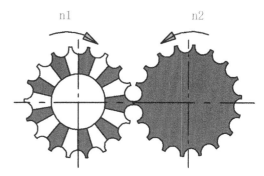

Figure 4. The front view of pill-making structure of Chinese traditional medicine automatism granulator.

Figure 6. Divided structure of auto-granulator.

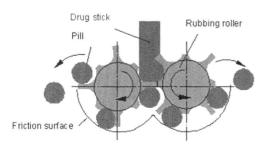

Figure 5. Rub-spheronization structure.

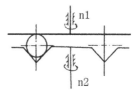

Figure 7. Co-shaft and bi-pan rubbing diagram.

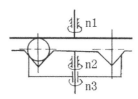

Figure 8. Co-shaft and tri-pan rubbing diagram.

The characteristic parameters of bad: pill shape is not good.

The descriptions of the technical conflict are available in 39 standard engineering parameters by parameters 12 and 32.

No. 12 Shape: Outside contour of the object, or appearance.

No. 32 Manufacturability: The simple and convenient degree of objects or systems in the manufacturing process.

The available invention principle number 1, 13, 27 and 28 can be found in conflict matrix, and the invention of the principle contents and specific solutions are as follows:

3.2.1 *Division*

The new scheme is shown in Figure 6, black and white part of roll n1 is independent to each other and together with roll n2, three axial velocities are different.

The structure principle of before segmentation (see Fig. 4) for Chinese traditional medicine system is similar to two coaxial plates lapping mode (see Fig. 7), and the segmented structure is similar to three coaxial plates lapping mode (see Fig. 8). This structure is good for pill rotation direction change, and strengthens the pelletizing effect.

3.2.2 *Reverse*

The original scheme (see Fig. 4) is external meshing scheme, if a pill rolling knife unchanged, another pill rolling cutter tooth number becomes infinite, then become rack and rack rolling balls, the pelletizing process schematic diagram is shown in Figure 9.

In Figure 9, two rubbing knives mold very good when the rack structure are rolling pill, and the pills in the cylindrical touch rub into pills. This structure can't do continuous pill production, pill rolling rack knife need complex rolling knife loop structures.

Rack reverse, then form inner gear, two pill rolling knife form Internal gear rolling structure (see Fig. 10), and the coincidence degree bigger than external engagement. Under the same conditions, internal gear rolling structure has longer pills rubbing time and better pill shape than external engagement. In Figure 10, two roller gears have the same linear velocity.

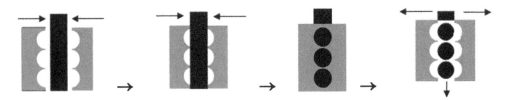

Figure 9. Rack rubbing diagram.

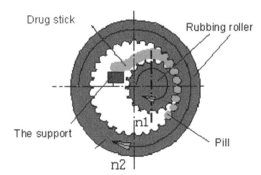

Figure 10. Internal gear rolling structure.

4 CONCLUSIONS

Granulating structure is a core of granulator system for Chinese traditional medicine granulator and need innovation. By simplified evolution pattern, Chinese traditional medicine granulator conceptual design result is as follows:

According to structure convolution, rub-spheronization structure (see Fig. 5) is formed. According to time convolution and invention principle number, divided structure of auto-granulator (see Fig. 6) is formed by division principle, rack rubbing structure (see Fig. 9) and internal gear rolling structure (see Fig. 10) are formed by reverse principle.

ACKNOWLEDGEMENTS

This study was financially supported by National Natural Science Foundation Key Items of P.R. China (70931001), National Natural Science Foundation (61273203), and The Natural Science Foundation Of Hebei Province (F2012501033) and the Technology Support Project of Northeastern University at Qinhuangdao (XNK201407).

REFERENCES

[1] Hyunseok Park & Ree, J.J. & Kwangsoo Kim. 2012. An SAO-based approach to patent evaluation using TRIZ evolution trends. Management of Innovation and Technology (ICMIT), 2012 IEEE International Conference on Digital Object Identifier: 594–598.
[2] Feihua Huang. 2013. Technology innovation and new product development process integrating QFD and TRIZ. Innovation Conference (SIIC), 2013 Suzhou-Silicon Valley-Beijing International: 127–131.
[3] Samoila, C. & Ursutiu, D. & Cotfas, P. & Cotfas, D. 2013. TRIZ method and remote engineering approach. Global Engineering Education Conference (EDUCON), 2013 IEEE: 1–4.
[4] Fei Yu & HuanGao Zhang & JianGuang Sun & RunHua Tan & GuoZhong Cao. 2012. Trimming-based conflict discovery and problem-solving process model. Management of Innovation and Technology (ICMIT), 2012 IEEE International Conference on: 647–651.

Future Information Engineering and Manufacturing Science – Zheng (Ed)
© 2015 Taylor & Francis Group, London, ISBN 978-1-138-02644-5

Risk management in commercial banks

Jianwen Zheng
Wuhan University of Technology, Hubei, China

ABSTRACT: Risk is the fundamental element that drives financial behaviour. Without risk, the financial system would be vastly simplified. However, risk is omnipresent in the real world. Financial Institutions, therefore, should manage the risk efficiently to survive in this highly uncertain world. The future of banking will undoubtedly rest on risk management dynamics. Only those banks that have efficient risk management system will survive in the market in the long run. The effective management of credit risk is a critical component of comprehensive risk management essential for long-term success of a banking institution. Credit risk is the oldest and biggest risk that bank, by virtue of its very nature of business, inherits. This has however, acquired a greater significance in the recent past for various reasons. Foremost among them is the wind of economic liberalization that is blowing across the globe. India is no exception to this swing towards market driven economy. Better credit portfolio diversification enhances the prospects of the reduced concentration credit risk as empirically evidenced by the direct relationship between concentration credit risk profile and NPAs of public sector banks.

Keywords: credit risk; management; commercial banks

1   INTRODUCTION

1.1   *Risk management*

The future of banking will undoubtedly rest on risk management dynamics. Only those banks that have the efficient risk management system will survive in the market in the long run. The effective management of credit risk is a critical component of comprehensive risk management essential for long-term success of a banking institution. Credit risk is the oldest and biggest risk that bank, by virtue of its very nature of business, inherits. This has however, acquired a greater significance in the recent past for various reasons. Foremost among them is the wind of economic liberalization that is blowing across the globe. India is no exception to this swing towards market driven economy. Competition from within and outside the country has intensified. This has resulted in multiplicity of risks both in number and volume resulting in volatile markets. A precursor to successful management of credit risk is a clear understanding about risks involved in lending, quantifications of risks within each item of the portfolio and reaching a conclusion as to the likely composite credit risk profile of a bank.

The corner stone of credit risk management is the establishment of a framework that defines corporate priorities, loan approval process, credit risk rating system, risk-adjusted pricing system, loan-review mechanism and comprehensive reporting system.

1.2   *Significance of the study*

The fundamental business of lending has brought trouble to individual banks and entire banking system. It is, therefore, imperative that the banks are adequate systems for credit assessment of individual projects and evaluating risk associated therewith as well as the industry as a whole. Generally, banks in India evaluate a proposal through the traditional tools of project financing, computing maximum permissible limits, assessing management capabilities and prescribing a ceiling for an industry exposure. As banks move in to a new high powered world of financial operations and trading, with new risks, the need is felt for more sophisticated and versatile instruments for risk assessment, monitoring and controlling risk exposures. It is, therefore, time that banks managements equip themselves fully to grapple with the demands of creating tools and systems capable of assessing, monitoring and controlling risk exposures in a more scientific manner.

Credit risk, that is, default by the borrower to repay lent money, remains the most important risk to manage till date. The predominance of credit risk is even reflected in the composition of economic capital, which banks are required to keep a side for protection against various risks.

According to one estimate, credit risk takes about 70% and 30% remaining is shared between the other two primary risks, namely market risk (change in the market price and operational risk i.e., failure of internal controls, and so on). Quality borrowers (Tier-I borrowers) were able to access the capital market directly without going through the debt route. Hence, the credit route is now more open to lesser mortals (Tier-II borrowers). With margin levels going down, banks are unable to absorb the level of loan losses. There has been a very little effort to develop a method where risks could be identified and measured. Most of the banks have developed internal rating systems for their borrowers, but there has been a very little study to compare such ratings with the final asset classification and also to fine-tune the rating system. Also risks peculiar to each industry are not identified and evaluated openly. Data collection is regular driven. Data on industry-wise, region-wise lending, industry-wise rehabilitated loan, can provide an insight into the future course to be adopted.

Better and effective strategic credit risk management process is a better way to manage portfolio credit risk. The process provides a framework to ensure consistency between strategy and implementation that reduces potential volatility in earnings and maximize shareholders wealth. Beyond and over riding the specifics of risk modeling issues, the challenge is moving towards improved credit risk management lies in addressing banks' readiness and openness to accept change to a more transparent system, to rapidly metamorphosing markets, to more effective and efficient ways of operating and to meet market requirements and increased answerability to stake holders.

The New Basel Capital Accord is scheduled to be implemented by the end of 2006. All the banking supervisors may have to join the accord. Even the domestic banks in addition to internationally active banks may have to conform to the accord principles in the coming decades. The RBI as the regulator of the Indian banking industry has shown keen interest in strengthening the system, and the individual banks have responded in good measure in orienting themselves towards global best practices.

1.3 *Credit Risk Management (CRM) dynamics*

The world over, credit risk has proved to be the most critical of all risks faced by a banking institution. A study of bank failures in New England found that, of the 62 banks in existence before 1984, which failed from 1989 to 1992, in 58 cases it was observed that loans and advances were not being repaid in time [5]. This signifies the role of credit

risk management and therefore it forms the basis of present research analysis.

Researchers and risk management practitioners have constantly tried to improve on current techniques and in recent years, enormous strides have been made in the art and science of credit risk measurement and management [6]. Much of the progress in this field has resulted form the limitations of traditional approaches to credit risk management and with the current Bank for International Settlement' (BIS) regulatory model. Even in banks which regularly fine-tune credit policies and streamline credit processes, it is a real challenge for credit risk managers to correctly identify pockets of risk concentration, quantify extent of risk carried, identify opportunities for diversification and balance the risk-return trade-off in their credit portfolio.

The two distinct dimensions of credit risk management can readily be identified as preventive measures and curative measures. Preventive measures include risk assessment, risk measurement and risk pricing, early warning system to pick early signals of future defaults and better credit portfolio diversification. The curative measures, on the other hand, aim at minimizing post-sanction loan losses through such steps as securitization, derivative trading, risk sharing, legal enforcement, and so on. It is widely believed that an ounce of prevention is worth a pound of cure. Therefore, the focus of the study is on preventive measures in tune with the norms prescribed by New Basel Capital Accord.

2 THE PROBLEM OF NON-PERFORMING

2.1 *Explanation*

Liberlization and Globalization ushered in by the government in the early 90 s have thrown open many challenges to the Indian financial sector. Banks, amongst other things, were set on a path to align their accounting standards with the International standards and by global players. They had to have a fresh look into their balance sheet and analyze them critically in the light of the prudential norms of income recognition and provisioning that were stipulated by the regulator, based on Narasimhan Committee recommendations.

Loans and advances as assets of the bank play an important part in gross earnings and net profits of banks. The share of advances in the total assets of the banks forms more than 60 percent [7] and as such it is the backbone of banking structure. Bank lending is very crucial for it make possible the financing of agricultural, industrial and commercial activities of the country.

The strength and soundness of the banking system primarily depends upon health of the advances. In other words, improvement in assets quality is fundamental to strengthening working of banks and improving their financial viability. Most domestic public sector banks in the country are expected to completely wipeout their outstanding NPAs between 2006 and 2008.

NPAs are an inevitable burden on the banking industry. Hence the success of a bank depends upon methods of managing NPAs and keeping them within tolerance level, of late, several institutional mechanisms have been developed in India to deal with NPAs and there has also been tightening of legal provisions. Perhaps more importantly, effective management of NPAs requires appropriate internal checks and balances system in a bank [9].

In this background, this chapter is designed to give an outline of trends in NPAs in Indian banking industry vis-à-vis other countries and highlight the importance of NPAs management. NPA is an advance where payment of interest or repayment of installment of principal (in case of Term loans) or both remains unpaid for a period of 90 days [10] (new norms with effect from 31st March, 2004) or more.

2.2 *Trends in NPA levels*

The study has been carried out using the RBI reports on banks (Annual Financial Reports), information/data obtained from the banks and discussion with bank officials. For assessing comparative position on CARR, NPAs and their recoveries in all scheduled banks namely, Public sector Banks, Private sector banks were perused to identify the level of NPAs.

The Table 2.1 lists the level of non-performing assets as percentage of advances of pubic sector banks and private sector banks. An analysis of NPAs of different banks groups indicates the public sector banks hold larger share of NPAs during the year 1993–94 and gradually decreased to 9.36 percent in the year 2003. On the contrary, the private sector banks show fluctuating trend with starting at 6.23 percent in the year 1994–95 rising upto 10.44 percent in year 1998 and decreased to 8.08 percent in the year 2002–03.

2.3 *International comparison of NPA levels*

Comparison of the problem loan levels in the Indian banking system vis-à-vis those in other countries, particularly those in developed economies, is often made, more so in the context of the opening up of our financial sector. The data in respect of NPAs level of banking system available for countries such as USA, Japan, Hong Kong, Korea, Taiwan & Malaysia reveal that it ranged from 1 percent to 8.1 percent during 1993–94, 0.9 percent to 5.5 percent during 1994–95, 0.6 to 3.0 percent during 2000 as against 23.6 percent, 19.5 percent and 14 percent respectively for Indian banks during this year [11]. The NPAs level in Japan, for example is at 3.3 percent of total loans, it is 3.1 percent in Hong Kong, 7.6 percent in Thailand, 11.2 percent in Indonesia, and 8.2 percent in Malaysia during 1994–95, whereas the corresponding figure for India is very high at 19.5 percent [12].

According to Ernst & Young [13], the actual level of NPAs of banks in India is around $40 billion, much higher than the government own estimates of $16.7 billion [14]. This difference is largely due to the discrepancy in the accounting of NPAs followed by India and rest of the world. According to Ernst & Young, the accounting norms in India are less stringent than those of the developed economies. Furthermore, Indian banks also have the tendency to extend past due loans. Considering India's GDP of around $ 470 billion, NPAs were around 8 percent of the GDP which was better than many Asian economic power houses. In China, NPAs were around 45 percent of GDP, while equivalent figure for Japan was around 28 percent and the level of NPAs for Malaysia was around 42 percent. On an aggregate level, Asia's NPAs have increased from $ 1.5 trillion in 2000 to $ 2 trillion in 2002 an increase of 33 percent. This accounts for 29 percent of the Asian's countries total GDP. As per the E & Y's Asian NPL report for 2002, the global slowdown, government heisting and inconsistency in dealing with the NPAs problem and lender complacency have caused the region's NPAs problem to increase. However looking from a positive angle, India's ordinance, on Securitization and Reconstruction of Financial Assets and Enforcements of Security Interest is a step in the right direction. This ordinance will help banks to concentrate on good business by eliminating the business of bad loans [15].

3 MANAGEMENT OF CREDIT RISK—A PROACTIVE APPROACH

3.1 *Explanation*

Risk is the potentiality that both the expected and unexpected events may have an adverse impact on the bank's capital or earnings. The expected loss is to be borne by the borrower and hence is taken care by adequately pricing the products through risk premium and reserves created out of the earnings. It is the amount expected to be lost due to changes in credit quality resulting in default. Whereas, the unexpected loss on account of individual exposure and the whole portfolio is entirely is to be borne by the bank itself and hence is to be taken care by the capital.

Banks are confronted with various kinds of financial and non-financial risks namely, credit, market, interest rate, foreign exchange, liquidity, equity price, legal, regulatory, reputation, operational, and so on. These risks are highly interdependent and events that affect one area of risk can have ramifications for a range of other risk categories. Thus, top management of banks should attach considerable importance to improve the ability to identify measure, monitor and control the overall level of risks undertaken.

3.2 Credit risk

The major risk banks face is credit risk. It follows that the major risk banks must measure, manage and accept is credit or default risk. It is the uncertainty associated with borrower's loan repayment. For most people in commercial banking, lending represents the heart of the Industry. Loans dominate asset holding at most banks and generate the largest share of operating income. Loans are the dominant asset in most banks' portfolios, comprising from 50 to 70 percent of total assets [17].

4 CONCLUSION

Credit risk management in today's deregulated market is a challenge. The very complexion of credit risk is likely to undergo a structural change in view of migration of Tier-I borrowers and, more particularly, the entry of new segments like retail lending in the credit portfolio. These developments are likely to contribute to the increased potential of credit risk and would range in their effects from inconvenience to disaster. To avoid being blindsided, banks must develop a competitive Early Warning System (EWS) which combines strategic planning, competitive intelligence and management action. EWS reveals how to change strategy to meet new realities, avoid common practices like

benchmarking and tell executives what they need to know—not what they want to hear. The reputation of a bank is very important for corporate clients. A corporation seeks to develop relationship with a reputable banking entity with a proven track record of high quality service and demonstrated history of safety and sound practices. Therefore, it is imperative to adopt the advanced Basel-II methodology for credit risk. The Basel Committee has acknowledged that the current uniform capital standards are not sensitive and suggested a Risk Based Capital approach. Reserve Bank of India's Risk Based Supervision reforms are a fore-runner to the Basel Capital Accord-II. For banks in India with the 'emerging markets' tag attached to them going down the Basel-II path could be an effective strategy to compete in very complex global banking environment. Indian banks need to prepare themselves to be competed among the world's largest banks. As our large banks consolidate their balance sheets size and peruse aspirations of large international presence, it is only expected that they adopt the international best practices in credit risk management.

REFERENCES

[1] Agarwal P. and Srikanth V., (2002), "A question of Reliability", Economic Times, July 24, 2002.
[2] Banking Bureau, (2002), "RBI Reports finds increase in NPAs of Commercial Banks", The Financial Express, November 16, 2000.
[3] Murthy E.N., (2002), "Managing Credit Risk", ICFAI Reader, Vol.2, February 2002, pp. 3.
[4] Murthy G.R.K., (2001), "Credit-Risk-Management in a market driven economy; The Acid Test for banks", IBA Bulletin, March 2001. pp. 105–123.
[5] Narasimham P.V., (1998), "Risk Management—Towards Sound & Strong Banking", presented at BECON' 98 conference, pp. 54–61.
[6] Narasimhan N., (2003), "Banking sledge hammers for NPA Files", Professional Bankers, March 2003, pp. 25–27.
[7] Pricewaterhouse Coopers, (2004), "Management of Non-Performing Assets by Indian Banks", IBA Bulletin, January 2004, Vol XXVI. No. 1., pp, 61–97.
[8] Rajeev A.S. (2004), "Basel II – Issues and Constraints", IBA Bulletin, June 2004, pp. 11.

Future Information Engineering and Manufacturing Science – Zheng (Ed)
© 2015 Taylor & Francis Group, London, ISBN 978-1-138-02644-5

Inference rules based on XML functional dependency and XML multi-valued dependency

Zhongping Zhang

The School of Information Science and Engineering, Yanshan University, Qinhuangdao, Hebei, China
The Key Laboratory for Computer Virtual Technology and System Integration of Hebei Province,
Qinhuangdao, Hebei, China

Shanshan Shi

The School of Information Science and Engineering, Yanshan University, Qinhuangdao, Hebei, China

ABSTRACT: There not only exists the relation of the functional dependency and multi-valued dependency in XML data constraint, but also exists the relation of data dependency obtained only from a certain existing relationship instance that is one-sided. For reducing data redundancy, in this paper, the concepts of XML functional dependency and XML multi-valued dependency based on the path expression in the Document Type Definition are given. And then the inference rules on the condition of the coexistence of functional dependency and multi-valued dependency in XML are proposed. Eventually, the validity and completeness of the inference rules are proved. The above-mentioned fundamentally solves the problems of implication of functional dependency and multi-valued dependency in XML data, laying a theoretical foundation for XML database design.

Keywords: Extensible Markup Language; inference rules; functional dependency; multi-valued dependency

1 INTRODUCTION

As a subset (Didier 2001) of SGML (standard generalized markup language) and XML (Extensible Markup Language) (Bray et al. 2008) collects the advantages of SGML and HTML (HyperText Markup Language). With the advantages of easy to manage, edit and query, XML becomes a bright spot in the development of network.

As the XML is more and more widely applied in many fields, more and more data are stored in XML form and the research of XML data dependency becomes an important research area. In recent years, experts and scholars study the problem of the standardization of the XML database only from XML functional dependency (Yin & Hao 2008, Yu & Jageadish 2008, Ahmad et al. 2008, Zhao 2009, Yin et al. 2012, Lv et al. 2012) or XML multi-valued dependency (Huang et al. 2008, Yin & Hao 2009, 2010, Yin et al. 2011), which also has made great achievements. In paper of Yu & Jageadish (2008) by using the thought of functional dependency in relational schema, they provide the definition of functional dependency in XML documents based on semantics. Based on the above paper, the authors take a new formalized definition on functional dependency in XML in paper of Ahmad et al. (2008). Based on the definition of the sub-tree information equivalence and

the sub-tree information consistency, XML strong multi-valued dependency and its inference rules are given in paper of Yin & Hao (2010). In the XML database, there exists the situation on the condition of the coexistence of functional dependency and multi-valued dependency in XML. Because of the complexity of the XML on the structure and the relatively late study, for the issues on the condition of the coexistence of functional dependency and multi-valued dependency in XML, there is no mature and perfect theory system (Cheng 2009). Data dependency obtained only from a certain existing relationship instance must be one-sided. Only give a set of complete inference rules, and we can solve fundamentally the problems of implication of functional dependency and multi-valued dependency, laying a foundation for the standardization of the XML database.

2 XML FUNCTIONAL DEPENDENCY AND XML MULTI-VALUED DEPENDENCY AND THEIR INFERENCE RULES

2.1 *XML functional dependency and XML multi-valued dependency and related definitions*

Definition 1: Let *T* be an XML tree that conforms to a Document Type Definition (DTD) *D*. A path instance in an XML tree T is a sequence such that $v_1 = v$, and for all v_i, $2 \leq i \leq n$, $v_i \in V$ and v_i is a child of v_{i-1}. A path instance $v_1 . v_2 . ... . v_{n-1} . v_n$ is said to be defined over the path $p_1 . p_2 . ... . p_{n-1} . p_n$; for all v_i, $1 \leq i \leq n$, $lab(v_i) = p_i$. The set of all path instances over a path p in a tree T is denoted by $Paths(p)$. The set of all path over D is denoted by $Paths(D)$.

Definition 2(XML Functional Dependency, XFD (Zhang & Liu 2012)): Let T be an XML tree that conforms to a DTD D and is complete (Millist 2003). An XFD is a statement of the form $p_1, ..., p_k \rightarrow q_1, ..., q_m$, $k \geq 1$, $m \geq 1$, $P = \{p_1, ..., p_k\}$ and $Q = \{q_1, ..., q_m\}$ where P and Q are the subsets in $Paths(D)$. There are every two distinct set of paths instances $V = \{v_1^1 . v_2^1 . ... . v_{l-1}^1 . v_l^1, ..., v_1^m . v_2^m . ... . v_{n-1}^m . v_n^m\}$ and $W = \{w_1^1 . w_2^1 . ... . w_{l-1}^1 . w_l^1, ..., w_1^m . w_2^m . ... . w_{n-1}^m . w_n^m\}$ in $Paths(Q)$, $n \geq 1$, $l \geq 2$, and T satisfies the XFD $p_1, ..., p_k \rightarrow q_1, ..., q_m$, where

(i) $Q \subset P$; or;
(ii) For every two distinct paths instances $v_1^i . v_2^i . ... . v_{t-1}^i . v_t^i$ and $w_1^i . w_2^i . ... . w_{t-1}^i . w_t^i$ in $Paths(q_i)$, if $Last(p_j) \in E_1$ and $x_{ij} = y_{ij}$, or $Last(p_j) \notin E_1$ and $val(Nodes(x_{ij}, p_j)) \cap val(Nodes(y_{ij}, p_j)) \neq \varnothing$ then $val(v^t) = val(w^t)$; where $x_{ij} = \{v|v \in \{\{v_1^i, ..., v_t^i\}$ and $v \in N(p_j \cap q_i)\}$, $y_{ij} = \{v|v \in \{\{w_1^i, ..., w_t^i\}$ and $v \in N(p_j \cap q_i)\}$, $1 \leq j \leq k$, $1 \leq i \leq m$, $t \geq 2$.

Definition 3(XML Multi-valued Dependency, XMVD (Zhang & Liu 2013)): Let T be an XML tree that conforms to a DTD D. An XMVD is a statement of the form $p_1, ..., p_k \rightarrow \rightarrow q_1, ..., q_m \mid r_1, ..., r_s$, $1 \leq k$, $1 \leq m$, $1 \leq s$, $P = \{p_1, ..., p_k\}$, $Q = \{q_1, ..., q_m\}$ and $R = \{r_1, ..., r_s\}$ where P, Q and R are the subsets in $Paths(D)$, $\{\{p_1, ..., p_k\} \cup \{q_1, ..., q_m\} \cup \{r_1, ..., r_s\}\} \subset Paths(D)$. The tree T satisfies the XMVD if whenever there exists a q_i, $1 \leq i \leq m$, and two distinct path instances $v_1^i . v_2^i . ... . v_{t-1}^i . v_t^i$ and $w_1^i . w_2^i . ... . w_{t-1}^i . w_t^i$ in $Paths(q_i)$, $2 \leq t$, such that:

i. $val(v^t) \neq val(w^t)$;
ii. There exists a r_j, $1 \leq j \leq s$, and two nodes z_1, z_2, where $z_1 \in Nodes(x_{ij}, r_j)$ and $z_2 \in Nodes(y_{ij}, r_j)$, such that $val(z_1) \neq val(z_2)$;
iii. For all p_h, $1 \leq h \leq k$, there exist two nodes z_3 and z_4, where $z_3 \in Nodes(x_{ijh}, p_h)$ and $z_4 \in Nodes(y_{ijh}, p_h)$, such that $val(z_3) = val(z_4)$;
iv. There exists a path instance $v_1'^i . v_2'^i . ... . v_{t-1}'^i . v_t'^i$ in $Paths(q_i)$ such that $val(v_t'^i) = val(v_t^i)$, and there exists a node z_1' in $Nodes(x_{ij}', r_j)$ such that $val(z_1') = val(z_2)$ and there exists a node z_3' in $Nodes(x_{ijh}', p_h)$ such that $val(z_3') = val(z_3)$;
v.   There exists a path instance $w_1'^i . w_2'^i . ... . w_{t-1}'^i . w_t'^i$ in $Paths(q_i)$ such that $val(w_t'^i) = val(w_t^i)$, and there exists a node z_2' in $Nodes(y_{ij}', r_j)$ such that $val(z_2') = val(z_1)$ and there exists a node z_4' in $Nodes(y_{ijh}', p_h)$ such that $val(z_4') = val(z_4)$.

Where, $x_{ij} = \{v|v \in \{v_1^i, ..., v_t^i\}$ and $v \in N(r_j \cap q_i)\}$, $y_{ij} = \{v|v \in \{w_1^i, ..., w_t^i\}$ and $v \in N(r_j \cap q_i)\}$, $x_{ijh} = \{v|v \in \{v_1^i, ..., v_t^i\}$ and $v \in N(p_h \cap r_j \cap q_i)\}$, $y_{ijh} = \{v|v \in \{w_1^i,$

$..., w_t^i\}$ and $v \in N(p_h \cap r_j \cap q_i)\}$; $x_{ij}' = \{v|v \in \{v_1'^i, ..., v_t'^i\}$ and $v \in N(r_j \cap q_i)\}$, $y_{ij}' = \{v|v \in \{w_1'^i, ..., w_t'^i\}$ and $v \in N(r_j \cap q_i)\}$, $x_{ijh}' = \{v|v \in \{v_1'^i, ..., v_t'^i\}$ and $v \in N(p_h \cap r_j \cap q_i)\}$, $y_{ijh}' = \{v|v \in \{w_1'^i, ..., w_t'^i\}$ and $v \in N(p_h \cap r_j \cap q_i)\}$.

Lemma 1: Let T be an XML tree that conforms to a DTD D and is complete. Let p_h, q_i and r_j be paths in $Paths(D)$, $1 \leq h \leq k$, $1 \leq i \leq m$, $1 \leq j \leq s$. If T satisfies the XFD $p_1, ..., p_k \rightarrow q_1, ..., q_m$ then it satisfies the XMVD $p_1, ..., p_k \rightarrow \rightarrow q_1, ..., q_m \mid r_1, ..., r_s$ for any path R, where $R = \{r_1, ..., r_s\} \in VPaths(D)$.

Definition 4 (Minimum base): Let T be an XML tree that conforms to a DTD D and path set $P = \{P_1, ..., P_k\}$. Assuming that $Paths(D) = P_1 \cup ... \cup P_k$, the minimum base set of P is a partition of $Paths(D)$, $S_1, ..., S_q$, where

(i) Every P_i is a set of certain S_j by union operation;
(ii) There not exists the partition that satisfies the condition (i) and the number of which is less than q, $1 \leq i \leq k$, $1 \leq j \leq q$.

Definition 5 (Dependency base): Σ is a set of XFD and XMVD in XML tree T conforming to DTD D. And $P \subseteq Paths(D)$. The minimum base set of P^+ is a path dependency base relative to Σ, denoted by $DEP(P)$.

2.2 Inference rules on the condition of the coexistence of XFD and XMVD

Solving decision problem of logical implication is that a known dependency set deduces other dependencies, which are correct. The introduction of inference rules is the main way to solve decision problem of logical implication. There are some inference rules on the condition of the coexistence of XFD and XMVD.

1. (XFD Reflexivity) if $Q \subseteq P$, $P \rightarrow Q$
2. (XFD Augmentation) if $P \rightarrow Q$ and $S \subseteq R \subseteq Paths(D)$, $PR \rightarrow QS$
3. (XFD Transitivity) if $P \rightarrow Q$ and $Q \rightarrow R$, $P \rightarrow R$
4. (XMVD Complementation) if $P \rightarrow \rightarrow Q$, $P \rightarrow \rightarrow R$ and $R = Paths(D) - P - Q$
5. (XMVD Reflexivity) if $Q \subseteq P$, $P \rightarrow \rightarrow Q$
6. (XMVD Augmentation) if $P \rightarrow \rightarrow Q$ and $V \subseteq W \subseteq Paths(D)$, $PW \rightarrow \rightarrow QV$
7. (XMVD Transitivity) if $P \rightarrow \rightarrow Q$ and $Q \rightarrow \rightarrow V$, $P \rightarrow \rightarrow V - Q$
8. (Composition) if $P \rightarrow Q$, $P \rightarrow \rightarrow Q$
9. (Coalescence) if $P \rightarrow \rightarrow Q$ and $R \subseteq Q$ and there exists $W \subseteq Paths(D)$ which satisfies $W \cap Q = \varnothing$ and $W \rightarrow R$, $P \rightarrow R$
10. (XMVD Union Rule) if $P \rightarrow \rightarrow Q$ and $P \rightarrow \rightarrow R$, $P \rightarrow \rightarrow QR$
11. (XMVD Difference Rule) if $P \rightarrow \rightarrow Q$ and $P \rightarrow \rightarrow R$, $P \rightarrow \rightarrow Q - R$, $P \rightarrow \rightarrow R - Q$
12. (XMVD Intersection Rule) if $P \rightarrow \rightarrow Q$ and $P \rightarrow \rightarrow R$, $P \rightarrow \rightarrow Q \cap R$

13. (XMVD Pseudo-Transitivity Rule) if $P\rightarrow\rightarrow Q$ and $WQ\rightarrow\rightarrow R$, $WX\rightarrow\rightarrow Z-WY$
14. (Mixed pseudo-transitivity rule) if $X\rightarrow\rightarrow Y$ and $X\rightarrow\rightarrow YZ$, $X\rightarrow\rightarrow Z-Y$.

3 CORRECTNESS AND COMPLETENESS OF THE INFERENCE RULES

The correctness of the inference rules is that "the dependency which is deduced by the inference rules in the dependency set Σ must be in Σ^+". And the completeness is that "the dependency in Σ^+ can be deduced by the inference rules in Σ". The correctness ensures all dependencies deduced are correct, and the completeness ensures we can deduce all dependencies logically implied.

3.1 *Proof of the correctness of inference rules*

Easy to prove, assume all XFD and XMVD are the dependencies based on the document tree T in the following proofs:

1. In accordance with the condition (i) of definition 2, $Q\subseteq P$ satisfies definition of XFD, so this rule holds.
2. Suppose XFD: $P\rightarrow Q$ holds, but $PR\rightarrow QS$ is false. Because $P\rightarrow Q$ holds, ① when the path q_i satisfies the condition (i) of definition 2, the left path extends $\{p_1, ..., p_k, r_1, ..., r_g\}$ which also satisfies the condition (i) of definition 2; ② if the path q_i satisfies the condition (ii) of definition 2 and $r_t = q_i$ satisfies the condition (i) of definition 2, the left path set extends $\{p_1, ..., p_k, r_1, ..., r_g\}$ which satisfies the condition (i) of definition 2 and $PR\rightarrow Q$ holds. If r_t satisfies the condition (ii) of definition 2 for a random t, namely $Last(r_t)\notin E_1$ and $val(Nodes(x_{it}, r_t))\cap val(Nodes(y_{it}, r_t))\neq\emptyset$ or $Last(r_t)\in E_1$ and $x_{it} = y_{it}$, $PR\rightarrow Q$ holds. Because of $S\subseteq R\subseteq Paths(D)$ and the path s_i which satisfies $r_i = s_i$, we know $PR\rightarrow QS$ holds according to definition 2. If there exists t that makes r_t not satisfy the condition (ii) of definition 2, all instance path sets corresponding to the left path sets are different. We can deduce $PR\rightarrow QS$ is true in this case, where $P = \{p_1, ..., p_k\}$, $1\leq j\leq k$; $Q = \{q_1, ..., q_m\}$, $1\leq i\leq m$; $R = \{r_1, ..., r_g\}$, $1\leq t\leq g$; $S = \{s_1, ..., s_h\}$, $1\leq d\leq h$.
3. We assume that $P\rightarrow Q$ and $Q\rightarrow R$ hold. If $P_{1\,val} = P_2$ for two path instance sets P_1 and P_2 according to definition 2, there is $Q_{1\,val} = Q_2$. And because $Q\rightarrow R$ holds, there exists $R_{1\,val} = R_2$ for two path instance sets Q_1 and Q_2 where $Q_{1\,val} = Q_2$. That is to say that if $P_{1\,val} = P_2$ for two path instance sets P_1 and P_2, there correspondingly exists $R_{1\,val} = R_2$, which indicates $P\rightarrow R$ holds.
4. This rule can be directly deduced by the definition of XMVD.
5. This rule can be directly deduced by the definition of XMVD.
6. We adopt proof by contradiction. We assume that $P\rightarrow\rightarrow Q$ holds but $PW\rightarrow\rightarrow QV$ does not hold. Due to $P\rightarrow\rightarrow Q$, according to the definition 3 there exists a path $v'^i_1.v'^i_2.....v'^i_{t-1}.v'^i_t$ in $Paths(q_i)$ to make $val(v'^i_t) = val(v^i_t)$. And there exists a node z'_1 in $Nodes(x'_{ij}, r_j)$ to make $val(z'_1) = val(z_2)$ and there exists a node z'_3 in $Nodes(x'_{ijh}, p_h)$ to make $val(z'_3) = val(z_3)$. Then there exists a path $w'^i_1.w'^i_2.....w'^i_{t-1}.w'^i_t$ in $Paths(q_i)$ to make $val(w'^i_t) = val(w^i_t)$. And there exists a node z'_2 in $Nodes(y'_{ij}, r_j)$ to make $val(z'_2) = val(z_1)$ and there exists a node z'_4 in $Nodes(y'_{ijh}, p_h)$ to make $val(z'_4) = val(z_4)$. According to the reflexivity rule, $W\rightarrow\rightarrow V$ holds. In conclusion, the path u in the PW satisfies that there exists a path $v'^i_1.v'^i_2.....v'^i_{t-1}.v'^i_t$ in $Paths(u_i)$ to make $val(v'^i_t) = val(v^i_t)$. And there exists a node z'_1 in $Nodes(x'_{ij}, r_j)$ to make $val(z'_1) = val(z_2)$ and there exists a node z'_3 in $Nodes(x'_{ijh}, p_h)$ to make $val(z'_3) = val(z_3)$, which satisfies the condition (iv) of definition 3. The other paths $P' = Path(D)-\{PQWV\}$ is a subset of $Paths(D)-PQ$ and is also a subset of $Path(D)-WV$, which explains any paths in P' satisfy the condition (v) of definition 3. Therefore $PW\rightarrow\rightarrow QV$ holds contradicted with the assumption.
7. Assume XMVD: $Q\rightarrow\rightarrow V$ holds. According to the definition 3, there is $Q\rightarrow\rightarrow V-Q$. Similarly, when $Q\rightarrow\rightarrow V-Q$ holds, $Q\rightarrow\rightarrow V$ also holds according to the definition 3. Hence, we consider the situation that there is no common path of the left and the right of $Q\rightarrow\rightarrow V$.

   For each $q\in Q_T(p)$ it always satisfies $V_T(p) = V_T(pq)$, where p and q are the path instances of P and Q, respectively. Obviously, there is $V_T(pq)\subset V_T(p)$. If there exists a path instance $v\in V_T(p)-V_T(pq)$ where v belongs to $V-Q$, it satisfies the path instances p and q. At the same time it exists in some tree tuples but exists not in the tree tuple t, where path instance sets p, q and v are in the tree tuples. Further known, $Q_T(pv)$ does not include the q.

   Because of $Q\rightarrow\rightarrow V$, V is only related to the path set Q and has nothing to do with the other paths according to the definition 3. Then we know $V(pq) = V(q)$. Due to $V_T(p) = V_T(pq)$, $V_T(p) = V_T(q)$ holds for $q\in Q_T(p)$. Hence we need to prove $V_T(pw) = V_T(p)$ for $q\in Q_T(p)$ where $w\in W$ and $W = Paths(D)-PV$. Because there is no common path of the left and the right of $Q\rightarrow\rightarrow V$, we get $Q\subset PW$. For a path instance set pw, there exists a mapping q' in Q. According to the definition of $Q_T(p)$, q' belongs to $Q_T(p)$. And because of $Q\rightarrow\rightarrow V$, there is $V_T(pw) = V_T(q')$, namely $P\rightarrow\rightarrow V$.
8. This rule can be deduced by the lemma 1.

9. We assume $P\rightarrow\rightarrow Q$ and $W\cap Q = \Phi$ are true. According to the definition 3 $Q_T(p) = Q_T(pw)$. Due to $R \subseteq Q$, there is $R_T(p) = R_T(pw)$. According to the definition 2 we can know $WP\rightarrow W$. And because of $W\rightarrow R$, we get $WP\rightarrow R$ according to XFD transitivity. It can be observed that $R_T(pw)$ contains only one element, and that is to say that $R_T(pw)$ contains only a value of R path set. And due to $R_T(p) = R_T(pw)$, $R_T(p)$ also contains only a value of R path set. According to the definition of XFD, we can know $P\rightarrow R$.

10. According to XMVD extension rules $P\rightarrow\rightarrow P\cup Q$ and $P\cup Q\rightarrow\rightarrow R\cup Q$ are true. And according to XMVD reflexivity $P\cup Q\rightarrow\rightarrow Paths(D)-P\cup Q\cup R$ is true. Then according to XMVD transitivity $P\rightarrow\rightarrow Paths(D)-P\cup Q\cup R$ holds. Lastly, according to XMVD reflexivity $P\rightarrow\rightarrow Q\cup R$ holds.

11. According to XMVD union rule we know that $P\rightarrow\rightarrow QR$ is true. And according to XMVD reflexivity $P\rightarrow\rightarrow Paths(D)-P\cup Q\cup R$ is also true. Then $P\rightarrow\rightarrow (Paths(D)-P\cup Q\cup R)\cup R$ holds according to $P\rightarrow\rightarrow R$ and $P\rightarrow\rightarrow Paths(D)-P\cup Q\cup R$ and XMVD union rule. Using XMVD reflexivity rule $P\rightarrow\rightarrow Paths(D)-P\cup(Paths(D)-P\cup Q\cup R)\cup R$ holds, namely $P\rightarrow\rightarrow (Q-R)-P$. According to XMVD reflexivity rule we know $P\rightarrow\rightarrow Paths(D)-P\cup((Q-R)-P)$, and that is to say that $P\rightarrow\rightarrow Paths(D)-(Q-R)$ is true. Once again to use XMVD reflexivity rule we know $P\rightarrow\rightarrow Q-R$ holds. We can prove $P\rightarrow\rightarrow R-Q$ in a similar way.

12. According to XMVD union rule we know that $P\rightarrow\rightarrow QR$ is true. And according to XMVD reflexivity $P\rightarrow\rightarrow Paths(D)-P\cup Q\cup R$ is also true. Then $P\rightarrow\rightarrow (Paths(D)-P\cup Q\cup R)\cup(Q-R)$ holds according to XMVD difference rule and XMVD union rule. Using XMVD reflexivity rule $P\rightarrow\rightarrow Paths(D)-P\cup(Paths(D)-PQR)\cup(Q-R)$ holds, and that is to say that $P\rightarrow\rightarrow (Q-R)-P$ holds. According to XMVD reflexivity rule we know $P\rightarrow\rightarrow Paths(D)-P\cup((Q\cap R)-P)$, and that is to say that $P\rightarrow\rightarrow Paths(D)-(Q\cap R)$ holds. Once again to use XMVD reflexivity rule we know $P\rightarrow\rightarrow Q\cap R$ holds.

13. We know that $W\cup P\rightarrow\rightarrow W\cup Q$ according to XMVD augmentation and $WP\rightarrow\rightarrow R-(W\cup Q)$ according to XMVD transitivity.

14. We know that $P\rightarrow\rightarrow PQ$ according to XMVD augmentation and $W\rightarrow W-PQ$ according to XMVD reflexivity. Because of $PQ\rightarrow W$, we know that $PQ\rightarrow W-PQ$ according to XFD transitivity and $P\rightarrow\rightarrow W-PQ$ according to XMVD transitivity. $P\rightarrow\rightarrow W-PQ$ and $PQ\rightarrow W-PQ$ are applied to the coalescence, and then we get $P\rightarrow W-P\cup Q$. According to XFD augmentation we know that $P\rightarrow P(W-PQ)$, and that is to say that $P\rightarrow P(W-Q)$ holds. According to XFD reflexivity $P\rightarrow W-Q$ holds.

3.2 *Proof of the correctness of inference rules*

The proof of the completeness is to prove that the XMVD and FD in Σ^+ can be deduced by the inference rules in Σ.

Let Σ be a set of functional dependency and multi-valued dependency over D. Set $P = \{p_1, ..., p_k\}$ and we know that $P\rightarrow P^+$ and $P^+\rightarrow P$. Let $P_1, ..., P_m$ be the dependency base of P over $Paths(D)-P^+$. So $P^+, P_1, ..., P_m$ are a partition based on $Paths(D)$.

We can adopt the way to construct T to prove them. We construct a document instance tree T, which consists of 2^m tree tuples. The value of paths of each tree tuple t in P^+ is 1and the value of the path instance in P_i is a_i $(a_i\in(0,1))$. The document tree has the following features:

Property 1: That each P_i is XMVD of the right path set holds in the T.

Property 2: If the right of a dependency is a non-void subset of P_i, we say this dependency holds in the T if and only if the intersection of the left of this dependency and P_i is non-empty, where $i = 1, ..., m$.

We discuss the situation about condition of satisfying completeness hypothesis in the T.

First of all, let f be a functional dependency in Σ^+. We need to prove f holds in the T. According to the definition 2 f can be formally expressed as $Q\rightarrow p$ ($Q\rightarrow PW$ can be expressed as $Q\rightarrow P$ and $Q\rightarrow W$), where p is a path. If $p\in P^+$, f holds in the T according to the definition 2 because P^+ has only a value. If $p\notin P^+$ ($p\in P_i$, $i = 1, ..., m$) and the intersection between Q and P_i is empty, there is $P\rightarrow\rightarrow P_i$ according to the definition 5. According to the coalescence rule $P\rightarrow p$ can be deduced by $P\rightarrow\rightarrow P_i$ and $Q\rightarrow p$ in conflict with $p\notin P^+$. Therefore, Q must intersect with P_i. According to the property 2 $Q\rightarrow p$ holds.

Second, let $g: Q\rightarrow\rightarrow V$ be a multi-valued dependency in Σ^+. We need to prove g holds in the T. From the above analysis it is known that $Q\rightarrow V\cap P^+$ holds in the T. And $Q\rightarrow\rightarrow V\cap P^+$ holds according to the composition. Then we need to prove $Q\rightarrow\rightarrow V\cap P_i$ holds in the T. For certain i, there is $V\cap P_i = \varnothing$ and $V\cap P_i = P_i$. According to the property 1 $Q\rightarrow\rightarrow P_i$ holds in the T and $Q\rightarrow\rightarrow\varnothing$ also holds. For certain i, there is $V\cap P_i\subseteq P_i$. If $Q\cap P_i = \varnothing$, $Paths(D)-P_i\rightarrow\rightarrow V$ holds according to the extension rules. According to the definition 5, $P\rightarrow\rightarrow P_i$ holds. According to the symmetry of the multi-valued dependency, $P\rightarrow\rightarrow Paths(D)-P_i$ holds. Then $P\rightarrow\rightarrow V-(Paths(D)-P_i)$ can be deduced by the transitivity of the multi-valued dependency. Further, $P\rightarrow\rightarrow V\cap P_i$ holds contradicted with $P_i\in DEP(P)$. Hence the intersection of Q and P_i is not empty. According to the property 2, $Q \rightarrow \rightarrow V \cap P_i$ holds in the T. So for i $Q \rightarrow \rightarrow V \cap P_i$ holds. According to $Q \rightarrow V \cap P^+$ and the union rule $Q \rightarrow \rightarrow V$ holds in the T.

In the end, we consider the left is the dependency of P not in Σ^+. Let f' be that the left is a functional dependency of P not in Σ^+, and it is formalized as $P \rightarrow Q$. If this dependency is a functional dependency, there is $Q \not\subset P^+$. According to the definition 5, Q must intersect with certain P_i. If $P \rightarrow Q$ holds, $P \rightarrow Q \cap P_i$ holds according to the definition 2. According to the definition 5 there is no intersection between P^+ and P_i, which is contradicted with the property 2. Hence $P \rightarrow Q$ does not hold in the T. Let g' be that the left is a multi-valued dependency of P not in Σ^+, and it is formalized as $P \rightarrow \rightarrow Q$. If this dependency is a multi-valued dependency, $P \rightarrow \rightarrow Q \cap P^+$ and $P \rightarrow \rightarrow Q \cap P_i$ are true in the T. So there is $Q \cap P_i \subseteq P_i$. According to the property 2 $P \rightarrow \rightarrow Q \cap P_i$ does not hold. Because $P \rightarrow \rightarrow P_i$ holds, according to the XMVD union rule $P \rightarrow \rightarrow Q \cap P_i$ holds in the T, which is contradicted with the assumption. Hence $P \rightarrow \rightarrow Q$ does not hold in the T.

Summing up the above parts, the inference rules are complete.

4 CONCLUSION AND FUTURE WORK

The paper studies the XML data constraint on the condition of the coexistence of functional dependency and multi-valued dependency in XML, laying a theoretical foundation for XML database design. We have introduced the data dependency constraints of XML based on DTD and the path expression and proposed the inference rules on the condition of the coexistence of functional dependency and multi-valued dependency in XML. Then we have proved the validity of and completeness of the inference rules. In the actual XML database, there exists situation of join dependency. Contents about concepts of join dependency and its inference rules need to be studied further.

REFERENCES

[1] Ahmad, Kamsuriah et al. 2008. Defining functional dependency for XML. *Journal of Information Systems, Research & Practices* 1(1): 26–34.

[2] Bray, Tim et al. 2008. *Extensible Markup Language (XML) 1.0*. http://www.w3.org/TR/xml.

[3] Cheng, Luqing 2009. Comparison of functional dependency of XML and relational database. *Information Technology and Informatization* 6(10): 52–54.

[4] Didier, Martin 2001. *XML Advanced Programming*: 2–19. Beijing: China Machine Press.

[5] Huang, Haiyan et al. 2008. Study on normalization algorithm of multi-valued dependency for XML documents. *Journal of Hunan University of Science and Engineering* 29(12): 126–129.

[6] Lv, Teng et al. 2012. Uncertain XML functional dependencies based on tree tuple models. *Web-Age Information Management*: 340–349.

[7] Millist, W.Vincent et al. 2003. Multivalued dependencies and a 4 NF for XML. *Proceedings of 15th International Conference of Advanced Information Systems Engineering*: 14–29.

[8] Yin, Lifeng et al. 2011. Inference rules for XML strong MVD and XML weak MVD. *Advanced Materials Research* 1289(268): 2009–2015.

[9] Yin, Lifeng et al. 2012. Inferior strong functional dependency and inference rules for XML. *Electronic Design Engineering* 1(20): 6–9.

[10] Yin, Lifeng & Hao, Zhongxiao 2008. Inference rules of XML strong functional dependency. *Computer Science* 35(9): 165–167.

[11] Yin, Lifeng & Hao, Zhongxiao 2009. Normalization of XML document with strong MVD under incomplete information circumstances. *Journal of Computer Research and Development* 46(7): 1226–1233.

[12] Yin, Lifeng & Hao, Zhongxiao 2010. Inference rules for XML strong multi-valued dependencies based on XML Schema. *Computer Engineering and Applications* 46(28): 152–156.

[13] Yu, C. & Jageadish, H.V. 2008. XML Schema refinement through redundancy detection and normalization. *The International Journal on Very Large Data Base* 17(2): 203–223.

[14] Zhang, Zhongping & Liu, Zhixiao 2012. Fourth normal form and free redundancy for XML document based on path expression. *2012 2nd International Conference on Computer Science and Network Technology*: 1818–1822.

[15] Zhang, Zhongping & Liu, Zhixiao 2013. Fourth normal form and no redundancy for XML document based on path expression. *Journal of Chinese Computer Systems* 34(5): 1091–1096.

[16] Zhao, Xiangguo et al. 2009. XML functional dependency and schema normalization. *Proceedings of the 9th International Conference on Hybrid Intelligent Systems*: 307–312.

Future Information Engineering and Manufacturing Science – Zheng (Ed)
© *2015 Taylor & Francis Group, London, ISBN 978-1-138-02644-5*

Design and analysis of horizontal-axis wind turbine main shaft

Yan Xia & Quan Liu

Mechanical and Electrical Engineering School, Beijing Information Science and Technology University, Beijing, China

ABSTRACT: Wind generator is will wind conversion for mechanical can eventually conversion for power of power mechanical, earth of wind potential very big, and level axis wind generator accounted for mainstream status, paper for level axis wind generator spindle and the growth box of drive for design, and using ANSYS for spindle strength check, analysis axis of maximum by force and the deformation, then on growth box for reasonable of match design calculation, coupled with on gear box of installation requirements, completed level axis wind generator spindle and the growth box drive of design.

Keywords: wind turbine machinery; the spindle; speed-up gearbox; ANSYS analysis

1 INTRODUCTION

Wind turbine machinery is to convert wind energy into mechanical energy [1], finally into electrical energy, hundredth of the earth's wind energy is equivalent to the global annual combined consumption of coal, oil and other fossil fuels, wind energy potential is tremendous.

Horizontal axis wind turbines are dominant position on the market at present. The main transmission mechanism of these wind turbines generally includes: low-speed shaft (spindle), gearbox, high-speed shaft, brake, bearings and coupling, and so on.

Before the design of the spindle, the transmission form of the gearbox is determined as the form of primary planetary gear train with secondary cylindrical gear, so as to determine the power parameters such as spindle drive [2].

2 THE DESIGN OF THE SPINDLE

2.1 *The parameters design of the spindle*

Usually, the overall transmission efficiency of planetary gear train is 0.98, of a pair of rolling bearings is 0.99, closed gear's is 0.97, the coupling's is 0.99, so the efficiency of secondary cylindrical gear transmission can be calculated as follows [3]:

$$\eta 1 = 0.99^3 \times 0.97^2 \times 0.99 = 0.905 \tag{1}$$

According to the generators power 2 kW, so the spindle power can be calculated as follows:

$$P_0 = \frac{P}{\eta} = \frac{2}{0.98 \times 0.905} = 2.255 \text{ kW} \tag{2}$$

Spindle's transmission torque

$$T = \frac{9.55 \times 10^6 \text{ P}}{n} = \frac{9.55 \times 10^6 \times 2.255 \text{ kw}}{20 \text{ r / min}}$$
$$= 1076800 \text{ N} \cdot \text{mm} = 1076.8 \text{ N} \cdot \text{m} \tag{3}$$

$$F = \frac{2T}{d} = \frac{2 \times 1076.8 \text{ N} \cdot \text{m}}{0.08 \text{ m}} = 26920 \text{ N}, \quad T = F\frac{d}{2} \tag{4}$$

The material of the spindle is chosen as steel grades 45#, elastic modulus EX is about 2×1011 Pa, the Poisson's ratio PRXY is 0.269, and the density is 7890 kg/m³.

Preliminary estimate the diameter of spindle as follows:

Calculated from the above equation, the minimum diameter of axle spindle is 53.14 mm, considering the spindle by the load and torque is bigger, in order to avoid some quality problem caused by fracture, deformation, and so on, the smallest diameter of spindle will be set as 80 mm.

2.2 *The structure design of the spindle*

The spindle is installed in front of the base at a certain elevation above [4], the head is connected

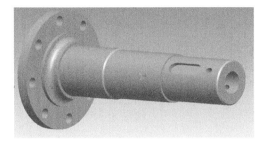

Figure 1.　The spindle.

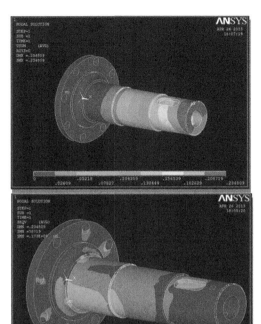

Figure 2.　Deformation and stress analysis of spindle.

with wheel hub, and the type is cantilever. Due to the spindle connected to the hub and gear box, so the design of the spindle should be considered. Wheel hub is an empty shell structure, a flange is connected by bolts, and also connected with gearbox with ordinary flat key is shown in Figure 1.

2.3　Checking spindle's intensity

In the whole transmission system, the spindle load bearing and the biggest impact, in order to guarantee the reliability of the transmission system, spindle's intensity must be performed on.

The spindle's material is 45# steel, the mechanical properties are as follows:

Density	Yield strength	Elasticity modulus	Poisson's ratio
7.89 g/cm³	355 MPa	200 MPa	0.33

Spindle with ANSYS intensity [3]. Spindle design loads had the gyroscopic torque, wheel, and so on. When you build solid models to ANSYS [5], some characters had a little effect on stress calculations, and can cause meshing difficulties, increasing the number of grid. Accordingly, some geometric simplification for details of the entity model. Analysis first identifies the main points or surfaces, the spindle force concentrated mainly on the surface in contact with flat keys, one of the main constraints on the bearing installation and connection hub flanges. Then import the spindle in ANSYS, set the material properties, mesh, imposing constraints and loads of final deformation analysis, analysis of stress and strain analysis of obtained results. Static analysis, static load imposed by the spindle rotation torque m. Handles the drawing of the adoption of the deformation and stress to withstand these loads [6] are shown in Figure 2.

3　SPEED-UP GEARBOX DESIGN

Wind turbine gearbox is an important transmission parts [7], due to the low speed wind turbine by wind to get, far less than generating the required speed is required under gearbox turbines in wind generated power is delivered to the engine and give it access to the appropriate speed. This design type gearbox with planetary-parallel structure, characterized by: wind turbine speed of spindle speed for 60 r/min and 2 pole generator takes about 1750 r/min of speed to achieve the rated power of the generator. [8]

Calculate gearbox ratio as follows:

$$i = \frac{n_H}{n} = \frac{60 \text{ r/min}}{1750 \text{ r/min}} = \frac{1}{29.16} \qquad (5)$$

3.1　Design of planetary gear train

First select the planet number $K = 3$.

Determination of the planetary gear, we must meet the following four conditions:

1. Transmission ratio conditions:

$$z_3/z_1 = i_1H-1, \ z_3 = (i_1H-1) z_1 \qquad (6)$$

2. Concentric condition:

$$z_2 = (z_3 - z_1)/2, z_2 = z_1(i_1H - 2)/2 \quad (9)$$

3. Uniform installation conditions:

$$(z_1 + z_3)/K = N \quad (8)$$

4. Adjacency condition: the addendum circle may not be a collision between two planets round

$$(z_1 + z_2) \sin (180°/K) > z_2 + 2ha* \quad (9)$$

By calculation, the number of teeth of the gear can be drawn: the center wheel $z_1 = 20$, planetary wheel $z_2 = 40$, inner gear ring $z_3 = 100$. Gear accuracy level 8 gear material: 20CrNiMoH, the hardness is above Hv740. Select gear module m = 2, addendum coefficient ha* = 1, top clearance factor c* = 0.25.

The addendum circle diameter of gear available is as follows:

Central gear: da = 44 mm,
Planet gear: da = 84 mm,
Gear ring: da = 195 mm.

3.2 *Structure design of cylindrical gear*

Gear Velocity Ratio Assignment: $i_1 = 1/5$
The shaft speed: epicyclic gear output speed: $i_2 = 360$ r/min
Cylindrical gear shaft speed:

$$n_{II} = \frac{n_1}{i_1} = \frac{360}{1/5} = 1800 \text{ r/min} \quad (10)$$

Generator speed:

$$n = n_{II} = 1800 \text{ r/min} \quad (11)$$

The shaft input torque:

$$T_I = 58 \text{ N} \cdot \text{m}, T_{II} = 10.72 \text{ N} \cdot \text{m}, T = 10.61 \text{ N} \cdot \text{m}$$

Low speed shaft, spur gear shaft, and motor.

3.3 *Design of cylindrical gear calculation*

Helical gearing, 40 Cr gear materials, hardness of 280 HBS, for 45 steel pinion materials, hardness of 240 HBS. Choose the number of large gears, $z_1 = 100$, $z_2 = 20$, primaries Helix angle $\beta = 14°$.

1. To calculate the stress cycles

$$N_1 = 60n_1jL_h = 9.33 \times 10^8 \quad (12)$$

$$N_2 = \frac{N_1}{i_1} = \frac{9.33 \times 10^8}{1/5} = 4.6 \times 10^9 \quad (13)$$

2. The calculation of contact fatigue allowable stress

Take the failure probability is 1%, the safety factor $S = 1$, so

$$[\sigma_H]_1 = \frac{K_{HN1}\sigma_{\lim1}}{S} = 0.94 \times 650 \text{ MPa} = 611 \text{ MPa}$$

$$[\sigma_H]_2 = \frac{K_{HN2}\sigma_{\lim2}}{S} = 0.9 \times 550 \text{ MPa} = 495 \text{ MPa}$$

$$(14)$$

Allowable contact stress:

$$[\sigma_H] = \frac{[\sigma_H]_1 + [\sigma_H]_2}{2} = 553 \text{ MPa} \quad (15)$$

3. Calculations
Big gear dividing circle diameter:

$$d_{1t} \geq 2.32\sqrt[3]{\frac{KT_1}{\varphi_d}\frac{u \pm 1}{u}\left(\frac{Z_E}{[\sigma_H]}\right)^2} = 119.13 \text{ mm} \quad (16)$$

Calculating circumferential velocity:

$$v = \frac{\pi d_{1t}n_1}{60 \times 1000} = 0.75 \text{ m/s} \quad (17)$$

Calculation of tooth width b:

$$d_{2t} = d_{1t} \cdot i_1 = 119.13 \times 1/5 = 23.826 \text{ mm} \quad (18)$$

$$b = \varphi_d \cdot d_{2t} = 23.826 \text{ mm} \quad (19)$$

Profile Section Modulus:

$$m_{nt} = \frac{d_{1t}\cos\beta}{z_1} = 1.156 \text{ mm} \quad (20)$$

Calculation of depth:

$$h = 2.25m_{nt} = 2.25 \times 1.156 = 2.601 \text{ mm} \quad (21)$$

According to actual load factor correction calculated pitch circle diameter:

$$d_1 = d_{1t}\sqrt[3]{\frac{K}{K_t}} = 119.13 \times \sqrt[3]{\frac{2.02}{1.6}} = 128.8 \text{ mm} \quad (22)$$

4. Calculation module:

$$m_n = \frac{d_1\cos14°}{z_1} = \frac{128.8 \times \cos14°}{100} = 1.25 \text{ mm} \quad (23)$$

5. According to the tooth root bending strength design. The design formula of bending strength is as follows:

$$m \geq \sqrt[3]{\frac{2KT_1 Y_\beta \cos^2 \beta}{\varphi_d z_1^2 \varepsilon_\alpha} \left(\frac{Y_{Fa} Y_{Sa}}{[\sigma_F]} \right)} \qquad (24)$$

Press contact strength calculated pitch circle diameter, and count the number of large gears.

$$z_1 = \frac{d_1 \cos 14°}{m} = \frac{128.8 \times \cos 14°}{1} \approx 125 \qquad (25)$$

Small gear teeth:

$$z_2 = z_1 \cdot i_1 = 125 \times 1/5 = 25 \qquad (26)$$

This design surface contact fatigue strength of gear not only meets the and meets the tooth root bending fatigue strength, compact.

6. Calculation of center distance:

$$a = \frac{(z_1 + z_2)m_n}{2\cos \beta} = \frac{(125 + 25) \times 1}{2 \times \cos 14°} = 77.3 \text{ mm} \qquad (27)$$

Round the center distance is 77 mm. After you press the rounded Center distance correction of spiral angles.

4 SUMMARY

ANSYS analysis in this paper, first identifies the main points or surfaces of the spindle end deformation analysis, analysis of stress and strain analysis of obtained results, and static strength well to ensure the overall reliability of the transmission system, and for the overall design of speed increase gearbox has laid a good foundation. Design and material selection play a vital role of the local and system-wide coordination in order to achieve good results, horizontal axis wind turbine speed increase gearbox transmission and spindle design.

REFERENCES

[1] Steinbuch M, Dynamic Modelling and Robust Control of a Wind Energy Conversion System [PhD Thesis], 1990.
[2] Pena r, Clare J.C, Asher G.M, Doubly fed induction generator using back-to-back PWM converters and its application to Variable-speed wind-energy generation, IEE Proceedings-Electric Power Applications, 1996, 143(3); 231–241.
[3] Patir, N., and Cheng, H.S., "Prediction of the Bulk Temperature in Spur Gears Based on Finite Element Temperature Analysis," ASLE Transactions, Vol. 22. pp. 25–36, 1979.
[4] Townsend, D.P., and Akin, L.S., "Analytical and Experimental Spur Gear Tooth Temperature as Affected by Operating Variable," ASME Journal of Mechanical Design, Vol. 103, pp. 216–226, 1981.
[5] Wang, K.L., and Cheng, H.S., "Numerical solution to the dynamic load, film thickness and surface temperature in spur gears—Part I, Analysis", ASME Journal of Mechanical Design, vol 103, pp177–187, 1981.
[6] Wang, K.L., and Cheng, H.S., "Numerical solution to the dynamic load, film thickness and surface temperature in spur gears—Part II, Results", ASME Journal of Mechanical Design, vol 103, pp188–194, 1981.
[7] Haruo Houjoh; Kiyohicko Umezawa; Shigeke Matsumura Vibration analysis for a pair of helical gears mounted on elastic shafts 1996.
[8] Tsuta T. Excitation force analysis of helical gear-pair with tolerance in their tooth shape and pitch, mounted on flexible shaft 1991.

Research on the design and application of test circuits for a digital voltmeter

Hao Chen, Weiwei Hu & Junliang Li
School of Reliability and Systems Engineering, Beihang University, Beijing, China

ABSTRACT: By providing a typical A/D circuit based on ICL7135 4-1/2 digit voltmeter, the paper focuses on the design of fault test circuit. The principle of digital voltmeter circuit and the main circuit failure modes and mechanisms with FMMEA will be presented, additional new test circuit designs will be demonstrated in the original A/D circuit and the feasibility will be analyzed and clarified. Finally, recommendations and some methods to design test circuit for a few main failure modes will be proposed.

Keywords: ICL7135; digital voltmeter; FMMEA; test circuit design

1 INTRODUCTION

With the increasing variety of large and complex system, people are paying more and more attention on reliability, fault diagnosis and prognostics, maintenance support and other problems. In the system maintainability, Condition Based Maintenance with benefits of small-scale logistical support, affordability, automation, high efficiency, and avoiding major catastrophic accidents has a very good prospect and requires the system itself with the ability to predict failures and manage their health status for achieving the goal of "affordability". At the same time, the concept of failure Prognostics and Health Management (prognostics and health management, PHM) is proposed through the use of kinds of sensors to acquire data and many of intelligent reasoning algorithms (such as physical models, neural networks, data fusion, fuzzy logic, expert systems, etc.) to assess the health status of the system. PHM predicts the failure before it occurs and collects information from various resources available to provide a range of maintenance and support measures for achieving the system's maintainability.

Along with the rapid development of digital technology, modern weapons and equipment take advantage of a lot of complex advanced electronic equipment, especially applications of the various microcircuits, which also give rise to serious problems and burdens for testing, maintenance and support, so that the fault monitoring and prediction of electronic devices have become major factors affecting combat readiness and cost.

Due to the features of failure in electronics, PHM is particularly difficult for electronic products.

Currently methods abroad can be divided into three categories when PHM is applied to electronic products, one of which is based on monitoring characteristic parameters. For example, by means of monitoring the electronic parameters such as current, voltage, resistor, the failure or the health status of equipment can be obtained.

In order to determine the representative properties' parameters for a typical A/D circuit accurately, the failure analysis is necessary to find the weaknesses which provide useful information for the targeted design of test circuit. The test circuit achieves the A/D circuit failures real-time monitoring by collecting data.

2 THE DIGITAL VOLTMETER BASED ON THE ICL7135

The most commonly used way for digital voltmeter circuit is to associate the A/D chip with a microcontroller which is used to accomplish the conversion from analog signals to digital signals and the digital display, whose advantages are the appliance of circuit design, the simplicity of components and the easiness to allow more extensions. The design of voltmeter circuit in the paper does not depend on a microcontroller, but the function of dual-integration and automatically zero-cleared of an ICL7135 chip itself.

2.1 Design of the digital voltmeter

The design as follows: the digital voltmeter is formed by an analog signal acquisition, an A/D conversion, a digital signal output display, a decoding display driver, a power supply, a clock, and other major functional modules. An ICL7135 chip is the core of module for achieving the conversion from analog voltage signals to digital voltage signals, the positive voltage of analog circuit is supplied by +5V DC, the clock is provided through the 555 timer, the reference voltage is offered by the use of a combination of potentiometers, resistors, capacitors and rectifiers, and the measured analog voltage signals which need to be filtered and rectified for acquisition is realized through a combination of capacitors and resistors. The CD4511 is adopted to decode and drive digital signal and display the output. The CD4511 is composed of a CMOS BCD latch/7 segment decoder/driver, which is able to perform the function of BCD conversion, blanking, latch control, seven segment decoder and driver and provide great pull-up currents that can directly drive digital tubes to display. The Driving Array MC1413 realizes the drive and display of digital tubes which are 0.3 inches common-cathode with the decimal point selector circuit. Basic tasks of the digital voltmeter are to measure and display the DC voltage digitally between −2V and +2V.

In the design above, it is proved that the first field will remain lit from the initial state. It is possible to make the circuit first blanking correct by designing a composite logical gate circuit which takes advantage of the blanking terminal of CD4511 decoding chip in the voltage meter circuit.

According to the design, the diagram of a digital voltmeter circuit is shown in Figure 1.

2.2 *The functional structure of digital voltmeter circuit*

Based on the design, the diagram of functional modules in a digital voltmeter is shown in Figure 2.

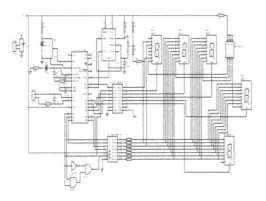

Figure 1. The diagram of a digital voltmeter circuit.

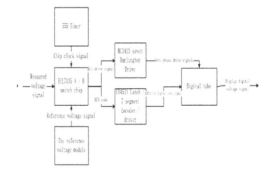

Figure 2. The functional block diagram of a digital voltmeter.

The analog voltage to be measured completes the digital conversion by the synergy of each module in the voltmeter and displays the voltage value in the digital tubes, as shown in the figure above.

3 THE FAILURE MODE, FAILURE MECHANISM AND EFFECT ANALYSIS

3.1 *Analysis overview*

FMMEA: failure modes, failure mechanisms and effect analysis. In this paper, FMMEA is applied to study failure modes, failure mechanisms of digital voltmeter circuit which may exist in the components and to identify the effects of various failure modes on each component of digital voltmeter.

For the digital voltmeter designed in this paper, the purpose to make use of FMMEA is to determine variety of potential failure modes, failure mechanisms, causes and effects, identify the degree of risks and provide important reference for designing targeted test circuit.

3.2 *Theoretical preparation*

In FMMEA, it is necessary for the product and its environmental conditions to describe in detail. This description is generally used to represent by the mission profile. Typical tasks of the circuit designed in this paper are to measure voltage signal and display it in digit. The mission profile is shown in Figure 3.

According to the knowledge of FMMEA, the features of digital voltmeter circuit and hardware configuration, we conduct FMMEA analysis about products and develop practical steps shown in Figure 4.

According to the description of digital voltmeter circuit function, we can determine several function modules of digital voltmeter: a digital voltmeter is constituted by the clock signal module (01),

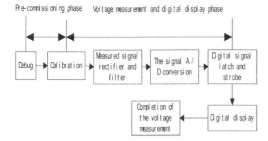

Figure 3. The mission profile.

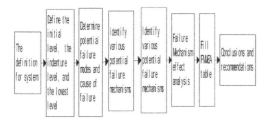

Figure 4. The process of FMMEA.

the reference voltage input module (02), the measured voltage input module (03), the analog to digital conversion module (04), and the digital voltage display module (05).

3.3 *The results of FMMEA*

From FMMEA the possible failures of the digital voltmeter can be found, including: clock generation capacitor C5, clock generation resistors R6/R7, divider resistor R5 are open or short; reference voltage potentiometer is open or mechanical damage; zener diode D21N4728A is open or short; rectifier capacitor C2 is short; rectifier resistor R1 is open; integrating capacitors C1/C3/C4 are open or short; integral diode D11N60 is open; decoding chip CD4511 has no output or output is high; limiting current resistors R8–R14 are open and inverting driver MC1413 chip is invalid etc.

The paper adds the test circuit to strengthen preventive monitoring of these failures to improve system reliability and reduce the potential failures, which may occur, to avoid the hazard of failures.

4 THE TEST CIRCUIT DESIGN FOR DIGITAL VOLTMETER

After the FMMEA for the designed circuit, we clearly understand main failure modes and failure mechanisms that the digital voltmeter circuit, even typical A/D circuit, may have. In the following

reliability growth tests, we are able to improve the design for a system whose severity is high or which is more prone to fail and enhance the reliability of products by increasing the number of redundant systems. It is effective of FMMEA to find the cause of failure, failure mechanisms and solutions timely, reduce time and costs for maintenance and improve products' testability.

The design of test circuit is a key and difficult point in the field of electronic products' reliability, which is proved quite valuable as people are paying more attention on the failure Prognostic and Health Management (PHM) technology.

This paper is designed for the A/D converter circuit, whose failures can be monitored by the specific test circuit.

4.1 *The test circuit design for the ICL7135 chip*

Analysis: The previous FMMEA shows that, when a failure from the internal circuit exists, the chip is unable to complete A/D conversion. When the signal output of 21-pin is BUSY and active high, the positive integral automatically becomes high at the start and the reverse integral automatically proves to be low at the end. Based on features about this pin, when the chip normally makes a double integral, the output of the 21th pin is sawtooth waveform which can be measured by an oscilloscope to determine working conditions of the chip.

Design: Wires from the 21th pin of ICL7135 chip as a test point when necessary are connected to an oscilloscope. (If the output is sawtooth waveform, the chip is working properly; if the waveform is linear, the failure in the chip circuit has occurred).

4.2 *The test circuit design for clock signal module 555*

Analysis: The output of clock signal module provides a clock pulse signal waveform for the 22th pin of ICL7135. This port can be measured to determine whether the clock module is working normally.

Design: Wires from the 22th pin of clock module or the third pin of 555 chip as a test port are connected directly to an oscilloscope to show the waveform of port output (If the waveform is continuous pulse, the module is normal; if the waveform is linear, the failure has occurred in the module).

4.3 *The test circuit design for reference voltage circuit modules*

Analysis: The reference voltage module whose normal output is a DC voltage of 1V is formed by a divider resistor R5, a potentiometer R4 and a zener

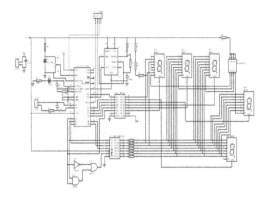

Figure 5. The circuit diagram for a digital voltmeter with test circuit interfaces.

diode D21N4728A. By the analysis of previous sections, when R5 is open, D2 has a short circuit or R4 is open or short, the output voltage is 0V; when R5 has a short circuit, the output voltage is 3.56V; when D2 is open, the output voltage varies slightly with fluctuations. Based on the above features, the kind of failure in the circuit can be determined substantially through testing the output of the reference voltage. If fluctuations are shown by a connected oscilloscope, we are able to determine the existence of D2-open.

Design: From the analysis above, wires from the second pin of ICL7135 REF input signal, which is also the reference voltage output signal, are connected to the test port and a multimeter is used to measure the voltage which is a criterion for possible failures.

4.4 The application of digital voltmeter test circuit

According to the analysis in the previous sections, several test circuit interfaces involving the modules

mentioned above are added based on the original circuit diagram, which makes it easy for digital voltmeter to detect and locate the fault. The circuit diagram for a digital voltmeter with test circuit interfaces is shown in Figure 5. By adding the test interfaces, the circuit is easier to locate the fault in a module, shorten the time of diagnosis and improve testability. It provides a great convenience for products' performance and health monitoring.

5 CONCLUSIONS

Based on the design of 4–1/2 digit voltmeter, the paper identifies main failure modes of the circuit through FMMEA analysis and explores the targeted circuit design for testing. This paper selects several typical circuit failure modes, analyses the corresponding test methods and judging criteria, designs the test interfaces, and finally forms the module for circuit test which can cover 60–70% basic failures of the entire circuit. In this paper, several typical failure modes selected for designing the test circuit are not enough for the whole circuit and deeper research is to be studied in the further.

REFERENCES

[1] Bo Sun & Rui Kang & Jinsong Xie. 2007. Research and application of Prognostics and Health Management System, *System Engineering and Electronic Technology*, 2007(10).
[2] Peiquan, Li. 2000. Principles and Applications of AD converter ICL7135, *Journal of Changwei Teachers College,* Vol. 19 No. 2 Apr. 2000.
[3] Yun Huang & Yunfei, En. 2007. Electronic Component Failure Mode Effect Analysis Technology, *Electronic Components and Materials*, Vol. 26 No. 4 Apr. 2007.

Future Information Engineering and Manufacturing Science – Zheng (Ed)
© 2015 Taylor & Francis Group, London, ISBN 978-1-138-02644-5

Research on water-entry impact and flow-fluid characteristics of tangent ogive structure

Ke Lin
School of Aeronautics and Astronautics Engineering, Air Force Engineering University, Xi'an, China
Xi'an Flight Academy of Air Force, Xi'an, China

Duo Qi, Hai-Yan Yang, Xiao-qiang Zhang & Ling Wang
School of Aeronautics and Astronautics Engineering, Air Force Engineering University, Xi'an, China

ABSTRACT: Authors of this paper are preceding the water-entry processes of three tangent ogives structures with different velocities and water-entry angles, based on moving grid technique and numerical simulation of two-phase model. Through analyzing the results of simulation, the load peak value influences are revealed when several ogives enter water with different velocities and water-entry angles. From the simulation, it can be concluded that the 30° tangent ogive does well on the axis performance and the 90° tangent ogive shows more stability on radial direction.

Keywords: water-entry; gas-water flow field; ogive; impact load; progress of force and velocity

1 INTRODUCTION

Many processes, such as seaplane's landing, spacecraft's recycling and torpedo's entry into the water, will face the problem of water-entry. Especially, spacecraft and torpedoes have a relatively large speed, which can produce huge impact load during the early time of entering water. The impact may cause structural damages, device failure, trajectory out of control and other issues. Analysis of the load distribution and flow field changes on the water-entry processes is the basis of shape optimization design and security assessment, also the core content of hydro-ballistic research.

Given the importance of water-entry, many scholars carried out much work on it for a long time. In early studies, flash photography techniques were mainly used to observe and analyze the phenomena [1]. Von Karman raised an added mass method to calculate the impact load. Taking the influence of fluid compressibility into account, V.A. Eroshin had done a great amount of experiments and theoretical researches [2]. By W.L. Mayo, the research is extended from impact load to various factors, such as the influence of waves [3]. With the development of computer technology and numerical analysis methods, the nonlinearity of three-phase coupling, fluid compressibility and structural distortion get to be analyzed effectively [4]. Although researchers above had made a lot of progress, it is not enough to study the whole process of water-entry. Selecting tangent ogive structure

as the research object, this paper studies the gas-water flow field evolution, impact load characteristics, force and movement in different velocities and angles when the structure enters water.

2 OGIVE STRUCTURE MODEL

In the early time of entering water, the head shape of structure will determine the load characteristics directly, so the structure can be simplified as slender body with the same aspect ratio and mass distribution. Tangent ogive structure has features of stable hydro-ballistic and small sailing resistance, and it is usually used as underwater weapon. A typical ogive structure shape is shown in Figure 1, which is described by apex angle α and scanning angle β. R_1 is scanning radius, R_0 is projectile radius, and β is the central angle corresponding to arc AB [5]. For tangent ogive structure, according

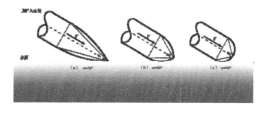

Figure 1. Ogives of three different apex angles enter water.

to the geometrical relationship, $\alpha = \beta$, $R_1 = R_0/\sin\alpha$. If R_0 is constant, R_1 is only determined by α, so the problem of head shape optimization becomes parameter optimization with variable α.

It is shown in Figure 1 that ogives of three different apex angles enter water in $30°$ water-entry angles. When the water-entry angle is smaller than apex angle, it is the tangent point of free surface and spherical head that contact with water first. Considering the axis symmetric characteristics and calculating works of unsteady moving grid, this paper simplifies 3D 6DOF movements to 2D 3DOF to study the movement and rotation of ogives in vertical symmetric plane.

3 SIMULATION MODEL

This paper uses multiphase flow technology and movement grid model of CFX to carry out numerical simulation, and the solution process is shown in Figure 3. Make some assumptions as follows: 1) ogive is a rigid body; 2) water cannot be compressed; 3) the attack angle is 0 at the beginning of water-entry; 4) the cavitation effect is out of consideration.

3.1 Multiphase flow technology

According to the simulation requirements of gas-water environment during the crossing process, it is necessary to introduce multiphase flow model in governing equation. Function r, which represents the ratio of each fluid phase's volume and the whole grid's volume, is defined as the volume fraction of fluid [6]. Use N-S equation weighted by r to describe gas-water movement in the total computational domain. The momentum equation is:

$$\frac{\partial}{\partial t}\left(r_\alpha \rho_\alpha U_\alpha\right) + \nabla \cdot \left(r_\alpha\left(\rho_\alpha U_\alpha \otimes U_\alpha\right)\right)$$
$$= -r_\alpha \nabla p_\alpha + \nabla \cdot \left(r_\alpha \mu_\alpha\left(\nabla U_\alpha + \left(\nabla U_\alpha\right)^T\right)\right)$$
$$+ \sum_{\beta=g,w}\left(\Gamma_{\alpha\beta}^+ U_\beta - \Gamma_{\beta\alpha}^+ U_\alpha\right) + S_{M\alpha} + M_\alpha \quad (1)$$

Different subscript α of function r represents different fluid. If $\alpha = g$, the fluid is air and if $\alpha = w$, the fluid is water. $S_{M\alpha}$ represents external forces or other momentum source. M_α represents internal force incremental of α caused by other fluids. $\Gamma_{\alpha\beta}$ represents mass conversion rate from β to α of unit volume. $\left(\Gamma_{\alpha\beta}^+ U_\beta - \Gamma_{\beta\alpha}^+ U_\alpha\right)$ represents momentum transfer caused by mass transfer between phases. Using function r to weigh k-ε model, the transport equation of turbulent energy k and turbulent dissipation rate ε is:

$$\frac{\partial}{\partial t}\left(r_\alpha \rho_\alpha k_\alpha\right) + \nabla \cdot \left(r_\alpha\left(\rho_\alpha U_\alpha k_\alpha - \left(\mu + \frac{\mu_{t\alpha}}{\sigma_k}\right)\nabla k_\alpha\right)\right)$$
$$= r_\alpha\left(P_\alpha - \rho_\alpha \varepsilon_\alpha\right) + T_{\alpha\beta}^{(k)} \quad (2)$$

$$\frac{\partial}{\partial t}\left(r_\alpha \rho_\alpha \varepsilon_\alpha\right) + \nabla \cdot \left(r_\alpha \rho_\alpha U_\alpha \varepsilon_\alpha - \left(\mu + \frac{\mu_{t\alpha}}{\sigma_\varepsilon}\right)\nabla \varepsilon_\alpha\right)$$
$$= r_\alpha \frac{\varepsilon_\alpha}{k_\alpha}\left(C_{\varepsilon 1}P_\alpha - C_{\varepsilon 2}\rho_\alpha \varepsilon_\alpha\right) + T_{\alpha\beta}^{(\varepsilon)} \quad (3)$$

$T_{\alpha\beta}^{(k)}$, $T_{\alpha\beta}^{(\varepsilon)}$ represent respectively the movement of k and ε between two fluids.

3.2 Moving grid model

Aiming at time varied characteristics of water-entry process, use moving grid technique to solve the problem of mesh movement and distortion.

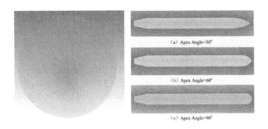

Figure 2. Computational domain mesh.

Figure 3. Pressure and free surface contour at −0.2 ms.

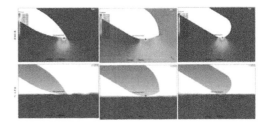

Figure 4. Pressure and free surface contour at 0.2 ms.

Mesh model, in which prism mesh is adopted to refine local mesh, can reduce numerical errors and improve the convergence speed, as is shown in Figure 4. Through changing initial position, interface direction and gravity projection direction of gas-water medium, it can complete numerical study of different water-entry angles.

4 THE RESULT AND ANALYSIS OF NUMERICAL SIMULATION

4.1 The structure and evolution of gas-water flow field

Ogives of different apex angles (30°, 60°, 90°) enter water with the 30° water-entry angle and 50 m/s velocity. The changing processes of initial flow field pressure and interface of gas and water are shown in Figure 3–Figure 6. Start time at the moment when head contacts water, for example, $t = -0.2$ ms represents 0.2 ms before contacting water.

As is known from above, because of large water-entry velocity, the air between ogive and free surface is compressed. At about 10 mm away from water ($t = -0.2$ ms), approximate 1.8 atm appears near the tangent point, which spreads to water by gas-water interface, and reduces gradually during the diffusion to surrounding.

The pressure increases with step after contacting water, at the moment of 0.2 ms, surface peak pressure of three ogives respectively reach 27.9 bar (30°), 28.8 bar (60°) and 34.3 bar (90°). The peak pressure appears at the vertex of 30° apex angle ogive. For the contact point of ogive with 60° or 90° apex angle is not the head vertex, so the peak point appears at the tangent point of head and water surface.

After the ogive head contacts the water surface, the peak pressure become moving. The 30° ogive moves backward along ogive surface, following with expansion and radiation of high-pressure area. At the time of 1.0 ms, a new peak pressure appears at the vertex of 30° ogive. For the ogives with 60° and 90° apex angle, the peak pressures are moving toward the vertex of the head.

During the process of water entry, the early peak pressure of the 30° ogive gradually disappeared near the water surface, and the second one still exists at the vertex of the head. However, the peak pressure of which decreases to 17.5 bars at the time of 1.8 ms.

Finally, the peak pressure on the three ogives stays below to the vertex slightly till it disappears.

4.2 Peak value analysis of overload

Given controllable speed and attitude constraints of torpedoes, study the numerical simulation of axial load and radial load peak value when the three ogives enter water at 100–200 m/s with water-entry angles of 30°, 45° and 60°.

As shown above, with the increase of water-entry velocity, peak values of overload and pressure become higher. With the increase of water-entry angle, the axial load increases, but the radial load decreases.

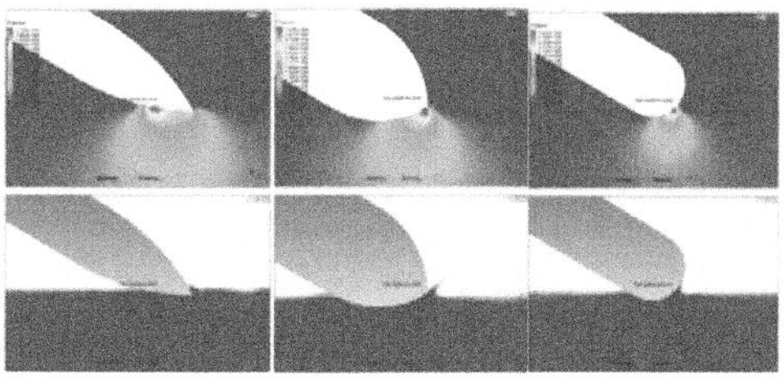

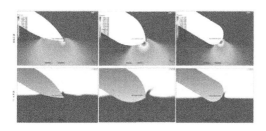

Figure 5. Pressure and free surface contour at 1.0 ms.

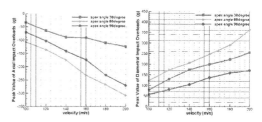

Figure 7. Change trends of impact load of 30° water-entry angle.

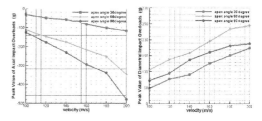

Figure 8. Change trends of impact load of 45° water-entry angle.

Figure 6. Pressure and free surface contour at 1.8 ms.

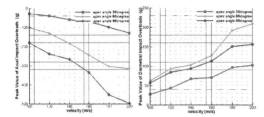
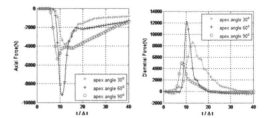

Figure 9. Change trends of impact load of 60° water-entry angle.

Figure 10. Force process at 50 m/s with water-entry angle of 30°.

Table 1. Comparison of impact load and pressure peak.

Water-entry	Peak value of axial overload (L_A)	Peak value of radial overload (L_D)
30°	$L_A(60°) > L_A(90°)$ $> L_A(30°)$	$L_D(60°) > L_D(30°)$ $> L_D(90°)$
45°	$L_A(90°) > L_A(60°)$ $> L_A(30°)$	$L_D(60°) > L_D(90°)$ $> L_D(30°)$
60°	$L_A(90°) > L_A(60°)$ $> L_A(30°)$	$L_D(60°) > L_D(90°)$ $> L_D(30°)$

Obviously, it is not a monotone relationship between apex angle and peak values above. Comparisons at the same water-entry velocity are shown in Table 1. Values in "( )" represent different apex angles, L_A represents peak value of axial overload, L_D represents peak value of radial overload, P_m represents peak value of pressure.

Make these peak values as evaluating indicator, and ogive of 30° apex angle has the smallest impact load and the best performance.

4.3 Process analysis of force and movement

Analyze the numerical simulation of the force and movement when the three ogives enter water at 50 m/s with water-entry angle of 30°. The simulation output is a series of discrete points with timestamps. Use B-spline to approximate and interpolate. The process of force is shown in Figure 10.

As shown above, for axial force, all the three ogives reach peak value at the fourth time step after contacting water; however, the decay processes are different. Ogive of 30° apex angle converges after about ten time steps, ogive of 60° apex angle converges after about six time steps, and ogive of 90° apex angle gradually decreases after reaching peak value.

For radial force, the smaller the apex angle is, the later the load reaches peak value. It is the latest for ogive of 30° apex angle to reach the peak value and the time is longest. It is obvious that ogive of 30° apex angle has the largest radial impulse by

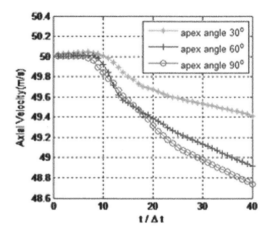

Figure 11. Change curves of water-entry axial velocity.

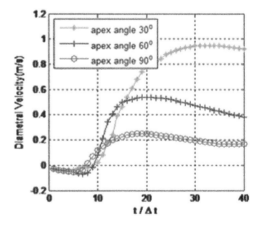

Figure 12. Change curves of water-entry diametral velocity.

integrating, which is disadvantageous to attitude control and structure design.

The velocity and attitude changes of three ogives during the water-entry process are shown in the Figures 11–13. After 12 ms of water-entry,

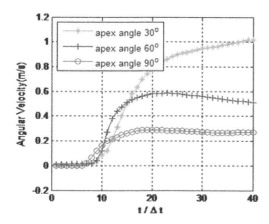

Figure 13. Change curves of water-entry angular velocity.

the reductions of axial velocity of 30°, 60°, and 90° apex angle ogives are 1.2%, 2.2% and 2.6% respectively. The radial velocity of 30° apex angle ogive increases from 0 m/s to 0.9 m/s and then converges. It increases from 0 m/s to 0.55 m/s and then decreases smoothly for the 60° ogive. It increases from 0 m/s to 0.25 m/s and then decreases for the 90° ogive. The change rules of rotating angular velocity are same as radial velocity. During the water-entry process, axial movement of 30° apex angle ogive is steady, but the radial movement and angular velocity have a big fluctuation, which corresponds to force analysis.

5 CONCLUSIONS

Through numerical simulation study on the process of water-entry of three typical tangent ogives, draw some conclusions as follows:

1. Acquire the effects of different apex angle having on pressure peak value distribution by studying change rules of two-phase interface during water-entry.
2. Study the load peak value influences when different ogives enter water with different velocities and water-entry angles. With the increasing of water-entry velocity, impact overload increases. With the increasing of water-entry angle, axial load increases but radial load decreases. There is no rigid monotone relationship between apex angle and load peak value. In comparison, the ogive of 30° apex angle has the smallest load peak value.
3. The simulation of force and movement during water-entry process shows that axial movement of 30° apex angle ogive is steady and its velocity reduction is smallest, but compared with 90° apex angle ogive, the radial movement and angular velocity of 30° apex angle ogive have a big fluctuation.

In summary, taking structure strength design into account, the ogive of smaller apex angle has a better performance. However, if the stability of hydro-ballistic is considered, the ogive of bigger apex angle is better. It is necessary to have a comprehensive optimum selection according to the requirements of trajectory planning and configuration design.

REFERENCES

[1] Watanabe. Resistance of impact on water surface. Inst Phys Chem Res, 1930(12): 251–267.
[2] Eroshin V.A, Romanenkov N.I, Serebryakov I.V, etal. Hydrodynamic forces produced when blunt bodies strike the surface of a compressible fluid. Mekhanika Zhidkostii Gaza, 1979, 6: 44–51.
[3] Mayo W.L. Hydrodynamic impact of a system with a single elastic mode. NACA 1074–1075, 1947.
[4] Wang Z, Yang J, Koo B, et al. A Coupled Level Set and Volume-of-Fluid Method for Sharp Interface Simulation of Plunging Breaking Waves. International Journal of Multiphase Flow, 2009, 35(3): 227–246.
[5] Sussman M, Smith K.M, Hussaini M.Y, et al. A Sharp Interface Method for Incompressible Two-Phase Flows. Journal of Computational Physics, 2007, 221(2): 469–505.
[6] J. Breil, S. Galera, P.H. Maire. A Two-Dimensional VOF Interface Reconstruction in A Multi-Material Cell-Centered ALE Scheme. International Journal for Numerical Methods in Fluids, 2011, 65(11): 1351–1364.

Future Information Engineering and Manufacturing Science – Zheng (Ed)
© 2015 Taylor & Francis Group, London, ISBN 978-1-138-02644-5

The analysis of the risk compensation in the insurance of the security of production in coal mine under government intervention

Chun Yan, Zhi-bo Liu, Chang-chao Chen & Yan-xing Li
College of Mathematics and System Science, Shandong University of Science and Technology, Qingdao, China

ABSTRACT: The issue on the security of production in coal mine is not only associated to the outlook of the industry, but also an important task in the construction of the harmonious society. In this paper, we firstly analyzed the government's function in the insurance of the security of production in coal mine, then we analyzed the government's game action in the risk compensation of the safe production of coal mine with utility theory and game theory, discussed the insurance model of the coal mine production security under three different conditions that government subsidizes insurance company, the miner owner and simultaneously for insurance company and the miner owner and put forward the risk prevention strategy that insurance participate in main part in coal mine production security.

Keywords: the security of production in coal mine; the government, insurance company; the miner owner; risk compensation

1 INTRODUCTION

The issue on the security of production in coal mine has been a major problem affecting the healthy development of industry and the image of China's coal industry. Therefore, in recent years, the academic circles pay close attention to the issues, appeared much of the related research. This paper thinks that the introduction of the insurance mechanism in the coal mine production safety regulation, introduction of economic factors, can fully mobilize enthusiasm, improve enterprise risk awareness, supervise and help coal enterprises to increase safety input, effectively reduce the probability of safety accidents. Even if the accident happens, the insurance compensation can also solve the funding problem of national security production loss partly to provide reasonable compensation for the victims and their families.

At present, coal mining enterprises in addition to participate in the national compulsory social insurance for industrial and commercial, can also buy accidental injury insurance for workers, insurance of employer liability in our country also is emerging. But the insurance is not perfect, there still exist many problems in the relevant mechanism. In addition, there still exist many problems, such as the high claim risk and loss ratio result from serious adverse selection, the insurancial liability of the commercial insurance and industrial injury insurance overlap, the reinsurance market and intermediary market don't participate in actively

and so on. All these problems make the role of insurance in the production safety of coal mine enterprise's role severely limited. Therefore, should establish a set of effective insurance mechanism and perfect, multi-level insurance system to standardize the process of coal production.

2 THE ROLE OF GOVERNMENT

2.1 *The government's advantage and forms of intervention*

Through the use of different forms, and intervention in the risk management in the security of production in coal mine on different levels, the government not only can intervene the operation of the market for the insurance the safety of mine workers, also can reduce the operation risk and loss control. The following summarizes the measures government can take in the coal mine safety production risk management and the corresponding advantages and disadvantages.

1. After the coal mining accident, the government provides direct assistance. This is the most original and the most commonly used process by the majority of developing countries in government participation. From a long-term perspective, the insurance plans beforehand is more efficient than rescue after the accident, it also can reduce the government's financial burden, but considering the social, public and other factors,

the government should play an important role after the accident.

2. Government should participate in taking the risk of the security of production in coal mine. [Q]Government can take to act as major insurers and reinsurers, which are the two ways to undertake the risk of the security of production in coal mine. Of course, due to the direct participation of the national body, the insurance company's moral hazard became a problem. Once the countries involved in reinsurance, the insurance company may relax underwriting, when the loss claim exceeds the limit, the insurance companies tend to relax the damage, and control the account of the loss, and even use excessive statistical way on the loss to profit from the government finances.

3. National Loan Scheme. Considering commercial insurance companies' ability to disperse intertemporal risk of coal mine production safety in terms of time, the state needs only to provide temporary liquidity loans to help the solvency of insurance companies is insufficient, does not need to take the initiative to undertake the risk of the security of production in coal mine for the insurance company. Of course, the problem of state loans lies in inhibiting innovation of the private sector, particularly capital markets, and the government side will face credit risk. Especially for developing countries, the financial markets stay bad debts more prominent, this model will make the government undertake the risk of the security of production in coal mine in disguise.

4. The direct financial subsidies to the private sector. Through financial subsidies, Market needs of the insurance of the security of production in coal mine will be expanded, the enthusiasm of the insurance companies' participation will also improve. In this process, the government will not bear the risk of the security of production in coal mine, but there may be some financial burden and may also lead to moral hazard of the insured and the insurance company.

5. Compulsory insurance and punishment mechanism. Administrative intervention measures taken by the Government and the implementation of compulsory insurance can help the Government insured can directly address the lack of willingness of insured and adverse selection problems.

These forms of participation are respectively for different purposes can play different roles, are not mutually exclusive. Government may take one or several comprehensive measures to compensate for the lack of private market. Government has the advantage of assisting the private sector to complete the risk diversification, punish those who damage market rules and address the lack of willingness on insurance and adverse selection problems on the demand side, so the government should choose the most economical and cost-effective manner with minimal negative impact of market intervention, which also become as the basic principles to build a successful mix of coal mine safety production risk management plans.

2.2 The effectiveness of government intervention

From the micro level, if in the case of government intervention, insurance policy holders get more than the government does not interfere in the case, then the insured gained more certainty and government intervention would not intervene. Based on the macro-level view, we will have to consider of the increased social welfare caused by the government intervention the contrast of economic costs resulting from it.

Due to the controversial evaluation system about whether the insurance of the security of production in coal mine is better under government intervention, the phenomenon of "government failure" needs to be further understanding, this article can not give a clear conclusion to the superiority of government intervention. This paper argues that the choice of government intervention methods depends on its purpose, it should minimize interference in the private market. Based on this, the following article will analyses whether different levels of government intervention form affect the private market operation and whether it can reduce the market failure areas, that is the cost of government intervention—benefit analysis.

3 THE GOVERNMENT'S GAME MODEL OF THE RISK COMPENSATION OF THE SECURITY OF PRODUCTION IN COAL MINE

National financial subsidies and policy support is necessary for the insurance of the security of production in coal mine, thus providing protection for the smooth implementation of the coal mine production safety insurance. In the insurance of the security of production in coal mine, if the government's dominant position and role is not clear, it will lead to the insurance runs in the low efficiency and the loss of social welfare. In this regard, this article will do further analysis on the risk compensation of the insurance of the security of production in coal mine with utility theory and game theory.

3.1 *The basic assumption of the model*

① Let mine owners' initial wealth is W, the probability of occurance of the insurance of the security of coal mine workers is P, the loss of wealth of the mine owners in the dangerous condition is L, the insurance payment of the insurance company after damaged is Y, assuming that the insurance rates for the π, mine owners should premiums paid shall πY.

② Let mine owners are risk averse insured persons, the utility function is $U_{(W)}$, according to the principle of welfare economics, $U'_{(w)} > 0$, $U'' < 0$. If the mine owners choose not to insure, expected utility for:

$$U_{S0} = P \times U(W - L) + (1 - P) \times U(W) \qquad (1)$$

On the other hand, expected utility is obtained as:

$$U_{S1} = P \times U(W+Y - L - \pi Y) + (1 - P) \times U(W - \pi Y) \qquad (2)$$

③ Let the insurance company is risk-neutral, if not danger, their income is πY; If the danger, which suffered losses as a $Y - \pi Y$. Also assume that the company's operating expenses is a constant K, then the insurance company's expected return is:

$$U_1 = P \times (Y - \pi Y) + (1 - P) \times \pi Y - K \qquad (3)$$

3.2 *Game analysis on the insurance mode of the safe production of coal mine under participation of government*

The government can take appropriate measures to make Pareto optimal (insurance underwriting) to become the Nash equilibrium, for example through the introduction of premium subsidies for mine owners, giving insurance companies operating subsidy or tax relief. The following three cases are discussed.

1. Government subsidies for insurance companies. If $U_{S1} > U_{S0}$, while $U_1 < 0$, mine owners driven by the interests willing to insure in this condition, but the insurance company earnings is negative, that is operating at a loss. If the government give subsidies to insurance companies for M, making the $U_1 + M > 0$, then (insurance underwriting) will be a Nash equilibrium of the game (as shown in Table 1).

2. Government subsidies for mine owners. If $U_{S0} > U_S$, and $U_1 > 0$, the insurance company is profitable at this time, but (insurance underwriting) strategy has damaged the interests of miners, the mine owners do not want to insure. If the government gives premium subsidies for mine owners is πYc (c for mine owners' premium subsidy rate provided by the government, $0 \leq c \leq 1$),

Table 1. The game analysis of government subsidies for insurance companies.

| | Insurance company | |
	Underwrite	Do not underwrite
Mine owners		
Insure	U_{S1}, $U_1 + M$	U_{S0}, 0
Do not insure	U_{S0}, $-K$	U_{S0}, 0

Table 2. The game analysis of government subsidies for mine owners.

| | Insurance company | |
	Underwrite	Do not underwrite
Mine owners		
Insure	$U_{S1} + \pi Yc$, U_1	U_{S0}, 0
Do not insure	U_{S0}, $-K$	U_{S0}, 0

Table 3. Game analysis of giving subsidies to the miners and the insurance companies at the same time.

| | Insurance company | |
	Underwrite	Do not underwrite
Mine owners		
Insure	$U_{S1} + Y$, $U_1 + M$	U_{S0}, 0
Do not insure	U_{S0}, $-K$	U_{S0}, 0

making $U_{S1} + \pi Yc > U_{S0}$, then (insurance underwriting) will be the Nash equilibrium of the game (as shown in Table 2).

3. The government offers subsidies to mine owners and the insurance company at the same time. If $U_{S1} < U_{S0}$, and $U_1 < 0$, at this time the mine owners do not want insurance, the insurance company is not willing to underwriting. If the government subsidies for insurance companies is M, premium subsidies given for mine owners is πYc, making the $U_1 + M > 0$, $U_{S1} + \pi Yc > U_{S0}$, then (insurance, underwriting) will be a Nash equilibrium of the game (as shown in Table 3).

Through the analysis of the current domestic situation, the first approach is unrealistic and undesirable. If the government wants to play a role in coal mine production safety insurance, the government needs to develop coal mine production safety insurance system and to give substantial support to allow mine owners to benefit. For the second way, if the government chooses to give premium subsidies to the mine owners, the government needs this part of the flow of funds to strengthen supervision, but also consider the problem of matching their investment and consumption. For the third way, if the government chooses to give subsidies to mine owners and insurance companies at the same time, the government needs to consider the adequacy of funds, and how to share other issues.

4 RISK PREVENTION STRATEGIES OF INSURANCE PARTICIPATION MAIN BODY OF SAFETY PRODUCTION IN COAL MINE

According to the foregoing analysis, under the participation of different subjects in game behavior, risk transfer and prevention model for coal mine production safety insurance is shown in Figure 1.

It can be found that if the feedback of the market regarded as the only guide for the transfer of risk among the governments at all levels and insurance companies and mine owners, then there is a risk of causing inefficient allocation of resources in the market, causing the insurance participants to make economic decision improperly, thus making losses of the whole social utility. All the participants in the different position and role in the insurance of the security of production in coal mine, and they are different in interests demands, so the risk prevention strategies they take should also be different.

The government, first of all should make laws on the insurance market of the security of production in coal mine. Secondly, improve financial support system to match the insurance of

the security of production in coal mine. Thirdly, improve regulatory system of the market for the insurance of the security of production in coal mine. Fourthly, perfect the insurance organization system of the insurance of the security of production in coal mine.

The insurance companies, first of all should design the categories of insurance, collate and stipulate the premium rates, and establish a hierarchical set loss claims system according to the extent of damage to occur, the amount of the claimed loss probability and payment and other factors of coal mine production safety accident. Secondly, in order to optimize governance structure of insurance companies, and do well in internal management of the risk and control the risk.

The mine owners, first of all should establish correct risk prevention awareness, respect and safeguard the legitimate rights and interests of miners, fully aware of the insured for hedging rather than profit. Secondly, insured object should be selected according to market needs and with their own advantages, and the miners should determine the breadth and depth to insure according to the insurance premium rate and the amount of the categories of insurance provided by insurance companies. Thirdly, the mine owners need to strengthen the management of production safety work, pay close attention to the occurrence of coal mine accidents, and control the risk in the lowest level.

5 CONCLUSIONS

This paper analyzes the risk compensation in the insurance of the security of production in coal mine under government intervention with game theory. Discussed the government's advantage, introduce and compare effectiveness of the five forms of intervention government intervention compulsory insurance and punishment mechanism; the government's game action in the risk compensation of the safe production of coal mine with utility theory and game theory, discussed three models under government's participation, the insurance of the security of production in coal mine; and this paper proposed the risk prevention strategies for participants of the insurance of the security of production in coal mine.

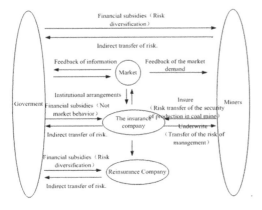

Figure 1. The model of risk transfer and prevention of coal mine safety production insurance.

ACKNOWLEDGMENTS

This work was financially supported by the China Postdoctoral Science Foundation funded project (2012M521362), the project of 2013 annual National

Statistics Scientific Research (2013LY037), the Postdoctoral Scientific Foundation in Shandong China (201303071), and the International Cooperation Training Project of Shandong Province Outstanding Young and Middle-aged Backbone Teachers.

REFERENCES

[1] Li R. Problems and Strategies of coal miners' industrial injury insurance. Coal mine safety. 2009(1):107–109.

[2] Yan C, Liu X.M. Review on the research of coal mine enterprise's safety production. China mining magazine 2009,18(3):19–22.

[3] Hu W.G, Liu L.Y. The Game Analysis on the Supervision of the Colliery Production Safety in China. The Journal of Quantitative & Technical Economics. 2008 (8):94–109.

[4] Lu, Y., Li, X. A study on a new hazard detecting and controlling method: The case of coal mining companies in China. Safety Science (2011), vol.49, pp.279–285.

[5] Shi. G.P. The status and improvement of technology of modern coal mine safety production management. Technology & Business (2013), vol.20, pp.41 (in Chinese).

Future Information Engineering and Manufacturing Science – Zheng (Ed)
© 2015 Taylor & Francis Group, London, ISBN 978-1-138-02644-5

Automatic segmentation of 3D point cloud using spectral clustering

Shoubin Liu & Zongwei Zhang

School of Mechanical Engineering and Automation, Harbin Institute of Technology Shenzhen Graduate School, Shenzhen, P.R. China

ABSTRACT: Point cloud segmentation is a difficult problem for the objects with topological branches. In this paper we propose a Spectral Clustering (SC) method to partition the point cloud into clusters by analyzing the eigenvalues of Laplacian matrix. The Laplacian matrix is constructed by using both topological information and local direction information of the points. Compared to the redundant clusters resulted by general spectral clustering method, the segmentation algorithm can generate accurate clusters for the point cloud. Experiment results show the effectiveness of the proposed algorithm.

Keywords: point cloud; segmentation; spectral clustering

1   INTRODUCTION

With the development of various applications of 3D point cloud, the partition of point cloud becomes a more urgent topic. Although a lot of works have been done on segmentation of point cloud data from 3D objects with single topology, it is still a difficult problem to segment of point cloud data from 3D objects with topological branches, such as the real world staffs. In this article we mainly use the spectral clustering algorithm to deal with the point cloud data. In addition our method tries to find a new construction of the similarity matrix and to find a new way to segment the point cloud automatically.

Spectral clustering algorithm is a powerful method for data analysis. Its principles can be found in the Chung's study (Chung 1997). The algorithm of spectral clustering can be expressed as follows (Ng et al. 2001).

Given a set of points $S = \{s_1, \ldots, s_n\}$ in R^l,

1. Form the similarity matrix $W \in R^{n \times n}$, where $w_{ij} \in [0\ 1]$ describes the similarity between s_i and s_j.
2. Define D to be a diagonal matrix $D = diag(d_{ij})$, whose (i,i) element is the sum of S's i-th row, $d_{ij} = \sum_{j=1}^{n} w_{ij}$, and construct Laplacian matrix $L = D - W$.
3. Define the normalized graph Laplacian $L_{rw} = D^{-1}W = I - D^{-1}W$, where I is the identity matrix.
4. Find $x_1, \ldots, x_k$, the k least eigenvectors of L_{rw}, and form the matrix $X = [x_1, x_2, \ldots, x_k] \in R^{n \times k}$, by stacking the eigenvectors in columns. The clustering number k can be obtained by analyzing the eigenvalues of the normalized Laplacian.
5. Treating each of X as a point in R^k, clustering them into k clusters via k-means or any other algorithms that attempt to minimize distortion.
6. Assign the original points S_i to cluster j if and only if row i of the matrix X was assigned to cluster j.

2   RELATED WORK

2.1   *Graph representation of points*

Given a set of data points $s_1, \ldots, s_n$, we can describes the relationship of the data points by using an undirected graph $G = (V,E)$, where V is the vertex set and E is the edge set. In this article we suppose that the graph is weighted, that is each edge in E carries a non-negative weights $w_{ij} \geq 0$. The weighted matrix is the similarity matrix $W = (w_{ij})$ $i, j = 1, \ldots, n$. Since the graph is undirected, the corresponding results are that $w_{ij} = w_{ji}$. If $w_{ij} = 0$, this means that the vertexes v_i and v_j are not connected. The degree of each vertex can be defined as $d_i = \sum_{j=1}^{n} w_{ij}$. The degree matrix D is defined as the diagonal matrix with the degrees $d_1, \ldots, d_n$ on the diagonal.

2.2   *Laplacian matrix of the graph*

The unnormalized Laplacian matrix of the graph is defined as:

$$L = D - W \tag{1}$$

The normalized graph Laplacian matrix is defined as two ways: $L_{sym} = D^{-1/2}W^{1/2}$ and $L_{rw} = D^{-1}L$. The former is a symmetric matrix and the latter is related to random walk (Liu & Feng 2013).

Theorem 1 Let G be an undirected graph with non-negative weights. Then the multiplicity k of the eigenvalues 0 of both L_{rw} and L_{sym} equal the number of connected components A_1, ..., A_k in the graph. For L_{rw}, the eigenspace of 0 is spanned by the indicator vectors of those components. For L_{sym}, the eigenspace of 0 is spanned by the vector $D^{1/2}I$, where I is the indicator vector of A.

The proof of this theorem can be find in the research of Luxburg (Luxburg 2007).

Given a subset of vertexes $A \subset V$, we denote its complement $V\backslash A$ by $\bar{A}$. Define $|A|$ as the number of vertexes in A and $vol(A) = \sum_{i \in A} d_i$. For two disjoint sets $A, B \subset V$, we define $W(A, B) = \sum_{i \in A, j \in B} w_{ij}$.

Spectral clustering is related to graph partitioning, which means choosing the clusters A_1, ..., A_k to minimize the object function:

$$cut(A_1, ..., A_k) = \frac{1}{2} \sum_{i=1}^{k} W(A_i, \bar{A}_i) \qquad (2)$$

The two most common graph cuts are the RationCut and the normalized Ncut:

$$RatioCut(A_1, ..., A_k) = \frac{1}{2} \sum_{i=1}^{k} \frac{W(A_i, \bar{A}_i)}{|A_i|}$$
$$= \frac{\sum_{i=1}^{k} cut(A_i, \bar{A}_i)}{|A_i|} \qquad (3)$$

$$Ncut(A_1, ..., A_k) = \frac{1}{2} \sum_{i=1}^{k} \frac{W(A_i, \bar{A}_i)}{vol(A_i)}$$
$$= \frac{\sum_{i=1}^{k} cut(A_i, \bar{A}_i)}{vol(A_i)} \qquad (4)$$

From Wagner (Wagner & Wagner 1991) we get that both the RationCut and the Ncut are NP hard. Fortunately, spectral clustering problems can be relaxed. Ncut leads to normalized spectral clustering, and RatioCut leads to unnormalized spectral clustering.

Theorum 2 (Luxburg 2007) Let $A, H \in R^{n \times n}$ be symmetric matrices, and let $\|.\|$ be the Frobenius norm or the two-norm of matrix. Consider $\tilde{A} = A + H$ as a perturbed version of A. Let $S_1 \subset \mathbb{R}$ be an interval. Denote $\sigma_{S_1}(A_1)$ as the set of eigenvalues of A in S_1, and denote V_1 as the eigenspace corresponding to all those eigenvalues. Denote $\sigma_{S_1}(\tilde{A}_1)$ and $(\tilde{V}_1)$ as the similar quantities for $\tilde{A}_1$. Define the distance between the spectrum of A inside S_1 and the spectrum of A outside of S_1 as:

$$\delta = \min\{| \lambda - s |; \lambda \text{ eigenvalue of } A, \lambda \notin S_1, s \in S_1\} \qquad (5)$$

Define $d(V_1, (\tilde{V}_1)) = \| \sin\theta(V_1, (\tilde{V}_1)) \|$ as the distance between the two subspaces V_1 and $(\tilde{V}_1)$, which is bounded by

$$d(V_1, (\tilde{V}_1)) \leq \frac{\| H \|}{\delta} \qquad (6)$$

This theorem tells us that the larger the eigengap δ is, the closer the eigenvector of the ideal case and the perturbed case are, the better spectral clustering works. The eigengap discussed here is mainly used to choose the clustering numbers k in the following process.

3 ALGORITHM CONSTRUCTION

3.1 *Similarity matrix construction*

We construct the similarity matrix by using the structure of the graph. If s_i and s_j are connected by an edge, then a positive weight $w_{ij} \in (01]$ is set to describe their similarities. If two points are disconnected, then we set $w_{ij} = 0$. When constructing similarity graph, the goal is to model the local neighborhood relationship between data points. The k-nearest neighbor graph is used to construct the similarity matrix. It has been shown that k should be greater than 6 to ensure stable computation (Pauly et al. 2003). In this paper we use the kd tree to calculate the k-nearest neighbors (Mount 1998).

The principle of constructing similarity matrix is to let the points with high weights if they are close enough. Considering the surfaces are not always a plane, we'd better take the direction information into account. If the direction of two points are close enough, we should give them high weights, otherwise 0 weights are given. The following describes an implementation of constructing a similarity matrix.

The Gaussian similarity function $w(s_i, s_j) = exp(-d^2(x_i, x_j)/(\sigma^2))$ is usually applied as a similarity function, where the parameter σ controls the width of the neighborhoods. In our algorithm, we take the surface direction information into consideration, and thus the corresponding similarity function can be express as:

$$w(s_i, s_j) = exp\left(-d^2(x_i, x_j)/(\sigma^2)\right) * exp(-angdis^2/(\delta^2)) \qquad (7)$$

where *angdis* is the angular distance between two points, and δ is the angular distance scaling.

We then introduce a local scaling to calculate the weight function (Zelnik-Manor & Perona 2004). Instead of selecting a single parameter σ, we create a local scaling parameter σ_i for every data points s_i. The distance from s_i to s_j as 'seen' by s_i

is $d(s_i,s_j)/\sigma_i$, while the converse is $d(s_i,s_j)/\sigma_j$. Therefore the square distance d^2 can be rewritten as $d(s_i,s_j)d(s_i,s_j)/\sigma_i\sigma_j = d^2(s_i,s_j)/\sigma_i\sigma_j$. Thus the weight function can be expressed as:

$$w(s_i,s_j) = exp\frac{-d^2(s_i,s_j)}{\sigma_i\sigma_j} * exp\left(\frac{-angdis^2}{\delta^2}\right) \qquad (8)$$

where $\sigma_i = d(s_i,s_k)$ and $\sigma_j = d(s_j,s_k)$. s_k is the k-th neighbor of points s_i or s_j. The selection of k is independent of scale and is a function of the data dimension of the embedding space. In our experiment, we set $k = 10$. The purpose of using σ_i and σ_j is for adaptive regulating the relative distance of two points s_i and s_j.

The distance between two points are not directly calculated. We first construct a matrix which contains topological information of each point. Given a point $s_i = [x_i, y_i, z_i]$, find its neighborhood $s_i' = [x_i', y_i', z_i']$. By subtracting its mean vector, we obtain the centralized point set $\bar{s}_i'$. Then we use the coordinates in $\bar{s}_i'$ to construct a topology matrix for each point as:

$$T = \begin{bmatrix} x_i^2 & x_iy_i & x_iz_i \\ y_ix_i & y_i^2 & y_iz_i \\ z_ix_i & z_iy_i & z_i^2 \end{bmatrix} = U\Sigma U' \qquad (9)$$

The unitary matrix U is the eigenvectors and Σ is the corresponding eigenvalues. Then we can get a mapping point with topological information by using the unitary matrix times the point vector:

$$P = U * S_i \qquad (10)$$

The *angdis* can be obtained by using the main direction of each point (Cheng et al. 2011). We calculate the surface direction information by using the principle component analysis (Smith 2002). Given a point s_i from V, we first find its k-nearest neighbors $S = [s_1, ..., s_k]$. Then principle component analysis is applied to calculate the main direction. Specifically, by singular value decomposition, we obtain

$$\bar{S} = \sum_{i=1}^{3} \lambda_i \zeta_i \xi_i^T, \quad \lambda_1 \geq \lambda_2 \geq \lambda_3 \qquad (11)$$

where $\bar{S}$ is the centralized S by subtracting the mean vector. λ_i is the singular value of $\bar{S}$. ζ_i and ξ_i are the left singular vector and right singular vector of $\bar{S}$, respectively. Then, we use ξ_3 as the surface direction of a point. We get the *angdis* of two points by their cosine similarities. Specifically, $angdis = 1 - \cos(\xi_i,\xi_j)$. As for the angular distance scaling, we use the same strategy as the distance

similarity. Considering that the same part of the surface has similar cosine similarity, we use the angular distance as criterion. Given two points s_i and s_j, if the angle between their surface direction is smaller than a threshold, we say that they have the same surface similarity, i.e. $s(i,j) = 1$. Thus the weight function can be expressed as:

$$w_{ij} = \begin{cases} \exp\left(\dfrac{-d^2(s_i,s_j)}{\sigma_i\sigma_j}\right)\exp\left(\dfrac{-angdis^2(s_i,s_j)}{\delta_i\delta_j}\right), & s(i,j)=1 \\ 0, & others \end{cases}$$

$$\qquad (12)$$

3.2 Estimating the number of clusters

From theorem 1 we find a way to discover the number of groups by analysis the eigenvalues of Laplacian matrix. Considering the group overlaps and noise, we do not take the number of 0 eigenvalues as the number of groups. According to theorem 2, we propose a new method to find the clustering number. First find the gap between each two adjacent eigenvalues. Then we can get the mean gap value. After getting the mean gap, we can test if each gap is greater than the mean gap. If it is, we set a flag for two clusters. Thus we can get accurate clustering numbers by using the gaps of eigenvalues. Finally we use the k-means method to cluster the founded eigenvectors.

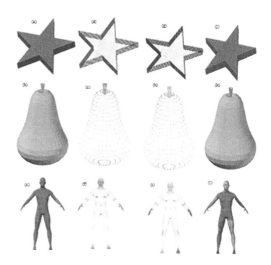

Figure 1. Partition of different point clouds using the proposed method: (d), (e) and (f) are original point clouds of a star, a pear and a human body. (a), (b) and (c) are original point clouds in shading. (g), (h) and (i) are corresponding segmentation results of point clouds. (j), (k) and (l) are segmentation results of point clouds in shading. Different colors indicate regions identified by the algorithm.

4 EXPERIMENT RESULTS

We use a star and a pear as two simple examples and a human model as one complex example to test the effectiveness of the proposed algorithm. The partition results are showed in Figure 1. The first row in Figure 1 shows that the star can be decomposed into three parts, i.e. the top and bottom planes and the vertical plane between them. The second row in Figure 1 shows the pear can be decomposed into two parts, i.e. the handle part and the body part. Since the handle of the pear is of a different topological structure, the algorithm identifies it successfully. The third row in Figure 1 shows the segmentation results of a human body. The algorithm successfully finds in total 10 clusters corresponding to the head, the torso, two arms with hands, two thumbs, two legs, and two feet.

5 CONCLUSIONS

This paper proposed an automatic algorithm for segmentation of 3D point cloud based on spectral clustering. The main work is concentrating on the estimation of clustering number by analyzing the gaps of eigenvalues. Experiment results show that this algorithm is effective for obtaining accurate clusters. Further work includes improvement of the settings of threshold parameters and the constructing of weight matrix with more topological structures.

ACKNOWLEDGMENT

This work is supported by the national natural science foundation of China under grant 60873141.

REFERENCES

[1] Cheng, J. et al. 2011. 3D human posture segmentation by spectral clustering with surface normal constraint. *Signal Processing* 91(9): 2204–2212.

[2] Chung, F. 1997. Spectral graph theory. *American Mathematical Soc.* 92: 1–214.

[3] Liu S. B. & Feng K. 2013. Point cloud segmentation based on moving probability. *Applied Mechanics and Materials.* 380–384: 1796–1799.

[4] Luxburg, U.V. 2007. A tutorial on spectral clustering. *Statistics and computing* (3): 1–28.

[5] Mount, D. 1998. ANN programming manual. *Technical report Dept. of Computer Science U. of Maryland*: 1–37.

[6] Ng, A. et al. 2001. On spectral clustering: Analysis and an algorithm. *Proceedings of Advances in Neural* 2:849–856.

[7] Pauly, M. et al. 2003. Shape modeling with point-sampled geometry. *SIGGRAPH'03* 22(3): 641–650.

[8] Smith, L. 2002. A tutorial on principal components analysis. *Cornell University* 51:52–79.

[9] Wagner, D. & Wagner, F. 1991. Between min cut and graph bisection. Springer Berlin Heidelberg: 744–750.

[10] Zelnik-Manor, L. & Perona, P. 2004. Self-tuning spectral clustering. *NIPS* 2: 1601–1608.

Future Information Engineering and Manufacturing Science – Zheng (Ed)
© 2015 Taylor & Francis Group, London, ISBN 978-1-138-02644-5

Hua chi block low permeability reservoir sensitivity evaluation

W.C. Qu & K.P. Song
Petroleum Engineering Institute of Northeast Petroleum University, China

B. Bi
Well Logging Company of Daqing Drilling Company, China

ABSTRACT: As a typical low permeability and water sensitivity reservoir, combined with Chang qing oilfield Hua chi block of reservoir geology, Hua chi block of injection water static scale experiments, research on temperature and ion concentration and PH value on the influence of scaling, through the whole core laboratory analysis of reservoir, and complete the sensitivity analysis. To prevent injection water damage to reservoir, based on project research results, in the implementation of operations, the selection and compatibility of formation fluid into the well and the application of clay stabilizer. Technology into damage to lift, at the same time, to enforce the protection of the reservoir has been hurt, to the scientific and efficient development of hua chi block.

Keywords: low permeability; scale; reservoir sensitivity; prediction

1 INTRODUCTION

Hua chi block is the southeast broken nose structure, located at the southern tip of the western slope in the western sag of Chang qing oilfield. For a set of sand shale interbed reservoir sedimentation and sedimentary types of alluvial fan, provenance from the south, change of sand body, eastern with western thick thin, the characteristics of the south north of thick thin. From the analysis of fluid distribution and reservoir occurrence as in thick layered edge water reservoirs, district a total of 180 oil wells, comprehensive water cut oil 24.5×10^4 t tired, recovery degree 10.8%, oil recovery rate 1.5%, at present formation pressure is 3.1 MPa. [1–3]

1.1 *Reservoir petrology characteristics*

Affected by the formation reservoir sedimentary types, reservoir sand body changes greatly. Sand body flat bout point for the south to north thick thin, thick thin east west. Reservoir lithology is mainly for red conglomeratic sandstone and conglomeratic anisometric sandstone, ranging from sand rock, siltstone, etc, which mainly contains gravel anisometric sandstone, reservoir rock grain sorting poor. Rock particles coarser, the median size is about 0.32 mm. Core analysis data, the reservoir porosity of 11.2%~22.4%. Average porosity was 14.2%, the permeability of 1.1×10^{-3}~$194.0 \times 10^{-3}\mu m^2$, for low porosity and low permeability reservoir. The body of the sand body parts are slightly better reservoir physical property[2–6]; Vertically, the reservoir physical properties change has no obvious regularity.

1.2 *Reservoir clay minerals*

Mesozoic reservoir clay mineral content is generally 3.1 ~ 13.5%, averages 8%, moderate levels; Which give priority to with smectite mineral composition, the average content of 69.6%, followed by illite, average content of 18.0%.

1.3 *The wettability of the reservoir rock*

Through the 5 block sample core laboratory analysis, relative water wet reservoir rocks were greater than 90%, maximum 99.5%, minimum 90.3%, shows that the reservoir rock wettability is strong hydrophilic type.

1.4 *Reservoir fluid properties*

Fault block reservoirs of 20 °C ground degassed oil density of 0.8461 g/cm³, 50 °C ground degassed oil viscosity of 8.171 mPa. S, wax content 10.05%, freezing point is 25 °C ~ 30 °C, colloid and asphaltene content 15.39%. For waxy crude oil and high freezing point thin oil.

2 THE RESERVOIR SENSITIVITY EXPERIMENTS

2.1 Water sensitivity evaluation experiments

Experimental data shows that hua chi block reservoir of strong water sensitive reservoir, index from 0.7 to 0.9, the total reservoir clay mineral is 67.9%, mostly expansive, the effect of montmorillonite mineral with the loss of the experimental fluid salinity, core permeability is decreased. The higher the fluid salinity of reservoir permeability damage.

2.2 Alkali sensitivity evaluation experiments

Hua chi block core permeability increased with PH, and the PH increased from 7 to 11, permeability was reduced by 38%, silicate deposits formed, when the PH is greater than 12, quartz and feldspar begins to dissolve, core permeability and bigger. Hua chi blocks contain alkali sensitivity, alkali sensitivity of the weak. PH greater than 11 after, with the increase of PH value, rock mineral and clay mineral dissolution, formation permeability increase.

2.3 Salt sensitivity evaluation

When the brine salinity is from 2000 mg/L to 1500 mg/L, hua chi block core permeability were significantly decreased, coring permeability significantly lower front salinity for the critical salinity, hua chi block critical salinity is 2000 mg/L. In order to prevent the formation of salt sensitivity reservoir, the well fluid salinity should be greater than 2000 mg/L.

2.4 Speed sensitivity evaluation

In the constant temperature box of 65 °C by low velocity to core injection of kerosene, measuring the velocity of the permeability; Then follow the same steps, and then according to certain traffic demand is gradually increasing level differential injection, observe the change of core permeability, the injection rate increases to find obvious changes of the critical flow velocity.

Learn from Figure 1 hua chi block core for weak velocity sensitive, speed sensitive permeability damage value is 15.0%. Core critical flow rate of 1.00 cm³/min, when the flow rate increased from 1.00 cm³/min to 1.50 cm³/min, the core permeability decreased obviously, fell by 11.69%. Hua chi block 1 hua chi well core for weak velocity sensitive, velocity sensitive permeability damage value is 13.5%.

2.5 Acid sensitivity evaluation experiments

Hua chi block medium to weak acid sensitivity, acid formation for soil acidification. Experimental

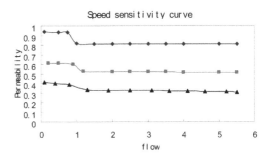

Figure 1. Reservoir core speed sensitivity experiment research.

data show that hydrochloric acid to the stratigraphic dissolution ability is weak, less than 10%, of stratum soil acid dissolution ability strong, up to 20%. So as a result of the existence of hematite and calcite, acidification need use hydrochloric acid formation processing first, and then to acid soil acidification. After the acid flow back. To prevent the secondary precipitation pollution formation.

3 THE STATIC SCALE PREDICTION EXPERIMENTS

This experiment is based on hua chi-19 sample data for water distribution. According to the ion concentration in water samples every 2 l water with drug volume are obtained.

3.1 Experimental instruments and medicines

This experiment mainly contains experimental instrument for electronic temperature control water bath pot, FA1004 A electronic balance, PH indicator and 101-3ab electrothermal blowing. Hob, iron ring, besides, there are long neck funnel, glass rod, filter paper, jar, spatulas, ears ball, measuring cylinder (500 ml), volumetric flask and, tweezers, pipette, etc. The main experimental drugs are calcium chloride (CaCNa2SO4), sodium chloride (Nacl), potassium chloride (Kcl).

3.2 The experimental steps

This experiment is mainly to discuss different mixing ratio, different temperature and different PH value on the scale of the impact. Experiments in total is divided into two parts, the first step is under ordinary conditions of different temperature and mixing ratio of scale experiments, the second part is under ordinary conditions of different PH value and the mixing ratio of scale experiments.

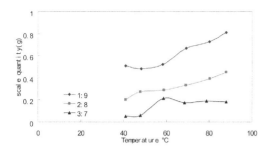

Figure 2. Atmospheric under different temperature different mixing ratio scale experiments.

3.3 *The influence of various factors on scaling*

Experimental temperature, mixing ratio, PH value are discussed the influence of three factors on scaling.

As shown in Figure 2, when mixing ratio constant, as the temperature rises, the quantity of scaling, however, change is not big. Thus, temperature may affect scale, but the effect was not significant. When temperature is constant, with the increase of mixture ratio, amount of obvious changes have taken place, and can be seen from the table, with the increase of Ca ion concentration, reduce scale quantity and this is due to the increase of Ca ion concentration increased scale, the dissolution of the generation of reduced scale.

By the above experiments can be concluded that for hua chi blocks, temperature on the scaling effect is not obvious, but the amount of mixing ratio on scaling has obvious effects, increases as the mixing ratio rises, the quantity and structure.

4 CONCLUSION

1. Hua chi block water sensitivity index is larger, strong water sensitivity damage. Hua chi block reservoir clay mineral amount was 67.9%, mainly composed of montmorillonite mineral, easy expansion, montmorillonite mineral water into the reservoir formation damage.
2. Hua chi reservoir belongs to the weak water sensitivity formation, the critical salinity is 4000 mg/L, all into the well fluid before injection formation should adjust its salinity in about 4000 mg/L, or add expansion agent, otherwise easy to damage the formation.
3. The sandstone reservoir optimal acid soluble loss rate is between 20% ~ 30%. Hydrochloric acid pool in China block reservoir core dissolution loss rate is smaller, less than 10%.

REFERENCES

[1] Allen M.B, Macdonald D.I.M, Zhao Xun, et al. Early Cenozoic two-phase extension and late Cenozoic thermal subsidence and inversion of the Bohai Basin, northern China. Marine and Petroleum Geology. 1997.
[2] Curiale J.A, Bromley B.W. Migration induced compositional changes in oils and condensates of a single field. Organic Geochemistry. 1996.
[3] Lisk W, Eadington P.J. Quantitative evaluation of the oil-leg potential in the Oliver gas field, Timor Sea, Australia. American Association of Petroleum Geologists Bulletin. 2002.
[4] Peters K.E, Moldowan JM. The Biomarker Guide: Interpreting Molecular Fossils in Petroleum and Ancient Sediments. 1993.
[5] Peters K.E, Walters C.C, Moldowan J.M. The Biomarkers Guide. 2005.
[6] Sinninghe Damste J.S, Kenig F, Koopmans M.P, et al. Evidence for gammacerane as an indicator of water column stratification. Geochimica et Cosmochimica Acta. 1995.

Future Information Engineering and Manufacturing Science – Zheng (Ed)
© 2015 Taylor & Francis Group, London, ISBN 978-1-138-02644-5

A new method for testing the environment adaptability of ordnance

Chun Hui Liu, Yun Tao Hao, Yu Chen & Hong Yi Liu
Baicheng Ordnance Test Center of China, Baicheng City, China

ABSTRACT: In this paper, according to the electromagnetic sensitivity mechanism of ordnance, a new method based on nonlinear frequency mixture theory to test the environment adaption of ordnance is proposed. In order to find the electromagnetic vulnerability in ordnance under test, the electromagnetic environment simulation is built according to the ordnance characteristic, which is different with those method presented in literature.

Keywords: ordnance; test; electromagnetic environment; environment adaptation

1 INTRODUCTION

In recent year, more and more ordnance comprises complex electronic circuits or equipments which radiate and receive electromagnetic energy intended and inadvertently. As a result, the electromagnetic environment in battlefield is worse and worse. How to evaluate the ordnance performance in complex electromagnetic environment, or in other words, how to test the ordnance electromagnetic environment adaptation is a very important question. So many countries in world take the ordnance electromagnetic adaptation as a main research object, and many achievement is reported [1~3]. In this paper, a cheap and effective method is proposed according to the electromagnetic sensitivity mechanism of ordnance to test the environment adaption.

2 THE METHOD AND THEORY FOR TEST ORDNANCE ELECTROMAGNETIC ENVIRONMENT ADAPTABILITY

2.1 The method for test ordnance electromagnetic environment adaptability

There are three steps in the proposed method as follow:

1. Application of the nonlinear characteristics of ordnance, construct a new kind of electromagnetic environment needed by the test of the ordnance complex environment adaptability. How to construct this kind of complex electromagnetic environment will be introduced in subsequent section.
2. Then in the constructed electromagnetic environment, operate the ordnance and observe its performance.
3. Accordance with the performance data achieved in step 2), evaluate the complex environment adaptability of the ordnance under test.

2.2 The method of construction of the complex electromagnetic environment

There are six steps in the method of construction of the complex electromagnetic environment as fellow:

1. Illuminate the ordnance under test with several (suppose three in subsequent section) electromagnetic generators which can step radiating frequency.
2. According to test requirements, extract hundreds or thousands of frequency in the the test frequency band (the number of the frequency points should be $3n$, n is a integer), and number these frequency from 1 to $3n$.
3. The first electromagnetic generator only radiates frequencies numbered $3k-1$, the second electromagnetic generator only radiates frequencies numbered $3k-2$, and the third electromagnetic generator only radiates frequencies numbered $3k$. Where k is an integer from 1 to n.
4. In test, three generators sweep frequency according to their work frequency sequences respectively.
5. During frequency sweep, the dwell time on each frequency is not equal for every generator, in order to form n^3 groups of radiation frequencies.
6. The average electronic field strength on the point where the ordnance is located from each generator should be 200 V/m.

There are two schemes for complex electromagnetic environment generator, that is, radiation

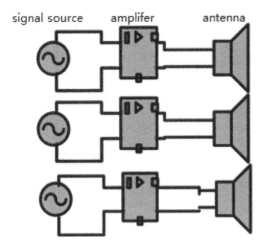

Figure 1.   The radiation generator.

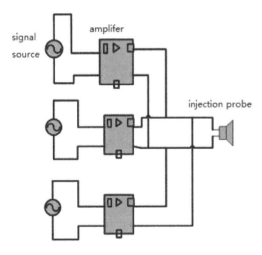

Figure 2.   The injection generator.

generator and injection generator, as shown in Figure 1 and 2.

3   THE PRINCIPLE OF ELECTROMAGNETIC ENVIRONMENT CONSTRUCTION SCHEME

3.1   *The principle of multi-frequency sensitivity*

According to the past engineering experience, there is a situation that when a few different frequency interference weapon system at the same time, sensitive phenomena occurs, but with these frequency interference weapon system respectively, does not take place any sensitive phenomena [4]. If several frequencies interfere the ordnance and sensitive phenomena occurs, in that each of these frequency makes different parts occur some kind of bad response respectively at the same time, and these bad responses overlap and result in performance degradation or loss. The principle can be introduced by Figure 3.

3.2   *The principle of frequency conversion due to the nonlinear characteristics*

Most of ordnance contain nonlinear circuits or equipment, therefore when multiple frequency go into the weapon system at the same time, a lot of new frequencies current will produce in its circuits or equipment [5~7]. Now assume that a ordnance is a fourth order nonlinear systems and three frequency go into the system at the same time. In the case that the number of the frequency combination is minimum, the number of sum frequency and difference frequency is as follows:

$$C_4^1 * C_4^1 * C_4^1 + C_3^2 * C_4^1 * C_4^1 + C_3^2 * C_4^1 * C_4^1 = 272$$

With 12 doubling frequency, a total of 284 high frequency current exist in the ordnance at the same time. If the 12 doubling frequency occur combination once again in the nonlinear system, there will be $C_{272}^1 * C_{271}^1 * C_{270}^1 * 4 = 79608960$ frequencies in the system, plus the 12 doubling frequency, altogether there are 79608972 high frequency circuit in the system. Therefore, if there is some electromagnetic vulnerability in ordnance, with so many group of frequency the electromagnetic vulnerability can be detected. Now we can get the conclusion that if we construct an electromagnetic environment based on nonlinear character, the electromagnetic environment adaptability of ordnance can be tested.

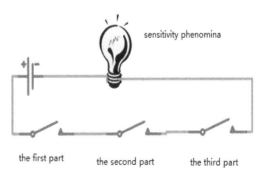

Figure 3.   The principle of multi-frequency sensitive.

4 THE SCHEME AND PRINCIPLE OF FREQUENCY SWEEP

In order to make the three generators can traverse n^3 frequency groups without repeat, the speeds of sweeping frequency of the three generators should meet some a condition. Obviously, if the first generator sweep frequency from first frequency to last frequency need time is T, the second generator sweep frequency time should be n^2T, and the third generator sweep frequency time should be n^3T. But when we use the sweep frequency scheme in wide frequency band, for example, in 10 K – 40 GHz, the trouble of the sweep frequency time would be several years is arise. It is too long to tolerate in test, so we should reform the scheme to make it practical. So we should select frequency points before test to reduce the value of n. According to the electromagnetic compatibility report afforded by the ordnance manufacturer, and the ordnance antenna character, shape dimension, cable length and so on, decide the sweep frequency band and points. Then, get rid of these points in the antenna working frequency band and we achieve the frequency table for sweep frequency in test. Number these frequency points and separate these frequencies in according to their number into three groups to afford three generators. The three generators sweep frequency according to their own frequency group respectively to construct complex electromagnetic environment in nonlinear ordnance. In all frequency, we set the electronic field strength to be 200 V/m submitting to GJB1389 A-2005 (equal to mil-std-464). Generally, the sweep frequency is less than 200, so n is about 70. If the frequency duration T is 1 s, the ordnance complex electromagnetic environment test can be finished within fifteen working days.

5 CONCLUSION

In this paper, according to the electromagnetic sensitivity mechanism of ordnance, a new method based on nonlinear frequency mixture theory to test the environment adaption of ordnance is proposed. With the new method, we have performed many ordnance tests effectively and cheaply. So the method proposed in this paper can be used for ordnance test and find the electromagnetic vulnerability.

REFERENCES

[1] Liu, Shanghe, and Guozhi SUN. "Analysis of the Concept and Effects of Complex Electromagnetic Environment." *Journal of the Academy of Equipment Command & Technology* 1 (2008): 000.
[2] Shang-he, L.I.U. "Effect of Electromagnetic Environment to Weaponry and Its Trend of Development." *Journal of Institute of Command and Technology* 1 (2005): 000.
[3] Rotkiewicz, Wilhelm. "Electromagnetic compatibility in radio engineering." *NASA STI/Recon Technical Report A* 84 (1982): 15397.
[4] Li, Nan, and Xue-fei Zhang. "Constitution Analysis of Complicated Electromagnetic Environment in Battlefield." *Equipment Environmental Engineering* 1 (2008): 003.
[5] Xu, Jinhua, et al. "A meta-synthesis of evaluating battlefield electromagnetic environment effects." *System Simulation and Scientific Computing, 2008. ICSC 2008. Asia Simulation Conference-7th International Conference on.* IEEE, 2008.
[6] Armstrong, Keith. "Validation, verification and immunity testing techniques for EMC for functional safety." *Electromagnetic Compatibility, 2007. EMC 2007. IEEE International Symposium on.* IEEE, 2007.
[7] Carter, Nigel J. "The past, present and future challenges of aircraft EMC." *Electromagnetic Compatibility, 2003. EMC'03. 2003 IEEE International Symposium on.* Vol. 1. IEEE, 2003.

Future Information Engineering and Manufacturing Science – Zheng (Ed)
© 2015 Taylor & Francis Group, London, ISBN 978-1-138-02644-5

Building Chinese polarity lexicons with co-occurrence using non-negative matrix factorization

C.M. Yang, H. Zhang & T.X. He
School of Computer Science and Technology, Southwest University of Science and Technology, Mianyang, Sichuan, P.R. China

ABSTRACT: This paper looks inside sentiment lexicons acquisition, one of the most significant tasks of Sentiment Analysis. Aimed at the problems of the Chinese polarity lexicons only reflects the language knowledge, lack of pragmatic knowledge, we present a novel approach to building Chinese polarity lexicons with co-occurrence from real data by a semi-supervised learning algorithm. Firstly, a relation matrix was constructed between the emotion word and the evaluation object from corpora by PMI (Point-wise Mutual Information). Secondly, use NMF (Non-negative Matrix Factorization) down to its co-occurrence matrix between emotion words, and new matrix of emotion words and evaluation object. Finally, the two relation matrix combined with the feature of synonymy and morpheme, and Label Propagation (LP) algorithm was used to run the relation map to distinguish the polarity of the emotion words. Experimental results show that the proposed method improves the accuracy and recall.

Keywords: polarity lexicons; pragmatic knowledge; co-occurrence; Label Propagation

1 INTRODUCTION

Emotion lexicons is the most crucial resource for most sentiment analysis algorithms. With high-quality emotion lexicons, systems using simple methods can achieve competitive performance (Xu et al. 2010; Taboada et al. 2011; Feldman 2013). There are three options for acquiring the sentiment lexicon: manual approaches in which people code the lexicon by hand, dictionary-based approaches in which a set of seed words is expanded by utilizing resources like WordNet, and corpus-based approaches in which a set of seed words is expanded by using a large corpus of documents from a single domain. However, to manually build emotion lexicon is time-consuming. Many research works focus on automatic construction of lexicons.

Emotion lexicons can be constructed automatically in a semi-supervised or unsupervised way. For example, more emotion words and polarity can be obtained by extending the semantic relationships (synonym, antonym) of a set of seed words utilize resources like WordNet or HowNet (Kim & Hovy 2006); Commendable and derogatory sentences can be identified from the massive Internet corpus by establishing language rules, and polarity can be estimated by the probability of words (Kaji & Kitsuregawa 2007); Polarity of words can be calculated by graph-based algorithm which ranks words according to a few seed emotion words (Xu et al. 2010).

The above methods reveal the application rules of languages from different perspectives, i.e. semantic relationship reveals the universal rules of human languages, representing priori knowledge; statistical relationship reveals the application habits of languages in specific fields, representing evidences. However, the analysis results might contradict with the actual situations because some words will have different meanings in specific fields or the context due to the ambiguity of words and non-standard Internet language. A lexicons that is constructed based on massive corpus statistics will reflect customs of language expressions, but there are many undesired results within it and it cannot cover all related emotion words.

In this paper, we adopted a combination of statistical and dictionary method to automatically build Chinese polarity lexicons with co-occurrence relation. First, the relation matrix between emotion words and evaluated objects is constructed based on the massive Internet corpus by PMI. Then, it will be decomposed by NMF into co-occurrence relationship matrix between emotion words, as well as the relationship matrix between emotion words and evaluated objects. Finally, the emotion words will be classified into positive or negative based on the two matrixes and the already constructed emotion lexicons.

The remainder of this paper is organized as follows. In Section 2, related works are introduced.

In Section 3, we describe a statistics-based method to acquiring co-occurrence relationship from multiple resources. We apply the method to some real data for checking its performance; the experimental result is shown in Section 4. The conclusions and some possible future work are presented in Section 5.

2 RELATED WORK

Open Mind Common Sense (Havasi et al. 2010), a common sense base, describes the objective world by the relationships between concept collections. It can be employed to conduct general emotional reasoning for events, actions and objects in the objective world. SentiWordNet (Baccianella et al. 2010) has been modified based on WordNet so that it assigns each word with specific polarity, namely, negative, positive and objective. It is constructed in a semi-supervised way.

The above mentioned emotion knowledge ontology mostly appears in the form of merely emotion lexicons. Enormous accurate linguistic knowledge has been summarized manually. However, pragmatic knowledge cannot be reflected effectively in the condition of dynamic Internet languages. At present, someone has put forward the approach of machine learning to learn knowledge from the Internet, among which the famous ones include FreeBase (Bollacker et al. 2008) of Google, Probase (Wu et al. 2012) of Microsoft. All their researches have the core idea of learning knowledge ontology that reflects concepts and the relationships, under certain regulations by means of extracting massive texts from the Internet in a semi-supervised way. Such ontology not only represents the grammar of language itself, but also adapts to the pragmatic habit of Internet language.

3 METHOD DESCRIPTION

Emotion polarity of text is determined all together by emotion words, evaluated objects and the context. In actual cases, one or several emotion words and evaluated objects both emerge in one sentence or paragraph. From statistical perspective, it is the co-occurrence relationship between words, namely, the relationship between distribution of emotion words and evaluated objects. In order words, to obtain the distribution of words, it is firstly to extract relationship between emotion words and evaluated objects from the text, and then, the correlation between emotion words, as well as between emotion words and evaluated objects is factorized.

3.1 Emotion words—evaluated objects relationship matrix build

The relationship between emotion words and evaluated objects is co-occurrence relation. It can be measured by PMI which is a method used in information theory and statistics for measuring correlation of two random variables (Xiaoming & Yan 2013), as indicated as

$$pmi(w_1, w_2) = \log_2 \frac{p(w_1, w_2)}{p(w_1)p(w_2)} \tag{1}$$

where w_1, w_2 are words in the linguistic data; $p(w_1)$ and $p(w_2)$ are occurrence ratios of words in the corpus. $p(w_1, w_2)$ are the co-occurrence ratios of words w_1, w_2. It can be calculated roughly by the times of co-occurrence of two words within certain distance.

According to the definition in (1), the correlation between emotion word s and evaluated object t in the text is calculated by (2)

$$Rel(s,t) = \log_2 \frac{Count(s,t)}{Count(s) \times Count(t)} \tag{2}$$

where $C(s,t)$ shows the times of co-occurrence of emotion word s and object t in certain sentence; $C(s)$ and $C(t)$ respectively shows the times of occurrence for emotion word s and object t in the corpus.

3.2 Relationship matrix of between emotion words build

Co-occurrence of two emotion words does not show that they have parallel relation, because they might be normal collocations. However, if two emotion words are closely related to an object, they are replaceable for each other with the same meaning or even the same emotion. Therefore, from the semantic perspective it is more reasonable to obtain the statistical relationship between emotion words via the co-occurrence relationship of emotion words and objects.

This paper constructs the co-occurrence matrix of emotion words through NMF. Matrix V can be decomposed into matrix W and H, as shown in (3):

$$nmf(V) = W \times H \tag{3}$$

For example, V is an $m*n$ dimensions relation matrix between random variable S and T. After decomposition, matrix W and H are respectively $m*k$ and $k*n$ dimensions matrix. W can be regarded as the distribution of random variable S

on k hidden feature sets. H^T can be regarded as the distribution of random variable T on the hidden feature set.

Suppose the relation matrix of emotion words and objects obtained is ST, decompose ST into matrix SF and FT by NMF, as shown in (4)

$$nmf(\mathbf{ST}) = \mathbf{SF} \times \mathbf{FT} \qquad (4)$$

where SF can be regarded as the distribution of emotion words on hidden feature set F. Thus, co-occurrence matrix SS of emotion words can be realized by formula (5)

$$\mathbf{SS} = \mathbf{SF} \times \mathbf{SF}^T \qquad (5)$$

In the case of large amount of data, matrix ST is enormous. Therefore, the decomposed matrix SF and FT are stored. As per the definition of NMF, the approximate matrix of ST is obtained from (6)

$$\mathbf{ST}' = \mathbf{SF} \times \mathbf{FT} \qquad (6)$$

Based on (4), (5) and (6), we can get the co-occurrence matrix SS of emotion words as well as the new relation matrix ST' of emotion words and evaluated objects.

4 EXPERIMENTS

In order to verify the influence that co-occurrence relation between emotion words and evaluated objects has on the emotion classification in actual corpus, the experiment firstly adopts the method stated in chapter two by construction of relation matrix between emotion words, as well as between emotion words and evaluated objects based on the enormous Internet data. Combining this relation matrix with semantic and morpheme features, the lexicons is classified into positive and negative categories by Label Propagation algorithm (Lu et al. 2010) based on the three datasets.

4.1 Corpus

During the experiment of construction of co-occurrence matrix of emotion words, we extracted large amount of comments data from product reviews on the Internet. Among the data, it is primarily the corpus organized by Songbo Tan (Du & Tan 2010). This corpus contains 12,000 pieces of comments in three different domains as shown in Table 1.

First, the tokenizers are used to carry out words segmentation and part-of-speech tagging for the

Table 1. Statistical description of the corpus.

Corpus	Positive	Negative	Total
Ctrip_htl	2000	2000	4000
Jingdong_NB	2000	2000	4000
Dangdang_Book	2000	2000	4000

Table 2. Statistical description of the polarity classification corpus.

Corpus	Positive	Negative	Total
Zhu	5045	3498	8543
HowNet	4566	4370	8936

corpus. Adjectives are extracted as emotion words and nouns or nominal phrases as evaluated objects. Before matrix calculation, candidate objects that have less than two emotion words will be filtered. Then, the emotion words network is constructed according to the method stated in chapter 2. NMF algorithm employs Chih-Jen Lin (Lin 2007) in version, the hidden feature dimension of which is 100. Relation matrix of emotion words and evaluated objects with 1652*6182 dimensions is obtained, and the co-occurrence relation matrix of emotion words with 1652*1652 obtained, too. Finally, normalization of row maximum value for the two matrixes is conducted.

In the experiment of classification of positive and negative words, three datasets, namely, Zhu's Chinese polarity lexicons, HowNet emotion words are used. In these lexicons, some words have both positive and negative meanings. Table 2 shows the detailed information of the two datasets. In the experiment, the datasets are classified into five parts randomly to conduct five-fold cross validation, four as training sets and one as testing set.

4.2 Evaluation metrics

This study adopts typical *Precision*, *Recall* and *F1* of the information retrieval to evaluate experimental results. Precision rate refers to the percentage of correct classification of words in category A. *F1* measures precision rate and recall rate as a whole. It can be calculated by (7)

$$F1 = \frac{2 \times Precision \times Recall}{Precision + Recall} \qquad (7)$$

As the positive and negative in the experiment are not balanced, precision, recall and *F1* are used for each category respectively. In addition, micro-average

Table 3. Comparisons of Zhu's Chinese polarity lexicons.

| Method | Positive | | | Negative | | | All | |
	Pre	Rec	F1	Pre	Rec	F1	Mac F1	Mic F1
A	92.4	95.8	94.0	94.1	88.0	90.9	92.6	92.8
B	**92.7**	**96.1**	**94.4**	**94.5**	**88.7**	**91.5**	**93.0**	**93.3**

Table 4. Comparisons of HowNet.

| Method | Positive | | | Negative | | | All | |
	Pre	Rec	F1	Pre	Rec	F1	Mac F1	Mic F1
A	91.3	88.6	90.0	88.8	90.3	89.5	89.8	89.7
B	**91.9**	**89.6**	**90.7**	**89.7**	**90.7**	**90.2**	**90.5**	**90.5**

F1 and macro-average *F1* are used to evaluate the overall result. Micro-average *F1* means that the total precision rate and recall rate are calculated directly by respective formula, and then calculate *F1*, while macro-average *F1* means that precision rate and recall rate of each category are calculated first, then, the results are averaged to get the total precision rate and recall rate, finally, calculate *F1*.

4.3 *Experimental results*

In order to verify the validity of co-occurrence relation method of emotion words, we carry out the experiment of classification of positive and negative words, which compare it with the method that only considers synonymous and morpheme (Method A). In our experiment, synonymous, morpheme and co-occurrence are all incorporated into the similarity model of words, and linear combination is conducted on the three relations in the weight of 2:2:1 (Method B). The results are shown in Tables 3 and 4.

From the above experimental results, we can see that the method that considers co-occurrence relation will achieve better results in the commendatory and derogatory lexicon and HowNet lexicon. It verifies that the proposed words co-occurrence relation assumption can reflect the modification relationship between words in the context and increase the precision rate of emotion words classification.

5 CONCLUSION

Emotion lexicons is an important corpus for emotion analysis. Regarding the problem of existing emotion dictionaries that linguistic knowledge is considered only while Internet language and pragmatic knowledge of context are hardly reflected, this paper proposes to extract co-occurrence relationship between words from the Internet corpus and incorporate them into existing dictionaries to construct a Chinese polarity lexicons with co-occurrence. First, the relation matrix of emotion words and evaluated objects is constructed based on the large amount of linguistic data on the Internet with the help of PMI. Then, the matrix is decomposed into co-occurrence relation matrix of emotion words as well as the relation matrix between emotion words and evaluated objects through the non-negative matrix factorization. Finally, combining the two matrixes with the already constructed emotion dictionary, the emotion words are classified into positive and negative category via LP algorithm. Based on the experimental results of Zhu's Chinese polarity lexicons, HowNet, this method has effectively improved the precision rate and recall rate of the dictionary that considers only morpheme and semantic.

ACKNOWLEDGEMENT

This work is supported by the Scientific Research Funds in education department of Sichuan Province (12ZB326), the Open Project Program of Engineering Lab of Network Fusion of Miyang (12ZXWK04), and the Research Fund of SWUST (12zx7116).

REFERENCES

[1] Baccianella, S., Esuli, A. & Sebastiani, F., 2010. SentiWordNet 3.0: An Enhanced Lexical Resource for Sentiment Analysis and Opinion Mining. LREC, 10, pp. 2200–2204.
[2] Bollacker, K. et al., 2008. Freebase: a collaboratively created graph database for structuring human knowledge. In Proceedings of the 2008 ACM SIGMOD international conference on Management of Data. pp. 1247–1250.
[3] Du, W. & Tan, S., 2010. Optimizing modularity to identify semantic orientation of Chinese words. Expert Systems With Applications, 37(7), pp. 5094–5100.
[4] Feldman, R., 2013. Techniques and applications for sentiment analysis. Communications of the ACM, 56(4), pp. 82–89.
[5] Havasi, C. et al., 2010. Open Mind Common Sense: Crowd-sourcing for Common Sense. In In Collaboratively-Built Knowledge Sources and AI. p. 51.
[6] Kaji, N. & Kitsuregawa, M., 2007. Building Lexicon for Sentiment Analysis from Massive Collection of HTML Documents. In EMNLP-CoNLL. pp. 1075–1083.

[7] Kim, S.-M. & Hovy, E., 2006. Identifying and analyzing judgment opinions. In Proceedings of the main conference on Human Language Technology Conference of the North American Chapter of the Association of Computational Linguistics. pp. 200–207.

[8] Lin, C.-J., 2007. Projected gradient methods for nonnegative matrix factorization. Neural computation, 19(10), pp. 2756–2779.

[9] Lu, B. et al., 2010. Learning Chinese polarity lexicons by integration of graph models and morphological features. In Proceedings of the 6th Asia Information Retrieval Societies Conference. pp. 466–477.

[10] Taboada, M. et al., 2011. Lexicon-Based Methods for Sentiment Analysis. Computational Linguistics, 37(2), pp. 267–307.

[11] Wu, W. et al., 2012. Probase: A probabilistic taxonomy for text understanding. In In Proceedings of the 2012 ACM SIGMOD International Conference on Management of Data. pp. 481–492.

[12] Xiaoming, D. & Yan, T., 2013. Improved mutual information method for text feature selection. In Computer Science & Education (ICCSE), 2013 8th International Conference on. pp. 163–166.

[13] Xu, G., Meng, X. & Wang, H., 2010. Build Chinese emotion lexicons using a graph-based algorithm and multiple resources. In Proceedings of the 23rd International Conference on Computational Linguistics. pp. 1209–1217.

Future Information Engineering and Manufacturing Science – Zheng (Ed)
© *2015 Taylor & Francis Group, London, ISBN 978-1-138-02644-5*

Application research of stochastic parallel gradient descent algorithm in adaptive optics system

Lei Zhang

Opto-Electronic Engineering School, Changchun University of Science and Technology, Changchun, P.R. China

ABSTRACT: With the expansion of the adaptive optics applications, such as optical communication in the atmosphere is no beacon of light scenarios, and there is a strong flicker of long-range atmospheric transport scenarios, based on the conventional adaptive optics wavefront sensor control technology the application is limited. In recent years, the sensor-based adaptive front optimize system performance without the direct wave optical control technology, due to the simplicity and complexity of flashes and other environmental conditions on the adaptability of its hardware implementation, more and more research in the field of adaptive optics attention. In the early development of adaptive optics to hill climbing and multi-frequency vibration technology as the representative of the search optimization algorithm was introduced adaptive optics control methods in the past. However, due to slow convergence and high bandwidth requirements of the system, the hardware complexity, low signal to noise ratio are difficult practical reasons. SPGD algorithm control parameters is simple, good convergence efficiency and stability, we have become the preferred method of optimization.

Keywords: adaptive optics; SPGD algorithm; wavefront correction

1 INTRODUCTION

The traditional use of adaptive optics systems is the wave front sensor to detect the wave front distortion of information, and then control the wavefront aberration corrector to compensate. With the expansion of the technology development and application in the field of conventional adaptive optics systems due to the difficulty of wave-front detection exhibit some limitations. With the theory of stochastic optimization algorithm development, before the use of optimization algorithms to optimize system performance indicators directly without wave adaptive optics control technology, more and more attention in the field of adaptive optics. At present, all kinds of new optimization algorithm introduced control adaptive optics system to make adaptive optics system optimized prior to imaging targets without wave sensors become an important class-based research. The main application of the method are Stochastic Parallel Gradient Descent Algorithm, Simulated Annealing, Genetic Algorithm, Algorithm of Pattern Extraction, Simplex Algorithm, Functional Approximation Method, SPGD algorithm applied research in the absence of adaptive optics wavefront sensor in the control system achieved rapid development, its correction effect is good, fast convergence characteristics more suitable for application to the strong turbulence, adaptive optics system multiunit.

2 MATHEMATICAL PRINCIPLES

Depending on the type of adaptive optics system, first determine the appropriate performance evaluation function J, and the relationship between the performance evaluation function control vector $u(r)$ between the unknown, $\nabla J[u(r)]$ by measuring only data to determine. SPGD algorithm derived from the theory of stochastic approximation parallel perturbation stochastic approximation (SPSA) algorithms and artificial neural network technology in the random error descent algorithm, using parallel perturbation gradient estimation techniques, and therefore belongs to no model optimization problems.

While applying a random disturbance to the system $\{\delta u_j\}(j = 1,2, ..., N)$, the change amount corresponding to the evaluation of system performance as a function of δJ

$$\delta J = J(u_1 + \delta u_1, ..., u_l + \delta u_l, ..., u_N + \delta u_N) \\ -J(u_1, ..., u_l, ..., u_N) \qquad (1)$$

For δJ Taylor series expansion, we have

$$\delta J = \sum_{j=1}^{N} \frac{\partial J}{\partial u_j} \delta u_j + \frac{1}{2} \sum_{j,i}^{N} \frac{\partial^2 J}{\partial u_j \partial u_i} \delta u_j \delta u_i + ... \qquad (2)$$

Now consider the variable u_l, known disturbances $\{\delta u_j\}$, δJ be measured, estimates the gradient

$$\frac{\delta J}{\delta u_l} = \frac{\partial J}{\partial u_l} + \varphi_l \tag{3}$$

where in the estimation error, its expression is

$$\varphi_l = \sum_{j \neq l}^{N} \frac{\partial J}{\partial u_j} \frac{\delta u_j}{\delta u_l} + \frac{1}{2} \sum_{j,i}^{N} \frac{\partial^2 J}{\partial u_j \partial u_i} \frac{\delta u_j \delta u_i}{\delta u_l} + \dots \tag{4}$$

Assume $\{\delta u_j\}$ is a random variable, equation $\{P(\delta u_j)\}$ on both sides to take the mathematical expectation,

$$\left\langle \frac{\delta J}{\delta u_l} \right\rangle = \frac{\partial J}{\partial u_l} + \sum_{j \neq l}^{N} \frac{\partial J}{\partial u_j} \left\langle \frac{\delta u_j}{\delta u_l} \right\rangle$$
$$+ \frac{1}{2} \sum_{j,i}^{N} \frac{\partial^2 J}{\partial u_j \partial u_i} \left\langle \frac{\delta u_j \delta u_i}{\delta u_l} \right\rangle + \dots \tag{5}$$

In SPGD algorithm, the disturbances $\{\delta u_j\}$ are statistically independent random variables, the average is zero, the variance is constant, the probability density function $\{P(\delta u_j)\}$ symmetrical about 0. When N is large, can be obtained

$$\left\langle \frac{\delta u_j}{\delta u_l} \right\rangle = 0, \left\langle \frac{\delta u_j \delta u_i}{\delta u_l} \right\rangle = 0, \forall j, i, l \tag{6}$$

Therefore, the statistical average sense, the random values can be $\{\delta J/\delta u_l\}$ as a estimate of the true gradient component of $\{J_l'\}$

$$J_l' = \frac{\partial J}{\partial u_l} \approx \frac{\delta J}{\delta u_l} \tag{7}$$

And traditional steepest descent method is similar to when the estimated gradient component, you can control the variables u_l iterative update

$$u_l^{(n+1)} = u_l^{(n)} - \gamma \frac{\delta J^{(n)}}{\delta u_l^{(n)}} \quad l = 1, 2, \dots N \quad n = 0, 1, 2, \dots \tag{8}$$

The basic formula is the iterative formula SPGD algorithm, also known as unilateral SPGD algorithm. In practical applications, can be used to increase the gradient of a bilateral disturbance estimation accuracy, bilateral SPGD known algorithm, which method

$$J_{\pm}^{(n)} = J\left(u_1^{(n)} \pm \frac{\delta u_1}{2}, \dots, u_j^{(n)} \pm \frac{\delta u_j}{2}, \dots, u_N^{(n)} \pm \frac{\delta u_N}{2}\right)$$
$$\delta J^{(n)} = J_+^{(n)} - J_-^{(n)} \tag{9}$$

3 BASIC PROCESSES

In the adaptive optics technology, the stochastic parallel gradient descent algorithm is a variation of the main and control voltage variation using system performance evaluation of the control voltage gradient estimate the amount and direction of the gradient search iterations, so that system performance evaluation indicators optimized.

Operation of the process: first, the respective variables are initialized, the disturbance by the disturbance voltage generator generates a random vector $\vec{\delta u}^k = (\delta u_1^k, \delta u_2^k, \dots, \delta u_{37}^k)$ Where in each of δu_i^k ($i = 1,2, \dots, 37$) are independent and the same $\delta u_i^k = \pm 0.5$ Bernoulli distribution $P\{\delta u_i^k = 0.5\} = P\{\delta u_i^k = -0.5\} = 0.5$. The Voltage $\vec{u}^{k-1} + \vec{\delta u}^k$ and $\vec{u}^{k-1} - \vec{\delta u}^k$ are supplied to the drive, $J_+^{(k)}$ and $J_-^{(k)}$ obtained by $\vec{u}^{(k)} = \vec{u}^{(k-1)} + \gamma \vec{\delta u}^{(k)} [J_+^{(k)} - J_-^{(k)}]$ to be calculated $\vec{u}^{(k)}$, so that the iteration loop, until the performance index value reaches the requirement.

SPGD iterative algorithms often iterates hundreds of times in order to fully converge, and get a lot of adaptive optics system performance evaluation should be on the deformable mirror when driving and the use of CCD to obtain spot image, so the convergence rate is largely being dependent on the speed of the image on the CCD and the response time of deformable mirror. Thus, for most non-model-based optimization algorithm adaptive optics system, CCD sampling frequency and speed of response deformable mirror to a large extent determines the bandwidth of the system.

If the number of each iteration of the algorithm can streamline time-consuming, you can make the convergence rate to increase, reducing each iteration algorithm to obtain the number of system performance evaluation function is one of them one way, thus reducing the CCD sampling and a deformable mirror drive algorithm number. Therefore, according to the usual experience, sum the minimum number of convergence of the algorithm, which was as an algorithm iteration threshold, when the number of iterations is less than the threshold value, no calculation of the current system performance evaluation, only the number of iterations is greater than the threshold value when the fishes current system performance evaluation function to judge, to see whether it conforms to the requirements of the system, in line with the iteration ends do not meet to continue the cycle. Thus, the number of iterations is less than the threshold value, for each iteration time, only two access performance evaluation function value, a certain amount of saving computation time.

SPGD correlation phase change random perturbation algorithm and a voltage caused by the phase distortion of the actual wavefront has a great influence on the convergence rate, the greater the correlation between the convergence faster. In the actual operation of the atmosphere communication, especially in strong turbulence channels, turbulence complex, difficult to obtain the actual phase of the wavefront aberration distribution. Wavefront sensor as a wavefront sensor element, in the case of strong turbulence cannot detect the real wavefront information, but the information and the actual wavefront information to detect if there is some correlation, you can use this information to optimize SPGD voltage perturbation algorithm to improve convergence speed.

4 SIMULATION ANALYSIS

Wavefront sensor in strong turbulence in the case of the channel cannot recover the correct wavefront, mainly because of fluctuations in the amplitude exceeds the dynamic range of light Hartmann wavefront sensor, some of the sub-light spot formed by the lens does not fall within its corresponding CCD within the range falls within the area adjacent to, during reconstruction of the wavefront, because of the slope of the wave front of the obtained information is not complete, the recovery of the actual wavefront and the wave front error is large, thereby wavefront correction cannot be achieved. However, in this case, if the information obtained by the wavefront sensor and the actual

wavefront aberration is still some correlation, then the correlation of these information processing can be applied to improve the algorithm SPGD.

To facilitate the calculation and processing of the simulation to simulate the Hartmann sensor 10 mm square aperture side, the sub-lens array is a square array of 10×10, wherein each side of the square sub-aperture 1 mm. Constructed from the phase distribution of the wave front aberration shown in Figure 1, the 500×500 dots, whose z-axis coordinate unit wavelength λ, x-axis and y-axis coordinate unit mm, a few additional units of the wavefront phase map.

Original two dimensional matrix, through the front lens sub-Hartmann sensor array, it is divided into 10×10 blocks Thus, in Figure 1, the required wavefront matrix into blocks, divided into 10×10 50×50 matrix bands, wherein each sub-matrix

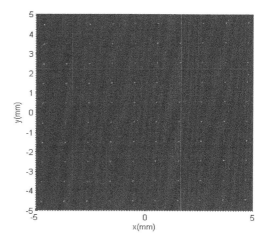

Figure 2.   Hartman dimensional graphics imaging lens.

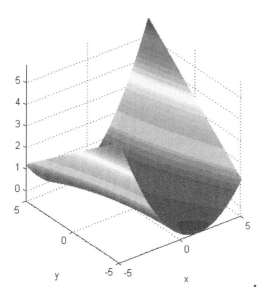

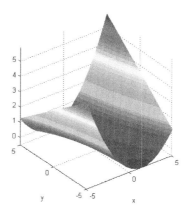

Figure 1.   Zernike polynomials constructed before the phase out of the original wave.

Figure 3.   The next phase of the reconstructed wavefront weak turbulence case.

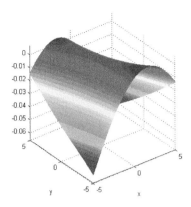

Figure 4. Under conditions of weak turbulence wavefront reconstruction errors.

corresponds to a sub-lens Hartman. Each sub-matrix of the FFT transform matrix is obtained after the image distortion of the spot matrix Hartmann wavefront through the lens array after the light patterns, i.e. the two-dimensional CCD image obtained, as shown in Figure 2. 100 sampling points resulting wavefront slope vector G has 200 components, after getting G, according to Zernike mode method Zernike polynomial coefficients calculated, and then reconstruct the wavefront.

In the case of weak turbulence, before each can be considered as a sub-lenses corresponding to the light spot falls on the CCD area, the reconstructed wave shown in Figure 3. Figure 4 is a reconstruction of the wave front error under weak turbulence.

5 CONCLUSIONS

Assuming the system uses high-speed DSP and FPGA to implement, using optical power system performance evaluation, then the circuit delay can be ignored, the performance indicators are not needed to obtain CCD imaging. 3000 Hz frequency response of the deformable mirror as an example to analyze the use of pre-and post-adoption optimize overall system optimization algorithm computing speed.

Before optimization, the deformable mirror response three times, for the first iteration of the algorithm, 150 iterations to meet the requirements (SPGD algorithm does not require complete convergence, assuming extremes can reach 80%), the algorithm converges once takes about 150 ms, far far less than ideal speed.

After optimization, the deformable mirror in response to two times for one algorithm iterates 60 times to meet the requirements, the completion of a wavefront correction algorithm takes about 40 ms, 10 ms still unable to meet the requirements. In this regard, on the one hand, hopes to improve the response speed of the hardware, doubling the speed of response, wavefront correction algorithm speed is doubled; hand to further improve the algorithm research.

REFERENCES

[1] Debbie Kedar, Shlomi Arnon. Optical Wireless Communication Through Fog in the Presence of Pointing Errors. Applied Optics, Vol. 42, Issue 24, pp. 4946–4954 (2003).
[2] Debbie Kedar, Shlomi Arnon. Evaluation of coherence interference in optical wireless communication through multiscattering channels. Applied Optics, Vol. 45, Issue 14, pp. 3263–3269 (2006).
[3] Marius Aharonovich, Shlomi Arnon. Performance improvement of optical wireless communication through fog with a decision feedback equalizer. JOSA A, Vol. 22, Issue 8, pp. 1646–1654 (2005).
[4] Heba Yuksel, Stuart Milner, and Christopher Davis. Aperture averaging for optimizing receiver design and system performance on free-space optical communication links. Journal of Optical Networking, Vol. 4, Issue 8, pp. 462–475 (2005).
[5] Kumar, A., Jain, V.K. Antenna Aperture Averaging and Power Budgeting for Uplink and Downlink Optical Satellite Communication. Signal Processing, Communications and Networking, 2008.:126–131.
[6] Jaime A. Anguita, Mark A. Neifeld, and Bane V. Vasic. Spatial correlation and irradiance statistics in a multiple-beam terrestrial free-space optical communication link. Applied Optics, Vol. 46, Issue 26, pp. 6561–6571 (2007).
[7] Jaime A. Anguita, Mark A. Neifeld, Björn Hildner, and Bane Vasic. Rateless Coding on Experimental Temporally Correlated FSO Channels. Journal of Lightwave Technology, Vol. 28, Issue 7, pp. 990–1002 (2010).

Future Information Engineering and Manufacturing Science – Zheng (Ed)
© 2015 Taylor & Francis Group, London, ISBN 978-1-138-02644-5

The FDTD method application in the problem of electromagnetic wave propagation in cylindrical coordinates

M. Levin & P. Zatsepin
Altai State University, Barnaul, Russia

ABSTRACT: The paper provides basic information about the FDTD method. The benefits of using the method in cylindrical geometry were described. The corresponding formulas for iterative calculations were determined. The article also contains an algorithm for the implementation of further numerical experiment.

Keywords: FDTD; cylindrical

1 INTRODUCTION

Some of the most prevalent problems in electrodynamics are propagation, reflection and attenuation of electromagnetic waves. The reason of it is the fact that sometimes analysis of an object just cannot be done directly. In this case, we can make a conclusion about the object's nature using parameters of electromagnetic wave reflected from it. However, this problem cannot always be solved analytically and, therefore, numerical methods must be used. The FDTD method (Finite Difference Time Domain) is a numerical analysis technique used for modeling computational electrodynamics. Its basis is the discretization of well-known Maxwell's equations written in differential form.

$$\nabla \times \bar{E} = -\mu \frac{\partial \bar{H}}{\partial t} \tag{1a}$$

$$\nabla \times \bar{H} = \varepsilon \frac{\partial \bar{E}}{\partial t} + \bar{J} \tag{1b}$$

According to these equations electric and magnetic fields are generated and altered by each other and by charges and currents. Difference equations allow us to determine the values of intensities on the current time step on the basis of known values of the fields obtained on the previous. The grids of fields are shifted from each other by the discretization time step, and for each of the spatial variables. The FDTD is attractive due to its generality and ease of implementation. It should be noted that the applicability of this method is independent of type of the source. The many other methods do not have this feature. Calculations carried out using the FDTD become much simpler by specifying used

geometric grid. The present article is focused on the cylindrical coordinates.

2 FORMULATION

This coordinate system is convenient to use considering objects symmetrical along one of the axes. Propagation of electromagnetic waves by an antenna may be taken as example. Using simple conversion

$$\begin{cases} x = \rho\sin\varphi \\ y = \rho\cos\varphi \\ z = z \end{cases} \tag{2}$$

We shall rewrite Maxwell's equations mentioned before

$$\frac{\partial E_z}{\rho \partial \varphi} - \frac{\partial E_\varphi}{\partial z} = -\mu_\rho \frac{\partial H_\rho}{\partial t} \tag{3a}$$

$$\frac{\partial E_p}{\partial z} - \frac{\partial E_z}{\partial \rho} = -\mu_\varphi \frac{\partial H_\varphi}{\partial t} \tag{3b}$$

$$\frac{\partial(\rho E_\varphi)}{\rho \partial \rho} - \frac{\partial E p}{\rho \partial \varphi} = -\mu_z \frac{\partial H_z}{\partial t} \tag{3c}$$

$$\frac{\partial H_z}{\rho \partial \varphi} - \frac{\partial H_\varphi}{\partial z} = \varepsilon_\rho \frac{\partial E_\rho}{\partial t} + J_\rho \tag{3d}$$

$$\frac{\partial H_p}{\partial z} - \frac{\partial H_z}{\partial \rho} = \varepsilon_\varphi \frac{\partial E_\varphi}{\partial t} + J_\varphi \tag{3e}$$

$$\frac{\partial(\rho H_\varphi)}{\rho \partial \rho} - \frac{\partial H_p}{\rho \partial \varphi} = \varepsilon_z \frac{\partial E_z}{\partial t} + J_z \tag{3f}$$

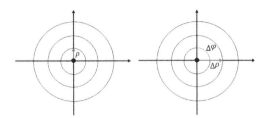

Figure 1.   Problem geometry.

Coming up it makes sense to move to a two-dimensional region taking the value of z constant and equal to zero.

We shall get

$$\frac{\Delta E_z}{\rho \Delta \varphi} = -\mu_\rho \frac{\Delta H_\rho}{\Delta t} \tag{4a}$$

$$-\frac{\Delta E_z}{\Delta \rho} = -\mu_\varphi \frac{\Delta H_\varphi}{\Delta t} \tag{4b}$$

$$\frac{\Delta (\rho E_\varphi)}{\rho \Delta \rho} - \frac{\Delta E_\rho}{\rho \Delta \varphi} = -\mu_z \frac{\Delta H_z}{\Delta t} \tag{4c}$$

$$\frac{\Delta H_z}{\rho \Delta \varphi} = \varepsilon_\rho \frac{\Delta E_\rho}{\Delta t} + J_\rho \tag{4d}$$

$$-\frac{\Delta H_z}{\Delta \rho} = \varepsilon_\varphi \frac{\Delta E_\varphi}{\Delta t} + J_\varphi \tag{4e}$$

$$\frac{\Delta (\rho H_\varphi)}{\rho \Delta \rho} - \frac{\Delta H_\rho}{\rho \Delta \varphi} = \varepsilon_z \frac{\Delta E_z}{\Delta t} + J_z. \tag{4f}$$

Now we present differentials as differences on the current and next step

$$\frac{E_z^i - E_z^{i+1}}{\rho \Delta \varphi} = -\mu_\rho \frac{H_\rho^i - H_\rho^{i+1}}{\Delta t} \tag{5a}$$

$$-\frac{E_z^i - E_z^{i+1}}{\Delta \rho} = -\mu_\varphi \frac{H_\varphi^i - H_\varphi^{i+1}}{\Delta t} \tag{5b}$$

$$\frac{E_\varphi^i - E_\varphi^{i+1}}{\Delta \rho} - \frac{E_\rho^i - E_\rho^{i+1}}{\rho \Delta \varphi} = -\mu_z \frac{H_z^i - H_z^{i+1}}{\Delta t} \tag{5c}$$

$$\frac{H_z^i - H_z^{i+1}}{\rho \Delta \varphi} = \varepsilon_\rho \frac{E_\rho^i - E_\rho^{i+1}}{\Delta t} + J_\rho \tag{5d}$$

$$-\frac{H_z^i - H_z^{i+1}}{\Delta \rho} = \varepsilon_\varphi \frac{E_\varphi^i - E_\varphi^{i+1}}{\Delta t} + J_\varphi \tag{5e}$$

$$\frac{H_\varphi^i - H_\varphi^{i+1}}{\Delta \rho} - \frac{H_p^i - H_p^{i+1}}{\rho \Delta \varphi} = \varepsilon_z \frac{E_z^i - E_z^{i+1}}{\Delta t} + J_z \tag{5f}$$

Basic formulas for iterative calculations are

$$H_\rho^{i+1} = H_\rho^i - \frac{J_z \rho \Delta \varphi \mu_\varphi \Delta t^2}{\varepsilon_z \mu_\rho \mu_\varphi - \Delta t^2 (\mu_\rho + \mu_\varphi)} \tag{6a}$$

$$E_\varphi^{i+1} = E_\varphi^i - \frac{J_\varphi \Delta t \Delta \rho s_\rho \rho^2 \Delta \varphi^2 + \Delta t^s (J_\varphi \Delta \rho + J_\rho \rho \Delta \varphi)}{(s_\rho \rho^2 \Delta \varphi^2 - s_\rho \Delta \rho) \Delta t^2 - \mu_z \Delta \rho s_\varphi} \tag{6b}$$

Further computation algorithm might look like this:

1. Definition of initial and boundary conditions.
2. Determination of φ and ρ coordinates step size and number of iterations.
3. While the number of remaining iterations is not equal to zero.
   a. Calculation of the basic formulas (6a-b).
   b. Finding the other variables by the formulas (5a-b).
4. End of the algorithm.

3   CONCLUSION

The resulting equations can be the basis for further numerical experiment. Their role, in addition to finding the values of fields for new iterations, is also to be a basis for making a report of electromagnetic wave propagation that can be very useful when constructing interactive graphics models. As for the specific practical application, the range of tasks that can be solved using this method, especially considering its specification for cylindrical coordinates, is very wide: from the simple transfer of information and to remote sensing of the Earth surface.

REFERENCE

[1] D.M. Sullivan. 2000. Electromagnetic simulation using the FDTD method. New York: The IEEE Inc.: 1.

Future Information Engineering and Manufacturing Science – Zheng (Ed)
© *2015 Taylor & Francis Group, London, ISBN 978-1-138-02644-5*

Finite-Difference Time Domain method for electromagnetic pulse scattering by dual layers structure in two-dimensional space

A.Y. Rykshin & P.M. Zatsepin
Altai State University, Barnaul, Russia

ABSTRACT: The Finite-Difference Time Domain (FDTD) method for the solution of electromagnetic pulse scattering by dual dielectric layers structure has been considered. An incident field was generated by infinite thread-like source of electric current located parallel to the boundaries of layers. The FDTD technique was applied to solve the Maxwell equations in the time domain by using the finite-difference analog. Electromagnetic fields characteristics in the region over the given structure were studied.

Keywords: FDTD; Pulse; Scattering; Two-Layer Structure

1 INTRODUCTION

Studying propagation and scattering of electromagnetic fields by local objects represents significant interest in problems of remote sensing, radar-location, geolocation [1,2]. But analytical or semi-analytical solution this kind of problems corresponds to some hardship [3]. While the FDTD method provides us the way simple enough to solve numerically partial differential equations [4].

In this paper, the FDTD technique is used to solve numerically Maxwell's time-dependent curl equations to obtain the value of the electromagnetic field density within a finite two-dimensional space that contains dielectric layered structure (figure 1).

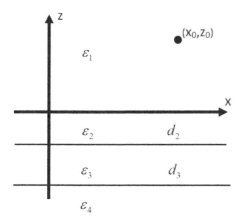

Figure 1. Geometry of the problem.

At the initial time infinite thread-like source of electric current is turned on, the source is located at point (x_0, z_0). The radiated electromagnetic wave propagates and scatters by dielectric layers of permittivity ε_2, ε_3 with thickness d_2, d_3 respectively. The permittivity of upper and lower half spaces is equal to 1.

2 FINITE-DIFFERENCE TIME DOMAIN METHOD IN TWO-DIMENSIONAL SPACE

Actually, the total space is discretized by a grid mesh, where the scattering structure is defined by properly assigning the permittivity over the grid points. Then the Maxwell's equations are discretized by using finite-difference approximations in both time and space. The region of calculations is limited, but it will be defined sufficiently large to have no influence from boundaries in the observation point (0,0).

The Maxwell's equations for our problem formulation can be written as follows:

$$\begin{cases} \dfrac{dE_y}{dz} = \mu_0 \dfrac{dH_x}{dt} \\[2mm] \dfrac{dE_y}{dx} = -\mu_0 \dfrac{dH_z}{dt} \\[2mm] \dfrac{dH_x}{dz} - \dfrac{dH_z}{dx} = \varepsilon_0 \varepsilon \dfrac{dE_y}{dt} + j_y^e \end{cases} \tag{1}$$

where j_y^e—electric current density with time dependence as

$$j(t) = -\frac{t-t_0}{\tau^2} e^{-\frac{(t-t_0)^2}{2\tau^2}} \qquad (2)$$

We can use equations (1) to construct the finite-difference analog of the Maxwell's curl equations, which involves only real calculations. To do this, space and time need to be discretized. We will use the following notation to indicate location of sampled in time and space electromagnetic fields:

$$E_y(x,z,t) = E_y(m\Delta x, n\Delta z, q\Delta t) = E_y^q[m,n]$$
$$H_{x,z}(x,z,t) = H_{x,z}(m\Delta x, n\Delta z, q\Delta t) = H_{x,z}^q[m,n] \quad (3)$$

where Δx and Δz are the spatial offsets between sample points and Δt is the temporal offset. The indices m, n correspond to the spatial steps by x-direction and z-direction respectively, while the superscript q corresponds to the temporal step. Using notations (3) we replace derivatives in equations (1) with finite differences:

$$\frac{E_y^q[m+1,n] - E_y^q[m,n]}{\Delta x}$$
$$= \left(H_z^{q+1/2}\left[m+\frac{1}{2},n\right] - H_z^{q-1/2}\left[m+\frac{1}{2},n\right] \right) \cdot \left(-\frac{\mu_0}{\Delta t} \right)$$
$$(4a)$$

$$\frac{E_y^q[m,n+1] - E_y^q[m,n]}{\Delta z}$$
$$= \left(H_x^{q+1/2}\left[m,n+\frac{1}{2}\right] - H_x^{q-1/2}\left[m,n+\frac{1}{2}\right] \right) \frac{\mu_0}{\Delta t} \quad (4b)$$

$$\frac{H_x^{q+1/2}\left[m,n+\frac{1}{2}\right] - H_x^{q+1/2}\left[m,n-\frac{1}{2}\right]}{\Delta z}$$
$$- \frac{H_z^{q+1/2}\left[m+\frac{1}{2},n\right] - H_z^{q+1/2}\left[m-\frac{1}{2},n\right]}{\Delta x}$$
$$= \varepsilon_0 \varepsilon \frac{E_y^{q+1}[m,n] - E_y^q[m,n]}{\Delta t} + j_y^{q+1/2}[m,n] \quad (4c)$$

Then we can obtain the update equations for electric and magnetic field densities:

$$H_z^{q+1/2}\left[m+\frac{1}{2},n\right]$$
$$= \left(E_y^q[m+1,n] - E_y^q[m,n] \right)\left(-\frac{\Delta t}{\mu_0 \Delta x} \right)$$
$$+ H_z^{q-1/2}\left[m+\frac{1}{2},n\right] \qquad (5a)$$

$$H_x^{q+1/2}\left[m,n+\frac{1}{2}\right]$$
$$= \left(E_y^q[m,n+1] - E_y^q[m,n] \right)\frac{\Delta t}{\mu_0 \Delta z} \qquad (5b)$$
$$+ H_x^{q-1/2}\left[m,n+\frac{1}{2}\right]$$

$$E_y^{q+1}[m,n]$$
$$= \left[\frac{1}{\Delta z}\left(H_x^{q+1/2}\left[m,n+\frac{1}{2}\right] - H_x^{q+1/2}\left[m,n-\frac{1}{2}\right] \right) \right.$$
$$- \frac{1}{\Delta x}\left(H_z^{q+1/2}\left[m+\frac{1}{2},n\right] - H_z^{q+1/2}\left[m-\frac{1}{2},n\right] \right)$$
$$\left. - j_y^{q+1/2}[m,n] \right]\frac{\Delta t}{\varepsilon_0 \varepsilon} + E_y^q[m,n]$$

$$(5c)$$

3 NUMERICAL RESULTS

Let's take advantage of obtained update equations (5) and make calculations of electric field density at the observation point (0,0). We place the source into the point (20,20), and equate half width of the pulse τ from the equation (2) to 10^{-9} s. On the figure 2 normalized to the maximum scattered electric field density for the case of $\varepsilon_1 = \varepsilon_4 = 1$, $\varepsilon_3 = 6$, $\varepsilon_2 = \{2;10;80\}$, $d_2 = d_3 = 0.1$ m is presented. It is noticeable on the represented graphs that the main maximums of the pulse are shifted simultaneously with relative permittivity of the first layer changing. Differences in amplitudes are the result of reflection coefficient distinctions. For the close values of relative permittivities of the layers, one

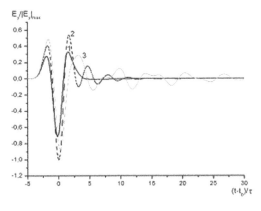

Figure 2. Normalized to the maximum scattered electric field density for the case of $\tau = 10^{-9}$ s, $\varepsilon_1 = \varepsilon_4 = 1$, $\varepsilon_3 = 6$, $d_2 = d_3 = 0.1$ m, (1) $\varepsilon_2 = 2$, (2) $\varepsilon_2 = 10$, (3) $\varepsilon_2 = 80$.

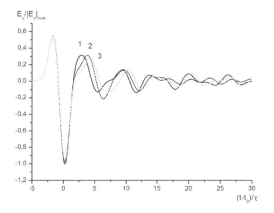

Figure 3.   Normalized to the maximum scattered electric field density for the case of $\varepsilon_1 = 1$, $\varepsilon_2 = 4$, $\varepsilon_3 = 80$, $\varepsilon_4 = 1$, $d_3 = 0.1$ m, (1) $d_2 = 0.1$ m, (2) $d_2 = 0.2$ m, (3) $d_2 = 0.3$ m.

can see the rapid damping of secondary reflections of the pulse.

On the figure 3 normalized to the maximum scattered electric field density for the case of $\varepsilon_1 = 1$, $\varepsilon_2 = 4$, $\varepsilon_3 = 80$, $\varepsilon_4 = 1$, $d_3 = 0.1$ m and $d_2 = \{0.1$ m; 0.2 m; 0.3 m$\}$ is presented. In comparison with figure 2, there are significant secondary reflections of the pulse on these graphs. In addition, it is noticeable that the influence of the reflected pulse within the layered structure to the third maximum of the basic reflected pulse.

These graphs are close to the one's that authors computed for this problem solved by semi-analytic method—with use of Fourier transform on spatial coordinates and wavelet transform on the time coordinate [3]. Hence, we may test for goodness of fit our theoretical researches.

By the way, one can see the convenience of the update equations (5) to the parallel computing, for example using OpenCL, and may have a good practice of such programs development.

ACKNOWLEDGMENTS

This study was supported by grant RFBR No 13–05–98037 r_sibir_a.

REFERENCES

[1] Fedorova L.L., Sokolov K.O. 2011 Georadiolokaciya massiva gornyh porod rossypnyh mestorozhdeniy kriolitozony, perekrytogo elektroprovodyaschim sloyem. Gorny inform.-analyt. bulleten. No. 8: 310–316. (rus).

[2] Rykshin A. Yu., Zatsepin P.M. 2009. Modelirovanie rasseyaniya elektromagnitnogo impulsa na dielektricheskom sloe. St. Petersburg State Polytechnical University Journal. Physics and mathematics. No. 4(88): 14–18. (rus).

[3] Rykshin A.Yu., Zatsepin P.M., Ulanov P.N. 2014. The modeling of a short electromagnetic pulse scattering by dual layers structure using wavelet transform. St. Petersburg State Polytechnical University Journal. Physics and mathematics. No. 1(189): 136–143. (rus).

[4] Schneider John. B. 2013. Understanding the Finite-Difference Time-Domain Method. www.eecs.wsu.edu/~schneidj/ufdtd.

Future Information Engineering and Manufacturing Science – Zheng (Ed)
© 2015 Taylor & Francis Group, London, ISBN 978-1-138-02644-5

Using DEMATEL to analyze the quality characteristics of mobile applications

Wen-Ming Han, Cheng-Hsien Hsu & Cheng-Yu Yeh
Department of Management Information System, Takming University of Science and Technology, Taipei, Taiwan

ABSTRACT: This study applied DEMATEL to analyze mobile App quality characteristics from the perspectives of developers and users. According to the results, users regard security as the most important quality characteristic, while functional suitability, usability and reliability are regarded as the most important quality characteristics by developers. The differences in viewpoints of users and developers may be the reason that the current mobile APP development quality has not been fully accepted by users. Therefore, developers need to rethink whether their quality priorities and considerations can meet user expectations to prevent waste of resources and resource crowding out effects.

Keywords: software quality; ISO/IEC 25010; DEMATEL; mobile applications

1 INTRODUCTION

Problems and destructive power arising from low-quality software are serious problems. For example, in 1960, due to the undetected FORTRAN syntax errors, the U.S. Venus probe spacecraft launching failed. In Canada, the software error resulted in the excessive radiation released by the radiotherapy unit to cause the death of many cancer patients (Jones, 2011). These cases prove that software quality is a serious and challenging issue. To address these challenges, researches have suggested that software quality management is the key factor to ensure customer satisfaction and affirmation of project delivery (Brown, 1991; Kan, 1991).

Mobile App is the software on smartphones, tablets and other mobile devices. The original application is only concerned with productivity and information search; however, with an increasing demand on everyday life quality, mobile App has been extended to games, order tracking, and ticket booking. According to ABI Research report, the mobile App market output may reach 27 billion USD in 2013. Despite many mobile Apps are creative, the negligence of the key factor, namely product quality, has led to usage problems and negative reviews.

The purposes of this study are the following: (1) to explore the mobile App software product quality characteristics to clearly grasp the facts about mobile App quality and provide mobile App developers with a clear quality vision for planned control; (2) to explore the cognitive differences regarding quality characteristics from the perspectives of developers and users. The developers can find out whether their products meet the quality expectation of the users, and carry out the proper control and quality improvement plan under limited resources.

2 ISO/IEC 25010

Weinberg (1991) suggested that quality management is the value in the minds of some people, and the quality is whether the needs are met. Quality refers to the degree that products provided by the producer meet the needs of the customers. By applying this definition to the field of software products, the software product quality can be defined as a set of element that allows customers to understand that software products meet or exceed their needs, such as maintainability and portability. Since the software product quality is the primary consideration of customers, how to measure, monitor and ensure the quality of software products in the software development life cycle has become a serious issue that the software industry must face.

The prerequisite to quantization control of software product quality to win customer satisfaction is to be able to clearly describe and define the elements of software product quality. Because the software product quality factors have multi-faceted elements, numerous studies have discussed software product quality factors to define a complete software product quality model. In other words, the unstructured software can be objectively

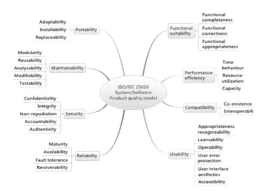

Figure 1.   ISO/IEC 25010.

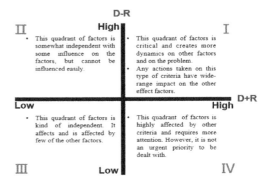

Figure 2.   The causal diagram.

quantified and compared in terms of software product quality across organizations.

ISO/IEC 25010 (2011) is the unit standard under ISO/IEC 2501n (2014). ISO/IEC 25010 renames the original internal and external quality point of views of ISO/IEC 9126 into software quality and system quality point of views. A total of 8 quality characteristics and 31 quality sub-characteristics are defined in ISO/IEC 25010. The 8 quality characteristics are functional suitability, performance efficiency, compatibility, usability, reliability, security, maintainability and portability. Each quality characteristic consists of quality sub-characteristics that are in line with their own meanings. Figure 1 presents this quality model.

3   DEMATEL

DEMATEL was developed by Battelle Memorial Institute in Geneva, Switzerland. The DEMATEL method was originally used to solve complex tangle of issues (e.g., race, hunger, environmental and energy issues) of the Science and Human Affairs Program. Its initial application area is for the study structural problems in the world, along with complex analysis of world problems to develop adaptation methods and review of existing research, models and data about the world's problems. Later, DEMATEL has been widely used in various research fields, such as agricultural development, female employment, environmental analysis, product surveys and medical practices, as it can affect factorial relationship by plane geometry so as to help understand the nature of the problem and then seek characteristic solutions (Gabus and Fontela, 1973; Fontela and Gabus, 1976). In general, DEMATEL calculation includes five steps: (1) to establish a direct relationship matrix; (2) to establish an initialized direct relationship matrix; (3) to calculate the

normalized direct relation matrix; (4) to calculate the total impact relationship matrix; (5) to establish a causal relationship diagram.

The drawing of the causal relationship diagram needs to add up horizontal and vertical sum of the total impact matrix generated by the fourth step. Usually, R represents the sum of each column of the total impact matrix, the degree of the criterion that directly or indirectly affects other criteria; D represents the sum of each row of the total impact matrix, which is the extent to which this factor is influenced by other factors. Centrality $(D + R)$ is on behalf of the strength of relationship between factors. Being closer to the right represents stronger strength whereas it is weaker if closer to the left. Causality $(D - R)$ represents the significance of the factor, which is the strength of influence or impact. When $(D - R) > 0$, D is relatively more significant; $(D - R) < 0$, then R is relatively more significant. In other words, factors in the first quadrant $(D - R)$ is positive and $D + R$ is very large) represent the driving factors to solve the core problems and thus should be the priorities for treatment; factors in the second quadrant $(D - R)$ is positive and $D + R$ is very small) are other factors of independence and little influence. Factors in the third quadrant $(D - R)$ is negative and $D + R$ is very small) represent other few factors that are affected or affect others. Factors in the fourth quadrant $(D - R)$ is negative and $D + R$ is large) represent the core problems to be addressed immediately but not the direct improvement of the quality characteristics. Figure 2 illustrates the management implications of each quadrant of the causal relationship diagram.

4   RESULTS

In this study, we designed the DEMATEL questionnaire based on the 8 quality characteristics of ISO/IEC 25010. The pairwise comparison scale of quality characteristics includes 0—no influence, 1—low influence, 2—medium influence and 3—high influence, the research questionnaire Cronbach α coefficient is 0.98, which is in line with the proposal by Wortzel (1979) that "Cronbach α coefficient in the range from 0.7 to 0.98 can be regarded as highly reliable", in addition the questionnaire content was reviewed and modified by advisor professor and the associated advisor professors to ensure Content Validity. Finally, the overall recovery rate of the survey samples was 53.18%.

4.1 User's perspective

Greater centrality (D + R) value represents that the quality characteristic is relatively important. According to Table 1, four quality characteristics are above the average in overall centrality (28.807) in the descending order of Functional Suitability, Security, Usability and Reliability.

The symbol of causality (D − R) represents whether the quality characteristics affects or is affected by others. According to Table 1, four quality characteristics are positive in D − R value (Performance efficiency, Maintainability, Compatibility and Security), and four quality characteristics are negative in D − R in the descending order of Functional Suitability, Usability, Reliability and Portability.

Figure 3 illustrates the Causal diagram described according to Table 1. The quality characteristics are evenly distributed in the four quadrants suggesting that Mobile APP users perceive that eight quality characteristics are different. Therefore, developers should consider the user perceptions of quality characteristics so that the Mobile APP can be interesting and win positive feedbacks.

Table 1. The direct and indirect effects of the quality characteristics under user's perspective.

Quality characteristics	D	R	D + R	D − R
Functional suitability	14.583	15.573	30.156	−0.990
Reliability	14.143	14.799	28.943	−0.656
Usability	14.301	15.144	29.445	−0.843
Performance efficiency	14.011	13.764	27.775	0.246
Portability	14.072	14.186	28.258	−0.114
Maintainability	14.325	13.282	27.607	1.043
Compatibility	14.512	13.612	28.124	0.900
Security	15.282	14.868	30.150	0.414

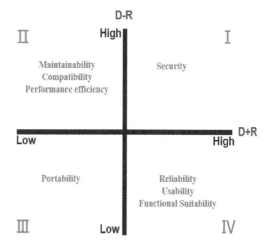

Figure 3. The Causal diagram under user's perspective.

Table 2. The direct and indirect effects of the quality characteristics under developer's perspective.

Quality characteristics	D	R	D + R	D − R
Functional Suitability	8.926	8.498	17.424	0.428
Reliability	8.810	7.969	16.779	0.841
Usability	8.770	8.346	17.116	0.424
Performance efficiency	8.000	7.799	15.798	0.201
Portability	6.777	8.130	14.907	−1.353
Maintainability	7.424	8.648	16.072	−1.224
Compatibility	7.321	7.553	14.874	−0.233
Security	8.486	7.570	16.055	0.916

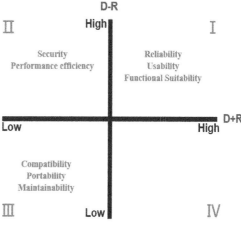

Figure 4. The causal diagram under developer's perspective.

133

4.2 *Developer's perspective*

Greater centrality (D + R) value means that the quality characteristic is relatively more important. According to Table 2, three quality characteristics are greater than the average centrality value (16.128) in the descending order from Functional Suitability, Usability and Reliability.

The symbol of causality (D − R) represents whether quality characteristics affects or is affected by others. According to Table 2, five quality characteristics have positive D − R values (Functional Suitability, Reliability, Usability, Performance efficiency and Security), 3 quality characteristics have negative D − R values in the descending order from Portability, Maintainability, and Compatibility.

Figure 4 illustrates the Causal diagram described according to Table 2, quality characteristics are clearly unevenly distributed in the four quadrants, in particular, there is no quality characteristic in the fourth quadrant.

5 CONCLUSIONS

User is the final accepter of the quality. Many developers failed by ignoring user experience, therefore, how to consider the quality characteristics perceived as important by users in the early product development period is an important issue that cannot be ignored because the Mobile APP quality control activities dominated by developers are to make the product quality acceptable to users.

From the perspective of D + R greater than centrality, Functional Suitability, Usability and Reliability are quality characteristics perceived as important by both the developer and the user. From the perspective of D − R below 0, Portability is the only quality characteristic that the user and the developer commonly agree. From the perspective of D − R above 0, both the developer and the user regard Performance efficiency and Security can affect other quality characteristics.

Based on the results of this study, there are three directions for future research and extended discussion: (1) to combine ISO/IEC 25010 system/software product quality architecture to support the use of non-hierarchical structure software product quality model; (2) to explore differences in the quality assessment results arising from different data collection methods to add the appropriate information to check the practical skills to make it more complete; (3) to analyze the Mobile APP product user quality elements by case study to verify the completeness of the proposed research method in this study.

REFERENCES

[1] Brown, M.G. 1991. Baldrige Award Winning Quality: How to Interpret the Malcom Baldrige Award Criteria. ASQC Quality Press, Milwaukee.
[2] Jones, C. 2011. The Economics of Software Quality, Addison Wesley Longman, Boston, MA.
[3] Fontela, E and Gabus, A. 1976. The DEMATEL Observer, DEMATEL 1976 Report, Battelle Geneva Research Center, Switzerland, Geneva.
[4] Gabus, A., & Fontela, E. 1973. Perceptions of the world problematique: communication procedure, communicating with those bearing collective responsibility. DEMATEL Report No. Geneva, Switzerland: Battelle Geneva Research Center.
[5] International Organization for Standardization, ISO/IEC 25000. 2014. Systems and software engineering—Systems and software Quality Requirements and Evaluation (SQuaRE)—Guide to SQuaRE.
[6] International Organization for Standardization, ISO/IEC 25010. 2011. Systems and software engineering—Systems and software Quality. Requirements and Evaluation. (SQuaRE)—System and software quality.
[7] Wortzel, L. 1979. Multivariate analysis. N. J., Prentice Hall.
[8] Weinberg, GM. (1991). Quality Software Management: Systems Thinking, New York: Dorset House.
[9] Kan, S.H. 1991. Modeling and Software Development Quality, *IBM Systems Journal* 30(3): 351–362.

Future Information Engineering and Manufacturing Science – Zheng (Ed)
© 2015 Taylor & Francis Group, London, ISBN 978-1-138-02644-5

Application of node based coincidence algorithm for solving order acceptance with multi-process capacity balancing problems

Watcharee Wattanapornprom & Tieke Li
Dongling School of Economics and Management, University of Science and Technology Beijing, Beijing, China

Warin Wattanapornprom & Prabhas Chongstitvatana
Department of Computer Engineering, Faculty of Engineering, Chulalongkorn University, Bangkok, Thailand

ABSTRACT: Over the past decade the strategic importance of order acceptance has been widely recognized in practice. This paper presents the application of node based coincidence algorithm to solve the order acceptance problem with multi-process capacity. The results show that Node Based Coincidence Algorithm (NB-COIN) is a potential algorithm which can maximize both profit and can maximize the capacity used at the same time.

Keywords: order acceptance; limited capacity; node based coincidence algorithm; genetic algorithm

1 INTRODUCTION

In economics, excess demand refers to excess of need over supply of products being offered to the market at a given price. This leads to upper prices along with the opportunity of employment. However, excess demand in Order Acceptance (OA) problems is the situation that orders arrival rate greater than service level of manufacturers. So far, under the high competition conditions, the manufacturers cannot straightforwardly raise their prices or just increase the temporary working capacity, subsequently, they need to select the most profitable set of the orders with regular employment cost and capacity.

OA is classified as a multi-dimensional knapsack problem which is a well-known NP hard problem. Additionally, there also exists the necessity of order sequencing which makes it much more difficult than the general knapsack problems. For example, the difference sequence of orders can result in difference profit level (Senju & Toyoda 1968, Kleywegt & Papastavrou 2001).

In 2011, Slotnick presented a recent overview of OA which addresses simultaneous order acceptance and scheduling decisions. From the literature it was found that most of the researches focus on accepting order in a single machine or process. However, by assuming that all the processes are grouped into a single process, the accepting consequences become inefficient in many real production

situations such as tardiness or over or under capacity utilization.

This paper presents a new technique to solve order acceptance or rejection in multi-process environments using Node Based Coincidence Algorithm (NB-COIN). The method is presented in section 2. The results are compared with Genetic Algorithm in section 3. Finally, the section 4 concludes the work.

2 METHODOLOGY

2.1 *The order acceptance model*

The set of order $i = (1, 2, ..., i)$, where i is one of the k product type and profit per unit is P_{ik}. Each order must be processed through set of production unit $N = (1, 2, ..., n)$. An order i is said to be early if finishing time t is equal or less than due date d, $t-D_i \leq 0$ and overdue if t is more than the due date $t-D_i > 0$. A product k consumes capacity CTP_{nt} as e_{ijknt} per unit, so the selected orders will occupy total production capacity $\sum_i e_k q_{ikt}$ for $\forall t$. Each production order consists of several jobs. The jobs have precedence (i.e., job $j + 1$ can start only if job j is completed). RT_n is the regular working time allowed in a day, which is assumed to be eight hours. The model can be defined as follow:

Capacity constraint
RT_n Total capacity of workstation n

CTP_{nt} Unassigned capacity of workstation n at period t $(t = 1, ..., T)$

e_{iknt} Consumption of CTP_{nt} for product k in order i

f_{ijknt} Time unit that workstation n utilize CTP_{nt} for product k in order i by job j at period t

g_n Cost of unassigned capacity of workstation n CTP_n per time unit

α_n Cost rate of leftover capacity at workstation n

d_{nt} Amount of leftover capacity at workstation n

Order constraint

p_{ik} Profit of product k in order i

q_{ik} Demand quantity of product k in order i

Decision constraint

$R_{ik} = 1$, if the order i for product k is accepted $= 0$, otherwise

$F_{ijktn} = 1$, if the order i for product k is produced at workstation n by job j at period $t = 0$, otherwise

Model objective

Maximize $Z = \sum_i \sum_k p_{tk} q_{tk} R_{tk} - \sum_t \sum_n \alpha_n d_{nt} g_n$ (1)

Subject to

Workstation-level activities constraint

$$\sum_t \sum_i \sum_k \sum_j e_{kijnt} q_{tkt} \times R_{tk} \leq \sum_t CTP_{nt} \quad \forall n \quad (2)$$

$$d_{nt} = RT_n - CTP_{nt} - \sum_t \sum_n e_{ijk} q_{ikt} f_{iktn} \quad \forall n \quad (3)$$

Order-level activities constraint

$$f_{ijknt} q_{ik} \geq F_{ijktn} \quad \forall i,j,k,n,t \quad (4)$$

$$f_{ijknt} q_{ik} \leq e_{kijnt} q_{ikt} F_{iktn} \quad \forall i,j,k,n,t \quad (5)$$

$$\sum_n t F_{i|j|knt} \leq D_i R_{ik} \quad \forall i,k,t \quad (6)$$

$$\sum_n \sum_t f_{ki(j-1)nt'} q_{ik} + \sum_n f_{iknt} q_{ik}$$
$$\geq \sum_n e_{i(j-1)knt} q_{ik} F_{ijkt} \quad \forall i, \frac{j}{\{1\}}, k, t \quad (7)$$

Binary and non-negativity constraint

$$R_{ik} = 0 \text{ or } 1 \quad \forall i,k \quad (8)$$

$$F_{ijknt} = 0 \text{ or } 1 \quad \forall i,j,k,n,t \quad (9)$$

$$f_{ijknt} \geq 0 \quad \forall i,j,k,n,t \quad (10)$$

This problem is considered to be a two objectives optimization problem. However, the two objectives are bind into one single objective. The objective function consists of two parts (i) to maximize the total profit and (ii) to minimize the leftover capacity. Generally speaking, the objective is to choose the set and sequence of the profitable orders using as much working capacity as possible. The leftover capacity is considered to have some certain penalty cost. The first set of constraints is established to ensure that the whole capacity of production plant is not disrupted. Constraint (3) was set to calculate the penalty of under capacity utilization. Constraints (4) and (5) set the F_{ijknt} decision variables to either 1 or 0. The F_{ijknt} is the indicator variable; it becomes 1 when $f_{ijknt} > 0$, indicating that job j of item i is being processed on station n in period t, otherwise it becomes 0. The F_{ijknt} variable is used to ensure the precedence relationship. The constraint set (6) ensures that when an order for an item is accepted, the completion time of the final job of that order does not exceed the order due date. The constraint set (7) imposes precedence restrictions to ensure that job j of item i can be processed in period t only after completing job j–1.

2.2 Solution procedures

This work compares the result of Node Based Coincidence Algorithm (NB-COIN) (Waiyapara et al. 2013) with Genetic Algorithm (GA) (Syswerda 1991). The algorithms are modified such that they would consider only the accepted sets of orders.

2.2.1 Node based coincidence algorithm

NB-COIN is a permutation based Estimation of Distribution Algorithm (EDA). It generates solution strings in sequences, ensuring that only valid permutations are sampled. NB-COIN is a variation of Coincidence Algorithm (COIN) proposed by Wattanapornprom and others (2013). It uses a data structure called coincidence matrix H to model substructures from absolute positions. The matrix H_{xy} represents the probability of y found in the absolute position x. The update equation of NB-COIN is

$$H_{xy}(t+1) = H_{xy}(t) + \frac{k}{(n)}\left(r_{xy}(t+1) - p_{xy}(t+1)\right)$$
$$+ \frac{k}{(n)^2}\left(\sum_{j=1}^{n} p_{xj}(t+1) - \sum_{j=1}^{n} r_{xj}(t+1)\right) \quad (11)$$

where k denotes the learning step, n is the problem size, r_{xy} is the number of xy found in the better-group, and p_{xy} is the number of xy found in the worse-group. The incremental and detrimental step is $k/(n-1)$, and the term $k/(n-1)^2(\sum_{j=1}^{n} p_{xj}(t+1) - \sum_{j=1}^{n} r_{xj}(t+1))$ represents the adjustment of all other H_{xj}, where $j \neq x$ and $j \neq y$.

136

After each population was evaluated and ranked, two groups of candidates are selected according to their fitness values: better-group and worse-group. The better-group is selected from the top $c\%$ of the rank and is used as a reward, and H_{xy} is increased for every pair of xy found in this group. The punishment is a decrease in H_{xy} for every pair of xy found in the worse group of the bottom c% of the population rank.

The pseudo code of NB-COIN is simplified as follows:

Step 1 Initialize the model
Step 2 Sample the population
Step 3 Evaluate the population
Step 4 Select candidates
Step 5 Update the model
Step 6 Repeat steps 2 to 5 until terminated.

2.2.2 *Genetic algorithm*

The GA used in this research is the permutation based GA with Position-based crossover (PBX) (Syswerda 1991). PBX preserves not only absolute order substructures but also relative order substructures from two parents. Figure 1 illustrates the steps and the example of PBX. The proto offspring 1 mimics the absolute order substructures from the parent 1 and then imitates the relative sequence order of the remaining substructures from the parent 2 and vice versa.

For this problem, the chromosomes are sequenced subsets of jobs. The diversity is maintained by ancestor replacement. If a new candidate is better than its ancestors it is used to replace one of its own parents. In this study, the local search is also applied to the new candidates with improvement. The swapping and insertion operations are randomly applied to the candidates until the candidates are no longer improved. The pseudo code of GA is as follows:

Step 1 Randomly generate the population.
Step 2 Evaluate the population.
Step 3 Perform crossover and mutation. If the newly generated candidate is better than its ancestors, then perform the local search until the candidate is no longer improved.
Step 4 Repeat Step 3 until the maximum number of generation is reached.

Although the encoded solution of GA is a full set of the jobs in the pool, the evaluation process considers only the accepted orders. The evaluation process not only evaluates the orders sequence, but also re-sorts the orders sequences to separate the accepted and rejected orders as illustrated in the Figure 2. The sequence of the accepted orders is kept in the accepted pool while the remaining orders are kept in the rejected pool. The candidate solution is re-sorted by concatenating the accepted pool with the rejected pool.

Even though, GA and NB-COIN are in the same group of evolutionary algorithms, however, the evaluation process and the updating process of NB-COIN for the order acceptance are slightly different. GA needs to maintain the genetic materials, therefore the whole set of orders need to be maintained. However, NB-COIN can reproduce the missing sequences by itself. In addition, the sequences of the rejected pool are considered to be the useless information, therefore, NB-

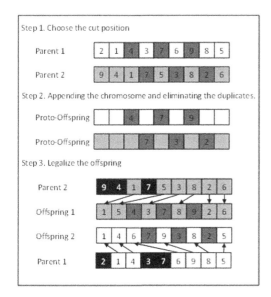

Figure 1. Position-based crossover (PBX).

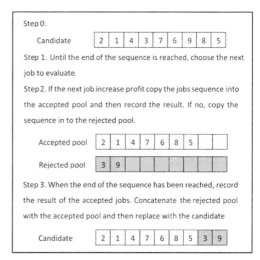

Figure 2. Evaluation with cutting off.

COIN only updates the models from the accepted sequences of orders. Consequently the evaluation process does not need to concatenate the rejected pool with the accepted pool. The evaluation processes in the Figure 2 simply use the accepted pool as the candidate for the NB-COIN.

2.3 *Test problems and experimental design*

A list of products and their profit per piece was randomly generated. The generated profits are ranged between 5 to 15 currency units per piece. Then these profit attributes were used to generate the capacity utilization for each product such that producing the least profitable product would utilize the most balance capacity in each process, while the random time were added according to their profits. The capacities used by each processes are ranged between 0.1 to 1 pieces per minute.

The ten problems of size 50, 75 and 100 were also randomly generated according to the products and their profits such that the less profitable products have more chance to be demanded. Each order was generated from a log-normal distribution with an underlying normal distribution with mean 0 and standard deviation 1. The quantity for each order was randomly generated using the range between 1×1000 pieces and 12×1000 pieces. Each product has to be processed through 5 parallel production units which mean that there are totally 5 processes $\times$ 5 parallel machines for each process. The maximum capacity was set to two weeks. The due dates of each order were generated from a uniform distribution plus calculated lead-time for each of the order. These parameters were imitated from the existent manufactures in Thailand. Therefore, the wage penalty for this problem was set to 300 baht per worker per one production unit per day.

To compare the results, both NB-COIN and GA were given the same population size and maximum number of generations which are equal to the problem size $\times$ 2. The probabilities of crossover and mutation of GA are equal to 0.8 and 0.2 respectively. The learning step, k, of NB-COIN is 0.05. The selection pressure of GA is 50% of the whole population, while NB-COIN uses 25% of the top ranks for rewards and 25% of the bottom ranks for punishment. Test programs were coded in Lazarus and ran on OS X 10.4 on Intel Pentium Core i5 2.50 GHz processor with 4 GB of RAM.

3 RESULTS

The performances of NB-COIN and GA are compared in terms of profit and capacity utilization. The performances are compared using the actual profit averaged from each of the best solutions out

Table 1. Performance of NB-COIN vs. GA in order acceptance with multi-process capacity balancing problem.

Problem size	NB-COIN		GA	
	Util.	Profit	Util.	Profit
50 orders	74.7%	18,5605	54.4%	14,4370
75 orders	80.7%	19,4074	57.5%	15,2562
100 orders	85.5%	21,0095	61.6%	16,1686

of ten runs. The capacity utilization is the wage penalty already deducted from the actual profit. The performance of NB-COIN to select from 50 orders is far better than GA that selected from 100 orders. The explanation is that the generated test problems were design such that the lowest profitable product utilizes the most balanced capacity. On the other hand, the most profitable product leaves more capacity leftover. The greedy profit maximization would results in the worse capacity utilization. NB-COIN is the algorithm that is good in solving multimodal and multi-objective problems (Waiyapara et al. 2013) as it tries to maintain the entire good substructures in order to recombine them.

4 CONCLUSION

This paper presents the application of NB-COIN to solve the order acceptance problem with multi-process capacity. The results show that NB-COIN is far better than GA for both profit and capacity utilization.

REFERENCES

[1] Kleywegt, A.J. & Papastavrou, J.D. 2001. *The dynamic and stochastic knapsack problem with random sized items*. Operations Research 49 (1): 26–41.
[2] Senju, S. & Toyoda, Y. 1968. *An approach to linear programming with 0–1 variables*. Management Science 15 (4), B196–B207.
[3] Slotnick, S.A. 2011. *Order acceptance and scheduling: A taxonomy and review*, European Journal of Operational Research 210(3): 527–536.
[4] Syswerda. 1991. A Handbook of Genetic Algorithms. *Schedule Optimization Using Genetic Algorithms.*
[5] Waiyapara, K. et al. 2013. *Solving Sudoku Puzzles with Node Based Coincidence Algorithm*; Proc. of International Joint Conference on Computer Science and Software Engineering (JCSSE 2013).
[6] Wattanapornprom, W. et al. 2013. *Application of Estimation of Distribution Algorithms for Solving Order Acceptance with Weighted Tardiness Problems*; Proc. of IEEE International Conference on Industrial Engineering and Engineering Management (IEEM13-P-0588).

Future Information Engineering and Manufacturing Science – Zheng (Ed)
© 2015 Taylor & Francis Group, London, ISBN 978-1-138-02644-5

Production and distribution plan for fresh produce under random fuzzy environment

Xuejie Bai
College of Management, Hebei University, Baoding, Hebei, China
College of Science, Agricultural University of Hebei, Baoding, Hebei, China

Lijun Xu
Bureau of Animal Husbandry and Fisheries of Baoding, Baoding, Hebei, China

ABSTRACT: This paper addresses the problem of production and distribution for fresh produce. A two stage minimum risk production and distribution planning model of agricultural products, which embeds the chance objective function, is proposed to handle the uncertainty of crops' yields that are assumed to be characterized by random fuzzy variables with known probability and possibility distributions. The two-stage random fuzzy minimum risk programming selected is that the decisions in a first stage are designed to meet the uncertain outcomes in a second stage. Since it is always difficult to handle with the random fuzzy model directly, we apply an Approximation Approach (AA) to evaluate the value of the objective function. Considering that the approximating model is neither linear nor convex, it can't be solved via the conventional optimization algorithm. Therefore, an approximate-based Hybrid Particle Swarm Optimization (PSO) algorithm is designed to solve the proposed model. Finally, an application example is presented to illustrate the significance of the random fuzzy production and distribution model as well as the effectiveness of the solution method.

Keywords: fresh produce; production and distribution plan; random fuzzy programming; mean chance; Particle Swarm Optimization

1 INTRODUCTION

The production and distribution plan of agricultural products, as an overall research field comprising of cultivation, harvest, storage, processing and distribution, plays an important role in the architecture of advanced planning systems (Akkerman et al., 2010; Shukla & Jharkharia, 2013). Recently, the activities from production to distribution has attracted many researchers' attention lately as a consequence of several reasons, such as the national focus on recent cases of agricultural produce contamination, the changing attitudes of a more health conscious and the preference of the better informed consumers (Ahumada & Villalobos, 2009). For example, Ahumada and Villalobos (2011) presented a mixed integer linear programming model used for production and distribution of agricultural produce with the objective of maximizing the revenues of a producer. Widodo et al. (2006) designed a dynamic programming approach to integrating the production, harvesting and inventory planning of flowers through the use of growth and loss functions for maximizing the demand satisfied.

In the production and distribution plan of agricultural products, a standard assumption of deterministic conditions may be questionable, because the variability of crops' yields and customers' demands are usually not known for certain. Some authors have modified traditional deterministic models to account for the uncertainty in most farm activities by modeling the unknown parameters as random variables with known probability distribution. For instance, Romero (2000) determined an efficient agricultural resource allocation using a multidimensional risk approach. Ahumada and Villalobos (2012) selected a two-stage stochastic program with risk level to build a tactical planning model for the production and distribution. Yu and Nagurney (2013) developed a network-based food supply chain model under oligopolistic competition and perishability with a focus on fresh produce and investigated a case study focused on the cantaloupe market.

In fact, the decision maker cannot obtain the perfect information of each parameter in the decision systems. Sometimes, the probability distribution function of the random variable may be partially known by estimation. In such environment, the random variable contains fuzzy information. Random fuzzy variable is an appropriate tool to describe this uncertain phenomenon (Liu, 2002). By means of random fuzzy variable, random fuzzy theory and random fuzzy programming have been well developed in the literature, such as the mean chance and equilibrium chance of a random fuzzy event (Liu & Liu, 2002), the expected value operator of a random fuzzy variable (Liu & Liu, 2003), the different modes of convergence (Liu & Dai, 2008; Liu & Wang, 2013), two-stage random fuzzy programming (Liu et al., 2008). The propose of this paper is to present a realistic two-stage random fuzzy model for production and distribution plan of agricultural products, in which the crops' yields are assumed to be uncertain and characterized as random fuzzy variables with known distributions. The objective function of the proposed model is to minimize the total costs incurred in the two stages. Then, we apply the AA (Liu et al., 2008) to evaluate the value of the objective function and discuss the convergence the AA. Since the approximating model is neither linear nor convex, it can't be solved via the conventional optimization algorithm. Therefore, an approximate-based hybrid PSO algorithm is designed to solve the proposed model. Finally, an application example is provided to illustrate the effectiveness of the hybrid PSO algorithm.

The rest of this paper is organized as follows. In Section 2, we develop a two-stage minimum risk random fuzzy production and distribution plan model. In Section 3, we design an approximation-based hybrid PSO algorithm to solve the new model. One numerical example is given in Section 4. Finally, Section 5 summarizes the main conclusions in our paper.

2 FORMULATION OF PRODUCTION AND DISTRIBUTION PLAN

In this section, we will construct a new type of two-stage random fuzzy programming model for the production and distribution plan of agricultural products.

The following index and parameters are used to describe the model throughout the paper (Table 1).

Based on the assumptions and notations, a two-stage random fuzzy production and distribution plan model for agricultural product can be established as follows:

$$
\begin{cases}
\min & Ch\left\{ \sum_{i=1}^{n} f_i x_i + Q\left(x, \xi_\gamma(\omega)\right) \le \varphi_0 \right\} \\
s.t. & \sum_{i=1}^{n} x_i \le M \\
& x_i \ge 0 \qquad i = 1, 2, \dots, n
\end{cases}
\tag{1}
$$

Table 1. List of notations.

Notation	Definitions
i	Index of crops, $i = 1, 2, 3 \dots, n$
M	The maximum possible acres of land
f_i	The nonnegative planting cost for different crop i
p_i	The unit purchase price for crop i
q_i	The unit selling price for crop i
d_i	The demand for different crop i
ξ_i	The random fuzzy yield for every crop i
x_i	Decision variable indicating acres of land devoted to crop i
y_{i1}	The amount purchased from market for crop i
y_{i2}	The amount sold for crop i

where $Q(x, \xi_\gamma(\omega))$ is the optimal value of the following linear programming

$$
\begin{cases}
\min & \sum_{i=1}^{n} p_i y_{i1} - \sum_{i=1}^{n} q_i y_{i2} \\
s.t. & \xi_i x_i + y_{i1} - y_{i2} \ge d_i \quad i = 1, 2, \dots, n \\
& y_{i1}, y_{i2} \ge 0 \qquad\qquad i = 1, 2, \dots, n.
\end{cases}
\tag{2}
$$

Expressions (1) and (2) are composed of our two-stage random fuzzy production and distribution plan model, where (1) is the first stage, and (2) corresponds to the second stage. The aim of the model is to minimize the risk that the total costs of a producer exceeds some given value. The costs is given by the combination of external conditions, which are out of the farmer's control, such as expected market prices, and those determined by the farmer himself, such as what and how much to plant in a given season. In the first stage, the constraint represents the main resources limiting the operations. Usually this resource is the result of strategic decisions, such as land available. The constraint makes sure that the resource used by a solution does not exceed the total availability of land. In the second stage, when the values taken by the random fuzzy variables are available, the customers' demands can be satisfied by corrective or recourse actions.

3 HYBRID PSO ALGORITHM

Particle swarm optimization algorithm, initially developed by Kennedy and Eberhart (1995), is a population-based algorithm, which is inspired by the social behavior of flocks of birds or schools of fish. Because of better intelligent background and theoretical framework, recently, the PSO algorithm has attracted much attention and been applied successfully in the fields of evolutionary computing, unconstrained continuous optimization problems and many others (Kennedy et al., 2001). For the subsequent discussion, we will give more detailed explanations about the hybrid PSO algorithm for solving the approximating random fuzzy production and distribution plan model (2).

Representation Structure: Suppose there are pop_size particles to form a colony. In the two-stage random fuzzy production and distribution plan model, we use a vector $X = (x_1, x_2, ..., x_n)$ as a particle to represent a decision.

Initialization Process: After that, we denote its current position of each particle by *pbest*, using P_k as abbreviation $P_k = (p_{k1}, p_{k2}, ..., p_{kn})$, which represents the personal smallest objective value so far at time. On the other hand, we set the global best particle of the colony as *gbest*, using P_g as abbreviation, $P_g = (p_{g1}, p_{g2}, ..., p_{gn})$ which represents the position of the best particle found so far at time t in the colony. Finally, we initialize the velocity V_k of the kth particle randomly, $V_k = (v_{k1}, v_{k2}, ..., v_{kn})$.

Updating Process: As mentioned above, the PSO algorithm is an evolutionary computation algorithm, and it searches for the optimal solution by renewing generations. Using the above notations, for the *pop_size* particles, the new velocity of the th particle is updated by

$$V_k(t+1) = \omega_t V_k(t) + c_1 r_1 \left(P_k(t) - X_k(t) \right) + c_2 r_2 \left(P_g(t) - X_k(t) \right) \tag{3}$$

for $k = 1, 2, ..., pop\_size$, where ω_t is the inertia weight that decreases linearly from 0.9 to 0.4; c_1 and c_2 are nonnegative constants, called the cognitive and social parameter, respectively; and r_1 and r_2 are two independent random numbers generated from the unit interval [0,1]. When the new velocity is obtained, we can update the position of the th particle by

$$X_k(t+1) = X_k(t) + V_k(t+1) \tag{4}$$

Summarizing the above process, we immediately yields the hybrid PSO flow as shown in Figure 1.

4 AN APPLICATION EXAMPLE

In this section, we propose an example to demonstrate the modeling idea. The example is described as follows.

Consider a farmer who specializes in raising grain, corn, and sugar beets on his 500 acres of land. During the winter, he wants to decide how much land to devote to each crop. The farmer knows that at least 200 tons (T) of wheat and 240 T of corn are needed for cattle feed. These amounts can be raised on the farm or bought from a wholesaler. The corresponding planting cost are $150, $230 and $260 per ton of wheat, corn and sugar beet, respectively. Any production in excess of the feeding requirement would be sold. Selling prices are $170 and $150 per ton of wheat and corn. The purchase prices are 40% more than this due to the wholesaler's margin and transportation costs, i.e., $238 and $210 per ton of wheat and corn. Another profitable crop is sugar beet which sells at $36/T; however, the European Commission imposes a quota on sugar beet production. Any amount in excess of the quota can be sold only at $10/T. The farmer's quota for next year is 6000T.

This problem was considered in (Birge & Louveaux, 1997) in which the crops' yields were assumed to be random variables. In this paper, we generalize the problem by assuming the yields are characterized by random fuzzy variables with known possibility and probability distributions.

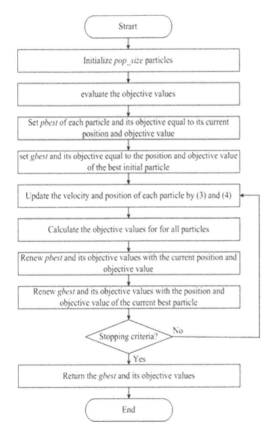

Figure 1. Hybrid PSO flow.

Assume that the yields of three crops $\xi = (\xi_1, \xi_2, \xi_3)$ is a random fuzzy vector defined as

$$\xi_1 \sim U(Z_1 - 0.4, Z_1 + 0.7), \quad Z_1 \sim (1.5, 2, 2.5);$$
$$\xi_2 \sim U(Z_2 - 0.5, Z_2 + 0.2), \quad Z_2 \sim (2.5, 3, 3.6);$$
$$\xi_3 \sim U(Z_3 - 3, Z_3 + 4), \quad Z_3 \sim (18, 22, 27).$$

As a consequence, we can set up the following model. Let and be the acres of land devoted to wheat, corn and sugar beets, and be tons of wheat sold and purchased, and be tons of corn sold and purchased, and be tons of sugar beets sold at the favorable price and lower price, respectively. The problem reads as follows:

$$\begin{cases} \min & Ch\{150x_1 + 230x_2 + 260x_3 + Q(x, \xi_\gamma(\omega)) \leq 120000\} \\ s.t. & x_1 + x_2 + x_3 \leq 500 \\ & x_i \geq 0, \qquad i = 1, 2, 3 \end{cases}$$

(5)

where $Q(x, \xi_\gamma(\omega))$ is the optimal value of the following linear programming

$$\begin{cases} \min & 238y_{11} - 170y_{12} + 210y_{21} - 150y_{22} - 36y_{31} - 10y_{32} \\ s.t. & \xi_1 x_1 + y_{11} - y_{12} \geq 200 \\ & \xi_2 x_2 + y_{21} - y_{22} \geq 240 \\ & y_{31} - y_{32} - \xi_3 x_3 \leq 0 \\ & y_{31} \leq 6000 \\ & y_{i1}, y_{i2} \geq 0, \qquad i = 1, 2, 3. \end{cases}$$

On the basis of the AA, we can turn the original two-stage production and distribution model of agricultural products into its approximating finite-dimensional programming. In order to solve the approximating problem, for each fixed first-stage decision variables, we generate 2000 sample points via the AA to calculate the value of the objective function. Then for each sample point, we can solve the second programming to obtain the optimal value as for $k = 1, 2, \ldots, 2000$. After that, the value of the objective function at x can be computed. If we set the population size in the implementation of the hybrid PSO algorithm is 50, then a run of the proposed algorithm with 800 generations gives the optimal solution $x^* = (179.5, 82, 235)$, with the mean chance 0.94862.

5 CONCLUSION

This purpose of the paper attempted to develop a new class minimum risk production and distribution model with random fuzzy yields and to discuss its solution method. The major results can be summarized as follows. Firstly, we proposed a two-stage random fuzzy production and distribution model, in which the crops' yields were uncertain and assumed to be random fuzzy variables. Secondly, we applied the AA to the proposed new model, and turned the original problem to an equivalent one. Considering that the approximating two-stage random fuzzy programming can't be solved by classical procedure, we designed an approximation-based hybrid PSO algorithm. Thirdly, we provided a numerical example to demonstrate the effectiveness of the proposed model. The computational results showed that the random fuzzy model had advantages over some existed modeling method for agricultural products.

ACKNOWLEDGMENT

This work is supported partially by the Natural Science Foundation of Agricultural University of Hebei (No. LG201306).

REFERENCES

[1] Ahumada, O. & Villalobos, J.R. 2009. Application of planning models in the agri-food supply chain: A review. *European Journal of Operational Research* 196(1): 1–20.
[2] Ahumada, O. & Villalobos, J.R. 2011. A tactical model for planning the production and distribution of fresh produce. *Annals of Operations Research* 190(1): 339–358.
[3] Ahumada, O., Villalobos, J.R. & Mason, A.N. 2012. Tactical planning of the production and distribution of fresh agricultural products under uncertainty. *Agricultural Systems* 112: 17–26.
[4] Akkerman, R., Farahani, P. & Grunow, M. 2010. Quality, safety and sustainability in food distribution: A review of quantitative operations management approaches and challenges. *OR Spectrum* 32(4): 863–904.
[5] Birge, J.R. & Louveaux, F. 1997. *Introduction to Stochastic Programming.* New York: Springer-Verlag.
[6] Kennedy, J. & Eberhart, R.C. 1995. Particle swarm optimization. Proc. of the IEEE International Conference on Neural Networks, 4: 1942–1948.
[7] Kennedy, J., Eberhart, R.C. & Shi, Y. 2001. *Swarm Intelligence.* San Francisco: Morgan Kaufmann Publishers.
[8] Liu, B. 2002. *Theory and Practice of Uncertain Programming.* Heidelberg: Physica-Verlag.
[9] Liu, Y. & Liu, B. 2002. Random fuzzy programming with chance measures defined by fuzzy integrals. *Mathematical and Computer Modelling* 36: 509–524.
[10] Liu, Y., Bai, X. & Hao, F. 2008. A class of random fuzzy programming and its hybrid PSO algorithm. *Lecture Notes in Artificial Intelligence* 5227: 308–315.
[11] Liu, Y. & Dai, X. 2008. The convergence modes in random fuzzy theory. *Thai Journal of Mathematics* 6(1): 37–47.

[12] Liu, Y. & Liu, B. 2003. Expected value operator of random fuzzy variable and random fuzzy expected value models. *International Journal of Uncertainty, Fuzziness and Knowledge-Based Systems* 11(2): 195–215.

[13] Liu, Y. & Wang, Y. 2013. Equilibrium mean value of random fuzzy variable and its convergence properties. *Journal of Uncertain Systems* 7(4): 243–253.

[14] Romero, C. 2000. Risk programming for agricultural resource allocation: A multidimensional risk approach. *Annals of Operations Research* 94(1): 57–68.

[15] Shukla, M. & Jharkharia, S. 2013. Agri-fresh produce supply chain management: A state-of-the-art literature review. *International Journal of Operations & Production Management* 33(2): 114–158.

[16] Widodo, K.H., Nagasawa, H., Morizawa, K. & Ota, M. 2006. A periodical flowering-harvesting model for delivering agricultural fresh products. *European Journal of Operational Research* 170(1): 24–43.

[17] Yu, M. & Nagurney, A. 2013. Competitive food supply chain networks with application to fresh produce. *European Journal of Operational Research* 224(2): 273–282.

Future Information Engineering and Manufacturing Science – Zheng (Ed)
© 2015 Taylor & Francis Group, London, ISBN 978-1-138-02644-5

Effect on group counseling for university freshmen after their psychological survey

Shuping Liu
Jiangxi Science and Technology Normal University, China

Lifeng Yang
Nanyang Normal University in Henan, China

ABSTRACT: *Objective*: In order to offer beneficial references for mental health education and consultation in college, we probe into a way of group psychological consulting which can overcome single interview defect after investigating the freshmen's psychological heath since there may exist various psychological problems and blocks among the new comers in college.
Methods: 372 freshmen, who were screened out through the survey of mental health, were taken as the objects of study, and their SCL-90 scores were taken as the research index to test the results of group counseling by statistical methods.
Results: Scores of interpersonal sensitivity, depression, anxiety and obsessive-compulsive dropped sharply among the new comers after group counseling, and T test reached the significant level.
Conclusion: Group counseling had the remarkable effect on students' Psychological adjustment.

Keywords: group counseling; freshmen psychological survey; SCL-90 etc

1 INTRODUCTION

Because of changes in life style, education and its social environments, freshmen often meet with all kinds of adaptive problems and disorders. On entering university, psychological survey should be done among freshmen, and students who were screened out from it were interviewed in a single method. That is, they were usually invited to counseling face to face. This method has two shortcomings, one is that the students who are invited have a doubt that they have psychological problems; the other is that only a small proportion of students with mental health problems can be treated effectively in this way. In addition, group counseling is mostly designed for interpersonal relationship problems, so that other psychological problems cannot be treated effectively. In order to overcome these defects, we use group counseling and design group activities according to psychological counseling theories to help the students.

Group counseling is a kind of psychological consultation in group situations, which can prompt the individual to improve their human relationships and learn new attitudes and behavior by means of observation, study, experience, self-knowledge, self-study and self-acceptance, and finally they can develop good adaptability [1]. The aim of this research is to explore the subject and effects of group counseling, so as to provide reference for mental health education and consulting work in colleges and universities.

2 SUBJECTS AND METHODS

2.1 Subjects

372 freshmen, who were screened out from 7012 freshmen through the survey of mental health, were taken as the object of study. These students were divided into four groups, which are "Harmonious Relationship" campsite (scores of interpersonal sensitivity ≥3, interpersonal disabilities), "Sunshine Mentality" campsite (scores of depressive ≥3, depressive mood) "Hearty Fly" workshop (scores of anxiety ≥3, Irritability nervousness), "Return workshop (scores of obsessive-compulsive ≥3, compulsive behavior) etc.

2.2 Tools

The scale of Self-reporting Inventory [2] was used, which has 90 items, including contents of psychotic symptom in feeling, affection, thought, consciousness behavior, lifestyle, interpersonal relationship and even eating and sleeping problems, and the factors of the scale reflect 10 kinds of psychological symptoms. Besides, the scale includes general information of students, such as name, gender, family and professional interests, etc.

2.3 Experimental process

2.3.1 Screening of homogeneous groups

Homogeneous group who were screened out through the survey of mental health means students who have same problems of moderate to severe symptoms, for example, scores of interpersonal sensitivity ≥ 3 (moderate to severe symptoms in interpersonal sensitivity), scores of depression ≥ 3, scores of depression ≥ 3, etc. And the homogeneous students were divided into experimental group and control group.

2.3.2 Implement group counseling activities

There was no significant difference between the experimental group and the control group before the group counseling. The control group did not take part in the group counseling, while the experimental group of 149 members took part in group counseling activities of 6 units. Each activity had 6 units, and each unit included the theme, target, enforcement and activity project. Each group counseling lasted for 6 weeks, and was held once a week. Let me tabulate the results as Table 1.

As shown in Table 1, each set of counseling program included 6 units, although the contents were different, the process of each program was the same. Unit 1 was an ice breaking game, which helped group members become familiar with each other quickly, and shortened the psychological distance; Unit 2 and Unit 3 were exploration part, which helped group members know themselves better, and understand their emotion disorders and behavior problems etc. Unit 4 and Unit 5 were behavior training part, which helped members

Table 1. Freshmen group counseling activities design and arrangement.

Activity title	Activity object (members screened out)	Unit content	Activity target
"Harmonious Relationship" campsite	Scores of interpersonal sensitivity ≥ 3	Unit 1: first encounter Unit 2: security and trust Unit 3: unity and cooperation Unit 4: warmly praise Unit 5: self-surpass Unit 6: happy reunion	(1) To eliminate strangeness and loneliness. (2) To cultivate interpersonal communication ability. (3) To enhance the sense of collective honor and develop co-operation.
"Sunshine Mentality" campsite	Scores of depression ≥ 3	Unit 1: first encounter Unit 2: first self-cognition Unit 3: multi-faceted ego Unit 4: interpersonal cognitive bias Unit 5: accept imperfection Unit 6: happy reunion	(1) To help members of high scores in depression to reduce depression. (2) To help members to improve the level of self-concept, self-acceptance and self-identity. (3) To enforce members' understanding of positive emotions.
"Hearty Fly" workshop	Scores of anxiety ≥ 3	Unit 1: contract and encounter Unit 2: trust and integration Unit 3: clearing off distress Unit 4: imagine relaxing training Unit 5: muscle relaxing training Unit 6: meeting the future	(1) To help members to realize moderate-intensity anxiety can improve learning efficiency. (2) To help members to master a set of methods for releasing tension before examinations. (3) To adjust the unreasonable idea and train self-confidence so that members can be accepted and cared. (4) To help others to find their own value and obtain satisfaction and confidence so that they can reduce examination anxiety.
"Return Nature" workshop	Scores of obsessive-compulsive ≥ 3	Unit 1: first encounter Unit 2: self-portrait Unit 3: to accept imperfection Unit 4: to come very naturally Unit 5: to recall desensitization Unit 6: to reach success	(1) To help compulsive members to realize obsessive compulsive is the result of perfection. (2) To change unreasonable ideas through adjusting the unreasonable cognitive. (3) To help members to master the therapy of compulsive behavior and compulsive ideas.

learn operation methods and skills of behavior therapy through various group games, so that they can relieve moods and change their behavior patterns. Unit 6 was the happy ending, and after the group counseling, their problems were solved. Also they were reacquainted themselves and acquired a new start. The director of group counseling was a full-time counselors, and the site for activities was mostly the group psychological training room.

2.3.3 *Group testing*

Both experimental groups and control groups were tested with SCL-90 before and after the group counseling, and papers were taken back several minutes later. In addition, we designed questionnaires for this research ourselves, and group members summed up and shared their satisfaction and comment showing satisfaction about their experience, which were also used as one of assessing factors for the effect of group counseling.

2.3.4 *Data processing*

SPSS17.0 software was used for data analysis

3 RESULTS AND ANALYSIS

Both experimental groups and control groups were T-tested with SCL-90, and results showed that there was significant difference in depression, anxiety and obsessive-compulsive, except interpersonal sensitivity, which means the mental health level of students who received system group counseling and training showed significant improvement than those who didn't receive group counseling (as shown in Table 2).

The scores of four control groups after the group counseling decreased a lot than before the group counseling, and T test reached a significant level, which showed that targeted group counseling can efficiently alleviate freshmen's psychological problems and distinctly promote the mental health level of freshmen. The differences of control groups in SCL-90 factors before and after group counseling are shown in Table 3.

Share and summary of experimental groups. The final report of experimental groups and views on behavioral changes were also the basis of assessing the effectiveness of group counseling. 90% of members said they liked group counseling activities and more than 84.6% of the members thought the effectiveness of the activity was good. At the same time more than 70% of the members thought they knew themselves better after the group consulting, while more than half of them said they liked the way of relaxation. One student said "I gained psychological sunshine and better mood", another member said "I found many people were

Table 2. Comparison between experimental group and control group in SCL-90.

Group	Number of students	Factors	X ± SD	T	Sig
"Harmonious Relationship" campsite control group	38 40	Interpersonal sensitivity	1.88 ± 0.63 2.15 ± 0.66	−1.85	0.069
"Sunshine Mentality" campsite control group	36 32	Depression	2.12 ± 0.69 2.58 ± 0.77	−2.60	0.012*
"Hearty Fly" workshop control group	37 34	Anxiety	1.98 ± 0.71 2.67 ± 0.88	−6.43	0.000**
"Return Nature" workshop control group	38 45	Obsessive-compulsive	2.31 ± 0.50 2.59 ± 0.53	−2.46	0.016*

Table 3. Comparison of experimental groups in SCL-90 before and after group counseling.

Group names	Number of students	Factors	Before group counseling X ± SD	After group counseling X ± SD	T	Sig
"Harmonious Relationship" campsite	38	Interpersonal sensitivity	3.15 ± 0.79	1.88 ± 0.63	9.94	0.0000**
"Sunshine Mentality" campsite	36	Depression	3.60 ± 0.88	2.12 ± 0.69	7.99	0.0000**
"Hearty Fly" workshop	37	Anxiety	3.17 ± 0.91	1.98 ± 0.71	6.27	0.0000**
"Return Nature" workshop	38	Obsessive-compulsive	3.44 ± 0.65	2.31 ± 0.50	8.49	0.0000**

living in confusion and pain, but we finally found out the outlet". We can see the effect of group counseling from the words of same participants mentioned above.

4 ANALYSIS AND DISCUSSION

4.1. *Necessity of group counseling*

After the pressure of college entrance examinations, some students bring former psychological problems into the university campus, such as examination anxiety, interpersonal communication disorder. These problems didn't disappear, on the contrary, adaptive problems, life style changes and changes in interpersonal relationships also continued to exist commonly. If all the psychological problems cannot be diagnosed immediately, they will affect the students' studies and life greatly.

Group counseling can help to explore oneself, change one's behavior, accept new ideas more readily, and eventually achieve the purpose of cognitive change and behavioral management. This activity will effectively prevent the occurrence of unfortunate events in school, and lay the foundation for better life.

4.2. *The results and reasons of group counseling*

Group counseling combines cognitive therapy with positive psychology. Cognitive behavior therapy, which is based on cognitive behavior technology, is formed from cognitive theories and behavior therapy, and systemic psychological therapy [3]. Moreover, positive psychology gives a new interpretation for many mental phenomena, so that it can arouse people's positive strengths and good qualities, and eventually it can help both ordinary people and talented people to exploit their potentialities and acquire a good life [4]. This research shows that group counseling which is designed on the basis of freshmen psychological survey can greatly improve students' adaptive capacity and mental health level, and these results are similar to the results of the such surveys research [5–8].

4.3. *Promoting group counseling in teachers and students*

Psychological health education highlights the importance and necessity of the work or college students. Psychological health education is not only the duty of professional psychology teachers but also the responsibility of all those who work with students. Psychological health education must be strengthened and promoted in all counselors, teachers and students. Psychological health education of college students is an important part of school moral education, and is also an important measure to implement quality education. Student counselors cannot solve various problems of college students in today's society. Group counseling, which is a new work principle and method, can help students learn from each other through psychological games and grow together; The student cadres can be good assistants for students counselor, and the Peer Counseling conducted by student cadres is a new mode, which has been an important supplement to the professional consulting in students psychological education. Psychological organization and class committee in psychology can work as carrier for psychological education in college, which extend activities like group consulting, scene play, psychological salon and Peer growth. So that professional team, students counselors and students cadres can cooperate with each other and form three-level network of mental health education in Colleges, which established a set of strict and effective working system for mental health and psychological crisis intervention, improved the effectiveness of student work.

REFERENCES

[1] Fumin Fan, Group Psychological Counseling, Beijing: Higher Education Press, 2005.
[2] Jihua Xiao, Jun Xu & Connell et, The Mental Health Scale (enlarged edition), Chinese Mental Health Journal, 1999.
[3] Ruolan Xu, On Theoretical Study and Application of Cognitive Behavioral Therapy, Journal of Chengdu University of Technology (Social Sciences), December 2006, Vol 14. No 4.
[4] Jun Ren & Haosheng Ye, Research on the Core Value of the Contemporary Psychology, Journal of Shanxi Normal University (Philosophy and Social Science Edition), July 2004, Vol 33. No 4.
[5] Kai zheng & Yuanqin He, Experimental Study of Group Counseling on College Students' Relationship, China Journal of Health Psychology, 2009, Vol 17. No 8.
[6] Yingli, Aishu Liu & Shouchen Zhang, Effect of Group Psychological Guidance on Interaction Anxiety of College Students, Chinese Journal of Health Psychology, 2005, Vol 13. No 4.
[7] Chunzi Peng, liang Shi Yan & XiaoHong Ma, Effect of Group Therapy on Social Anxiety of College Students, Chinese Mental Health Journal, 2003, Vol 17. No. 4.
[8] Cuihong Zhang, Zheng Xie & Zuoming Chen, Effects of Group Counseling on Social Anxiety of College Students, China Journal of Health Psychology, 2007, Vol 15. No 4.

Future Information Engineering and Manufacturing Science – Zheng (Ed)
© 2015 Taylor & Francis Group, London, ISBN 978-1-138-02644-5

Effect of the countervailing power on quality innovation of the upstream manufacturer

Kai Li, Wei-hua Chen, Hui-qing Su & Zhi-hui Liu
School of College of Business Administration, Northeastern University, P.R. China

ABSTRACT: A three-stage dynamic game model is built to study the effect of the countervailing power on quality innovation of upstream manufacturer. Through the analysis of best quality choice of supplier with the retailers' stackelberg competition, the results showed that when the leader retailer has the countervailing power, the quality innovation of manufacturer was inhibited with the enhancement of its countervailing power which reduces the market equilibrium price, increases the output further and causes the non-monotonic change of retailer's sales. In addition, the consumer welfare increases at first and then decreases with the countervailing power enhancing. Finally, the existence of waterbed effect for the wholesale price is confirmed under a certain condition.

Keywords: supply chain; quality innovation; stackelberg competition; countervailing power; waterbed effect

1 INTRODUCTION

Recently, the relation of traditional retail-supply industry has changed greatly with the increase of concentration degree for retail business. The role played by big retailer is becoming more and more important which also affects the competition action of upstream and downstream. In reality, retail organizations often carry out the sales promotion activity to attract consumers on the one hand and thereby put forward more beneficial supply contract terms to suppliers on the other hand which fully uses its scale advantage. This new type of vertical relations has attracted attention of many scholars and the regulation department. Especially, the research of the countervailing power for the big retailer has gradually become the focus of the study.

In 2008, the UK competition commission firstly investigated grocery retailers' Countervailing Power, especially for the upstream innovation investment cutting caused by retailers' Countervailing Power, and the 'waterbed effect' of the wholesale price[1]. Actually, the countervailing power of the large retailers will inevitably influence on the upstream of the competition strategy. In the Chinese dairy market, milk producers strong Countervailing Power forcing the farmers to provide low quality of raw milk is exposed repeatedly[2]. Similarly, many industries have the phenomenon of the strong downstream countervailing power which will inevitably cause some new industrial problems via this new type of vertical relations. One of the most direct problem is that the downstream countervailing power influence on upstream innovation quality, which is also a starting point of this study.

In theory, a lot of attention is paid on the influence of the countervailing power on the market price, products and consumer welfare. Research on Countervailing Power can be traced back to Galbraith's' Countervailing power Hypothesis'[3]. Recently, von-Ungern-Sternberg, Dobson & Waterson, Chen analyzed the 'Countervailing power Hypothesis'. von-Ungern-Sternberg and Dobson & Waterson (1997), respectively, to compete for the buyer forces formed downstream market concentration were studied, and that only when the retailer fierce competition "Countervailing power hypothesis" will be established. Chen considered that the difference of retailers scale exists and the dominant retailers has the countervailing power, and the edge retailer is the premise for 'Countervailing power Hypothesis'.

---

[1]UK Competition Commission (2008), Market Investigation into the Supply of Groceries in the UK. http://www.competition-commission.org.uk/rep_pub/reports/2008/538 grocery.htm.

[2]HealthNet, China's raw milk quality is the worst, http://health.msn.com.cn/Info/20100627/06371018563.shtml.
[3]Galbraith believes that the big retailer can get better supply contract than the competitor and save the cost of consumer.

Erutku, FU and Wu also concerned about the direct impact of the buyer countervailing force of the retailer and got some conclusions. These studies do not only deducted theoretically to a certain extent, but also lay a theoretical foundation for the study of indirect effects.

In addition, some scholars also concerned about the indirect effects. Chen, Inderst & Shaffer investigated the impact of the countervailing power on the upstream product based on the assumption of different sources of the countervailing power. Inderst & Wey concerned the incentive of the downstream Countervailing Power on the upstream product innovation and process innovation. Domestically, Qi Y. studied the impact of the countervailing power on the upstream product differentiation, and held that the increased Countervailing Power incented upstream to select complementary product supply. Some other scholars studied the indirect effect from the Purchaser Group, technology selection. However, few studies have focused on the effects of the countervailing power on the quality innovation of manufacturers, which is one of the major innovation of this research.

Researches on Quality Innovation were mostly undertaken basing on the vertical differentiation of products. Mussa and Rosen examined the consumer preferences, vertical differentiation and quality monopolist choice, and believed that the monopolist preferred to supply low-quality product for the first time. Economids studied the impact of vertical integration on the downstream manufacturers quality and innovation when product had the network externality, and believed that vertical integration would encourage to merger quality innovation of downstream manufacturers and suppress quality improvement of competitors. Similarly, Farrell & Katz examined incentives of Quality Innovation for upstream manufacturers by the vertical integration via building different upstream market structures (monopoly and competition), and believed that vertical integration promotes the innovation of merged firms, and inhibit the innovation of competitive manufacturers. Through assuming that the quality was exogenous and upstream and downstream carried non-cooperative negotiations, Battigalli & Fumagalli studied incentives of quality choice for upstream manufacturers by Countervailing Power, and believed that the Countervailing Power motivate providers to provide low quality products which harmed the consumer welfare, and vendors will choose to supply higher quality products when the long-run equilibrium existed. Zhang F. et al. investigated the quality improvement strategies in duopoly market enterprise and found that the quality innovation depended on the quality of the original product for the short-term equilibrium and exogenous quality.

In fact, the increase of the countervailing power of the retailer will inevitably affects the manufacturer quality innovation. Therefore, by assuming that quality innovation of the manufacturers requires a certain cost and the dominant retailer has the countervailing power, the influence of the countervailing power on the quality innovation of the manufacturer was researched, and the market equilibrium and consumer welfare under the competition structure are analyzed.

The remaining contents are arranged as following: The first is the literature review; The second part is set as the basic model assumption; The third part analyzes the countervailing power and Influence on the quality innovation of the manufacturer; The fourth part is Balance and extended discussion; The last is the conclusion.

2 THE BENCHMARK MODEL

A two-tier market structure is built in order to investigate the impact of the countervailing power on the quality innovation of the manufacturer. The upstream monopoly manufacturer M provides specific level s of quality products, and the downstream retailers of R_1 and R_2 carried the Stackelberg competition. Assuming polarization of the retailers market, the lead force R_1 of the retailers have the countervailing power, and the follow R_2 is the recipient of the contract (take it or leave it). So the R_1 can obtain a certain number of discounts in the procurement process $\gamma(0 < \gamma < 1)$. γ represents the countervailing power of R_1. The counter power of the R_1 increased with the augment of γ. The similar assuming can be found in the research of Erutku and Fu[4]. w_1 and w_2 are the wholesale prices of the retailers R_1 and R_2 respectively with $w_1 = w_2(1-\gamma)$.

It is assumed that consumers are uniformly distributed on the segment [0 1] and there is a minimum quality preference θ which also obey the segment [0, 1]. Each consumer can (or not) buy one unit of a maximum of commodity. Thus, a single consumer utility function can be described as follows

$$U = \begin{cases} \theta s - p, & \text{consumer buys one unit} \\ & \text{product of quality } s, \\ 0 & \text{not buy} \end{cases} \quad (1)$$

[4]Erutku and Fu investigate the impact of countervailing power under the assumption of non-cooperation game of upstream and downstream. Erutku (2005) check the effect of the buyer countervailing power on the market balance with alternative products and retailers' price competition, while Fu examine the influence of the buyer power on the market balance with the state-owned chain enterprise of the buying power.

where U indicates the level of the individual consumer's utility and p represents the product's retail price. Of course, consumer utility also depends on other factors like retail level of service, convenience, etc. In order to avoid interference of these factors and to simplify the study, consumer utility depends only on the product quality and price. When $\theta s - p = 0$, the preference of consumers for the purchase of products or not has no difference, and the market demand function can get

$$p = s - s(q_1 + q_2) \qquad (2)$$

Assume the marginal cost of the manufacturer M as c and standardize $c = 0$. In addition, improvements in the quality of the projection will also be required to pay the marginal cost of innovation $MC(s) = 1/2\,s^2$. So the manufacturer's total cost function can be described as follows:

$$C(q_1, q_2, s) = \left(c + \frac{1}{2}s^2\right)(q_1 + q_2) = \frac{1}{2}s^2(q_1 + q_2) \quad (3)$$

$C(\bullet)$ represents the total production cost of the manufacturer M. From the Eq. (4), the total cost includes the manufacturing cost and the innovation cost two parts, and total cost and the marginal cost of the manufacturer are both raised when the manufacturer improve product quality, that is $\partial C/\partial s > 0$ and $\partial^2 C/\partial s^2 > 0$.

Based on the above conditions, the present study examines the change of best quality level with the enhanced countervailing power, and the influence of countervailing power of the dominant retailer on the market equilibrium and consumer welfare.

3 COUNTERVAILING POWER AND THE QUALITY INNOVATION OF MANUFACTURER

To check the independent effect of the countervailing power on the quality innovation of the manufacturer, the complete information dynamic game model is chosen to analyze the whole process, while the situation of incomplete information can be treated as a new direction for the future. The specific game sequence is described below. The first stage, the manufacturer observes the countervailing power γ of dominant retailer R_1 and produces the product of quality level s. The second stage, the wholesale price for follow retailer w_2 is made based on the benefit maximization principle by the manufacturer and the price of dominant retailer $w_1 = w_2(1 - \gamma)$. The last stage, retailers carry on the stackelberg competition. The dominant retailer R_1 enter the market first and determine its market capacity q_1, and the follow retailer R_2 observe the

size of the R_1 market, to enter the market and its sales q_2 is determined.

Based on the above game process, the profit function of the manufacturer M can be written as:

$$\begin{aligned}\pi_M &= w_1 q_1 + w_2 q_2 - C(x, s) \\ &= w_2(1 - \gamma)q_1 + w_2 q_2 - \frac{1}{2}s^2(q_1 + q_2)\end{aligned} \qquad (4)$$

The profit functions of R_1 and R_2 can be got similarly

$$\pi_{R_2} = (p - w_2)q_2 = [s - s(q_1 + q_2) - w_2]q_2 \qquad (5)$$

$$\pi_{R_1} = (p - w_1)q_1 = [s - s(q_1 + q_2) - w_1]q_1 \qquad (6)$$

The game process can be solved by backward induction. Firstly, the retailer R_1 and R_2 perform the Stackelberg competition to ensure market balance in the third stage of the game. The retailer R_2 determines its sales volume q_2 after the observation of sales volume of R_1:

$$q_2 = \frac{s - sq_1 - w_2}{2s} \qquad (7)$$

The retailer R_1 enter into the market and determine its market scale q_1. The retail price $p = s - s(q_1 - (w_2 - s + sq_1/2s))$ is got after the substitution of Eq. (7) into the Eq. (2). Combined with Eq. (6) (First derivative of q_1), the sale of R_1 is:

$$q_1 = \frac{s - 2w_1 + w_2}{2s} \qquad (8)$$

The manufacturer decide the wholesale price w_2 for the follow retailer. The dominant retailer gets the wholesale price $w_1 = (1 - \gamma)w_2$ according to its countervailing power. the wholesale price of R_2 is

$$w_2 = \frac{s(3 - 2\gamma)(s + 2)}{16\gamma^2 - 16\gamma + 12} \qquad (9)$$

$$w_1 = (1 - \gamma)w_2 = \frac{s(2\gamma - 3)(\gamma - 1)(s + 2)}{4(4\gamma^2 - 4\gamma + 3)} \qquad (10)$$

the balance price and production of retail market are given as follows:

$$q_1 = \frac{\gamma(s + 2)}{2(4\gamma^2 - 4\gamma + 3)} - \frac{s}{8} + \frac{1}{4},$$

$$q_2 = \frac{s}{16} - \frac{(6s + 12)}{16(4\gamma^2 - 4\gamma + 3)} + \frac{3}{8} \qquad (11)$$

$$p = \frac{s(s + 6)}{16} - \frac{s(4\gamma - 3)(s + 2)}{8(4\gamma^2 - 4\gamma + 3)} \qquad (12)$$

In the first stage, the manufacturer decide the quality level on the basis of competition between retailers. Substitute the Esq. (9), (10), (11), (12) into Eq. (4), the best quality level of the manufacturer for the market balance is got according to the benefit maximization condition $(\partial \pi_M / \partial s) = 0$

$$s^* = \frac{2(40\gamma^2 + 18 - 24\gamma - \Delta)}{3(4\gamma^2 - 12\gamma + 9)} \tag{13}$$

where $\Delta = \sqrt{1552\gamma^4 - 1632\gamma^3 + 1368\gamma^2 - 216\gamma + 81}$ [5]. After the check of the effect of the countervailing power of the retailer on the quality innovation of the manufacturer, the conclusions are obtained as follows.

Proposition 1: With the augmentation of the countervailing power of the dominant retailer, the manufacturer tends to choose supply products of low quality.

Proof, the effect of the countervailing power is checked according to Eq. (14).

$$\frac{\partial s^*}{\partial \gamma} = \frac{128\gamma(24\gamma + \Delta - 40\gamma^2 - 18)}{(3 - 2\gamma)^3 \Delta} < 0 \tag{14}$$

Q.E.D

The quality innovation is restrained by the countervailing power because the best quality level decreased with the strengthen of the countervailing power of the dominant retailer under the balance circumstance from Eq. (14). The reason is that the wholesale price discount rate rises as the enhancement of the countervailing power, resulting in a decline in the profitability of the manufacturers supply. In order to compensate for the loss of profit, manufacturers lower the product quality reduce its total marginal cost to increase profit on the one hand and by adjusting the wholesale price to improve profit margin on the other hand. But there needs to be pointed out that manufacturers in the adjustment of product quality and supply decisions, need to guarantee all retailers are active in the market.

4 BALANCE AND EXTENDED DISCUSSION

4.1 Countervailing power and 'waterbed effect'

In reality, the countervailing power lies in its impact mostly on the trading conditions of upstream and downstream, that is, countervailing power prompts the retailer R_1 to get lower wholesale price, while the wholesale price of the competitor R_2 will also change. That is also a study focus of countervailing power, which is also examined in this paper.

Substitute Eq. (14) into (9) and (10) to give the retailer wholesale price R_1 and R_2 under the market equilibrium condition.

$$w_2^* = \frac{(40\gamma^2 - 24\gamma - \Delta + 18)(60\gamma + \Delta - 52\gamma^2 - 45)}{9(2\gamma - 3)^3(4\gamma^2 - 4\gamma + 3)} \tag{15}$$

$$w_1^* = \frac{(\gamma - 1)(24\gamma + \Delta - 40\gamma^2 - 18)(60\gamma + \Delta - 52\gamma^2 - 45)}{9(2\gamma - 3)^3(4\gamma^2 - 4\gamma + 3)} \tag{16}$$

Examine the relationship between countervailing power of retailer R_1 and the wholesale price ('waterbed effect'), the following conclusion is got.

Proposition 2: With Enhanced countervailing power of retailer of the dominant retailer, the wholesale price of the dominant retailer reduces while the wholesale price of the follow retailer rises first and falls then. When the countervailing power is weak, the 'waterbed effect' happens.

Refer to the Appendix $A-1$ for the proof.

The countervailing power of R_1 in this study reflects its forces on the ability to get volume discounts. Along with the enhanced countervailing power and the erode ability of profit margins for manufacturers, manufacturers is forced to raise the wholesale price for retailer R_2 to make up for the lost. However, w_2^* cannot increase all the time for keeping R_2 active in the market. So the manufacturer will reduce the wholesale price for all retailers to improve procurement amount and compensate for their profit loss, when $\gamma > 0.059$. This conclusion differs from viewpoint of Chen[1] who assumes that the products are of the same quality and the countervailing power of dominant retailer is the ability to negotiate. Their result shows that the countervailing power rises while reducing the wholesale price of dominant retailer and edge retailer, that is, there is not 'waterbed effect'. Our research is also unlike Inderst & Valletti [14] studies who assume that the external choosing value of retailer is its Countervailing Power. It is believed that the increasing of the Countervailing Power affects the wholesale price of the retailer and the countervailing power causes the 'waterbed effect' of the wholesale price market. In this article by assuming that supplier endogenously choose the product quality levels, and the dominant retailer has the countervailing power, the results show that only when the countervailing power is relatively low, the 'waterbed effect' of the wholesale price

---

[5]To ensure that all manufacturers and retailers are active in the market (profit level is positive)under the assumption, a corner solution s is rejected here and the manufacturer's optimal quality strategy s^* is got finally.

market happens. The relation of w_1^*, w_2^*, and γ is simulated results shown in Figure 1.

4.2 *Countervailing power and market equilibrium*

As consumers utility level strictly depends on the level of product quality and price, and countervailing power of R_1 affects the quality choice of the manufacturer, countervailing power is bound to affect the purchase decisions of the consumer. The effect of the countervailing power of R_1 on the market equilibrium is studied.

Retail price and sales volume is got via substitution Eq. (14) into Eqs. (11) and (12)

$$p^* = \frac{40}{2\gamma - 3} + \frac{287}{36} + \frac{8\gamma + \Delta - 6}{36(4\gamma^2 - 4\gamma + 3)} - \frac{5\Delta - 360}{6(2\gamma - 3)^2}$$

(17)

$$q_1^* = \frac{42\gamma + \Delta - 52\gamma^2 + 56\gamma^3 - 2\gamma\Delta + 9}{108 - 96\gamma^3 + 240\gamma^2 - 216\gamma}$$

(18)

$$q_2^* = \frac{\Delta - 138\gamma + 2\gamma\Delta + 188\gamma^2 - 152\gamma^3 + 9}{24(3 - 2\gamma)(4\gamma^2 - 4\gamma + 3)}$$

(19)

To investigate the effect of the countervailing power of R_1 on the p^*, q_1^* and q_2^*. We got proposition 3:

Proposition 3 As enhancement of the countervailing power against leader retailers, market equilibrium price is falling; the sales of dominant retailer increases firstly and then decreases; the sales of follow retailer decreases firstly and then increases.

Refer to the Appendix *A*–2 for the proof.

About the relationship research of countervailing power and consumer welfare, Chen thinks countervailing power improves consumer welfare

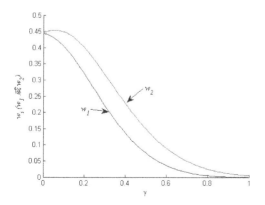

Figure 1. The change of wholesale prices with countervailing power enhancing.

when fringe retailers exist, but the investigation of Inderst & Shaffer (2007) states that countervailing power can reduce the product diversity, and then harm consumer welfare. The research of Wang also shows that the enhanced countervailing power of chain retail will weaken the welfare level of the consumer. Different from these researches, this paper believes that enhancement of countervailing power of leader retailers can affect non-monotonically on consumer welfare under the background of endogenous manufacturers quality control. This means weak countervailing power is good at improvement of consumer welfare while the strong countervailing power can make it worse.

5 CONCLUSION

The effect of the countervailing power for dominant retailer on the upstream quality innovation is analyzed and the market effect is also studied through the constructing upstream monopoly manufacturer, downstream duopoly retailer and the market structure of Stackelberg competition and assuming the endogenous choice to the quality of the product level for the manufacturer. The results show that the enhanced countervailing power suppresses the manufacturer's quality innovation; the countervailing power results in a decline of the equilibrium price and rise of output; the influence on the intermediate product wholesale price is not consistent. To be specific, the enhancement of the countervailing power deduces the wholesale price of dominant retailer, while the wholesale price of the follow retailer rises first and falls after. Namely, the 'waterbed effect' of the wholesale price happens when the countervailing power is weak. In the end, this paper argues that the increase of the countervailing power for the leader retailer results in a decline of consumer welfare level after rise first.

In fact, many studies have focused on the influence of the countervailing power on the competition strategy of the supplier. Inderst & Wey argue that the countervailing power suppresses the product innovation of the manufacturer and promote the technological innovation. Chen argues that the countervailing power can reduce the supply types. Battigalli *et al.* believe that the countervailing power helps improve the level of quality upstream under the long-term equilibrium. Different from those studies, it is accepted in this paper the countervailing power of the dominant retailer can inspire the manufacturers to produce products of low quality based on the assumption of the non-cooperative retailer-supplier relationship the conclusion of which offers a new theory to explain some current phenomena, such as the middle and upper dairy farmers supplying low quality raw milk in the dairy market, the makers providing low quality capsule products in the over-the-counter drug capsules market, etc. Of course, in other cases like the follow retailer has the countervailing power, the conclusion of this article is not changed.

Because of the complexity of the retailer-supplier relationship, the differences between the upstream and downstream way to trade, etc., this paper does not give all the situation of influence of the countervailing power on the upstream quality innovation. In addition, the upstream quality innovation is also affected by the regulation factors like taxes, subsidies and information asymmetry in reality. Therefore, extended study on the above problems can be developed in the future.

6 APPENDIX

A–1 Proposition 2 proving

According to Eqs. (A1) and (A2), the effect of the strengthen for the countervailing power of dominant retailer is studied.

$$\frac{\partial w_1}{\partial \gamma} = \frac{\begin{bmatrix} (21\Delta^3 - 4374\Delta - 746496\gamma \\ -236088\gamma^2\Delta + 495648\gamma^3\Delta \\ +76\gamma^2\Delta^3 - 645408\gamma^4\Delta - 32\gamma^3\Delta^3 \\ +531968\gamma^5\Delta - 258688\gamma^6\Delta + 49664\gamma^7\Delta \\ +3761640\gamma^2 - 10904544\gamma^3 + 20279808\gamma^4 \\ -25807104\gamma^5 + 22273920\gamma^6 - 12123648\gamma^7 \\ +3496704\gamma^8 + 63504\gamma\Delta - 68\gamma\Delta^3 + 15309) \end{bmatrix}}{9\Delta(4\gamma^2 - 4\gamma + 3)^2(2\gamma - 3)^4}$$

$$< 0 \tag{A1}$$

$$\frac{\partial w_2^*}{\partial \gamma} = \frac{2(24\gamma + \Delta - 40\gamma^2 - 18)(60\gamma + \Delta - 52\gamma^2 - 45)}{3(2\gamma - 3)^4(4\gamma^2 - 4\gamma + 3)}$$

$$- \frac{\begin{matrix}(3104\gamma^3 - 2448\gamma^2 + 1368\gamma - 108 - \Delta80\gamma + 24\Delta) \\ \times (60\gamma + \Delta - 52\gamma^2 - 45)\end{matrix}}{9\Delta(2\gamma - 3)^3(4\gamma^2 - 4\gamma + 3)}$$

$$- \frac{\begin{matrix}((3104\gamma^3 - 2448\gamma^2 + 1368\gamma - 108 - 104\Delta\gamma + 60\Delta) \\ \times (24\gamma + \Delta - 40\gamma^2 - 18))\end{matrix}}{9\Delta(2\gamma - 3)^3(4\gamma^2 - 4\gamma + 3)}$$

$$+ \frac{(8\gamma - 4)(24\gamma + \Delta - 40\gamma^2 - 18)(60\gamma + \Delta - 52\gamma^2 - 45)}{9(2\gamma - 3)^3(4\gamma^2 - 4\gamma + 3)^2}$$

$$\begin{cases} >0 & \gamma \le 0.059 \\ <0 & \gamma > 0.059 \end{cases} \tag{A2}$$

Q.E.D

A–2 Proposition 3 proving

From Esq. (A3)–(A5), the first derivatives of p^*, q_1^* and q_2^* are solved.

$$\frac{\partial p^*}{\partial \gamma} = \frac{776\gamma^3 - 612\gamma^2 + 342\gamma - 27 + 2\Delta}{9\Delta(4\gamma^2 - 4\gamma + 3)}$$

$$+ \frac{10\Delta - 720}{3(2\gamma - 3)^3} - \frac{80}{(2\gamma - 3)^2} - \frac{(8\gamma - 4)(8\gamma + \Delta - 6)}{36(4\gamma^2 - 4\gamma + 3)^2}$$

$$- \frac{7760\gamma^3 - 6120\gamma^2 + 3420\gamma - 270}{3\Delta(2\gamma - 3)^2} < 0 \tag{A3}$$

$$\frac{\partial q_1^*}{\partial \gamma} = \frac{\begin{matrix}\gamma(104\Delta + 5148\gamma - 8000\gamma^2 + 6208\gamma^3 - 1548) \\ +\Delta(2\Delta - 168\gamma^2 - 42) + 108\end{matrix}}{\Delta(96\gamma^3 - 240\gamma^2 + 216\gamma - 108)}$$

$$+ \frac{\begin{matrix}(288\gamma^2 - 480\gamma + 216) \\ \times (42\gamma + \Delta - 52\gamma^2 + 56\gamma^3 - 2\gamma\Delta + 9)\end{matrix}}{(96\gamma^3 - 240\gamma^2 + 216\gamma - 108)^2}$$

$$\begin{cases} >0 & \gamma < 0.854 \\ <0 & \gamma \ge 0.854 \end{cases} \tag{A4}$$

$$\frac{\partial q_2^*}{\partial \gamma} = \frac{\begin{matrix}(54\gamma - 45\Delta - 188\gamma^2\Delta + 136\gamma^3\Delta - 64\gamma^4\Delta \\ +126\gamma\Delta + 81 + 1512\gamma^2 - 4368\gamma^3 + 7376\gamma^4 \\ -5920\gamma^5 + 2560\gamma^6)\end{matrix}}{(8\gamma^3 - 20\gamma^2 + 18\gamma - 9)^2\Delta}$$

$$\begin{cases} <0 & \gamma < 0.429 \\ >0 & \gamma \ge 0.429 \end{cases} \tag{A5}$$

The change of retail marginal income is showed as follows from Eqs. (A6)–(A9).

$$MR_1 = p^* - w_1^*$$

$$= \frac{\begin{matrix}918\gamma + 9\Delta + 140\gamma^2\Delta - 136\gamma^3\Delta - 3384\gamma^2 \\ +8304\gamma^3 - 8240\gamma^4 + 5344\gamma^5 - 102\gamma\Delta + 81\end{matrix}}{36(3 - 2\gamma)^3(4\gamma^2 - 4\gamma + 3)} \tag{A6}$$

$$MR_2 = p^* - w_2^*$$

$$= \frac{\begin{matrix}9\Delta - 2646\gamma - 196\gamma^2\Delta + 232\gamma^3\Delta + 6120\gamma^2 \\ -13872\gamma^3 + 12880\gamma^4 - 9184\gamma^5 + 150\gamma\Delta + 81\end{matrix}}{36(3 - 2\gamma)^3(4\gamma^2 - 4\gamma + 3)} \tag{A7}$$

$$\frac{\partial MR_1}{\partial \gamma} = 11988\gamma + 297\Delta + 5964\gamma^2\Delta$$

$$-11456\gamma^3\Delta + 14960\gamma^4\Delta$$

$$-\frac{\left(32128\gamma^5\frac{\Delta}{3} + 42304\gamma^6\frac{\Delta}{9} - 86832\gamma^2 + 279792\gamma^3 - 618912\gamma^4 + 871616\gamma^5 - 865536\gamma^6 + 518400\gamma^7 - \frac{1665280\gamma^8}{9} - 1512\gamma\Delta - 1701\right)}{(2\gamma-3)^4(4\gamma^2-4\gamma+3)^2\Delta}$$

$$\begin{cases} \geq 0 & \gamma \leq 0.28 \\ < 0 & \gamma > 0.28 \end{cases} \qquad (A8)$$

$$\frac{\partial MR_2}{\partial \gamma} = -2\frac{\left(22842\gamma + 891\Delta + 13032\gamma^2\Delta - 25584\gamma^3\Delta + 35952\gamma^4\Delta - 26592\gamma^5\Delta + 12544\gamma^6\Delta - 199260\gamma^2 + 644328\gamma^3 - 1478736\gamma^4 + 2109408\gamma^5 - 2160960\gamma^6 + 1319808\gamma^7 - 495616\gamma^8 - 3078\gamma\Delta - 5103\right)}{3\Delta(2\gamma-3)^4(4\gamma^2-4\gamma+3)^2}$$

$$\begin{cases} < 0 & \gamma < 0.37 \\ \geq 0 & \gamma \geq 0.37 \end{cases} \qquad (A9)$$

Q.E.D

ACKNOWLEDGEMENTS

We are grateful to the funding of the National Natural Science Foundation "The formation mechanism and impact studies of countervailing power based on consumer preferences" (71172150).

REFERENCES

[1] Chen Z. Dominant retailers and the Countervailing Power Hypothesis. Rand Journal of Economics, 2003, 34(4): 612–625.

[2] Inderst R. & Wey C. How Strong Buyers Spur Upstream Innovation. London: London School of Economics, 2005: 1–36.

[3] Wang H. Countervailing Power, Transport Cost and Welfare. Journal of Industry Competition and Trade, 2010, (10): 41–53.

[4] Battigalli P. & Fumagalli C. Buyers Power and Quality Improvements. Milan: Bocconi University, 2007: 1–34.

[5] Galbraith, J.K. American Capitalism: The Concept of Countervailing Power. New York: Houghton Mifflin, 1952: 119–121.

[6] Hunter A. Notes on countervailing power. Economics Journal, 1958 (68): 89–103.

[7] Von-Ungern-Sternberg T. Countervailing Power Revisited. International Journal of Industrial Organization, 1996, 14(4): 507–519.

[8] Dobson P. W. & Waterson M. Countervailing Power & Consumer Prices. Economic Journal, 1997 (107): 418–430.

[9] Erutku C. Buying Power & Strategic Interactions. Canadian Journal of Economics, 2005, 38(4): 1160–1172.

[10] Fu H. & Li C. The Analysis of Countervailing Power Welfare for the State-owned Chain Enterprises. Industrial Economic Review, 2009, 8(4): 1–11.

[11] Wu X., Vertical Market Structure and Countervailing power Research. Industrial Economic Review, 2010, (1):39–47.

[12] Chen Z. Monopoly & Product Diversity: The Role of Retailer Countervailing Power. Carleton Economic Papers, 2004(1):4–19.

[13] Inderst R. & Shaffer G. Retail Mergers, Countervailing Power, and Product Variety. Economic Journal, 2007(117): 45–67

[14] Inderst R. & Valletti, T. M. Countervailing Power and the "waterbed effect". The Journal of Industrial Economics, 2011, 59(1): 1–20.

[15] Inderst R. & Wey C. Countervailing Power & Supplier Incentives. European Economic Review, 2007(51): 647–667.

[16] Qi Y. & Li K. & Liu Z. The Effect of the Countervailing power on the Competition Strategy of the Upstream Enterprise-Analysis Based on the Bargaining Balance. Operations Research and Management Science, 2012, 21(5): 217–222.

[17] Mussa M. & Rosen S. Monopoly & Product Quality. Journal of Economic Theory, 1978, 18:301–317.

[18] Economides N. Quality choice and vertical integration. International Journal of Industrial Organization, 1999, 17(6): 903–914.

[19] Farrell J. & Katz M. Innovation, Rent Extraction, and Integration in Systems Markets, The Journal of Industrial Economics, 2000, 48(2): 413–432.

Future Information Engineering and Manufacturing Science – Zheng (Ed)
© 2015 Taylor & Francis Group, London, ISBN 978-1-138-02644-5

To estimate a phase via two-entangled photons with multiple passes

L.X. Xia, R.Q. Liu, B. Xie, S.Q. Yu & X.L. Zhou
Department of Physics, Kashgar Teachers College, Kashgar, Xinjiang, China

ABSTRACT: In this paper, we use two-entangled photons to estimate an unknown phase by a super-stable optical interferometer with multiple passes. It is shown that, under a certain condition, our scheme enables considerable high precision to beat the standard quantum limit.

Keywords: entangled state; squeezed coherent state; cavity QED

1 INTRODUCTION

For optical phase measurements, at the fundamental level, its precision is limited by the number of quantum resources (photons or ions) that are used, leads to measure a phase with a precision remarkably improving the precision to $1/N$—the Heisenberg limit (HBL) (Nagata et al. 2007, Rarity et al. 1990). Whereas, using each resource independently, i.e., without entanglement, leads to measure a phase with a precision $1/\sqrt{N}$—the Standard Quantum Limit (SQL) (Kuzmich et al. 1998). It is shown by experiment that quantum entanglement enables higher precision to estimate an unknown phase. As a result, recent work in quantum interferometer has focused on N-photon $N00N$ states (Nagata et al. 2007, Rarity et al. 1990, Kuzmich et al. 1998, Walther et al. 2004, Mitchell et al. 2004, Resch et al. 2005), i.e., $(|N0\rangle + |0N\rangle)/\sqrt{2}$.

Unfortunately, such a conspicuously unconventional state is very difficult to produce. Up to now, the maximal number of photons or ions have been performed is only $N = 6$, where the SQL has been beaten only with two photons or four photons (Nagata et al. 2007, Resch et al. 2005). How to heighten precision is an important subject of a physical scientist in the measurement of a physical quantity.

Recently, Nagata *etc* demonstrated an optical phase measurement, with an entangled four-photon interference visibility greater than the threshold to beat the SQL (Nagata et al. 2007). Here they demonstrated an optical phase measurement with an entangled four photon interference visibility greater than the threshold to beat the SQL. And Sun *etc* propose and demonstrate experimentally a projection scheme to measure the quantum phase with a precision beating the SQL (Sun et al. 2008).

In this paper, we present a super-stable optical interferometer with multiple passes, and show that we use two-entangled photons to estimate an unknown phase; our scheme enables considerable high precision to beat the SQL.

This paper is organized as follows. In Section 2, we describe the super stable optical interferometer with multiple passes. In Section 3 and 4, we use single photon and two entangled photons to estimate an unknown phase, respectively. Our scheme enables considerable high precision to beat the SQL. And conclusions and remarks are made in Section 4.

2 A SUPERSTABLE INTERFEROMETER

In this section, we first give a brief introduction of the super-stable optical interferometer with multi-pass. This interferometer is consisting of one beam splitters (BS, 50:50) and three single-face plane mirrors, by 1, 2 and 3, in a horizontal plane (Nagata et al. 2007, Higgins et al. 2007). See Figure 1, it is $q = 2$, i.e., the mode goes d through the Phase Shift (PS) for two times, where A is a double-face plane mirror, and four single-face plane mirrors B, C, D and E. All of the five plane mirrors are fixed in a vertical plane. This interferometer can keep the optical path lengths in modes c and d to be sub-wave-length (nm) stable. The photons are input in modes a and/or b, and detected in modes e and/or f, after a PS, ϕ is applied to mode d for q times (here $q = 2$). For the other q, such as $q = 4; 8; 16; 32$, we can also get them by using the similar technique.

Due to a multi-photon interference effect, leads to a multi-photon detection probability (Nagata et al. 2007) as follows

$$P = \eta(1 - \cos qN\phi)/2, \tag{1}$$

where η is the intrinsic efficiency, N is the number of detecting the photons in the modes e and f.

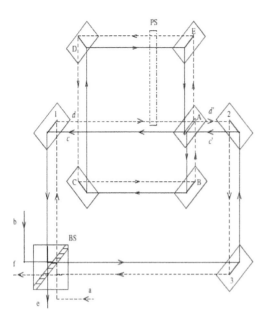

Figure 1.   A schematic of a super-stable optical interferometer with multiple passes, where $q = 2$.

We can obtain the corresponding SQL (Nagata et al. 2007) is

$$(\Delta\phi)_S = \sqrt{\eta/qN}, \tag{2}$$

And if the visibility of the interference is set V, then its precision (Nagata et al. 2007) is

$$\Delta\phi = 1/VqN \tag{3}$$

To beat the SQL, we <u>must make</u> $\Delta\phi < (\Delta\phi)_S$ to be held, i.e., $1/VqN < \sqrt{\eta/qN}$. <u>We find that its</u> threshold of visibility is $V_{th} = 1/\sqrt{\eta qN}$. If $V_{th} \geq 1$, then the precision can't beat the SQL, because the visibility of an experiment is not higher than 1; If $V_{th} < 1$ and $V > V_{th}$, then $\Delta\phi < (\Delta\phi)_S$, and its precision can beat the SQL (Okamoto et al. 2008).

In section 3 and section 4, by the super-stable interferometer with multiple passes, we use single photon and two-entangled photons to estimate an unknown phase and show that our scheme enables considerable high precision and can beat the SQL.

3   USING SINGLE PHOTON

In the following, we put a single photon in the input a of the interferometer, and put no photon in the input b, then the input state is $|10\rangle_{ab}$, after the BS, the state is

$$(|10\rangle + |01\rangle)/\sqrt{2} \tag{4}$$

After the mode d goes through the PS with q times, this state evolves to

$$(|10\rangle + e^{iq\phi}|01\rangle)/\sqrt{2}. \tag{5}$$

And after the same BS, the probability of detecting one photon in the mode e is

$$P_{1e} = (1 - \cos q\phi)/2, \tag{6}$$

where $\eta = 1$ and $N = 1$. According to Eq. (2), the corresponding SQL is $(\Delta\phi)_S = 1$. For the single-pass, $q = 1$, then its threshold of visibility is $V_{th} = 1$. Thus, using single photon with single-pass, its precision can't beat the SQL (Nagata et al. 2007). For the multi-pass, $q \geq 2$, then $V_{th} \leq \sqrt{2}/2 \approx 70.7\%$. In the literatures (Nagata et al. 2007, Higgins et al. 2007), it is shown that if $q = 2, 32$, the corresponding visibility can be $96.7\% \pm 0.5\%$ and $92\% \pm 0.5\%$ for single photon, respectively. So that $V > V_{th}$ is held, and it can beat the SQL to use single photon with multi-pass in experiment.

4   USING TWO-ENTANGLED PHOTONS

Next we put a single photon in each input of the interferometer, and then the input state is $|11\rangle_{ab}$. And which goes through the BS, the state is

$$(|20\rangle_{cd} + |02\rangle)/\sqrt{2} \tag{7}$$

The quantum interference of the two-photon amplitudes cancels the $|11\rangle_{ab}$ term. After the mode d goes through the PS with q times, this state evolves to (Hong et al. 1987)

$$(|20\rangle_{cd} + e^{i2q\phi}|02\rangle)/\sqrt{2} \tag{8}$$

The probability of detecting two photons in the modes e and f is

$$P_{1e1f} = (1 - \cos 2q\phi)/2 \tag{9}$$

where $\eta = 1$ and $N = 2$. For the single-pass, $q = 1$, then $V_{th} = 1/\sqrt{2} \approx 70.7\%$. If $q = 2$, then $V_{th} < 1/\sqrt{2}q = 50.0\%$.

It is necessary to address the experimental implementation of the proposed scheme. In the literatures [1, 8] show that if $q = 2, 32$, the corresponding visibility V can be $96\% \pm 0.5\%$ and $92\% \pm 0.5\%$ for single photon, respectively. Therefore, $V > V_{th}$ is held for an appropriate q, and the SQL can be beaten by using two-entangled photons with single-pass or multi-pass in experiment.

5 CONCLUSION

In conclusion, we propose a scheme to estimate a phase via single photon or two-entangled photons with multiple passes by two-entangled photons. And it is shown that, using single photon and two-entangled photons, we could estimate an unknown phase with a high precision than that to beat the SQL. Especially, our scheme is also very feasible by current technology. This is the reason that an increasing q can decrease the V_{th}, but V is almost unchanged. Furthermore, using N-entangled photons ($N > 2$), our scheme can attain higher precision than using single photons or two entangled photons under the same q passes through a phase shift.

This scheme has more promise for a wide range of metrology tasks and many important applications, especially in the light of continued development of high-flux single-photon or multiple-photon sources and efficient detectors.

ACKNOWLEDGMENT

This work is supported by the National Science Foundation of China (Grant No. 11064006, 11347194), the Person with Ability Study Foundation of Kashgar Teachers College (No. 15 JJ05018). One of the authors (X.L. Xia) would like to thank Professors X.L. Zhou and L. Guo, and Doctors R.Q. Liu.

REFERENCES

[1] Nagata, T. Okamoto, R. O'Brien, J.L. Sasaki K. Takeuchi, S. 2007. Beating the Standard Quantum Limit with Four-Entangled Photons. *Science* 316: 726–729.

[2] Rarity, J.G. Tapster, P.R. Jakeman, E. Larchuk, T. Campos, R.A. Teich M.C. Saleh, B.E. 1990. Two-photon interference in a Mavh-Zehnder interferometer. *Phys. Rev. Lett.* 65: 1348–1351.

[3] Kuzmich, A. Mandel L. 1998. Sub-shot-noise interferometric measurements with two-photon states. *Quant. Semiclass. Opt.* 10: 493–500.

[4] Walther, P. Pan, J.W. Aspelmeyer, M. Ursin, R. Gasparoni, S. Zeilinger, A. 2004. De Broglie wavelength of a non-local four-photon state. *Nature* 429: 158–161.

[5] Mitchell, M.W. Lundeen, J.S. Steinberg, A.M. 2004. Super-resolving phase measurements with a multiphoton entangled state. *Nature* 429, 161–.

[6] Resch, K.J. Pregnell, K.L. Prevedel, R. Gilchrist, A. Pryde, G.J. O'Brien1, J.L. White, A.G. 2005. Time-reversal and super-resolving phase measurements. *quant-ph* 0511214.

[7] Sun, F.W. Liu, B.H. Gong, Y.X. Huang, Y.F. Ou, Z.Y. Guo, G.C. 2008. Experimental demonstration of phase measurement precision beating standard quantum limit by projection measurement, *EPL* 82: 24001-1-24001-5.

[8] Higgins, B.L. Berry, D.W. Bartlett, S.D. Wiseman, H.M. Pryde, G.J. 2007. Entanglement-free Heisenberg-limited phase estimation. *Nature* 450: 393–396.

[9] Hong, C.K. Ou, Z.Y. Mandel, L. 1987. Measurement of sub-pico-second time intervals between two photons by interference. *Phys. Rev. Lett.* 59, 2044–2046.

[10] Okamoto, R. Hofmann, H.F. Nagataet, T. Obrien, J.L. Sasaki, K. Takeuchi, S. 2008. Beating the standard quantum limit, *quant-ph* 0804.0087.

Future Information Engineering and Manufacturing Science – Zheng (Ed)
© 2015 Taylor & Francis Group, London, ISBN 978-1-138-02644-5

Unsupervised feature selection algorithm based on spectral clustering and analysis of variance

Qing Qiang Wu, Yi Dong Wang, Zhi Yan Wang & Kun Hong Liu
Software School, Xiamen University, Xiamen, China

ABSTRACT: Now many data mining tasks have to deal with high dimension problems. Feature selection is put forward to solve such problems, the original data set is handled with feature selection, which can facilitate clustering, classification and other process of data mining by the relevant feature subset. In reality, the label of data set is difficult to obtain or cost relatively expensive in many cases, so unsupervised feature selection algorithm were proposed. This paper mainly focus on the unsupervised feature selection algorithm based on spectral clustering of improvements, the analysis of variance is introduced to the eigenvectors space by Spectrum mapping. Experiments show that the improved strategy can effectively increase the effect of unsupervised feature selection algorithm based on spectral clustering.

Keywords: feature selection; unsupervised; spectral cluster; variance

1 INTRODUCTION

In many computer applications there are high dimensional datasets such as massive web, database and so on, which put forward challenges in statistics, pattern recognition, artificial intelligence, data mining and other related disciplines. Handling high dimensional data usually increases time and space complexity and interferes with the analysis results simultaneously. In order to overcome these problems, feature selection algorithms are proposed, Which can realize the dimension reduction of high dimensional data. after that the feature subset can better reflect the nature of the data structure and improve the efficiency of data analysis and processing.

Feature selection algorithm can be divided into two categories: supervised and unsupervised. Now typical supervised feature selection algorithm includes Pearson Correlation coefficients [1], Fisher score [2] and Information gain[3]. In practice, many data sets lack label or have no label, in that the label generally is difficult to obtain or costs hugely. Therefore, it is necessary to study unsupervised feature selection algorithm, this paper will focus on the unsupervised feature selection based on spectral clustering analysis[4] and propose improving strategies.

This paper is bases on the multiple classification problems using unsupervised feature selection algorithm (MCFS), which uses Spectral Embedding for Cluster Analysis [5] and Learning Sparse Coefficient Vectors [6] to realize unsupervised feature selection algorithm, and propose the optimization

by fusioning the traditional feature selection and MCFS.

The rest of the paper is organized as follows: Section 2 is provided a brief review of the related work; Section 3 is introduced the principle and the improved method; in Section 4, the related experimental results are presented; Section 5 is provided the concluding remarks.

2 RELATED WORK

Feature selection methods can be classified into "wrapper" methods and "filter" methods [7, 8]. Wrapper methods through data mining algorithm to evaluate features, feature selection process depends on data mining algorithm such as classification and clustering. Using filter method mainly examines intrinsic properties of data to evaluate features prior to the mining tasks.

For unsupervised wrapper method, the mining algorithm which depends on is usually cluster [9, 10]; These feature selection methods will be associated with the process of feature selection and clustering results, which use the cluster results to guide the feature selection process, and then improve cluster effect. These methods tend to have high computational complexity, and are not suitable for large scale data set. This is why the unsupervised feature selection algorithm based on spectral clustering (MCFS) [4] chooses filter method.

Maximum variance may be the simplest but effective unsupervised feature selection method.

this method is assessed using variance of feature as the basis, MCFS is improved by this strategy in this paper, the only difference is that this article will use variance of square of feature as the assessment. Because cluster distance measure adopts Euclidean distance in this paper, so we choose to adopt the square of feature. the smaller variance of feature, the less information it contains, if the range of variation of feature value changes is not big, it will have no meaning to distinguish among various conditions. On the other hand, the greater variance of feature, the more information it contains. Therefore, by selecting features with larger variance, Maximum variance can gain some features which contains more information.

Laplacian Score[11] is a unsupervised feature selection algorithm, which is aimed at improving the effect of maximum variance. it can well reflect the data manifold structure. the goal is to obtain features which have low Laplacian score, these features have higher information and strong local structure expressing ability of the sample space.

NDFS[12] is another one, to exploit the discriminative information in unsupervised scenarios, performing spectral clustering to learn the cluster labels of the input samples, during which the feature selection is performed simultaneously. To reduce the redundant or even noisy features, $l_{2,1}$-norm minimization constraint is added into the objective function, but it uses grid search in the parameter selection, which has high computational complexity.

RUFS has been put forward recently [13], it utilizes $l_{2,1}$-norm minimization on processes of both label learning and feature learning, outliers and noise could be effectively handled and redundant or noisy features could be effectively reduced. But the algorithm still exist difficulties in parameter selection.

UDFS[14] is also unsupervised, which assumes that the class labels of input data can be predicted by linear classifier, different from the existing unsupervised feature selection algorithms, it chooses the most discriminative feature subset in batch mode. But it also has difficulties in parameter selection.

3 PRINCIPLE AND METHOD

The general problem of unsupervised feature selection can be modeled as: given a set of data $X = [x_1, x_2, \dots x_n]$, $x_i \in R^b$, find a feature subset which has d dimensions and contains the most valuable features. In other words, $\{x_1', x_2', \dots x_n'\}$ denotes the d-dimensional space R^d, can well preserve the geometric structure of the data represented in the original b-dimensional space R^b.

This paper will firstly introduce the principle of MCFS. This paper will use the embedded spectral clustering analysis to implement dimensionality reduction, then using maximum variance to calculate the weight information by the new space of eigenvectors, applying the weight information in learning sparse coefficient vectors at last.

The flowchart of MCFS-VS is given below, as shown in Figure 1.

3.1 Spectral embedding for cluster analysis

Spectral clustering does well in detecting the cluster structure of data,[15, 16]. The algorithm firstly according to the given sample data set defines a description of the affinity of paired sample points similarity matrix, and computes eigenvalue and eigenvector of the matrix, and then choose the right eigenvector for clustering.

Spectral clustering algorithm treat a data point as a vertex V of a graph, regarding corresponding weight E of connected vertices as the similarity among the vertices, then it will get a undirected weighted graph G (V, E) based on the similarity, so the clustering problem can be converted into graph partition problem.

Spectral clustering and manifold learning have a close ties [17], it assumes that the data is uniform sampling from a low dimensional manifold in the high dimensional Euclidean space, manifold learning is recovering low dimensional manifold structure from high dimensional sampling data. It is from the observed phenomenon to find the essence, to find the inherent law of data. Spectral clustering can be roughly divided into the following steps (assuming to be divided into K classes):

a. Establish similarity graph and set W as the weighted adjacency matrix of similarity graph;
b. Calculate unnormalized graph Laplacian matrix L (L = D−W, where D is the degree matrix).
c. Calculate the smallest feature K vectors of L;

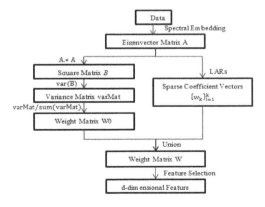

Figure 1. The flowchart of MCFS-VS.

d. The K eigenvectors are arranged together to form A, namely N * K matrix A, then executing kmeans clustering algorithm.

In this paper, the spectral clustering step (d) the N * K matrix A is handled as the following:

$$B = A.*A \qquad (1)$$

Getting N * K variance matrix B, each column of which get varMat variance matrix:

$$varMat = var(B) \qquad (2)$$

Calculating weighting matrix W_0 of each column in matrix A according to the income of the covariance matrix varMat of matrix B:

$$W_0 = varMat/sum(varMat) \qquad (3)$$

Matrix A vector calculate the square before variance respectively, considering using the eigenvector matrix obtained from the spectral clustering using Euclidean distance measure.

3.2 *Learning sparse coefficient vectors*

Getting "flat" embedded matrix A, it can measure the importance of each feature in the original data sets based on internal dimension column vector of A, and distinguish contribution of each feature of different label in this way.

Given a_k, namely the k column in A, by minimizing the fitting error, we can find a relevant feature subsets:

$$min_{wk}\|a_k - X^T W_k\|^2 + \beta|W_k| \qquad (4)$$

where w_k is a b-dimensional vector, $|w_k| = \sum_{i=1}^{b}|W_{k,i}|$ denotes the L1-norm of w_k, w_k essentially contains the combination coefficients for different features in approximating a_k, because of the nature of L1-norm, if β is big enough, some coefficients will be shrunk to exact zero, in this way, we can choose a subset which contains the most valuable features, MCFS utilize LARs[6] to solve LASSO[18] problem.

It is likely that some characteristics are related in high dimensional data set, several weak features' combination can distinguish among different classes. Some supervised feature selection algorithms have been designed to solve this problem, MCFS utilizes L1-regularized regression model to find the feature subset rather than through calculating the contribution of individual characteristics just based on considering this aspect.

3.3 *Feature selection strategy*

To select d-dimensional feature subset from original b-dimensional data set, we can choose from $\{w_k\}_{i=1}^{k} \in R^b$. The cardinality of each w_k is d and each entry in w_k corresponds to a feature. To choose d-dimensional feature subset, the strategy of MCFS is:

$$MCFS(j) = max_k|w_{k,j}| \qquad (5)$$

where $w_{k,j}$ is the j-th element of vector w_k, according to the type b to calculate MCFS score of the original feature, selecting top d features as a feature subset. In this paper, we will make a process that multiply variance matrix of reproduction clustering W and $\{w_k\}_{i=1}^{k} \in R^b$, generating a new weighting matrix $\{w_k\}_{i=1}^{k} \in R^b$ Handling after these are the same as MCFS, we called our algorithm MCFS-VS, VS means variance score.

4 EXPERIMENTS

In this section, we use two standard data sets to compare MCFS-VS and MCFS under the same conditions evaluating whether MCFS-VS improves the effect of MCFS, mainly using clustering to implement by assessment of NMI score and Accuracy[12].

4.1 *Data sets*

This experiment chooses two standard data sets, a statistical information of these data sets is given below:

- The first data set is Isolet Data Set, It contains 150 subjects who speak the name of each letter of the alphabet twice. The speakers are grouped into sets of 30 speakers each, and are referred to as isolet 1 through isolet 5. In our experiment, we use isolet 1 which contains 1560 examples with 617 features.
- The second data set is COIL20 Data Set, It contains 20 objects. Images of each object are taken 5 degrees apart as the object is rotated on a turntable and each object has 72 images. The size of each image is 32 × 32 pixels, with 256 grey levels per pixel. Thus, each image is represented by a 1024-dimensional vector. which contains 1440 examples with 1024 features.

4.2 *Clustering results*

Through the improved MCFS, experimental data set apply kmeans clustering by feature selection, we use the normalized mutual information metric to obtain NMI scoring again, and compute accuracy, to study whether the improved algorithm has effects on the improvement.

Figure 2 is Clustering performance vs. the number of selected features on Isolet Data Set.

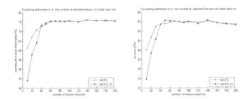

Figure 2. Clustering performance (%) vs. the number of selected features on Isolet data set.

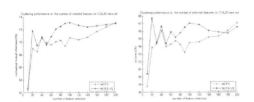

Figure 3. Clustering performance (%) vs. the number of selected features on COIL20 data set.

Experiment using MCFS-VS selected features to execute kmeans clustering, NMI score equals the average value which obtained from NMI algorithm by the results of kmeans clustering by iterativing 300 times. From Figure 2, we can see when choosing number of features between 40 to 60, the effect of MCFS-VS is better than MCFS, which NMI score is increased about 0.6%, and choosing a little features to execute kmeans clustering can get a good NMI score. So in practice, MCFS-VS has its value. and based on accuracy figure, we can get similar conclusion.

Figure 3 is Clustering performance vs. the number of selected features on COIL20 Data Set.

The experimental data processing is the same as Isolet data set, From Figure 3 we can see when choosing number of features between 60–90, the effect of MCFS-VS is better than MCFS, and choosing few features to execute kmeans clustering can get the a good NMI score and accuracy.

But from the two data set in our experiment, it also gets a problem that MCFS-VS still need to consider in choosing to select right number of features.

5 CONCLUSIONS

MCFS-VS is presented in this paper, on the basis of embedded spectrum analysis which are used to get the dimension reduction feature vector. it combines with the Maximum variance method to get the feature vector weight information. These weights is added to the sparse coefficient regression study, for as much as possible to extract the more important information of the original features, so it can effectively improve effectiveness of feature selection. But through the experiment, this approach also has a restriction which needs to select the right number of features.

REFERENCES

[1] Lee Rodgers, J. and W.A. Nicewander, *Thirteen ways to look at the correlation coefficient*. The American Statistician, 1988. 42(1): p. 59–66.
[2] Duda, R.O., P.E. Hart, and D.G. Stork, *Pattern classification*. 2012: John Wiley & Sons.
[3] Cover, T.M. and J.A. Thomas, *Elements of information theory*. 2012: John Wiley & Sons.
[4] Cai, D., C. Zhang, and X. He. *Unsupervised feature selection for multi-cluster data. in Proceedings of the 16th ACM SIGKDD international conference on Knowledge discovery and data mining*. 2010. ACM.
[5] Von Luxburg, U., *A tutorial on spectral clustering*. Statistics and computing, 2007. 17(4): p. 395–416.
[6] Efron, B., et al., *Least angle regression*. The Annals of statistics, 2004. 32(2): p. 407–499.
[7] Kohavi, R. and G.H. John, *Wrappers for feature subset selection*. Artificial intelligence, 1997. 97(1): p. 273–324.
[8] Law, M.H., M.A. Figueiredo, and A.K. Jain, *Simultaneous feature selection and clustering using mixture models*. Pattern Analysis and Machine Intelligence, IEEE Transactions on, 2004. 26(9): p. 1154–1166.
[9] Dy, J.G. and C.E. Brodley, *Feature selection for unsupervised learning*. The Journal of Machine Learning Research, 2004. 5: p. 845–889.
[10] Roth, V. and T. Lange. *Feature Selection in Clustering Problems in NIPS*. 2003.
[11] Zhang, D., S. Chen, and Z.-H. Zhou, *Constraint Score: A new filter method for feature selection with pairwise constraints*. Pattern Recognition, 2008. 41(5): p. 1440–1451.
[12] Li, Z., et al. *Unsupervised Feature Selection Using Nonnegative Spectral Analysis in AAAI*. 2012.
[13] Qian, M. and C. Zhai. *Robust unsupervised feature selection. in Proceedings of the Twenty-Third international joint conference on Artificial Intelligence*. 2013. AAAI Press.
[14] Yang, Y., et al. *l 2, 1-norm regularized discriminative feature selection for unsupervised learning. in Proceedings of the Twenty-Second international joint conference on Artificial Intelligence-Volume Volume Two*. 2011. AAAI Press.
[15] Dhillon, I.S., Y. Guan, and B. Kulis. *Kernel k-means: spectral clustering and normalized cuts. in Proceedings of the tenth ACM SIGKDD international conference on Knowledge discovery and data mining*. 2004. ACM.
[16] Yu, S.X. and J. Shi. *Multiclass spectral clustering. in Computer Vision, 2003. Proceedings. Ninth IEEE International Conference on*. 2003. IEEE.
[17] Belkin, M. and P. Niyogi. *Laplacian eigenmaps and spectral techniques for embedding and clustering. in NIPS*. 2001.
[18] Hastie, T., et al., *The elements of statistical learning*. Vol. 2. 2009: Springer.

Future Information Engineering and Manufacturing Science – Zheng (Ed)
© 2015 Taylor & Francis Group, London, ISBN 978-1-138-02644-5

Differences on the development of regional education under the geospatial perspective in China

Hongyu Xie
College of Travelism and Geography Sciences, Yunnan Normal University, Kunming, China

Jian-hou Gan
Key Laboratory of Educational Informalization for Nationalities, Ministry of Education, Yunnan Normal University, Kunming, China

Zhilin Zhao
College of Travelism and Geography Sciences, Yunnan Normal University, Kunming, China

ABSTRACT: In China, the development of regional education is unbalanced; there are many factors that affect the educational development. So this paper analyzed the differences on the regional education development from 6 main aspects in China: ① differences on regional location ② differences on the characteristics of the regional topography and terrain ③ differences on the characteristics of the regional demography ④ differences on the level of the regional economical development. And finally the conclusions are that this paper proposed 3 strategic countermeasures that are the following: countermeasure of national level, countermeasure of social dimension, and countermeasure of changing the backward education in the western and central region.

Keywords: Regional education; Differences; Coordinating; Geospatial

1 INTRODUCTION

The education had been a huge development since the New China found especially the reforming and opening up began in China. The compulsory education has become universal basically, high school gross enrollment rate has reached 82% since the year of 2010, and the gross enrollment rate reached 26.9% in 2011, the higher education has entered the popular stage. It can be seen that Chinese education has reached the level of developed countries from enrollment indicators in the world. However, there exists extremely uneven economic and social development in China and the regional quality of education, educational opportunities and the level of educational equity uneven accordingly. So it has great significance to reduce the regional differences of education, promote a balanced education, coordinate development, and find out the reasons of impacting the regional disparities so that it can promote Chinese eastern, central and western education to be balanced and coordinated development, and accelerate to change the educational backwardness in the Midwest. Through analyzed the different factors which the geographical features and cultural features affect the educational

development under the geospatial perspective, this paper attempted to find a way which promote Chinese education to be balanced and coordinated development, and change educational plight of backward areas in the western of China.

2 SIX FACTORS OF IMPACTING THE DIFFERENCES ON THE DEVELOPMENT OF REGIONAL EDUCATION

Natural environment is a basic condition for human survive and development as well as the human cultural and educational activities also are affected by it, at the same time; it has varying degrees of impact on the formation and improvement of school conditions and students' physical and mental development. China has a majority of the land area, population, and each region has their own characteristics and complex interlacing of the geographical features, which form a complex topography, rich and colorful cultural characteristics. On the one hand, those complex geographical areas offer a variety of favorable conditions for the development of education; on the other hand, it also exists more difficult for the development of education.

2.1 Factor 1—differences on the regional location

In the "human geography", as Huilin Chen said, it always exists a core area in the place of the cultural district, namely, the place that the performance and standard of cultural characteristics are most prominent and typical. It is the part of weaken culture from the outside of core area which it will be completely gone when the distance is far away. [1] Similarly, in the geographic space, the location of each area is an objective fact that is the political, economic and cultural center; it has an important impact on the development of culture and education. The center of political, economic and cultural is the core area of culture and education in this region, and it is also the best areas of development. Closer the core area, better the education, farther away from the core area, worse the education. As the Chinese famous economist Mr. Shanmai Wang did, he divided more than 30 provinces (city) district into four categories on the basis of the level of economic development: Beijing, Tianjin, Shanghai are the first region where they are the most economically developed regions, the area of Liaoning, Jiangsu, Zhejiang, Shandong, Guangdong belong to the second type. the third region are Hebei, Shanxi, Jilin, Heilongjiang, Anhui, Fujian, Henan, Hunan, Sichuan, new and other underdeveloped regions, the remaining provinces and other least developed regions belong to the fourth region in China. [2] According to statistics, the education in Beijing, Shanghai and other similar areas is always 2–3 times than other provinces. Coupled with the local cultural heritage is more intense, longer history of educational development, the level of educational development in these areas is the highest level in China. Since the economic development is very high, which results in the local investment in education makes bigger difference than other areas, so the quality and level of development will be greatly affected. Higher education represents the level of regional development. Unbalanced regional economic development level corresponds to the distribution to the higher education. It has shown that the number of eastern universities is more than the western and central region; as well as the quality of the College education. In addition, Mr. Shanmai Wang has done that educational quality and equitable education level which measure the various regions of China in 2009, As it can be seen from Figure 1, in 2009, the highest level of educational development is in Shanghai, Beijing, Zhejiang and other economically developed areas, and the lowest is located in Guizhou, Yunnan, Tibet and other

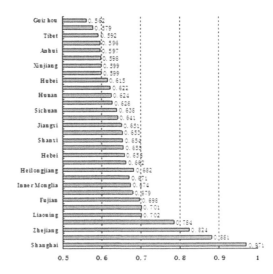

Figure 1. Chinese eductional development index and rank in 2009.

remote location economically underdeveloped regions.

2.2 Influence factor 2—regional topography and terrain characteristics differences

Chinese terrain is very complex, which can be divided into five types: plateaus, mountains, plains, basins, hills, of which the plateau is 26% of total area, the mountain is 33% of the total area, plains is 12% of the total area, 19% of the total area is the basin, hills is 10% of the total area. It can be judged from the terrain that the topography of China is rise gradually from east to west, which can be divided into the Yangtze River and the northern plains of the third step; Loess Plateau, the Sichuan Basin of the dominated second step and the first step in the Tibetan Plateau. Within each step, it is also divided into highland areas, mountain areas, hilly areas, basins and plains areas.

In the first ladder area, it is characterized by the plateau where the Tibetan plateau is the main plateau. This area is the most backward regions of educational development. On the one hand, since this region is remoteness, it is far from the core area of culture, education and developed coastal areas, so the historical and practical reasons that caused economic and cultural backwardness in the region, correspond to the educational development is restricted; on the other hand, because the exchange and traffic are inconvenience, so the educational information, educational content, teaching methods and other aspects of educational thought and external communication are all limited. And at last, the harsh natural environment is not conducive to the development of education.

In the second ladder area, it is characterized by mountain, basin where there are mainly in the Loess Plateau and the Sichuan Basin. Because it is rather nearer to the core of cultural and educational areas and developed coastal areas, so the educational development is in high levels relatively. However, it is not conducive to exchange because of the characteristics of the mountainous region, so that the development of education is a little difficult.

In the third ladder area, it is characterized by plains, hills where they are mainly in the Yangtze River and North China plains. The area is nearer to the core area of education, and the economic development is very high level so it is also convenient to develop education that provide favorable conditions for the development of exchanges, so that this region is the best level on the development of Chinese education.

Above all, the complex terrain brings a wealth material for the development of education, but also brings a variety of adverse conditions for the development of education.

2.3 Influence factor 3—regional climate characteristics differences

The ancient Greek historian (before 384 BC—325 BC) Aristotle thought that people in the cold regions are brave, but lack of wisdom and technology; the Asian people are wise, but the lack of courage; and the Greeks are able to self-reliance. He believes that the reason why the Greek nation has both excellent characters, because the Greece lies in between cold and hot climate apparently. [3] In the book of "Climate and Civilization," the author who was American scholar Huntington (Ellsworth Huntington, 1876–1974 years) thought if the nation is in the absence of climate contributing factor, it cannot reach the pinnacle of culture regardless of ancient or modern, i.e. [3] Although climate characteristics does not play a decision role on the form of the human personality, but it is the important factor of favoring for or districting the development of education. The warm climate provided favorable conditions for the people who are in warm and comfortable climatic conditions conducive to improve learning efficiency, which can create higher educational outcomes. The hot and humid areas are not conducive to learning, and to create relatively low educational outcomes. While the cold climate is not conducive learning and it will create the lowest educational outcomes. China is the vast area, and the climate types distribute widely. So this paper divided the characteristics of climate into three regions in China: the warm zone, cold zone and humid areas so that it will facilitate the study.

There are three different climatic characteristics, that is the hot and humid region where is mainly in The Yangtze River and the eastern coast of the tropical and subtropical regions, the warm region is mainly in North China Plain and the Yellow River Basin; the cold climate zone is mainly in alpine. In these three different climatic characteristics, the education level of the warm region is the highest which is represented by Beijing, and the education in the hot and humid region is higher which represented by Shanghai, Wuhan, Guangzhou, and the education level is backward in the clod region where mainly in Lhasa alpine. In addition to politics, history, geography, culture and other factors on the regional education level, the climatic characteristics has an important impact on the level of educational development undoubtedly.

2.4 Influence factor 6—economic development levels differences

There is a close relationship between education and economic, which the education provide spiritual motivation and intellectual support for the economic development while the economic development support education material foundation and basic conditions. So on the one hand, the economic development is inseparable from the support of the development of education which is also inseparable from the support of the economy. The regional economic development provide a basis for the development of regional education conditions, propose personnel training needs, pull the education scale and raise the level and adjustment of economic structure, determines the structure, size and efficiency of the education development. On the other hand, the regional educational development is a basic mean and way to improve worker productivity and innovation, the introduction and dissemination of technology, accelerate regional technological progress.

Figures 2–3 is a provincial GDP in the year of 2011. As it can be seen that the first layer of the best cities on the economic development strength are Shanghai, Beijing, Tianjin; the second layer of the better cities on the economic development strength are Zhejiang, Jiangsu and Guangdong.

The economic strength in Guangdong province is much better than the fifth floor of Hubei and Chongqing, but the education index level in Guangdong only 0.598 which is much lower than in Hubei and Chongqing where the educational indexes are 0.658 and 0.662. The phenomenon shows that although the economic development plays an important role in the education, it is not the decisive role. So it can be said that the economic

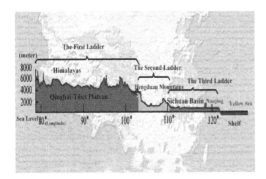

Figure 2. Chinese Terrain schematic profiles (from latitude 32°).

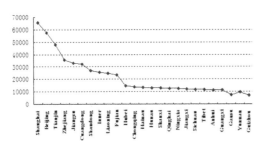

Figure 3. The economic rank of Chinese province in 2012.

development does not necessarily bring the development of education. However, the education in the Inner Mongolia of northern China has maintained a high level which the educational development index and economic strength has always been considered to be better than education and economy city—Hubei and Chongqing. This shows that the backward western regions can also be break through the education level.

3 COUNTERMEASURES OF PROMOTING THE BALANCED AND COORDINATED REGIONAL EDUCATION

The uncoordinated development of regional education has brought many economic and social problems. It has not only hampered Chinese sustainable economic development and social progress severely, but also resulted in educational opportunities and the educational process are not fair, and posed a potential threat on social cohesion, political stability, at the same time, it further exacerbated the brain drain in the Midwest. Therefore, it is particularly important to promote the regions coordinated development of education in the east, middle and west, and it is also to measure whether

it can make reasonable and effective means to promote the educational development.

3.1 Countermeasure of national level

Education is the foundation of the country's development which the fortunes raise and fall hinges on education. The world politics is changing now, the international competition is becoming fierce increasingly, and the science and technology is developing rapidly. In fact, the worldwide economic competition is the science and technology competition. The national science and technology innovation and cannot be separated from improving the quality of education. In this sense, who mastered the education in the 21st century, who will be able have the international competition in the 21st century? Therefore, the state should shift economic construction to the scientific and technological progress and improve the quality of workers, and put the development of science and technology and education in the first place. So if the state wants to become a comprehensive, coordinated and balanced country, it has to develop the education in the western region where it has the backward educational career.

If the state want to change the backward educational situation in the western region, the first thing is to change regions lagging traffic situation in the west and middle region where the education is more backwardness than the eastern region, so the state offer funding on the preferential policies, financial support, investment guidance and other traditional, it is an important factor to exchange the lagging traffic in the remote location the state, it should change the traffic condition in of the road, rail, water, air, etc., and create favorable traffic conditions in the central and western regions.

3.2 Countermeasure of the social dimension

It has an important practical significance to be a balanced and coordinated regional education which will promote social progress and achieve common prosperity and harmonious society. China has proposed the implementation of the strategy, which requires people to fully understand the importance of education, fully aware of the pilot of the education in the economic and social development. So people should do the following: on the one hand, people should strengthen propaganda and support joint development of education in the Midwest; on the other hand, develop the area of manpower, material, and intellectual, financial support backward areas. For example: people can support each other from region to region, province against province, city-to-city counterparts, if possible, it can even form a school counterpart support to schools to achieve the goal of "first drive after the rich, eventually reach common prosperity of the whole society, "the goal. And let backward areas achieve the balanced and coordinated development of education with the developed regions.

3.3 *Countermeasure of changing the backward education in the central and western regions*

Firstly, people should change the concept of "wait, rely on, ask for" in the central and western regions where the education is so backward, and explore the potential to develop economy in this region, so that it can provide strong support for educational development. Secondly, people in the central and western regions should fully exploit the educational backwardness of national characteristics, rich ethnic educational content, and develop the characteristics of national education, so that it can walk a new way of education. Third, the educational experts should change the backward education positively in the central and western regions, strengthen regional and extra-regional exchanges and cooperation, enrich the content of education, change the backward educational management methods, so that let the education suitable for the formation of the central and western regions, and at last training a large number of suitable candidates for the western region's economic and social development. Finally, people should change the traditional education concept, intensify propaganda and formulate preferential policies to create favorable conditions so that it will improve the investment environment to attract capital to the region, economic development and education.

4 CONCLUSION

In the "Education Planning Programs", it put forward that China should make the compulsory education gross enrollment rate of 95%, high school education gross enrollment rate of 90% and higher education gross enrollment rate to reach 40% by the year of 2020, which if is both an opportunity and a challenge for Chinese regional education development. So people should fully develop, and promote Chinese education to become a balanced and coordinated education.

ACKNOWLEDGEMENTS

This work was supported by Soft Science Program of China (2013GXS4D149), Ministry of Education, Humanities and Social Sciences Project (12YJCZH053), Minority Native Culture of Quality Engineering Projects in Yunnan.

REFERENCES

[1] Huilin Chen: "Human Geography". Science Press, 2007, (6): 137.
[2] Wanbing Shi: Higher Education and Social Justice Configuration. Shenyang Institute of Technology, 2004, (1):35–36.
[3] Kelly: Human Geography. Science Press, 2007, (6):9–10.

Future Information Engineering and Manufacturing Science – Zheng (Ed)
© 2015 Taylor & Francis Group, London, ISBN 978-1-138-02644-5

Study on the heritage and development of ethnical culture under the educational perspective

Jian-hou Gan
Key Laboratory of Educational Informalization for Nationalities, Ministry of Education, Yunnan Normal University, Kunming, China

Hongyu Xie
College of Travelism and Geography Sciences, Yunnan Normal University, Kunming, China

ABSTRACT: As an important route of human cultural heritage, education plays an important role in the process of inheriting the ethnical cultural. Education is a mechanism for cultural life which is an indispensable part of the cultural emergence and development. The existence and development of culture is achieved through education. So this paper tries to reveal and mine the unique feature of education which can inherit the culture through using the method of cultural anthropology of education, in-depth study on the relationship between cultural heritage and ethnical articles. And finally it will further explore educational possibilities for ethnical cultural heritage.

Keywords: ethnical culture; culture heritage; education

1 CONNOTATION AND SIGNIFICANT OF THE ETHNICAL CULTURAL HERITAGE

Cultural Heritage is actually a process of the enculturation which the anthropologist always said; it refers to the process of vertical handover like the race of baton among people in the same community, which has the characteristics of mandatory requirements and mechanism because of the restriction in the backgrounds of living and culture. [1] so that human culture has stability, integrity, continuity and other features in the development of human society. In the restricting inheritance mechanism, all cultural heritages are learned, not inherited from the biological aspects. There have the narrow and broad connotations of ethnical cultural heritage, of which the broad ethnical cultural heritage is the heritage of ethnical cultural, and the narrow ethnical cultural heritage refers to the heritage of a single ethnical minority. [2] Therefore, the heritage of ethnical culture in this article is the narrow connation, especially the cultural heritage of ethnic minorities.

Ethnical culture is the root of a nation, which is an important symbol of ethnical existence, and it also is an important driving force for Ethnical development. Inheriting the unique ethnical culture is the need for an ethnical survival and development, and also the common wealth of protecting

human being. The rich history and culture is created in the production of the unique and colorful life practice of humankind in China, which is the treasure of human culture, so it must be inherited and promoted by the people forever.

2 EDUCATIONS AND THE HERITAGE OF THE ETHNICAL CULTURE

As an important route to inherit ethnical culture, education plays an indispensable role in the process of the ethnical cultural heritage. The existence and development of culture is achieved through education. If there is no education, there won't have culture, and even the culture is difficult to show its vitality and history length. Thus, "education is a mechanism for cultural life" [3], and also it is an indispensable part in the process of the emergence and development of cultural factors. [4] Education is an activity to train people in a society, which is also the basic way of the social and cultural heritage and the social experience of passing production.

How to play an important role of education in the heritage of our excellent ethnical culture? After many years of studies on the educational theory and multicultural educational theory of domestic and foreign Countries, the Chinese ethnical education experts proposed the "theory of multiple integrated educations" under the background of diversity and unity. That is, the education in the multi-nations not only shoulders the transmission function of human common culture, but also has the function of passing the outstanding ethnical culture, while it also has the function of taking us the excellent traditional ethnical culture. [5] At the same time, education not only has the function of training students to have the good attitude of the ethnical culture, but also let them to have a positive attitude towards the ethnical culture.

3 THEORY BASIS OF EDUCATION ON THE HERITAGE AND DEVELOPMENT OF THE ETHNICAL CULTURE

Educational Anthropology is a comprehensive discipline between pedagogy and anthropology, which is a new emerging discipline used by anthropological perspectives and the various methods of social education.

Educational Anthropology originated in the social science research in the early 20th century in the western areas, which began to become common throughout the world in the 1970s. In the early 20th century, the anthropologists Headwaiter and Malinowski criticized that the cultural background and ethnical education were ignored by school, and they proposed that educators should be made use of to educate students in the ethnical culture. In the 1970s, researchers began to study the educational anthropology from the culture and personality to take the education as the issue of anthropological research, they recognized that the nature of education is the spread of culture, and schooling is a special form of cultural submission.

At present, educational anthropology is divided into two schools, that is the cultural educational anthropology which is represented by the Anglo-American, and the philosophy of educational anthropology which is represented by the Germany and Australia in the European countries, the former mainly emphasis on the theory of cultural perspective, and focuses on the field investigation and laboratory-based research and case studies, to research theory and practice in education; while the latter studies the issue from the perspective of human nature, and focuses on the principles of philosophical anthropology, abstract, speculative, deductive, inductive methods study to examine the images of people, the relationship between human nature molded and sound, researchers and education. [6]

The content of this paper adopt the vision and methods of cultural anthropology of education through the various elements of the relationship between ethnical cultural heritage and educational

investigation, trying to reveal and mine the unique function of education on the ethnical cultural heritage, then the possibilities of education on the ethnical cultural heritage will be further explored.

4 HERITAGE AND DEVELOPMENT OF THE ETHNICAL CULTURE UNDER THE PERSPECTIVE OF EDUCATION

4.1 Inheriting function of education on the ethnical culture

There is an inherent and natural link between Culture and education, such that they has the relationships of mutual restraint, interdependent and mutually means. Like some scholars believed imagery, without the culture, education like machine will has no raw materials to produce; at the same time, without education, the heritage and development of culture cannot be conducted effectively. Education can save, transfer and even create human culture; so the cultural function of education is reflected in the following aspects.

First, the cultural inheriting functions of education. Culture is created in the process of human activities, which is learned by individual, so it cannot be inherited by heredity, only be developed and renewed by the way of transmission. There has 2 ways of inheriting culture; they are the portrait of cultural heritage, which represents the cultural heritage in time, and the horizontal cultural transmission, which refers to the performance of cultural heritage in the geographical space. As an activity of educating humankind, education makes the culture as an intermediary, and has the important functions of spreading and passing culture. And compared to other means of cultural transmission, the cultures were inherited by education is the most basic of human culture, culture can be preserved by people through the master of them. Therefore, education is an effective way of cultural transmission. Because of education, culture can be passed from generation to generation, and the human culture can be able to accumulated and passed.

Second, the cultural optional functions of education. It is the starting part of cultural change and development for cultural option, which expresses it, selects and excludes the culture automatically. Although education has the function of inheriting culture, it is not equivalent to cultural transmission, because education is not spread all cultures, and it spread culture through some choices. Without the choice to transmit culture, it cannot be called education, especially the school education. So the educational standards of optioning are as following: on the one hand, valuable culture would be chose by education, the invaluable culture would be rejected. On the other hand, culture which fits to the social mainstream would be chosen. At last, according to the characteristics of ages and the rules of education, the educational cultural process would be chosen.

Third, cultural innovation and creation function of education. Culture was created by human beings, so education not only has the function of inheriting the cultures which have existed in the world, but also should have the function of updating and creating the cultural development. Firstly, the social culture is always based on the critique and selection of the education, and certain culture would be formed by the human needs, the process of alternative, form, and submission also is the process of the reformed and recreated of culture, so education has become a factor of new social culture. Secondly, cultural achievement can be created by the scientific research, the production of new ideas and concepts through education. Finally, all sorts of talented people would be evolved and transported in the process of education, and then those people would create new culture again, so that school will become the place of cultural creation.

Fourth, the openness of modern education also promotes cultural exchange and integration. Culture is the minds and common behaviors of people in a specific area with certain period, in the sense; any culture has regional and closed characteristics. However, the development of modern social productive forces and the formation of a market economy have broken the pattern of the politics, economy, so that the cultural exchange would be gradual integrated in the process of openness.

4.2 Soles of education on the ethnical cultural heritage

4.2.1 The role of education in promoting ethnical cultural heritage

Educational role in promoting ethnical cultural heritage is the vertical transmission function of ethnical cultural education, manifested in the following aspects:

First, the ethnical culture can be preserved by education. The process of inheriting the ethnical culture is the process of preserving culture, and education plays an important role on the promoting the preserve ethnical culture. At first, education has the function of surviving and developing ethnical culture, in the process of inheriting the ethnical culture, the cultural subjects who are the educators and educated people pass and undertake the ethnical culture through educational activities, so that the ethnical culture can be reserved and developed. And the next is that education can promote ethnical cultural heritage. "Cultural heritage

is results of cultural accumulation and dissemination. Culture can be prolonged preserved and disseminated because of the education, and it finally resulted in the basic core of culture "[7], thus an ethnical cultural tradition is formed.

Second, education has the role of promoting ethnical cultural alternative. Education not simply replicates the ethnical culture because of the restrictions of era development and social needs, the traditional culture and the developing level of educated people, it is an alternative selectively. The education in the school is a process of training people purposely and planned, which is a process of select, blend, and finish ethnical culture, so that let the ethnical culture into the every kinds of education.

4.2.2 The role of education on the restricting the ethnical culture

Education not only has the role of promoting the ethnical culture, but also restraining it. The main restricting ethnical culture in the following 2 aspects:

First, choosing the content of education restricts the ethnical cultural heritage. Culture is formed by mainstream culture and non-mainstream culture, which the ethnical culture is a part of the latter. Education is an important carrier of inheriting the ethnical culture. Education is not simply to inherit the ethnical culture, which is passed by education alternatively. Generally, in the process of education, how much quality of the role on the education, it restricts the quantity of ethnical cultural heritage; whether the good and positive ethnical culture is passed by education, it restricts the quality of the ethnical culture. Therefore, in a sense, choosing the content of education affects the quality of inheriting the ethnical culture.

Second, using the methods of education restrict the level of ethnical cultural heritage. The method is the sum of the main rules in order to achieve a certain purpose and established the means, procedures and approaches. ethnical cultural heritage is always expanded by a certain method in the process of education, whether the methods of education is appropriate, whether it is loved and accepted by the younger generation, it will affect the quality of the ethnical cultural heritage, which restricts the levels of ethnical cultural heritage.

4.3 Pathways of education on the ethnical cultural heritage

4.3.1 Types of education

According to the object, content and form of education, the education can be divided into Family education, school education and social education while the community education belongs to a part of social education, which has been discussing in recent years. Compared with school education, community education has its own function, it broken the restriction of education, and expand the education into the social grassroots, that is, community education has the organizing function of learning for lifelong and let individual to adapt to the social background to learn. Based on the characteristics of community education, it has more applicability and operability of inheriting the ethnical cultural heritage than the social education. Therefore, this paper focuses on the community education, family education and school education.

4.3.2 *Ethnical cultural heritage of community education, family education and school education*

4.3.2.1 Ethnical cultural heritage in the family education

Some scholars believe that the family is the basic social cells, which is the most important and most central organization of economy, social and spiritual home. In addition to the natural character, the upbringing of a person has a close relationship with his environment, and the environment has a good function of impacting and educating people. The education in the family has a deep impaction for a person, and has the subtle affection on a person.

Home is an important place to inheriting the ethnical culture, while the family education is a traditional means to continue and develop ethnical culture. Home is not only the production organization, but also the growth habitat for a person. From the beginning of born, people start to learn and inherit the ethnical culture, and this type of inhering culture is the oldest and most robust way. People accept the ethnical culture in the family; especially some of the folk specific industries, like the religions of ethnical culture, witchcraft, folk craft skills, some ethnical medicine skills, and those ethnical cultures have an important influence in the family education.

4.3.2.2 Ethnical cultural heritage in the school education

The characteristics of educational purpose, planning and organization determine education must have the unmatched function on the heritage of the ethnical culture. People cannot find a more effective way to train persons of inheriting the ethnical culture than the ways of outside. If the ethnical minority wants to preserve their own culture, they must be educated by the school to inherit the ethnical culture, and let people learn and apply the ethnical culture repeatedly, and the life of ethnical culture will be activated in the school. Therefore, the school education should and must be the import way to inherit the ethnical culture.

However, the school education need to be further improved for inheriting the ethnical culture. Some scholars believe that the knowledge content of one dollar, subject, and standard make the traditional ethnical culture marginalized, which are far away from the cultural background of students, and let them not to adapt to the environment, so that this phenomenon resulted in students is indifferent alienation to study, the methods of standardized evaluation lead minority students to lack the means to make learning traditional culture initiative. So measures must be improved to make the school education an important way to the development of ethnical cultural heritage.

4.3.2.3 Ethnical cultural heritage in the community education

Community education mainly represented the cultural activities in the village community like temples and other cultural activities of ethnical minority areas, those activities are ways of inheriting the ethnical culture. The temple education is the most complete and most advanced way to inherit ethnical culture before the school education emerged. Because it has the witches, priests, monks and other cultural transmitters, so it has the relatively fixed field and rules, and the unity of doctrine and classics, so it has huge impaction of the common psychological quality of ethnical minority.

5 CONCLUSION

In short, there is a close relationship between education and the heritage of ethnical culture, education has the important mission of disseminating the ethnical culture and affecting the heritage and development of ethnical culture, even affect the unity and stability of the Country. In the new era of socialism, it is an important significant to understand and hander the relationship between education and the heritage of ethnical culture, and make full use of various types of education.

ACKNOWLEDGEMENTS

This work was supported by Soft Science Program of China (2013GXS4D149), Ministry of Education, Humanities and Social Sciences Project (12YJCZH053), Minority Native Culture of Quality Engineering Projects in Yunnan.

REFERENCES

[1] Zhao Shilin, Yunnan minority cultural heritage Outline Kunming: Yunnan Ethicalities Publishing House, 2002:17.

[2] Cao Nengxiu Wang Ling. Minority areas of school education and Ethnical cultural heritage. Yunnan Normal University (Philosophy and Social Science), 2007, (2).

[3] General Education in Educational Anthropology, Nanjing: Jiangsu Education Press, 1998:160.

[4] Sun Yajuan, Cao Liushao Township can show. Xundian County as an example to talk about the cultural heritage of minority issues. Yunnan Agricultural University Volume 4, No. 1, 2010, (2).

[5] Fei Xiaotong, Pattern of Chinese Yuan integration Beijing: Central University for Ethicalities Press, 1999:28–32.

[6] General Education in Educational Anthropology tutorial Beijing: People's Education Press, 2005.

[7] Zheng Jinzhou, School of Education and Culture Beijing: People's Education Press, 2000:111–112.

[8] Feng Zengjun, General Education in Educational Anthropology tutorial Beijing: People's Education Press, 2005:211.

Future Information Engineering and Manufacturing Science – Zheng (Ed)
© 2015 Taylor & Francis Group, London, ISBN 978-1-138-02644-5

Exploring and development—summary of ethnical cultural education

Hongyu Xie
College of Travelism and Geography Sciences, Yunnan Normal University, Kunming, China

Wenlin Yan
College of Foreign Studies, Yunnan University of Finance and Economics, Kunming, China

Jidong Yi
Key Laboratory of Educational Informalization for Nationalities, Ministry of Education, Yunnan Normal University, Kunming, China

ABSTRACT: In recent years, along with the reform and opening up of society, and the development of the ethnical areas, the ethnical culture faced the phenomenon of serious loss, so the discussion about ethnical cultural education is becoming an increasing topic today. As for the ethnical education, it has been studied by many scholars from every aspects, however, because the ethnical cultural education is an important aspect of ethnical education, and those studies still had some restrictions, the problems was still relatively scattered which had not formed the same understanding, especially the basic theoretical issues has not yet conducted a systematic analysis. This paper contempt to summarizes the relative research in the past two decades, and trying to find some clues, which hope to make some useful exploration of the development of our ethnical cultural education.

Keywords: ethnical cultural education; multicultural integration education; EFA; exploration and development

1 INTRODUCTION

As for the study of ethnical cultural education, there are three questions must be figured out, what is the ethnical cultural education? Why we must to do the ethnical cultural education in society today? And what is the significance of doing the ethnical cultural education or what is the value of doing the ethnical cultural education? As a ethnical prosperity of transforming the nature and society, culture presented variety of forms and natures. The different culture was created by different ethnical people. More and more people in many countries began to realize that the ethnical culture is not only a matter of a ethnical prosperity, but also has an close relationship with the survival and development of a ethnical minority. So it has great strategic significance to understand and use the ethnical culture, and strengthen the ethnical cultural education.

1.1 Definition of the ethnical cultural education

As for the exact definition of the ethnical cultural education, there has been not a common conclusion among the scholars, and it has few discussions on this matter. The definition of ethnical cultural education was defined by broad and narrow definitions, while the former refers to the cultural education happened in the same ethnical minority, because they want to adapt to the mainstream of society to live and develop better and better, and on the other hand, the ethnical cultural education is an activity of inheriting and promoting their outstanding cultural heritage. And the latter also is known as the education of ethnical minority, which refers to the education, happened in the area of ethnical minority [1]. Some scholars believe that the ethnical cultural education includes ethnical language, ethnical culture courses, and some subjects in the integration of ethnical culture from the broadly speaking. And in a narrow sense, it is the course of ethnical culture specifically [2].

Combined with the views of those scholars, this paper thinks that the definition of ethnical cultural education has broad and narrow senses. From the broadly speaking, it refers to the education is conducted in the members of a nation who are have the common culture in a state. And the narrow definition is education is happened in the ethnical minority in a state.

1.2 Values of the ethnical cultural education

Chinese traditional culture is the sum of material culture and spiritual culture which has been created by people for thousands of years; it embodies the carrier of ethnical spirit. The precious value, and rich content has profound implication for the development of the ethnical culture, which is the important cohesion and sign of human civilization [3]. Combing views of scholars of the ethnical culture, the values of ethnical cultural education as follows:

Some scholars believe that ethnical culture is a fundamental symbol to distinguish their own culture with the other culture. In recent years, along with the reforming and opening up, the strong economy and globalization has come into our country, and the western culture also has strong impaction on the ethnical culture which lead to the ethnical culture is in jeopardy. So it has become an imperative task for us to inherit the ethnical culture, and education has an important role in this major event [1]. Ethnical cultural education are the needs of the great rejuvenation of the Chinese nation and uphold ethnical cultural identity and safeguard ethnical cultural security, but also the need to promote the scientific development of higher education [4]. Strengthen ethnical cultural education is the need to strengthen ethnical unity, social harmony and stability, which also is an important guarantee to achieve cultural diversity and rich the connotation of Chinese culture [5].

From the above views of scholars, the author believes that the values of conducting ethnical cultural education has 2 important aspects, it not only conducive to the maintenance of ethnical unity, social harmony and stability, but also help enhance ethnical comprehensive interethnical competitiveness from the ethnical level; and on the other hand, it also is conducive to enrich the spiritual life of the people, but also conducive to the cultural heritage of each nation, and enhance ethnical identity from the community level. At the same time, for the development of the ethnical cultural education, it is in favor of inhering, preserving, innovating and developing the ethnical culture, it also has an important practical significance to develop and revitalize the ethnical culture in China.

2 DESCRIPTION OF THE RESEARCH ACHIEVEMENTS ON THE ETHNICAL CULTURAL EDUCATION

2.1 Targets of the ethnical cultural education

The target of ethnical cultural education also the goals of ethnical cultural education, which refers to the standard of educating people, that is, what are the types of people because of the education. It is determined according to the situation for political, economic, production, cultural and scientific and technological development of society and educated certain physical and mental development, reflecting certain social demands of educating people, which also is the starting and ending of the educated work, but also determine the content of education, alternative education methods, inspection and evaluation of educational results. Accordance with the "Education Law", the educational goals of this stage is "to educate the moral, intellectual, physical, and can fully develop communist successors." So what kind of effect achieve through the cultivation of education. The views as following:

One view is that the goal of educate the ethnical culture for the young generation is to improve the ideological and moral qualities of the ethnical people, and let them have the moral qualities, scientific and cultural qualities [3]. Another view is that the major value of educating student the ethnical culture is to let the have the emotions, attitudes and values on theirs ethnical culture while the formation of the ethnical culture [6]. And to improve the overall quality of students and cultural conservation and viability [7]. Some other views thought the fundamental purpose of educating the ethnical culture is recreate the new culture of theirs ethnics[8], the ultimate goal is to train the ethnical people become a part of his nation [9].

It can be defined the values of ethnical cultural education in the following: through the cultivation of the educated to understand and master the nation's outstanding cultural achievements, while understanding and mastering the knowledge of Chinese nation and ethnical culture in the world, so that let people form a correct outlook on life and values and love the socialist motherland, and learn to live together in harmony, finally they will become useful for the development of the society.

2.2 Objects of the ethnical cultural education

The object of education is also known as the educated people. Education is an activity of educating people. The broad definition of educated people is the object of the educators exerts influence on them in the process of education. And the narrow one is the social members are affect by the educating activity, which are trying to shape and educate people through the systematic education. Therefore, different educational activities have different objects. So what objects of the ethnical cultural education? Or which part of people belongs to the object of the ethnical cultural education? The views as the following:

Tian Jingzheng thought it is the first step to educate the children in the process of the educational system, which also is very necessary and urgent [10]. Ma Xianglin considered that the heritage of outstanding ethnical culture must start from early childhood [9]. Lu Jiping thought the college students should inherit and carry forward the ethnical culture [11]. Chen Xinggui believed that the ethnical cultural education should be a universal learning and education, which not only be learned by minorities, the main ethnical people must learn [12]. Zhu Xiaoming noted that the ethnical cultural education is education for individual among all ethnical people; in particular, it is an education to strengthen the guidance of ethnical cultural education for students [8].

It can be understood from the above views, the objects of ethnical cultural education not only includes students, but also has the adult students, including both main ethnic minorities, and ultimately it can form the universal learning and life-long learning so that it will promote our excellent ethnical culture, let the ethnical culture stand in the world.

2.3 Content of the ethnical cultural education

What material can educate students to understand and master the knowledge? So the following are the views about the content of ethnical cultural education:

Some scholars believed that ethnical cultural education is the activity of taking the excellent traditional culture as the main content. [13], the excellent culture is something the west want to learn, so people should learn it [14]. Some scholars believe that the content of inheriting the ethnical culture in the school are the knowledge, feeling, attitudes and values of ethnical culture [2]. Another scholar took survive, life, adaptation, ethics, labor as the basic content of the ethnical cultural education, and they also emphasized multicultural education [9].

This paper can borrow the content of ethnical cultural education form Mr. Nan Wenyuan, he believes that ethnical culture can be divided into three levels of education: education of ethnical minority, education of Chinese culture and education in the world. Three educations interrelated and mutually exchanged, mutual influenced, which are the contents of the ethnical education is the first and foremost part of education, it is also the basic starting for the educated to study the ethnical culture.

2.4 Ways of the ethnical cultural education

As for the Ways of ethnical cultural education, that is, what manner or by what channels to educate people the ethnical culture. The existence and development of culture is achieved through education.

If there is no education, the culture will not be happened. Thus, "education is a mechanism for cultural life," which is an indispensable part of the emergence and development of cultural heritage; it is also the fundamental way to pass the experience of social life. However, how to play the important role of education in our ethnical cultural heritage? The ways of inheriting the ethnical culture mainly in the following areas:

Firstly, the ways of ethnical cultural education should be take the school education as the main way, supplemented by other education. On the one hand, it should begin the public school education from the type of ethnical cultural education, which includes social education, community education, vocational and technical education, family education, private school education, religious temples education, adult continuing education [15]. On the other hand, the traditional cultural programs should be conducted through the school education, in school; educators should strengthen the construction of campus culture, social practice, and use the advanced network technology and other ways to broaden the ethnical cultural education. In the family education, parents should pay attention to both the "persuasion" as well as emphasis on "teaching by example"; and in the community education, people should attach importance to carry out the construction of the cultural market and district activities, but also the media coverage of public sector should be opened [3]. Meanwhile, the importance of ethnical culture and modern educational should be integrated [14].

Secondly, community education should be developed rely on the community resources. Some scholars proposed that the sterility mechanism of ethnical cultural education will be formed through relying on the community resources [16]. And some other expert proposed that it is a stable way to realize the ethnical cultural education through the foundation of the University of Regional Ethnical Culture [17]. Another scholar proposed a way of "standing on the ground," that is, the ethnical culture can be accumulated in the practice of the ethnical festivals [9]. Those views of "On the ground", "rooted into the local culture" and "Regional University for Ethnicalities Culture" all have emphasized the ethnical education can not be divorced from the facts of the ethnical cultural areas, even can not live form life.

Thirdly, another way of educating ethnical culture is to strengthen higher education and innovate the style of ethnical cultural education. Zhao Lei, She Yuxia discussed the ethnical cultural education in the university under the background of information technology. Zhao Lei believed that it the favorable environment of ethnical cultural education can be created by strengthening educational team and determining reasonable goals of ethnical cultural, choosing positive content [18]. She Yuxia proposed that the ethnical cultural education can be made by using a variety of media and other communication platforms, the association of students, and the ethnical cultural environment [5]. Huang Jingjing thought the ethnical cultural education should make full use of the modern information technology to rich the ethnical culture, and to open us the wide field of the ethnical cultural education.

From the above studies, they mainly focused on the family education, social education, community education and school education, and emphasize the theory of "education is life, life is education," the ethnical cultural education can not be separated from the actual situation in ethnical areas, which should be innovated in practice activities. However, Kan Jun from the Southwest University said, in the background of the college entrance examination system, it is more obvious of family education and social education in the process of inheriting the ethnical culture [19].

3 PROBLEMS OF THE ETHNICAL CULTURAL EDUCATION

The importance and urgency of ethnical cultural education is recognized and valued increasingly recently, however, because of the various restrictions, the ethnical cultural education is still in the status of awkward position, and its development is still struggling. So analyzing the problems and reasons exist in the ethnical cultural education is important, mainly represented in the following:

First, the ethnical cultural education did not deserve its attention on every level of education. On the one hand, because of the concept of exam-oriented education, the goals of learning are to respond to a variety of exams. So the goals, principles, teaching methods and textbooks of the ethnical cultural education have not been strengthened; on the other hand, parents are keen to cultivate the child as "talent" and ignored as a "himself" in the family education, so the correct world outlook, life outlook and cultural values are ignored; at last, it exists some regime aspects on the social advocacy, the outstanding cultural book, and the cultural events held in terms of social education [20].

Secondly, the ethnical cultural education is divorced from the real life in the ethnical minority areas. The government emphasis on the enrollment and the quality of education, and the ethnical cultural education to students is ignored by educators. There has 2 main reasons, one the one hand, the education lose its ethnical charac-

ter because of the impaction on model of education in the mainstream society. On the other hand, education has become an activity of industrial production, and has divorced from the real life in the ethnical minority areas. The last is the value of ethnical cultural education has not been recognized through the judgment of mainstream cultural standard [10].

Thirdly, the development of ethnical cultural education has been restricted by the mainstream culture, exam-oriented education and the backward status of ethnic areas, on the one hand, the necessity and independence of ethnical cultural education was weakened by the involvement of mainstream culture, and the multicultural curriculum was ignored in the ethnical minority areas. One the other hand, as for the context and curriculum design of ethnical culture, the mainstream culture is the dominant one, and the minority culture become a part of the reference content [21]. Meanwhile, there existed varying degrees of western culture which affect the people in the ethnical minority area, and the ethnical culture was neglected. So because of those reasons, it has resulted in the loss and mutation of language and words [22].

4 DEVELOPING TREND OF THE ETHNICAL CULTURAL EDUCATION

How to develop the ethnical cultural education in the future period? How to handle the relationship between ethnical cultural education and multicultural education? Those are the questions we must to think. Professor Teng Xing who is form the Central University for Ethnicalities proposed a "theory of multi-cultural integration education" based on research experiences for many years. That is, the education should try his best to pass the favorite traditional culture, at the same time, it must has the function of inheriting the common cultural achievements of human beings, even has the function of passing the outstanding culture of ethnical minority This theory provided a theoretical foundation for the conflict between education and multicultural education, and give a direction to the development of ethnical culture education. That is, the ethnical culture is not retro which can not be xenophobic [1].

5 CONCLUSION

In short, because of the globalization and information technology, the economical development has been promoted, and the consciousness of people on the society, politics, and culture also has been changed. In the field of culture, every nation also has the challenge of cultural independence when they understand and identity the world culture. So it has become a urgent need to carry out and strengthen ethnical cultural education.[18] in order to explore a suitable way of ethnical cultural education, the content setting, improvement of teaching methods, and improvement of educational system are all need to do.

ACKNOWLEDGEMENTS

This work was supported by Soft Science Program of China (2013GXS4D149), Ministry of Education, Humanities and Social Sciences Project (12YJCZH053), Minority Native Culture of Quality Engineering Projects in Yunnan

REFERENCES

[1] Cao Kewu, Multicultural education in our philosophy of Ethnical Culture Education Education and Teaching Research, 2007 (2): 143.

[2] Shen Chunshan, Cultural Selection and Construction of Ethnical Cultural Courses—A Case Study of Yanbian Central University for Ethnicalities PhD thesis, 2012 (5): 28, 8, 8.

[3] Gong Liyan, excellent traditional culture of Chinese education and cultivate the ethnical spirit Northeast Forestry University, a master's degree thesis 2006 (6): 1, 1, 8–9, 32–39, 22–24.

[4] Huang Jingjing, College Students in Ethnic Cultural Education Exploration Educational Technology, 2012 (9): 33, 34.

[5] SHE Yuxia, Study on College Minority Cultural Education in China context of globalization, Higher Agricultural Education, 2012 (9):12, 12 –13.

[6] Shen Chunshan, Cultural Selection and Construction of Ethnical Cultural Courses—A Case Study of Yanbian Central University for Ethnicalities PhD thesis, 2012 (5): 28, 8, 8.

[7] Yanbian State Board of Education. Guidance on carrying out ethnic Korean Korean cultural education in primary and secondary schools statewide [DB/OL]. Http://www.ybedu.net/web1/show.aspx?id=587, 2007-10-10.

[8] Zhu Xiaoming, Liu Jun. Teach new period of ethnical culture. Heilongjiang Ethnical Series, 2004 (4):81, 80, 80–83, 81.

[9] Ma Xianglin. Sense of ethnical culture and education "on the ground" in teaching and research materials courses, 2012 (6): 82–84.

[10] Tian Jingzheng, Zhou Duanyun, on the plight of children ethnic minority areas of culture and education Contemporary Education Forum, 2009 (5): 22, 23.

[11] Lu Jiping, Zhang Chen, Thoughts on College Student Ethnical Culture Education for Governance Digest, 2010 (7): 133.

[12] Chen Xinggui multicultural education and minority cultural heritage Yunnan University for Ethnicalities (Philosophy and Social Sciences), 2005 (5): 33.

[13] Gong Liyan, excellent traditional culture of Chinese education and cultivate the ethnical spirit Northeast Forestry University, a master's degree thesis 2006 (6): 1,1,8–9,32–39,22–24.

[14] Zhu Xiaoming, Liu Jun. Teach new period of ethnical culture. Heilongjiang Ethnical Series, 2004 (4):81, 80, 80–83, 81.

[15] Nan Wenyuan, ethnical culture and ethnical education. Qinghai University for Ethnicalities (Social Science Edition), 1994 (4):92–93, 93.

[16] Pan Hung ethnical cultural education research and practice. Modern teaching, 2012 (12):53–54.

[17] Huuang Longguang, Ethnic Culture School: University of intangible cultural heritage of regional new model Heritage, 2012 (1):23–28.

[18] Zhao Lei. Ethnical Cultural Education in the Era of Globalization. Chongqing Vocational & Technical College (Comprehensive Edition), 2003 (10): 60–61, 59.

[19] Kan Jun, The three regional cultural heritage Southwest type of legal protection and education Southwest University PhD thesis, 2010 (4): 98.

[20] Gong Liyan, excellent traditional culture of Chinese education and cultivate the ethnical spirit Northeast Forestry University, a master's degree thesis 2006 (6): 1,1,8–9,32–39,22–24.

[21] Zhao Rong, Theoretical Perspective explores Cultural Minority Education in China. Sichuan Cadre College, 2011 (3):86–87.

[22] Wang Jinliang, ancient quiet environment and the conservation of biological diversity awareness in Yunnan Ethnic Culture Yunnan Normal University (Philosophy and Social Sciences), 2009 (1): 39.

Future Information Engineering and Manufacturing Science – Zheng (Ed)
© *2015 Taylor & Francis Group, London, ISBN 978-1-138-02644-5*

Study on e-education in the "security analysis and investment" course teaching practice

Ping Guo, Zhifen Xu & Xinning Zhang
Shijiazhuang Tiedao University, Shijiazhuang, China

ABSTRACT: With the rapid development of Chinese securities market, the general public is increasingly involved in the investment securities, and this trend highly requires improving practical ability in securities investment teaching. "Security Analysis and Investment" as a financial professional core courses and general courses for the economy and management majors, must to meet the needs of the community. This paper analyzes the defects of traditional methods of teaching, based on the experience of years of teaching, innovate teaching methods with computer and network technology such as multimedia teaching, software simulation teaching, experiment teaching, competition teaching and simulated trading teaching, then redesign the "Security Analysis and Investment" course evaluation system, finally summarize the effect of these innovation.

Keywords: network; multimedia; security; teaching; innovation

1 INTRODUCTION

Along with the rapid development of Chinese capital markets, the financial position of securities and securities markets has been upgraded constantly, which require the emplo yees in securities industry not only to have a solid theoretical foundation, but also to have a wealth of practical ability. The course of Security Analysis and Investment is currently a required course for the student of modern economy and management majors. In order to deepen students' understanding of what they have learned, and have some practical ability, it is necessary to present technical analysis of securities in the classroom, to introduce the securities investment analysis and operation software and related investment website, and to explore its application in the teaching process.

2 DEFECTS IN THE TRADITIONAL TEACHING METHODS

2.1 *Emphasis on theory than practice*

Traditional teaching of securities investment programs, most have stayed at the level of concepts and theories, only enable students to understand, and difficult to achieve using. In fact, there is a great difference between the theory and the practice of securities investment, as in the teaching methods of the securities investment analysis, macroeconomic analysis, industry analysis and technical analysis, the theory is very abstract and difficult to understand, only combining with the companies analysis and technical analysis, it is more vivid and imaged, to improve students' interest in learning.

2.2 *Isolation from actual situation in China*

Some books of Security Analysis and Investment introduce the basic theory and methods, but most of foreign theory and practice, though these can allow students to understand, but there are quite different with Chinese reality, and let the students have a sense of distance and difficult to understand.

2.3 *Lack of operational case*

Whether from the definition of stocks and bonds, or Security Analysis and Investment point of view, the textbooks are lacking the advancing cases, especially in the period of rapid development of China's stock market in recent years, and from the implementation of the non-tradable share reform to the introduction of options and stock index futures, and so on. In this series of changes the study in order to be able to keep up with the times, we need to research the realistic case taking place, to enable students to integrate into the society immediately.

3 THE INNOVATIVE TEACHING METHODS WITH THE APPLICATION OF COMPUTER NETWORK TECHNOLOGY

3.1 Multimedia teaching

The teaching of Security Analysis and Investment involves a lot of graphics, formulas, technical indicators, data such as the K-line theory, the gap theory, shape theory, wave theory, and the traditional teaching methods such as writing on the blackboard are difficult to obtain good results. The use of multimedia teaching tools, synthetically applying sounds, images, text, and color, turning the abstract into image, to reproduce the similar scene of content in teaching, so that students have an immersive experience. For example, showing the pictures of stocks and bonds samples, the students can have an intuitive feel through these financial investment instruments. The form of diagrams to show the interaction of price, volume, time and space vividly explains the securities investment theory. It's intuitive, flexible, real-time, and three-dimensional features bring a great convenience for Security Analysis and Investment teaching.

Multimedia teaching is an auxiliary practice teaching model, although this form of teaching has a certain degree of image and intuitiveness, but generally speaking, this practice of teaching is still a static model, relative to the changes of the diversity of the securities market and the requirements for the Security Analysis and Investment teaching to close contact with the practice, this form of teaching still has a certain degree of disparity.

3.2 Software simulation teaching

Software simulation teaching refers to the teaching form of connecting the internet to computer in the class, downloading and installing the securities investment software, and using the projection equipment to display the real-time data of all kinds of securities to students. Software simulation teaching is given by using a variety of market analysis software, through real-time quotes which are linked into the Stock Exchange by internet, have a dynamic observation for securities trading. This teaching mode integrates with the analysis of the macro economic situation, the economic policy

changes, the relevant industrial development trends, business background, brokers information as well as technical index changes, the teacher can help students to judge the stock market trend and the trend of stocks and then to be confirmed in the subsequent teaching process.

Software simulation teaching requires teachers to analyze securities trading in the classroom, and need teachers have the higher comprehensive quality, not only theoretical knowledge but also more abundant practical experience. For one thing, they should have a thorough understanding for the modern portfolio theory. Updating their knowledge constantly, and adding up new ideas into the teaching process appropriately. So students can feel the current trends of the securities market development, and can be attracted to pay more attention to this course after class. For another, teachers need to strengthen practices. Apart from participating in securities investment actively, the teachers regularly study in the related departments in securities industry, learning about the latest progress of the securities industry, and to teach the fresh practice experiences to students.

3.3 Experiment teaching

As a discipline with strong practice and operability the experiment training is an indispensable part of Security Analysis and Investment course. Taking advantage of network resource and securities trading software students can observe various securities investment tools, combine the theory and practice, analyze the market quotations and simulate the securities investment. Technical analysis is the unique analysis methods of stock quotes, but it is not easy to master since it's content is much more and complex. Students should have an intimate knowledge of various technical analysis theories and methods, e.g. to find the typical Head-and-shoulders Top and M Top graphics, to research the technical index (such as KDJ, RSI, MACD, and so on). After the simulation training, students should record the situation of the training and what they have learned from the training as a part of experiment report.

3.4 Competition teaching

"Security analysis and Investment" is a much practical and operational course. The introduction of lasting and healthy competitive activities in classroom teaching can improve the teaching effect, turn the students from passively listening to actively actual combat exercising, enhance the study enthusiasm of the students and make them have a solid grasp of the theory and practice operation abilities of securities investment.

Figure 1. Software simulation teaching mode.

During the classroom teaching, the students grouped themselves freely by dividing into 5–7 groups. Each group of students select their group name, configured of virtual capital of 100,000 Yuan RMB. Every week each student take turns in their groups to choose a stock and explain that at least three reasons, next week do settlement, and at the end of the semester do settlement of balances for each group with total amount of profit and loss, and record the usual results according to the ranking.

Competition teaching combines the teaching contents with students' practice, students feel what they learned in the classroom is very informative and useful, so they are absorbed in each class, hoping the teachers can teach more skills. Teachers do timely comment in the classroom, guiding the students to solve new problems found in the practical operation. And students can truly entered the stock market from the view of investors, feeling the price changes in the securities exchange in a simulated trading environment.

But the competition rules also have defects. Due to the limited time of classroom teaching, class time is not necessarily on the Chinese stock exchange time of 9:30–11:30 a.m. and 1:00–3:00 p.m., and the teacher forces the student to close the positions during lessons and students are not able to choose the time to sell, which making "luck" components increase in the profit and loss settlement.

3.5 *Simulated trading teaching*

It is divided into two steps for simulated practical operation of securities transactions. The teacher first introduces the entire transaction process in laboratory by opening an account (shareholder accounts and funds accounts), entrusting, dealing, to settling and delivering, to enable students to master the operational processes of Securities Transactions. Then let the students make trading operation in the analog securities system. The analog system is a professional website for simulated dealing of stocks, such as www.cofool.com; it also can be a simulated stock market of the securities trading software, namely the students use the simulated trading functions of the securities trading software to register voluntarily and then to obtain simulated funds. In addition, various websites such as Sina, Net Ease, Stock Star also organized simulated stock contest. These analogs are approximately the same with real stock trading process. The operation, transaction, price, volume and enterprise data are all real in the market, except for the funds. The students are completely independent to choose trading time, and can choose more stocks to form portfolios.

In technical analysis segment, teachers can let the students freely choose the securities according

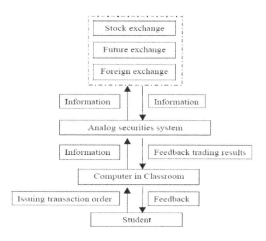

Figure 2.   Simulated trading teaching.

to market quotation by their own analysis and make trading decisions combining with various technical analysis and technical indicators. Then they adjust their portfolios according to the judgment on investment and the performance of each security in the market, pay attention to collect information in the market, and analyze and evaluate their investment portfolios every day to modify their own investment portfolios with ample grounds.

After completing the simulated transaction, the teacher makes a synthesized evaluation to the achievements for the investment in the period of time, and makes the evaluation report. Finally, the teacher should combine the teaching method and the simulated teaching together, in order to make the students truly feel the atmosphere of the securities market, broaden their horizon and improve their ability of integrating theory with practice. During this stage, teachers should request the students to experience transactions in the business hall of a securities company and consult the staff of securities department to introduce how to account, sale, delivery, settlement, and so on.

4   THE EVALUATION SYSTEM DESIGN OF "SECURITY ANALYSIS AND INVESTMENT" COURSE

In the design process of teaching evaluation system for the course of "Security Analysis and Investment", teachers should emphasize the students learning situation and knowledge mastering, adopting diversification of evaluation methods, in this way teachers can relatively and accurately judge the students' study effect, and promote the innovation of the teaching and learning.

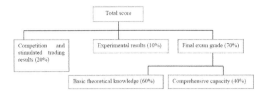

Figure 3. The evaluation system of "Security Analysis and Investment" course.

Therefore, the curriculum should adopt composite examination mode, namely the combination of usual evaluation and final exam, the combination of the quantitative indicators and non quantitative indicators, and the combination of professional evaluation and quality assessment, to use a variety of evaluation methods for comprehensive evaluation of course grades. The examination mode makes different from the past single appraisal model that through a final written examination results, and take "composite–whole process-diversity" evaluation system, as shown in Figure 3.

The course of "Security Analysis and Investment" is closely related with the times, if students are only satisfied in the classroom, they hardly comprehend the true meaning of this course. With the market economy changes rapidly, the teaching effect hardly synchronize with the change of investment mode, while the securities investment software simulation support real market environment and make the students keep up with market changes, and help students to gain effective investment experience.

5 TEACHING EFFECT ANALYSIS

5.1 Inspiring the students learning interest

By the improvement of traditional teaching way, students can make the organic integration of theoretical study and practice. Especially the multimedia teaching, simulation investment, sales practice, involving such as investment analysis, the practice opportunity to grasp the skills, investment psychological change, investment philosophy, and so on. These all make students theoretical knowledge of the textbook specification, thus deepening the understanding and grasp of theoretical knowledge. Some students even involved in stock investment after learning the course of the "Security Analysis and Investment".

5.2 Cultivating the team spirit

Through simulation and practice of investment in the securities sales department, students' active participation in earnest, contributing ideas and exert efforts, change the past situation that teachers and students, students and students part ways with each other, and greatly promote the feeling of love and helping each other between the students.

5.3 Improving the students' ability in practical analysis

Through years of teaching reform exploration of securities investment analysis course, teachers put the emphasis of the course on the cultivation of student's actual analysis, summarize the theoretical framework, such as stock market analysis as well as a variety of reliable patterns, main technical index analysis method, and carry on the practice into the classroom instruction, draw inferences about other cases from one instance, to make student deepen understanding and apply their knowledge to meet practical needs, greatly improving the students' interest in learning operation, wins students acclaim and achieves better teaching effect.

5.4 Cultivating students' health investment psychology

By teachers and students' discussion and summary on the success or failure of the simulation trade, professional managers lecture, students can understand the psychology of the sudden wealth, gambling and jealousy will cause the failure of the securities investment. At the same time, teachers should strengthen the education of students' morality and law, and guide students to have healthy and peaceful investment mentality in the process of securities investment.

ACKNOWLEDGMENT

The paper is the section of the research results for the project "The Independent innovation research on promoting development through evaluation for economy and management colleges under teaching assessment of domestic universities", which is funded by Department of Educational Science Planning Project in Hebei Province in 2014.

REFERENCES

[1] Yang yuangui. 2013. Exploration on Teaching Reform of securities investment. Chinese township enterprises accounting 2013(11): 203–204.
[2] Wang Hongling. Application of simulation in securities investment Teaching. Friends of Accounting 2010(3):116–118.
[3] Wu Yaoyou. Laboratories Application in the "Securities Investment" Teaching, Exam week 2009(9): 239–240.

Future Information Engineering and Manufacturing Science – Zheng (Ed)
© 2015 Taylor & Francis Group, London, ISBN 978-1-138-02644-5

Philosophical thinking on the essence of the contemporary youth*

Qing-gui Huang

Fujian Polytechnic of Information Technology, Fuzhou, Fujian, China

ABSTRACT: Marx's theory on human nature is "ideological and moral cultivation and legal basis" course solid theoretical foundation and the core problem of human nature, objective evaluation of historical development, analysis of modern western mainstream economics is the basic premise and the analysis tool of "economic man", theoretically beneficial to deepening of Marx's human nature research and development in the based on the Marx doctrine of human nature theory of nature; said further understanding of main body of socialist market economy from the practice, the realization of social transformation and reform goal, realize the transformation of teaching material system to teaching system, "play the function of moral education" course, need to realize the contemporary young students to grow up and become adult.

Keywords: human nature; the theory of human nature; contemporary youth; philosophical thinking

Marxist theories about the nature of people are "ideological and moral cultivation and legal basis" curriculum solid theoretical foundation and core, according to the socio-economic development and the needs of students, scientific materialism elaborate nature of the problem of people, the East and West, traditional Chinese culture traditional culture and Western ideas in the important differences between Marxist and non-Marxist, is the focus of contemporary pluralistic values of the intersection, is the focus of contemporary young students and hot, objective evaluation of the historical development of human nature problem, analyze the modern mainstream Western the basic premise of economics and analysis tools "economic man" theory, benefit from the deepening of Marxism Research and development in human nature adhere to Marxist theory on the basis of people; from the practice, said a better understanding of the socialist market economy the essential characteristics of the subject, to achieve social transformation and reform of health objectives for the realization of the teaching material system changes to the system, to play a "basic" moral function curriculum, adult college students need in order to achieve growth and taught. What is man, the meaning of life, attitude towards life and the purpose of life

and so we can not avoid the issue, one can recognize the worldly things, but then "can not properly understand their own" reality unacceptable; If we can properly understand their own helps to understand the enormous sublimation of life and cognitive ability for worldly things will get a huge boost. Thus, the ancient Greek temple of Apollo wrote a doorway: "Know thyself." Contemporary young students to know themselves, to pursue their own meaning of life is the eternal theme.

1 CHINESE ANCIENT UNDERSTANDING OF HUMAN NATURE AS THE ANGLE OF VIEW

In China, the era of the Yellow Emperor Yandi's attempt to answer the question "what is human". "There are seven foot bones, and difference, Daifahanchi, leaning and interest, that the people." [1] 83 this is the difference between the human and animal from the human features.

The difference between the spring and autumn period mainly focus on human and animal, and the understanding of human nature, realized the qualitative leap. Confucius proposed that "Their natures are much the same also, accustomed to far also" [2], 220 to put forward to discuss and study the basic problems of life, for us to point out the direction of life. This definition: "while humanity inborn nature." "Food, color, or." Thought is born with is called, the content is "food". Mencius advocates that human nature is good content is: man is different from animals is that man has, benevolence,

─────────

*This work is partially supported by Fujian Province Education Science "Twelfth Five-Year Plan" in 2011 research project (FJCGGJ11-089) and Fujian Polytechnic of Information 2012 education reform research project (Y12204).

righteousness, propriety, wisdom, morality and ethics. Xunzi advocated that human nature is evil. Xunzi said that the "people" that people have the understanding of "father and son" dear, "men and women don't" capability, this capability is provided by the "ceremony" of the ethics code of ethics. The core problem China ancient humanity relates to Philosophy and the humanities spirit to provide valuable reference for the contemporary youth self-cultivation, but also we accurately understand the essence theory of modern western human to make accurate judgments, and provide the basis for scientific understanding of human nature theory.

2 THE WESTERN UNDERSTANDING OF HUMAN NATURE AS THE ANGLE OF VIEW

The ancient Greek myth "J Fink Faith mystery" is the "what is" to start, the general Taigela that man is the measure of all things, Aristotle believes that man is a political animal, Socrates, that man is a rational animal is the first western the first people the difference between man and animal apart.

The modern bourgeois thinkers and scientists on the human as the system of fruitful, Copernicus "heliocentrism" ousted the long ruling "geocentric theory", shocked and shaken "God made man", began to liberate human nature free from the medieval religious theology, human itself the environment for human survival and to explain the nature and human. Because of the modern western philosophers in human nature on the mechanical materialism and idealism of the understanding, the understanding of human problems exist defects of mechanical materialism and idealism, providing the reference data for the essence of Marx's scientific understanding of.

3 MARX'S HUMAN NATURE THEORY

Marx in the "outline of Feuerbach" pointed out that "the essence of man is no abstraction inherent in a single person. In fact, it is the sum of all social relations." [3] 18 adhere to human society, it must be from the social practice, the social relations of human nature and its change rule, is the construction of "ideological and moral cultivation and legal basis" solid theoretical foundation and core.

3.1 *The nature of people must take the reality and concrete foundation*

Materialist dialectics, which generally exist in the individual, and the individual performance of the general. The social nature of man of the general provisions of the basic proposition is the foundation of theory on the understanding, is the philosophical premise of economic theory and economics. "People here are realistic, people engaged in activities, their communication to adapt to a certain development of their productivity, and this development (the most distant form until it) constraints." [4] 29, therefore, the reality person is engaged in material production; reality is the specific person, we examine the nature of human to the social production mode to find, that is living, realistic, specific people. Marx said that the "human nature is not the beard, blood, abstract physical nature, but human society". [5] 56 this is people will certainly become the premise of man in society, and society completely out of his description of the event the individual person is not likely to be the person, may not have the basic characteristics of labor, thinking, language, culture, may not have a real person, not Felba, said that "the single abstract—" personal "essence", the essence of man must be from the specific historical conditions and specific social practice to examine.

3.2 *The nature of people must be based on the social nature of man as the foundation*

Social human nature, Marx is common on the nature of human being based on dialectical analysis, from the perspective of human nature and its changing rules to person as social. Human nature is "the sum of all social relations", is the core of Marx's theory on human nature.

1. People are also the connection from the perspective of the social network. Everyone is inherited from the past and future development of the essence of contact point, the essence of man includes political capital, economic capital, relationship capital, health capital, capital (including self learning ability and competition ability) and resources are scarce opportunity cost. In today's market economy, we must re adjust the relationship between righteousness and interest and personal, others and society's relations, has a important practical significance. Einstein's special brain does not explain the nature of biological characteristics can not be individual to illustrate.

Other features to human nature embody the social relationship and decide the human perspective. Materialist dialectics, everything is connected, the social relationship is also widespread contact. "Social relationship means many people's cooperation", [6] 33 social relations including production, distribution, exchange, consumption, the relationship of material social relations and includes political, legal, moral, religious relations, the idea of social relationship and family, region, industry, class, ethnic relations margin complicated and difficult to deal with, cross entangled, mutual penetration, interaction of material social relations and ideological social relations. "No matter how detached personal relationships in the subjective, he in the social sense is always the product of these relationships". [7] 12 from human origin, other features all the characteristics include the characteristics and status of people with natural ability, social different people and people; from the process of personal growth, people must be socialized to make biological people into all the features with human society.

3.3 Study of the essence of human development must be to changes in historical perspective, is the theoretical foundation and core to build "Ideological and moral cultivation and legal basis"

Materialist dialectics said that everything is the development of movement and change, so the human nature "the sum total of social relations" in social relations are also changing, so the human nature is changed, the change is based on the social productive forces and production relations of social contradictions movement based on the development of changing. Marx said that the "human nature" is "in each historical era changes [8]" 669 "the history is nothing more than human nature change." [9], 142 from the possession of property system, the basic historical order nature of human evolution: the privatization—nationalization—socialization, private ownership society, few people understand the main resources, the majority rule, there is a difference, grade, level between people, such as the possession of land, exploitation of the peasants, reflecting the relations of exploitation of feudal landlord class; the capitalist productive resources exploitation of workers, reflecting the relations of exploitation of bourgeois socialism; to eradicate poverty, improve people's livelihood, realize common prosperity, conform to the socialist essential requirement, also conforms to the essential requirements of subjective and objective conditions more, human nature will be given more abundant display. The Communist society, human nature will be conscious, rich, comprehensive and free development.

3.4 The nature of people must adhere to the unity of generality and individuality as the angle of view

Materialist dialectics, common in personality, the personality characteristics. Marx pointed out that "the general nature of people first, and then to study in each historical era has changed the nature of man." [10] including 669 common people: collaborative human; different psychological and a different language, text and information exchange are decoded using; people of different cultural backgrounds, different types of each other, under-stand and understanding; human consciousness is the moral, legal and form the basis of common sense and necessary condition to meet their own needs and the needs of the society. Emphasize reality, human nature of the concrete, must certainly personality. Personality is the common to the performance, from the personality, nature is the abstraction. "People—not the abstract concept, but as a real, live, special person—is this kind of existence. These individuals is how, this kind of social relation itself is what". [11] 25 on Marx's theory of human nature is not only for all mankind, but also to each individual, so as to realize the unity of generality and individuality.

3.5 The nature of research must adhere to the people who are unified natural and social properties of perspective, is to build "ideological and moral cultivation and legal basis" of the foundation and core

From the perspective of Marxist argument essentially human contradiction between the productive forces and production relations campaign materialism is the sum of all social relations theory, but also from the perspective of dialectical materialism unified opposition about everything is scientifically explain who is contradictions natural and social properties unity.

1. To the basic properties of the human animal, the natural properties of the same sex with animals perspective. From individual human point of view, people are animals, and animals contain the same terms. Constitute the basic elements of our senior animal body with carbon, oxygen, nitrogen, hydrogen, and so on. Basically the same, the same basic genetic information is proved by modern science, overlooking the internal organs, limbs are basically the same physical structure as well as physiological processes and reflex food, color, and so on. the same basic functions. Differences in animal or human with a degree from the look on people, animals or animal who is the person's physical attributes and other natural instincts and desires, selfish, greedy, cruel, jungle, from the perspective of human freedom, equality, love is hypocritical. The class nature of human society is to follow the animal world jungle, jungle survival of the fittest, there is a ruling class and the ruling class. Capitalism will best meet human animal jungle, the ruling class has the final authority to determine the distribution of benefits rule, their liberty, equality, fraternity evolved killing and eating people, such as treatment of the Indians, to treat the people in backward countries, the individual who is creatures who, starting from the individual person to clarify the nature of

the human animal is. In a market economy, people's independence is an illusion, the fate of the need to master, wealth, economic status on the increase, followed by upgrading the political status will change their destiny, between individuals interdependence is a fact.

2. To distinguish between human and animal perspective. People are thinking of language and culture and wisdom of animals, and thinking associated with wisdom. Person is conscious (including thoughts, emotions, will, and other mental phenomena), rational animal, is in his own consciousness dominated activities; person can create text, art, plants, and so the things of nature there is no; ancient Chinese believe that people have understand morality, decency, love.

3. Work is a fundamental difference between man and animal is perspective. Labor is that people use tools to transform nature, to get the product to meet their own needs. Labor is the difference between humans and animals with the most fundamental boundaries. So Engels said: labor created man. All others are characteristic of the formation on the basis of Labor. First, with the root causes of labor, for the people is primary. Who must work to survive, relying on social production, distribution, exchange and consumption to achieve. Simply animal survival instinct. Second, labor people together in social work. As a combination of individuals who do not have animals in the form of social, human sexuality, for example, men and women who love sex is based on the formation of many social relationships of marriage, family, relatives, and so on. By way of civilization on the basis of, straight combined with distinctive and strong social. The animal is only determined by the instinct of group. Third, labor formed human thinking and consciousness, forming part of human language and culture, and human morality, ethics and so on. So labor is the most basic of human nature, based on the intrinsic reason why people are people and people with other attributes. Binding mode and binding relationship between labor is the essence of man, because a single animal person, through this combination, it has a user-friendly; This combination means or how the relationship determines all aspects of an individual's other psychological, social and so on. Are selfish, whether benevolent and other characteristics. Practice has proved that: the essence of man is the sum of all social relations. Any isolated since childhood, from the existence of social relations, even with the natural properties of the individual can not develop into a real human, human nature does not have.

Only eliminate the contradiction between man and nature, man and society, between people, in order to achieve true harmony between man and nature, man and society, between people, the real essence of man. In a market economy and the realization of the dream of today's China, the pursuit of fairness and justice, democracy and the rule of law is the target value of the basic requirements of human nature to achieve, and social systems in order to construct a free and comprehensive development, the development of our ability to improve our realm, is the direction of the efforts of everyone.

In short, as the nature of social beings who are unable to biological and psychological characteristics to accurately express the virtue of tolerance must be nurtured in order to be of value in life, to love life makes sense watering of tolerance, meaningful life can be wonderful, wonderful life to be happy. This requires tolerance and mutual understanding, harmony and happiness of life will not worry and hatred, when we eliminate people, man and nature, human and social contradictions and conflicts can really build people, man and nature, man and harmonious society, human nature will actually be achieved. Marxist theory human nature provides us with a clear theory of patriotism, ideals and beliefs and other major practical problems foundation also elaborated collectivism and socialist core value system, to solve the philosophical basis of various non—proletarian ideology and a wrong thought, is to establish a scientific world outlook, outlook on life and values based on the logical structure of the theory is the objective requirement and theoretical convincing.

REFERENCES

[1] Yang Bojun subset of rows speech [M] Beijing: Zhonghua Book Company, 1979.

[2] Xu Zhigang Annotation Analects • XVII Analects General Theory Beijing: People's Literature Publishing House, 2000.

[3] Marx and Engels, Selected Works Volume 1 Beijing: People's Publishing House, 1960.

[4] Marx and Engels, Selected Works Volume 3 Beijing: People's Publishing House, 1960.

[5] Marx and Engels, Selected Works Volume 1 Beijing: People's Publishing House, 1995.

[6] Marx and Engels, Selected Works, Volume 3 Beijing: People's Publishing House, 1960.

[7] Marx and Engels Collected Works, Volume 23 Beijing: People's Publishing House, 1972.

[8] Marx and Engels Collected Works, Volume 23 Beijing: People's Publishing House, 1972.

[9] Marx and Engels, Selected Works, Volume 3 Beijing: People's Publishing House, 1960.

[10] Marx and Engels Collected Works, Volume 23 Beijing: People's Publishing House, 1972.

[11] Marx and Engels Collected Works, Volume 42 Beijing: People's Publishing House, 1979.

Future Information Engineering and Manufacturing Science – Zheng (Ed)
© 2015 Taylor & Francis Group, London, ISBN 978-1-138-02644-5

Financial service innovation of 3PL—logistics finance

Y.Q. Wang & D.Y. Zhao
Three Gorges University, Yichang, Hubei, China

ABSTRACT: In recent years, logistics finance developing rapidly in China as an innovation business of logistics service, has become an important area to expand space for development and enhance competitiveness for the third Party Logistics (3PL) enterprises and financial institutions. This paper expounded the connotation, significance and background of logistics finance, conducted an analysis of several main business modes of logistics finance with flow charts and put forward some prevention measures to the risks that may exist in logistics financial services for 3PL enterprises, financing enterprises and financial institutions. The aim of this paper is that logistics finance business can be promoted and it will be helpful for participants to operate the business successfully.

Keywords: logistics finance; Finance-Transportation Warehouse (FTW); bonded warehouse; logistics factoring; risk prevention

1   INTRODUCTION

Modern logistics, called "the third profit source of enterprises", is attached great importance to its development in China. However, the rapid development of modern logistics cannot do without the support of financial services. Along with the globalization and networking of the modern logistics industry, logistics financial services such as clearing, payment and so on, have emerged and continue to develop and be improved. Logistics finance is a product under a certain stage of the development for the logistics industry and the financial sector. Logistics finance is becoming a new financial business, which 3PL enterprises treat as an innovation service for high-end competition and promote financial institutions to participate in. The development and innovation of logistics financial business in modes are the need for financial institutions to expand the business, but how logistics finance controls financial risk and makes the logistics industry, the financial industry and the financing enterprise healthy and rapid growth to achieve "win-win", is a permanent topic and deserves our in-depth research. Awareness and effective risks aversion is the key to carry out logistics finance smoothly or not.

2   OVERVIEW OF LOGISTICS FINANCE

2.1   *Connotation of logistics finance*

Logistics finance is only in recent years popular in China, with regard to the concept of logistics finance, at present there are three main academic point of views: (1) logistics finance in a broad sense, is a series of business activities, which refer to application of all kinds of financial products and effective integration logistics, business flow, capital flow with information flow as well as the organization and regulation of the operation efficiency of funds for the whole process of the logistics operation. In a narrow sense, it is the settlement and financing service, with which logistics suppliers provide customers, which often requires banks to participate in. (2) Logistics finance refers to financing activities in which financial instruments make logistics value-added in the supply chain business activity. (3) Logistics finance refers to the combination of the logistics industry and the financial industry as well as the combination of financial capital and the logistics commercial capital, which is a new business area of the financial industry [1].

Three main subjects involved in logistics financial service: 3PL enterprises and financial institutions, as well as small and medium-sized enterprises (financing enterprises). Cooperation with financial institutions, logistics enterprises provide financing services for Small and Medium Sized Enterprises (SMEs).

2.2   *The background of logistics finance*

2.2.1   *3PL services revolution*
The rapid development of modern logistics cannot do without the support of financial services.

On the one hand, logistics finance can help 3PL expand their scope of operation and enhance competitiveness; on the other hand the service can help SMEs to expand the financing channels, reduce the cost of financing. Logistics finance is the need to accelerate the development for modern logistics, which is a revolution of the third party logistics services.

2.2.2 *The financing difficulties of SMEs*
SMEs financing has been a worldwide problem. Due to their credit deficiencies for SMEs, lack of mortgaged real property, there are few financing channels, high financing cost and huge pressure on the development funding in the production operation. Logistics financial service emerged can revitalize current assets such as temporarily idle raw materials and finished products, and effectively support SMEs financing activities.

2.2.3 *Supply chain win-win goal*
For the purposes of modern 3PL enterprises, logistics finance can enhance the competitive ability of enterprises, expand their business scale, increase high-value-added services and increase their operating profits; for the supply chain enterprises, logistics finance can reduce the cost of corporate financing, broaden the financing channels for enterprises. In the case of financial institutions, logistics financial service can help financial institutions to expand the size of loans, reduce credit risks, and even help financial institutions with the disposal of certain non-performing assets [2].

2.2.4 *Enhanced innovation consciousness of financial institutions*
Currently, financial institutions are facing increasingly fierce competition. In order to obtain the advantage in the competition, financial institutions continue to carry out business innovation, which prompted the birth of logistics finance. Logistics finance can help banks attract and stabilize customers, expand business scale and enhance the competitiveness of banks; can solve the pledge evaluation, asset disposal and other services faced in the pledge and loans of banks [3].

2.2.5 *The implementation of "Property law"*
The implementation of "Property law" provides policy and legal basis for the logistics finance. The law provides that the movable property and goods have the right to pledge, which resolves the problem that the dealer has no fixed assets as collateral.

3 SEVERAL MAIN OPERATION MODES OF LOGISTICS FINANCIAL SERVICE

3.1 *The pledge of warehouse receipts mode*
The pledge of warehouse receipts refers to that financing enterprises store collateral in the 3PL enterprise's warehouse designated by banks and depends on special warehouse receipts pledge or pledge list issued by the 3PL enterprise to apply for bank loans, then were provided a certain percentage of loans by banks according to the use, liquidity and price volatility of the pledge. In this process, the third party logistics enterprise is responsible for the storage and regulation of the pledge and plays the role of the middleman and guarantor. Specific process is shown in Figure 1.

3.2 *The FTW mode*
Finance-Transportation Warehouse (FTW) is an integrated platform of 3PL service, which is at the core of warehousing and supervision of the collateral, evaluation, public warehousing, logistics distribution and the auction. Not only is FTW the framework of a new bridge for the cooperation between the bank and the enterprise, it will also be well integrated into the enterprise supply chain system, as important 3PL service provider for SMEs [4]. Financial institutions, according to the scale of 3PL enterprises, the performance of operation, the proportion of assets and liabilities as well as the creditworthiness, granted 3PL enterprises a line of credit. 3PL enterprises can provide related enterprises with flexible mortgage loans depending on the line of credit, directly monitoring the whole process of mortgage loans business, in which financial institutions are basically not involved in the actual operation of the mortgage loan project. Specific process is shown in Figure 2.

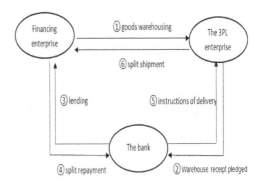

Figure 1.   The pledge of warehouse receipts mode.

192

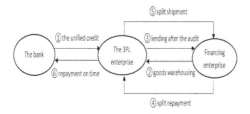

Figure 2.   The FTW mode.

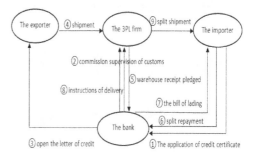

Figure 4.   Logistics factoring mode.

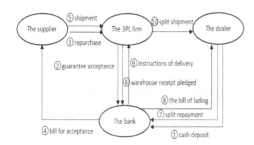

Figure 3.   Bonded warehouse mode.

3.3   *Bonded warehouse mode*

Bonded warehouse, known as buyer's credit, is the extension of warehouse receipts pledge business. It refers to that (1) financing enterprises (dealers) pay a certain proportion of deposits to the bank, (2) the suppliers receives acceptance bills which the bank issued and deliver goods to logistics enterprises designated by the bank (warehouse), (3) according to the bank's instruction, logistics enterprises deliver goods to the dealer. In this mode, the supplier must promise to buy back the pledged goods when dealers are unable to repay loans expiring. Specific business process is shown in Figure 3.

3.4   *Logistics factoring mode*

Logistics factoring, also called credit guarantees, refers to that (1) the bank issued credit certificates to the export enterprises in foreign countries after the domestic import enterprises paid a certain percentage of deposits, (2) export enterprises deliver goods to the warehouse of logistics enterprises designated by the bank, (3) logistics enterprises were responsible for storage and supervision of goods, controlled the ownership of the goods and delivered goods in batches to import enterprises

according to the bank's instructions, and finally the logistics companies and the import enterprises settle the purchase price in common.

In this kind of mode, the logistics company mainly played a role of factor: (1) to provide financing services to foreign export enterprises, (2) to provide credit risk guarantee and confidence evaluation and other financial services for domestic importers. Specific process is shown in Figure 4.

4   RISK PREVENTION AND CONTROL OF LOGISTICS FINANCIAL BUSINESS

4.1   *Risk prevention of financial institutions involved in the logistics financial business*

1. Establish a comprehensive information management system. Financial institutions establish a reliable and complete data information processing system, which can make an in-depth analysis of the data provided by 3PL enterprises to evaluate the authenticity and conduct an analysis and processing of the data offered by the financing enterprises to provide the basis for the decision whether to finance or not.
2. Strengthen credit management and to establish strategic alliance. To establish a strategic alliance means that the commercial banks and logistics enterprises establish a stable long-term cooperative partnerships, joint standardize operation, carry out their duties and maximize their strengths in order to reduce risks and increase revenue.
3. Strengthen the identification of warehouse receipt validity. Warehouse receipt has the special function in logistics financial services that can be pledged as a kind of securities to achieve financing, so banks as lenders must strengthen

the identification of the authenticity and legitimacy of warehouse receipt [5].

4.2 Risk prevention of logistics enterprises in the logistics financial business

The risks of logistics enterprises mainly come from internal operations of enterprises, customer credit, selection of collateral and storage of goods, etc. Effective measures to control risk are put forward with the combination of sources of risks below.

1. Establish a sound corporate management system for logistics enterprises. Logistics enterprises in the process of the development of the logistics financial business shall establish a sound corporate management system, strengthen internal management and avoid risks because of the imperfect supervision system and the decision-making errors management layers generated, etc.
2. Strengthen the investigation of financing enterprises credit. The third party logistics enterprises dealing with financing enterprises frequently, in addition to master the transactions of their goods, provide related information to the commercial banks, but also to investigate the ability to operation, in order to prevent the risk that the financing enterprises cannot pay back the loans on time.
3. Select the pledge with caution. The collateral selected has features of the function in a huge range and the price fluctuation in a tight range, should be easy-handing and non-perishable [6].

4.3 Risk prevention of financing enterprises participating in logistics financial business

Risks of financing enterprises to participate in logistics financial business mainly come from the high cost of financing and loans repayment pressure.

1. Select 3PL enterprises correctly. The third party logistics enterprises as the partners of financing enterprises can provide financing enterprises with effective regulation and storage, which may reduce cost and accelerate to obtain financing from banks.
2. Improve the viability of enterprises and strengthen integrity efforts. The problem of financing for financing enterprises is mainly due to lack of real property, poor credit and other reasons. The pressure to loans can be alleviated by improving the viability of enterprises and strengthening good faith efforts

that may promote the long-term cooperation between the logistics companies with financial institutions.
3. Specification and execution management of contract signed. The contract is one of the key aspects to prevent risks in logistics financial business. In execution of the contract phase, the implementation of 3PL enterprises should be supervised irregularly to guard against the occurrence of risks [5].

5 CONCLUSIONS

Logistics financial business with rapid development and innovation, not only improved the development mode of traditional logistics business and achieve the effective integration of logistics, information flow and capital flow, but also has great significance for enterprises to optimize the allocation of resources and enhance the attractiveness and competitiveness of the market. Logistics finance with increasingly rapid development is paid more and more attention, but in China it is still a relatively new financial business mode and remains to be further progress. I hope my study on this business is helpful for financial institutions, logistics companies and financial enterprises to strengthen the prevention of risks in logistics financial services and the promotion of more rapid and healthy development of logistics financial business.

REFERENCES

[1] Guangchun Li. Study on Evaluation of Non-systematic Risks of Innovative Mode of Logistics Finance. Zheng zhou: Zhengzhou institute of aeronautical industry management, 2013, 3.
[2] Pengxia Shi. Research on the Service Pattern and Risk Control of Logistics Finance in TPL Enterprise Xi'an: Chang'an University, 2008, 4.
[3] Xiangfeng Chen. Innovation of Logistics Finance Service—Finance Logistics Shanghai: Fudan University, 2004, 11.
[4] Xingli Wang, Wu Jian. Discussion on Logistics Financial Implications, Prospect, Modes, Practice and Research Hotspot. Financial Accounting, 2013, 10: 128–131.
[5] Shanshan Liu. Research on Risks Management of Logistics Finance—an Analysis on Demonstration of Qingdao Branch of Zhong Chu Development Stock co. LTD Qingdao: Ocean University of China, 2009, 6.
[6] Ying Jiang. Financial Logistics Risk Measure and Control Hunan: Hunan University, 2009, 10.

Future Information Engineering and Manufacturing Science – Zheng (Ed)
© 2015 Taylor & Francis Group, London, ISBN 978-1-138-02644-5

Research on the circuit of the fifth-order CNN hyper-chaotic system

Jie Xu, Juan Chang & Lian Zhang
Science Institute, Air Force Engineering University, Xian, China

Qinglin Xu
School of Technology Physics, Xidian University, Xian, China

ABSTRACT: The thesis has studied the simulation of the operation amplifier in Cellular Neural Networks nerve cell, which has realized the design of the fifth-order CNN hyper-chaos circuits. The simulations prove that the circuits are feasible and can be the same with higher CNN systems for its expansibility.

Keywords: cellular neural networks; hyper-chaos; circuit design

1 INTRODUCTION

Cellular Neural Networks[1-3] originates in Hopfield Neural Networks and Cellular Automata[4] were first proposed by Neumann. CNN reveals the superiorities of Neural Networks and Cellular Automata. Those are having the function of high speed parallel processing, having the characteristic of continuous time dynamics[5-6], and having the characteristic of local joints between cells, of piecewise linear output functions, and of continuous real-time signal processing, which can realize large scale IC, high speed parallel process, improve running speed, and have double-valued output. Presently, CNN as a flexible and effective neural networks model is attached important to the image processing, pattern recognition, control, secure communication, physics and so on[7-9]. In this paper, the fifth-order CNN hyper-chaos system is achieved with operational amplifiers, the process of the realization is analysed and explained. Finally, the simulations are accomplished with Pspice, and the feasibility of the conclusion is proved.

2 THE FIFTH-ORDER CNN STATE EQUATION

For the absolutely interconnected fifth-order CNN, the dynamic characteristic of its every cell is expressed by the state equation:

$$\frac{dx_j}{dt} = -x_j + \alpha_j y_j + \sum_{\substack{k=1 \\ k \neq j}}^{5} a_{jk} y_k + \sum_{k=1}^{5} S_{jk} x_k + i_j$$

$$(j = 1, \ldots, 5) \quad (1)$$

The parameter of the five cells are written as:

$$\alpha_k = 0 \, (k = 1, 2, 3, 4),$$

$$a_{jk} = 0 \, (j, k = 1, \ldots 5 \, (j \neq k)), i_j = 0 \, (j = 1, \ldots, 5)$$

$$S_{11} = S_{12} = S_{13} = S_{15} = S_{21} = S_{24} = S_{25}$$
$$= S_{31} = S_{35} = S_{42} = S_{51} = S_{52} = S_{53} = 0$$

$$S_{14} = S_{23} = S_{33} = S_{44} = 1, S_{41} = S_{43} = S_{45} = -1, S_{22} = 3$$

$$S_{34} = 14, S_{32} = -14, \alpha_5 = 200, S_{54} = 100, S_{55} = -99$$

Then the state equation of the fifth-order CNN system is written as:

$$\begin{cases} \dot{x}_1 = -x_1 + x_4 \\ \dot{x}_2 = 2x_2 + x_3 \\ \dot{x}_3 = -14x_2 + 14x_4 \\ \dot{x}_4 = -x_1 - x_3 - x_5 \\ \dot{x}_5 = 100x_4 - 100x_5 + 200\left[0.5\left(|x_5 + 1| - |x_5 - 1|\right)\right] \end{cases}$$

$$(2)$$

3 DESIGN OF THE UNIVERSAL CNN CHAOTIC CIRCUIT

The typical inner cells state equation of the two-dimensional CNN is as follows:

$$c\frac{dv_{xij}(t)}{dt} = -\frac{v_{xij}(t)}{R_x} + \sum_{c_{kl} \in N_r(ij)} A(i,j;k,l)v_{ykl}(t)$$

$$+ \sum_{c_{kl} \in N_r(ij)} B(i,j;k,l)v_{ukl}(t) + I$$

$$(1 \leq i \leq M; 1 \leq j \leq N) \quad (3)$$

The output equation is as follows:

$$v_{yij}(t) = \frac{1}{2}\left(\left|v_{xij}(t)+1\right|-\left|v_{xij}(t)-1\right|\right) \quad (1 \le i \le M; 1 \le j \le N)$$

$$(4)$$

Given $\tau = t/(R_x C)$, the scale transformation from equations (3) and (4) are

$$\frac{dv_{xij}(\tau)}{d\tau} = -v_{xij}(\tau) + R\left[\sum_{c_{kl}\in N_r(ij)} A(i,j;k,l)v_{ykl}(\tau)\right.$$

$$\left. + \sum_{c_{kl}\in N_r(ij)} B(i,j;k,l)v_{ukl}(\tau) + I\right]$$

$$(1 \le i \le M; 1 \le j \le N) \qquad (5)$$

$$v_{yij}(\tau) = \frac{1}{2}\left(\left|v_{xij}(\tau)+1\right|-\left|v_{xij}(\tau)-1\right|\right)$$

$$(1 \le i \le M; 1 \le j \le N) \qquad (6)$$

Extend the conception of the neighbor cells to general form to obtain the state equation of the following non-dimensional universal CNN:

$$\dot{x}_j = -x_j + \sum_{k=1}^{n} a_{jk}y_k + \sum_{k=1}^{n} S_{jk}x_k + v_j \qquad (7)$$

where n is quantity of the neighbor cells. The output equation is as follows:

$$y_j = \frac{1}{2}\left(\left|x_j+1\right|-\left|x_j-1\right|\right) \qquad (8)$$

In the equation (7), given $\tau = t/(R_b C_j)$, we can derive:

$$C_j \frac{dx_j(t)}{dt} = -\frac{x_j}{R_b} + \frac{1}{R_b}\left[\sum_{k=1}^{n} a_{jk}y_k(t) + \sum_{k=1}^{n} S_{jk}x_k(t) + v_j\right]$$

$$(9)$$

A possible circuit with operational amplifiers of the equation (9) omitted by Figure 1, which can prove:

$$S_{jk} = \frac{R_a}{R_k}, a_{jk} = \frac{R_a'}{R_k} \qquad (10)$$

4 CIRCUIT DESIGN AND ANALYSIS TO SIMULATIONS OF THE FIFTH-ORDER CNN

Select LF347N as the operational amplifier, and the voltage supplied by the power is $\pm 15V$. The composed CNN hyper-chaos circuit is omitted in Figure 2. Every cell's original parameters are as follows:

$R_{a1} = R_{a2} = R_{a5} = 30k\Omega, R_{a3} = 91k\Omega, R_{a4} = 10k\Omega,$

$R_{b1} = 0.5k\Omega, R_{b2} = 1k\Omega, R_{b3} = 1.5k\Omega, R_{b4} = 0.7k\Omega,$

$R_{b5} = 0.3k\Omega, R_{ci} = R_{di} = 91k\Omega, C_i = 0.047uF,$

$$(i = 1,\dots 5),$$

$R_{11} = R_{23} = 30k\Omega, R_{22} = 10k\Omega,$

$R_{32} = 6.5k\Omega, R_{33} = 91k\Omega,$

$R_{41} = R_{43} = R_{44} = R_{45} = 10k\Omega,$

$R_{34} = 6.5k\Omega, R_{54} = 0.3k\Omega,$

$R_{55} = 0.303k\Omega, R_{55}' = 0.15k\Omega.$

$R_e = 75k\Omega, R_f = 910k\Omega, R_g = 28k\Omega, R_h = 2k\Omega.$

Simulate the circuit of Figure 2 with Pspice. Figure 3 shows that the state variables x1, x3, x4, x5 of CNN hyper-chaos circuit change with time. The chaotic attractors' phases of the hyper-chaos circuit's state variables are omitted in Figure 4.

The experiment's results of Figures 3 and 4 show that the circuit can generate chaos signal with the

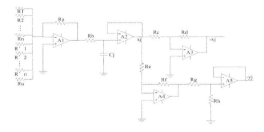

Figure 1. The circuit with CNN operational amplifiers.

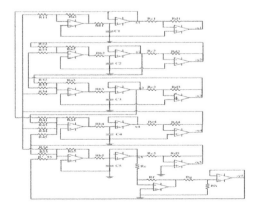

Figure 2. The fifth-order hyper-chaos circuit.

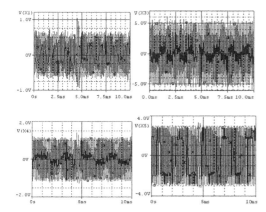

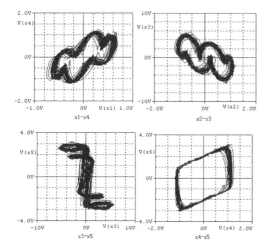

Figure 3. The state variables' time curves of the fifth-order CNN hyper-chaos circuit.

Figure 4. The chaotic attractors' phases of the fifth-order CNN hyper-chaos circuit.

characteristics of the chaos motion, such as global attraction, bounded, local repulsion, non-periodic and so on, and prove the feasibility of the design. The circuit's design uses fewer components, which has a simple construction, so it can be transformed into a practical circuit easily, and also has a good extendibility, so it can construct higher dimensional chaos circuit by adding cell module.

5 CONCLUSION

So much attention to the research of dynamical behavior between CNN cells and the realization of various dynamical behaviors with simple circuit is paid. In this paper, the state equation of CNN cells is given, the current integrated chips are utilized to construct elementary cell needed by the design of chaos circuit, and the fifth-order CNN hyper-chaos circuit is realized. And this plan is apt to accomplish the integration.

REFERENCES

[1] Chua L.O., Yang L. IEEE Transactions on Circuits and System, 1988, 35(10):1257–1272.
[2] N.E. Schoppa. Neuron, 2006, 49(2):271–283.
[3] Zeng Z., Wang J. Neural Computation, 2006, 18(4): 848–870.
[4] Fang-yue Chen, "CNN consilient dynamic system", Ph. D dissertation, Univ. Shanghai, 2003.
[5] Huang T., Chan A., Huang Y., et al. Neural Networks, 2007, 20(8):868–873.
[6] Cao J., Li X. Physica D, 2005, 212(1/2):54–65.
[7] Cao J., Feng G., Wang Y. Physica D, 2008, 237(13):1734–1749.
[8] Lou X., Cui B.J of Mathematical Analysis and Applications, 2007, 330(1):144–158.
[9] Gu H., Jiang H., Teng Z. Nonlinear Analysis: Theory, Methods & Applications, 2008, 68(10):3189–3200.

Future Information Engineering and Manufacturing Science – Zheng (Ed)
© 2015 Taylor & Francis Group, London, ISBN 978-1-138-02644-5

Green supply chain management system of modern enterprise construction strategy

Xiao-yuan Shi & Wei-hua Zhang
Zhejiang Yuexiu Foreign Languages College, Shaoxing, Zhejiang, China

ABSTRACT: With the continuous development of China's logistics industry, green supply chain management is becoming a hot issue for the current research. Enterprises can't succeed in the fierce market competition to occupy a favorable position to achieve long-term survival and development of the traditional management model for innovation and development, more good to adapt to changes in the market development demand. This paper first described the basic characteristics of the green supply chain management, and then analyze the current state of green supply chain management in China's enterprises, the final focus of the green supply chain management strategies and implementing measures were discussed.

Keywords: green supply chain; modern enterprise; basic characteristics; build strategy

1 INTRODUCTION

With the continuous development of China's logistics industry, green supply chain management is becoming a hot issue for the current research. With the continued enhancement of people's environmental awareness, a growing shift in the mode of economic growth, enterprises can't succeed in the fierce market competition to occupy a favorable position to achieve long-term survival and development if they do not make innovations in traditional management pattern and adapt to various needs of the changing market. This paper first describes the basic characteristics of the green supply chain management, and then analyzes the status quo of green supply chain management in China's enterprises, and finally focuses on the strategy for the green supply chain management and the measures. [1]

2 THE FEATURES OF GREEN SUPPLY CHAIN MANAGEMENT MODERN ENTERPRISES

2.1 *Features of green supply chain management*

Modern enterprises should consider the environmental benefits in their operation and management activities so as to avoid negative effects on ecological environment as much as possible when they implement green supply chain management. Therefore, enterprises must take into account the most effective use of management programs to the resources, the lowest emissions of pollutants or scientific recycling and utilization of waste, and finally, enterprises should make scientific and rational evaluation of the various effects on the ecological environment. If these programs fail to meet relevant requirements, they will not be implemented. The traditional supply chain management is to emphasize a customer-centered method and provide the best quality service, without taking into account environmental factors, and it apparently has been difficult to adapt to the latest requirements of the time.

2.2 *The closed-loop operation*

The factors of production and logistics elements in the implementation process of green supply chain management are not raw materials, semi-finished or final products in the traditional sense, and instead, they emphasize the green movement and circulation of material elements within the supply chain system. For example, waste generated in the production, or defective products, and expired products in the process of logistics and sales are needed to be recycling and reuse so as to reuse resources or resale products. Modern enterprises implement green supply chain management process, which is a scientific system of closed-loop operation, without a termination point in the green supply chain system.

2.3 *Emphasis on the application of information network technology and data sharing*

Development of information technology for the enterprise supply chain management system of green construction and implementation provides a convenient, can help companies in product design and production processes to seek suitable partners to achieve further optimization and sharing of social resources, to achieve resource conservation, inventory savings and production efficiency and other targets; Secondly, enterprises can grasp cutting-edge information about market supply and demand through e-commerce platform, thus achieving the high-quality procurement of raw materials and broadening of sales channels. In this way, enterprises reduce the expenditures in the procurement and marketing chain; Thirdly, with the help of modern network technology, companies can also improve the logistics efficiency, reduce waste of logistics resources, and reduce the adverse impact on the surrounding environment in the process of logistics and transport. Meanwhile, data sharing in green supply chain management system is beneficial to optimizing the raw material selection, product design, green production, green logistics, waste recycling and other aspects.

2.4 *With the idea of concurrent engineering*

Green supply chain management emphasizes its concern about the product life cycle, whether it is the production of raw materials, or manufacturing and recycling of products. All these need "green" concept. Concurrent engineering thinking also attaches importance to the product life cycle. It requires comprehensive consideration of all aspects in the design process, and takes into account the recycling and reuse of products and materials. Obviously, the ideal of green supply chain management and concurrent engineering thinking have the same purpose. [2]

3 FACTORS INFLUENCING GREEN SUPPLY CHAIN MANAGEMENT QUALITY

3.1 *Lack of "green" awareness*

Due to the influence of the traditional business management model, many companies blindly pursue the maximization of corporate profits, while ignoring the possible adverse impact on the ecological environment in the process of supply chain management. Many entrepreneurs even believe that only when companies get maximum economic profits, they can in turn make a greater contribution to society. They also think that excessive focus on environmental issues will increase the production and operating costs, which is not conducive to the improvement of economic benefits of enterprises. At the same time, if they want to guarantee the

quality of the implementation of green supply chain management, they have to reach consensus among all company members relevant to the whole supply chain and make joint efforts to achieve the desired structure and objectives. If the managers of many companies in China do not have the basic modern concept of green supply chain management, it can't fully guarantee the effective implementation of green supply chain management.

3.2 *The tradition patterns has formed a barrier*

The traditional supply chain management model lays stress on factors such as time, cost, quality and service, without taking into account the demand of environmental protection. Making suppliers as a starting point and the user as the end point, it is a one-way flow, whether it be capital flow, information flow or logistics. It does not pay attention to the closed-loop operation in all sectors of the supply chain management. Obviously, it is in conflict with the green supply chain management in many ways. It is precisely because there are a lot of traditional factors of the supply chain management in modern enterprises and they, to a large extent, hindered the effective implementation of green supply chain management strategies in modern enterprises. [3]

3.3 *Environmental policy and enterprise system is not perfect*

The government should play a guiding role in the implementation of the green supply chain management, actively carry out and build a sound environmental policy and system, and do a good job in supervision and management for the implementation of green supply chain to create a favorable policy environment and social environment. However, many local governments lay too much emphasis on economic growth, making inadequate efforts in fighting against environmental pollution. Some even damage the environment to obtain performance, which is shortsighted. Some governments even support the mode of production at the expense of ecological environment by some enterprises, which make many companies ignore the protection of ecological environment, and excessively pursue economic interests. Implementation of green supply chain management strategy, if there is no perfect system and policies for environmental protection, it is difficult to enter the in-depth and comprehensive implementation.

3.4 *Consideration of input and output by enterprises*

In the process of implementation of green supply chain management strategy, enterprises do not only invest operating costs in the production and management links, such as human resources, information, logistics, and capital, but also require all members to consider the social costs of the entire supply chain, which makes enterprises doubt their proportion of their input and output, and fear that they may not have the expected benefits, which to some extent limits the implementation of green supply chain management strategies in modern enterprises.

4 EFFECTIVE MEASURES TO BUILD GREEN SUPPLY CHAIN MANAGEMENT SYSTEM

4.1 *Green design system*

Researches indicate that the integrated performance of the product depends largely on the product design, which accounts for only a small percentage of the total cost. Therefore, modern enterprises must pay full attention to green design system, giving consideration to the possible impact on the environment by products in the product life cycle while designing products. Modern enterprises should also ensure the environmental benefits of the product throughout the life cycle, so that they can make full use of social resources to their lowest consumption. [4] First, the modular design can help companies quickly develop products, facilitate the optimization and maintenance of the product structure, conducive to scientific recycling and reuse of products; second, standardized design can reduce the workload of the production and processing, reduce the difficulty of the work and resource consumption; third, removable design and recyclable design are also important methods in green design system, reducing the difficulty of the work, and increasing the reuse of parts.

4.2 *Green supply system*

In order to ensure the effect of the implementation of the green supply chain management in modern enterprises, enterprises should ensure the supply of raw materials on the one hand, and they also need to build a scientific system of green supply. First, the strict selection of suppliers. In addition to considering the traditional factors in supply chain management, enterprises should look into the efforts made by suppliers in environmental protection, to see if it can meet the objectives of enterprises in supply chain management to promote the production of green products and energy saving and consumption reduction; second, implementing a green logistics policy. [5] For example, during the packaging process, we should try to

choose a biodegradable, pollution-free packaging materials; in the handling process, we should reduce the breakage rate of the goods, and reduce noise pollution; in the process of transport operations, we should reduce energy consumption and harmful emissions.

4.3 *Green production system*

In organizing production activities, modern enterprises should fully take into account the adverse effects on ecological environment by raw material input, product output, and a variety of resource consumption. Their aim is to build a scientific production system for environmental protection. First of all, in the choice of production technology and programs, scientific analysis should be given to the factors that may adversely affect the green supply chain management, such as processing methods, processing machinery efficiency and resource consumption. Modern enterprises should try to make production technology easier, lower the consumption of resources on the basis of production needs, and reduce pollution caused by production technology. Secondly, with the improvement of production and processing, enterprises should minimize the leftover of processing, which contributes to the conservation of materials and waste generation. Third, when selecting production and processing machinery and equipment, enterprises should try to choose high-performance, small emissions machining equipment, repair and maintenance to develop scientific system. And they should also draw up scientific rules about maintenance and repair.

4.4 *Construction of green marketing and green circulation system*

First of all, in the sales sector, enterprises should try to choose "green" marketing methods, and adopt ecological management method. For example, they can make use of e-commerce platform to carry out online marketing; in choosing sales promotion, enterprises should pay attention to the publicity of the green features of companies and products, select the most economical and most environmentally friendly ways of publicity; in the selection of suppliers or brokers, their green image should also be taking into account. [5]

Secondly, in the circulation, it is necessary to select the green, environmentally-friendly, and safe packaging materials or forms of packaging, as packaging can easily be discarded by the user, both a waste of resources and also causing damage to the ecology; besides, it is important to select the green mode of transport, such as centralized distribution, optimization of transportation routes, etc. Furthermore, a comprehensive evaluation should be given to the service life of products and the extent of their recycling when products are in service.

5 STRATEGIES TO IMPLEMENT AT GREEN SUPPLY CHAIN MANAGEMENT

5.1 *Create a strategic partnership with members*

Green supply chain management emphasizes the full attention of the entire product life cycle, requiring the active participation of each member in the supply chain. In view of this, enterprises should actively guide the consumers, encourage the public to actively participate in the process of consumption of green products, and recycle and reuse packaging or old products to achieve the desired objective of protecting the environment. At the same time, in the implementation of green supply chain management strategy, enterprises should develop a strategic partnership with suppliers and partners on the entire supply chain, which is very important. Because only by doing so can they realize sharing of information and resources, improving market competitiveness on the whole and making sure the implementation of management measures on green supply chain.

5.2 *Building of green system*

Each industry or enterprise, has its own characteristics and practical. Therefore, enterprises in the construction of green supply chain management system, must be in-depth analysis of their production operations and the market environment, and develop scientific management goals. Enterprises should take targeted management measures in product design, supply, packaging, manufacturing, sales, recycling and other sectors to enhance their green supply chain management system.

5.3 *To build a good corporate green culture*

Modern enterprises can effectively raise green awareness by building the "green culture", so that all personnel are aware of the importance of environmental protection. Strengthening the guidance of the concept on scientific development and the concept of sustainable development is beneficial to the implementation of green supply chain management. [6] First, enterprises should strengthen in-house promotion and training, improve corporate rules and regulations, strengthen the

establishment and penetration of green culture, and gradually build a green culture, so that completely change the traditional concepts of supply chain management, and form green atmosphere within the enterprise; Second, companies should carry out a series of forums, lectures and other activities relevant to green supply chain management in the region or within the industry, sharing green ideas and successful experiences; Third, to realize harmony between operating activities and performance appraisal on environment, enterprises should emphasize green business culture, and promote the implementation of management thinking about green supply chain.

6 CONCLUSION

To sum up, there is much room to improve the implementation of the green supply chain management in China's enterprises. Only when environmental awareness is raised among enterprises and the public, relevant laws and regulations and policies are improved and green supply chain management is more recognized and accepted, enterprises can be more active to implement green chain supply management. Under those circumstances, enterprises are able to build a sound management system of green supply chain, guarantee the implementation of relevant management measures and finally contribute to the social and economic harmony and sustainable development.

REFERENCES

[1] Zhang Min. 2011. Study on green supply chain performance evaluation based on circulation economy, *Oriental Enterprise Culture* (22).
[2] Xie Xiao-jin. 2008. Analysis of the Current Situation and Countermeasures of Green Supply Chain Management in China, *Journal of Sinopec Management Institute*. 10 (3).
[3] Nagel M.H. 2001. Environmental Supply Chain Management Versus Green Procurement in the Scope of Business and Leadership Perspective, *Electronics and the Environment* (3).
[4] Gilbert S. 2001. Greening Supply Chain: Enhancing Competitiveness through Green Productivity. Taiwan: *Asian Productivity Organization*.
[5] Nie Qing, Wang Jing. 2010. The Development of Green Supply Chain in Europe and Its Implications, *Ecological Economy*, (01).
[6] Hong Xiaoqing. 2010. Analysis of green supply chain management in Enterprises, *China Business & Trade*, (27).

Future Information Engineering and Manufacturing Science – Zheng (Ed)
© 2015 Taylor & Francis Group, London, ISBN 978-1-138-02644-5

Research of modern logistics enterprise financial management

Wei-hua Zhang
Zhejiang Yuexiu Foreign Languages College, Shaoxing, Zhejiang, China

ABSTRACT: One of the modern logistics industry as a pillar industry to promote the development of China's national economy, its importance, no doubt, has been seen as a key factor for determining a country's comprehensive competitive strength and the degree of modernization. Financial management as a core part of the management of modern logistics enterprises, the management level is good or bad will have a significant impact on normal business operations and decision-making of logistics. In this paper, the issues related to the financial management of modern logistics enterprises, financial management strategy of the new era of modern logistics enterprises are discussed.

Keywords: modern enterprise; logistics; financial management; financial risk

1   INTRODUCTION

With the development of China's market economy, the logistics industry as an emerging industry of China, gained unprecedented development opportunities and the degree of concern in the development process of China's national economy is playing an increasingly important role. However, because many of our logistics enterprises from the traditional storage companies or transport companies to shift from the management model are relatively backward, economic efficiency and competition in comprehensive strength are not strong. For these companies how to use the follow-up and implementation of modern financial management system, further address the corporate finance, capital operation, the quality of financial information is not higher aspects of the problem is particularly important, this is a modern logistics enterprises to improve their market competitiveness, better to deal with the inevitable choice of domestic and international competitive environment.

2   GOAL OF MODERN LOGISTICS ENTERPRISES TO DEVELOP FINANCIAL MANAGEMENT ACTIVITIES

The financial management of the logistics enterprises is to follow relevant laws and regulations, financial management principles based on the specific conduct of the enterprise financial activities, financial management process, effective treatment, thus achieving logistics enterprises in the financial management in a variety of goals.[1] Among them, the logistics enterprises to carry out financial management, the most fundamental goal is to achieve the maximization of corporate profits and minimize management costs. Logistics enterprises have to go through effective financial management measures to reduce the cost of the target customers, to help target customers more value, in order to better ensure their own development potential and market space. In this regard, the modern logistics enterprises to achieve the established goal of financial management and investment program for scientific analysis and decision-making and improve efficiency in the use of existing funds, increase in the financing, regulatory and other aspects of efforts to improve financial quality of information processing.

3   THE MAIN PROBLEM OF LOGISTICS ENTERPRISES IN THE FINANCIAL MANAGEMENT PROCESS

3.1   *Quality of financial information is not high*

At present, the cost of logistics enterprises constitute more and more complex, but many companies still follow the traditional cost accounting methods, which is difficult to ensure the accuracy and usefulness of corporate financial information, it is difficult for the investment and decision-making of logistics companies to provide scientific basis. Transportation enterprises, for example, the object of its cost accounting, according to the run line to be divided, be related to the warehousing costs, management fees and operating costs, from the perspective of operational tools, will involve air-conditioned vehicles, refrigerated trucks, container cars, bulk trucks, and so on, from a business point of view can be divided for the loading and unloading operations and cargo operations. Although under normal circumstances, the company can transport operating costs and warehousing costs point of view, the accounting of the cost of logistics enterprises, but it is difficult to segment the costs of different services, customers can not help to provide for future decision-making of enterprise managers. Business managers on the demand for financial information, often to different business and customer cost information for analysis and thus to optimize and configure the company's existing resources, to achieve increased market share and improve the management objectives of the economic benefits. The financing environment for logistics companies needs to be improved.

With the continuous increase of the degree of market opening in China, foreign logistics enterprises to enter China's logistics market, in large measure to increase the competitive pressure of China's logistics industry, an urgent need to increase investment in the domestic logistics enterprises, innovation and development of the logistics management mode as soon as possible. Many companies rely solely on their financial strength, it is difficult to achieve a rapid transformation, and need to raise more money outside the enterprise.[2] But as a whole, China's financial markets, inadequate logistics enterprises in the process of financing difficulties, the direction of the financing of domestic enterprises mainly from state-owned commercial banks, often the need for asset-backed or has a good bank credit, which is a lot of small and medium-sized logistics enterprises are not available, which greatly limits the pace of China's logistics enterprises from the traditional business transformation to a modern large-scale logistics enterprises.

3.2 Fails to establish and improve the financial regulatory system

Whether it is the small and medium-sized logistics enterprises, or large group logistics companies, to varying degrees, financial and lax supervision of specific performance in a weak enterprise basis of accounting, accounting information, financial accounting norms, failure to strictly abide by the relevant rules and regulations and other issues. This is mainly because a lot of logistics enterprises in the financial sector are not practical for corporate financial management, build a sound, a sound financial regulatory system, severely limits the quality of corporate financial regulation.[3] For example, some enterprises not related to financial management system to thoroughly implement

the branch financial management control is not strong; corporate accounting system and internal control system is relatively backward, it is difficult to obtain proper financial supervision effect; the overall quality of the financial management of logistics enterprises is not high, the lack of effective ways and means of financial regulators.

3.3 Backward mode of fund management, capital utilization is not high

Most logistics companies have suffered financial supervision and other issues, in varying degrees, specifically in the enterprise basic accounting work is weak, accounting information distortion, financial accounting irregularities, failure to strictly comply with the relevant rules and regulations and other issues. This is mainly because a lot of the logistics company's financial department did the actual corporate financial management, building a sound, a sound financial regulatory system, severely limits the quality of financial regulation.[4]

4 CAUSES OF IMPACT THE QUALITY OF FINANCIAL MANAGEMENT OF LOGISTICS ENTERPRISES IN CHINA

4.1 Quality of financial information is not high

Overall, China's rapid pace of development of the logistics industry has made unprecedented achievements in terms of scale of development and economic benefits, the study of logistics enterprises in financial management is not deep enough, not keeping up with the times forward pace, limiting the enterprise quality of service and maximum economic efficiency.[5] Logistics industry, for example, related to finance the administrative rules and regulations, not perfect, is not sufficiently standardized to the enterprise to carry out the supervision of the business and accounting, accounting more false information, the lack of guidance of scientific theories, and other factors are to some extent, limit in-depth study of the theory is not conducive to enhance the overall level of financial management of logistics enterprises.

4.2 Backwardness of concept managers

As many logistics companies are in transition from the traditional transport companies or storage companies from not very high quality of enterprise managers, not really understanding the various concepts associated with the logistics and financial management, lack of national macropolicy in-depth understanding, failed to update their knowledge structure, acceptance logistics expertise in financial management training and timely to establish a modern financial management concepts, this conduct high-quality financial management activities apparently is very unfavorable. For example, some managers only understand the basic knowledge of finance and accounting, asset management, capital operation, investment analysis and planning skills are quite lacking; concern of financial indicators, to focus only on profits and business income, rather than focus on a comprehensive analysis of the current financial situation and limits of corporate financial management functions effectively.

4.3 The lack of a favorable policy environment

With the continuous improvement of the logistics industry in the national economy status, countries and governments at all levels to give more and more attention of the logistics enterprises, but from the development of the current situation of China's logistics industry, the degree of concern and support of the government is still far enough. For example, China's road transport costs long-term high, lagging behind financial management policies and regulations related to logistics companies, logistics enterprises in the financing environment be further improved. Government has no particularly good policy environment for logistics companies to create, there will be restrictions in the financial management aspects of innovation and development of logistics enterprises.

5 IMPROVE FINANCIAL MANAGEMENT OF LOGISTICS STRATEGY

5.1 Strengthen the internal controls and reduce financial risk

For the various problems of modern logistics enterprises in the financial management process are essential to strengthen the company's internal control. Good corporate internal control system can effectively reduce or avoid the risk of various types of financial, adequate attention should cause each of logistics enterprises. Logistics enterprises in the management process there is always a variety of fraudulent behavior that affects the normal development of the corporate financial management activities. Logistics enterprises in the concrete practice of the process, according to the needs of corporate financial management and long-term development of the internal control system by professionals to build, and then in practice the process of constant adjustment; Second, enterprises should strengthen the relevant content advocacy

and training, so that each employee can recognize their responsibilities to improve the internal control attention from ideological; Finally, the internal control system to improve the effective integration of various internal control measures in-depth implement and enforce, and thus minimizes the likelihood of financial risk.[6]

5.2 Improve the overall quality of corporate financial practitioners

High diathesis financial management personnel is an important prerequisite to ensure the quality of corporate accounting information is a key factor to determine the quality and level of corporate financial management. Therefore, improving the business skills and overall quality of corporate financial management personnel can not be ignored. First, the logistics enterprises to maintain the stability of the accounting personnel, to avoid the accounting information distortion caused by the accounting change frequently or lack of professional skills;, logistics companies to finance staff to create a favorable employment environment and competitive salary system, and fully mobilize the enthusiasm and initiative of the accounting and financial personnel; again, logistics companies to invest enough time and effort to the development of scientific training programs, at the same time improve the professional skills of finance staff, should pay attention to the financial officer of logistics enterprises the cultivation of moral qualities.

5.3 Actively expand financing channels for small and medium-sized logistics enterprises

Better promote the development of China's logistics industry, to help logistics companies as soon as possible to establish the modern competitive business profit model, the relevant state departments should increase policy support for logistics companies, and appropriate relaxation of the loan conditions of the logistics enterprises, together with the financial institutions to give preferential interest rates to help the logistics enterprises in a number of qualified and has the potential to take the lead in achieving the development goals of the listing, in order to stimulate the development and progress of the industry as a whole. At the same time, logistics companies also own facing financial difficulties, and actively seek effective financing channels. For example, enterprises can actively introduce foreign investment, or through the diversification of forms of cooperation to achieve the expansion needs of the project, and then expand the business scale and enhance competitive strength.

5.4 To increase the intensity and scope of logistics enterprises in financial regulation

First, the logistics enterprises to reasonable set of financial regulatory organization, do a good job in all aspects of the organization and preparation work and focused on financial regulatory personnel training, to enhance the sense of responsibility and job skills of the financial regulatory personnel, as the deepening and expansion of the financial scope of regulation to lay a good foundation for the organization.[7]

Second, within the logistics companies, to form a good corporate financial monitoring of the atmosphere, the internal financial staff are actively involved in the financial regulatory activities in strict compliance with the relevant state laws and regulations and financial rules and regulations further optimize the workflow of enterprise financial management and financial control, enhance financial supervision of the normative, and thus enhance the efficiency of corporate financial management, and continuously improve the financial control of logistics enterprises.

Finally, logistics enterprises should use science-based budgeting and other means to effectively supervise the activities of corporate financial management. Budgeting can decompose the goal of corporate financial management and implementation, specifically related to the organization and personnel responsibilities and powers to effectively reduce the internal division of responsibilities. The same time, it can also be as a business, an important basis for performance evaluation and adjustment, timely detection and processing enterprises in the accounting treatment of aspects of the problems, and reduce financial risk.

5.5 Great importance to the regulation and use of enterprise funds

Logistics enterprises in the financial management process, we must strengthen the regulation and use of enterprise funds, as long as the operational efficiency and quality of the enterprise funds, in order to give full play to the proper value of the enterprise funds, to avoid the adverse capital idle or poor supervision impact. Occupies a very large proportion of liquidity in the capital structure of the logistics enterprises, enterprises must first improve the liquidity of the turnover rate, improve the utilization efficiency of corporate liquidity. At the same time, logistics companies in the financial management process, logistics enterprises must strive to improve the enterprise's own funds operating efficiency. In addition, companies should develop and improve the financial regulatory system, tracking the flow of information of the enterprise funds in a timely manner, to avoid the emergence of the problem of corruption of the internal black-box operation, to ensure the safety of the enterprise funds.

6 CONCLUSION

In summary, with the development of China's market economy and the changing competitive environment at home and abroad, the modern logistics enterprises must fully recognize the importance of financial management, a careful analysis of their own various problems encountered in the financial management process, to take targeted, scientific financial management measures to promote the logistics and financial management level to a higher level, enhance the market competitiveness of the modern logistics enterprises, to better serve the socio-economic development of our country.

REFERENCES

[1] Tian Fen, Lv Yongbo, Liuling Xiao, Chen Li. 2009. Financial risk early warning of logistics enterprises, Logistics Technology, vol. 28(11):69–72.

[2] Young G.O. 2009. "Synthetic structure of industrial plastics" in Wu Bo, Logistics Engineering and Management, 31(7):88–89.

[3] Peng Maohua. 2008. Suggestions of China's logistics enterprises in financial Management, Era of economic and trade (EDITION), 6(9):159–160.

[4] Wang, Q., K. Zantow, F. Lai, and X. Wang, 2006. Strategic Postures of Third-Party Logistics Providers in Mainland China, International Journal of Physical Distribution & Logistics Management, (36):793–819.

[5] Berson A., Smith S. and Thearling K. 2000. Building data mining applieations for CRM New York: MeGraw-Hill.

[6] Ding Zhaomei. 2010. Modern logistics enterprise financial management problems and countermeasures, The Chinese market, (19):27–29.

Future Information Engineering and Manufacturing Science – Zheng (Ed)
© 2015 Taylor & Francis Group, London, ISBN 978-1-138-02644-5

Research on information technology promotes chemical effective teaching

F. Yue & X.Z. Liu
Northeast Normal University, Changchun, Jilin, China

X.L. Yang
Huazhong Normal University, Wuhan, Hubei, China

ABSTRACT: In this paper, under the guidance of the chemistry curriculum standard in compulsory education phase, we have discussed how to utilize information technology to promote the effective teaching of chemistry through analyzing chemistry discipline ontology and the ways of integrating information technology into chemistry. Simultaneously, the text carries on a case analysis to demonstrate the effective application of information technology in chemistry teaching. Through putting forward the idea, we hope to provide reference for transforming and innovating chemistry teaching methods, and finally achieve the goal of cultivating the students' ability.

Keywords: information technology; effective teaching; chemistry teaching

1 INTRODUCTION

Curriculum standard requires chemistry teaching should stimulate the students' curiosity of learning chemistry, guide the students to understand the changing rule of the material word, form the basic concept of chemistry and guide the students to experience the process of scientific inquiry, enlighten students' scientific thinking and cultivate the students' practical ability. As is known to us all, chemistry is a discipline which based on the experiment, in the teaching, we should create the activities that give priority to experiment in order to stimulate student interest in science and guide students to obtain chemical knowledge in observation, experiment and discussion. However, the traditional teaching method cannot satisfy the demand of chemical teaching, and cannot match up with the progress of science and society. With the development of education informationization and teaching informationization, information technology will exert revolutionary influence on teaching.

2 CHEMISTRY DISCIPLINE ONTOLOGY ANALYSIS

2.1 The classification of chemical knowledge

Modern cognitive psychology divides knowledge into declarative knowledge, procedural knowledge

and strategic knowledge. In the implementation of the new curriculum in high school chemistry, because of the influence of the traditional knowledge concept, there exist a phenomenon that give more attention to declarative knowledge than procedural knowledge in chemistry teaching. Study and strengthen the teaching of procedural knowledge is an important issue of improving the quality of chemistry teaching.

Chemical declarative knowledge: such as, compounds, chemical terms, chemical basic concepts and theory.

Chemical procedural knowledge: such as, understanding the nature of the chemical reaction, doing the chemical experiment according to the chemical principle, explain the relevant chemical phenomena by using the related chemical principle.

Chemical strategic knowledge: with this, you can acquire new chemical knowledge, construct chemical knowledge system and manage the declarative knowledge as well as the procedural knowledge.

2.2 The feature of chemistry course

As a basic natural science, chemistry researches the composition, structure, nature and application of substance at the molecular and atomic level and its trait is to research and create substance. In compulsory education phase, the chemistry course is an important part of science education and characterized in basic. We should provide the students with basic chemical knowledge and skills for the sake of their future development, and we need to make students understand the material world from the perspective of chemical, enhance the students' ability of analyzing and solve simple problems by using chemical knowledge and scientific method. In addition, chemistry and other disciplines always intersect and permeate with each other, such as, physics, biology, geography, and so on. From this perspective, in our teaching, we should do our best to expand students' knowledge and horizons.

2.3 *Students' ability of chemistry*

Chemistry teaching should foster the students' basic concept of chemistry, enlighten students' scientific thinking and cultivate the students' practical ability which combines chemical thinking and practice. Chemical thinking means the construction of chemical knowledge system and mastering the knowledge skillfully and chemical practical ability is skill that means solving practical problems by the integrated use of chemical knowledge.

3 INTEGRATION WAYS OF INFORMATION TECHNOLOGY AND TEACHING

From the perspective of education, information technology can be seen as part of the educational environment or educational means, we can take advantage of the information technology to improve education and teaching thus promoting the reform and development of education. Integration is a constructive and procedural process that means making several relevant parts or factors become a new unified whole. The consequent of integration is to enable the factors of the system to coordinate together and interpenetrate together, and finally maximize the effectiveness of the system. Information technology and curriculum integration gradually emerges with the development of computer-aided instruction and information technology education. With respect to the definition of integrating information technology into curriculum, different experts hold different definitions, but they generally agree that integrating information technology into curriculum is to transform teaching methods, improve education quality and better teaching effectiveness.

According to professor Zhong S.C. who comes from northeast normal university of China, integration points means any one teaching steps in a lesson that information technology supports its implementation is better than conventional teaching methods. Professor Zhong S.C. consider that the application of information technology in primary and secondary education can be roughly

divided into three levels: The first is empirical use, which means teachers use information technology according to their own teaching design and arrangement of a lesson empirically. AS a result, teachers often lack study and reflection on whether it is necessary to use information technology and how to make use of it to support teaching efficiently. The second level is targeted application, that is to say, under the teaching design and arrangement of a lesson, analyzing every teaching steps and then find out the integration points accurately, and thus selecting the appropriate way of technical support according to integration points. The third level is innovative applications. This requires teachers firstly research the best way to learn a lesson, establish ideal learning process (explain later), find out integration points and finally carry out teaching by using appropriate technical support mode and teaching software. Currently, most teachers stuck in the first case, rarely reached the second level and hardly any reached the third due to lacking scientific ideas and methods.

4 INFORMATION TECHNOLOGY PROMOTES THE EFFECTIVE TEACHING OF CHEMISTRY

4.1 *Definition of effective teaching*

We think effective teaching includes three aspects: effect, efficiency and benefit. Effect means students can develop themselves comprehensively in the three-dimensional teaching objectives area. Efficiency means reducing the burden on students and making their learning relaxed, happy and active. Benefit means promoting the overall development of students, including internal and external, current and lifetime, even meet the social demand for high quality talents. The center of effective teaching is how to promote students "learning", and then the teacher's "teaching".

4.2 *Information technology and chemistry effective teaching*

In order to achieve the goal of "effect" and "efficiency", we have to conquer the difficult point of teaching and learning. The difficulties in chemistry teaching mainly include: First of all, there are many highly abstract knowledge as well as difficult to understand, such as, spatial structure of chemical bonds, primary cell reaction, spatial structure of organic and some chemical reaction phenomena. Secondly, expansion and extension of the experiment is another difficult. Chemistry is a discipline based on experiments. However, conventional laboratory conditions are not able to fully meet the needs of students' learning and inquiry. Plus, some chemistry experiments are dangerous that discourage students' learning and exploration. Thirdly, some of the knowledge is not closely linked with daily life and not easy to apply. That is students often can't find a link between life and the knowledge they learned or even discover the knowledge can be applied but cannot be verified.

Information technology can help and support students' learning for the above three cases. We can imagine utilizing information technology to build a teaching and learning platform that possesses the following features: On one hand, integrating a wealth of academic resources to render abstract knowledge, such as, various kinds of pictures, video and some other multimedia material. On the other hand, providing a virtual lab environment. Thus, students can do virtual experiments according to the chemical principle they have learned and doing exploration accordingly. In addition, this platform can offer typical cases that are closed to real life. This can deepen students' understanding of chemical knowledge. Plus, knowledge system construction tools can be integrated in the platform to help students collate and summarize knowledge. Rational use of information technology to build support tools and pedagogical disciplines platform can largely contribute to students' declarative knowledge and procedural knowledge mastery and greatly facilitate teachers' teaching and students' learning.

Information technology can provide strong support for the "benefit" of effective teaching. Current applications of information technology, such as, educational apps, massive open online courses, flipped classrooms, mobile learning, micro-courses, and so on. These can promote students' lifelong and all-round development.

4.3 Thought analysis of information technology using

First of all, teachers need to analyze the system of chemistry knowledge. That is to say, teachers must have a systematic understanding of the knowledge what they teach, clear the position of every knowledge point in the discipline and the connection between the knowledge points, ensure different knowledge point can cultivate students' various abilities. Importantly, teacher must grasp the important and difficult knowledge points. Secondly, on the basis of analyzing knowledge system, teacher should choose different teaching methods for different types of knowledge, such as, teacher guidance, cooperative learning, autonomous learning, and so on. Generally speaking, chemistry declarative knowledge is required to be presented visually. As a result, teacher can use

multimedia material to present abstract chemistry knowledge. Contrastively, chemical procedural knowledge focus on the application of knowledge and the experience of process. In this case, information technology can create situation, stimulate students' interest in learning meanwhile giving students access to solve problems by what they have learned. Chemical strategic knowledge is used for managing of the above two types of knowledge and constructing knowledge systems. Third is the design of ideal learning process. This means designing a lesson assuming that the teaching conditions can fulfill whatever you think. In this case, teachers cannot be restricted by conventional teaching conditions and teaching approaches, thus avoiding the empirical use of information technology and innovating teaching methods to complete the teaching process effectively.

5 CASE DESIGN AND ANALYSIS

In the case, the content is selected in junior high school chemistry book: *Oxygen making and properties*. The case focuses on the effective application of information technology.

5.1 Case design oxygen making and properties

Knowledge system analysis: Mainly includes the following points: before learning this lesson, the students have learned some basic properties of oxygen and the principle of making oxygen in the laboratory; this lesson is to understand the nature of oxygen through experiment operation; from the point of the whole set of teaching material, this experiment is the only gas making experiment. Therefore, students should summarize the general idea and method of making other gases; knowledge of this lesson can lay the foundation for the later experiment study and operation; in this lesson, the key point is grasping the nature of oxygen and the laboratory, and the difficult point is the knowledge of catalyst.

Teaching objectives: a) understand the laboratory instruments for making oxygen and assemble the instrument correctly in order to operate the experiment perfectly. b) Master a variety of methods for laboratory preparation of oxygen and take a preliminary understanding of the new material prepared by chemical methods. c) Understand the concept, function of a catalyst and grasp the concept, application of decomposition reaction. d) Clear the matters that need attention for making oxygen in the lab.

Teaching methods choice and innovations: According to the teaching objectives, we can use multimedia teaching methods, relying on pedagogical platform, constructing an open interactive learning environment, providing students with a platform for self-exploration, cooperative learning and teachers' guidance. These technical supports will contribute to students' understanding of knowledge and improve learning effectiveness.

Teaching environment: The class will be shown in the network classroom, because traditional laboratory teaching require students to go to the operating table to watch the teacher's experimental operation. Traditional method has the drawbacks: specific experimental details, experimental methods of operation and experimental considerations cannot be mastered merely by the teacher' teaching; Additionally, sometimes teachers need to demonstrate several times during the experiment which will spend a lot of time, and bring poor teaching effectiveness. By contrast, teaching in the network classroom, students can take advantage of the chemistry teaching platform to experiment. In this way, students can learn the specific experimental process and announcements independently, and finally improving the efficiency of classroom teaching.

Teaching preparations: This class will be carried out in the network classroom according to the teaching content, teaching objective and difficult teaching point in teaching. In the teaching, various methods are used, such as, demonstration method, group cooperation investigation and autonomous learning with the learning platform. Before class, teachers and students need to prepare something. For teachers: video collection for oxygen making in the lab, searching information of the use of oxygen and the related content of catalyst. For students: previewing the relevant learning content, collecting literature and video information for making oxygen through the network.

Teaching process: Teaching process mainly consists of four parts, that is, a) reviewing the old knowledge and introducing the new knowledge. Reviewing includes: nature of oxygen, test methods of oxygen, the use of oxygen; Introducing means raising a question: how can we prepare oxygen since oxygen is so important. b) exploring the new knowledge. This section, students can use the teaching platform proceed as follows: exploring the collection method of oxygen, recognizing laboratory instruments and medicines, grouping inquiry, report the results. In addition, students are able to explore in depth: The potassium carbonate solution instead of manganese dioxide experiments, platform to showcase potassium permanganate preparation of oxygen. c) consolidating the new knowledge. This means students do some exercises and relevant experiments rely on the platform. d) class summary. For this part, the teacher can use

the teaching platform to show students the relevant knowledge summary and knowledge system. Thus, students can master the knowledge better.

5.2 Case analysis

In this case, according to the previous proposed ideas, we firstly analyze the knowledge system, that is, the importance of the knowledge, the relationship between other knowledge and the key and difficult points. Then according to the teaching objectives, we simply introduce the choice and innovation of teaching methods, meanwhile providing teaching environment and preparation analysis to support the teaching methods. Finally, we design the teaching process. Basing on the thought of ideal teaching process, we break through the traditional teaching method and design some teaching steps that can come out through the support of information technology. This case is a macro design, and the details of how to apply information technology is not in-depth list and analysis. Teachers can use some ideas from the case to study and practice deeply and we will also give micro designs in the later research.

6 CONCLUSIONS

Information technology has a revolutionary impact on the development of education and must be highly valued. There are many teaching problems that cannot be solved by the traditional teaching methods. In this case, we had better take advantage of information technology to promote our teaching, thus can enable the students to study easily, joyfully, forwardly with high quality and efficient. However, the rise of information technology and its continuous development cannot enable all teachers to accept and make good use of it. Therefore, we need to study the theory and practice aspects deeply as well as other related research for the sake of cultivating teacher's information technology accomplishment and finally improving the teachers' application level of information technology. We expect the teacher's ability as to the application of information technology can reach the third level just as professor Zhong S.C. said.

In this paper, we take chemistry for example to put forward some suggestions and methods on the application of information technology in subject teaching and there are many shortcomings and the insufficiency. In the future, we who working in the field of education informatization need to do further research on the use of information technology in education.

REFERENCES

[1] Zhong, S.C. & Wang, W. 2013. Thinking about information technology to promote teaching method innovation. *China Educational Technology* 2013(02):106–110.

[2] Zhong, S.C. 2007. Information technology and curriculum integration method and practice effectively. *China Educational Technology* 2007(10):74–77.

[3] Liu, X.Z. 2011. *Internet-based Creative Classroom Teaching.* Changchun: Jilin education press.

[4] Wang, Y.S. 2014. Procedural knowledge and its teaching in middle school chemistry course. *Chemistry teaching* 2014(01):12–14.

[5] Yue, F. & Liu, X. 2014. Research of information technology to promote the effective teaching of English. *Information technology education of primary and secondary schools* 2014(01):27–29.

Future Information Engineering and Manufacturing Science – Zheng (Ed)
© 2015 Taylor & Francis Group, London, ISBN 978-1-138-02644-5

The double threshold method based on Kinect depth map

Guangsong Li & Jiehong Luo
Department of Information Engineering, Guangdong Polytechnic, Foshan, Guangdong, China

ABSTRACT: According to the poor segmentation effect of Kinect depth image, the paper puts forward the dual threshold method based on the optimal threshold, in which the object model of the depth image is extracted between the two thresholds. This method has the advantages that the depth information is not affected by the illumination and the surrounding environment, and also can obtain better segmentation of edge effect.

Keywords: threshold; double threshold method; Kinect; depth image

1 INTRODUCTION

Threshold refers to the data processing method that the actual received data is compared with the same type of data set in advance, according to the result of the comparison, and decide to keep or delete in a given interval. Image segmentation is the digital image processing technology, which is based on different parts of the image features, such as pixel gray of gray image, color of the color image, object texture and so on, and the image is divided into different regions to extract the required target [1–3]. In this process, usually the part need to be extracted called object, and the other parts are called background. In order to further use of object, it is need to be separated accurately from the background. L. Zhiguo et al. [4] put forward the hierarchical optimization method to improve the efficiency of the sampling particles from the state space, but the extracted image features in the complex light conditions, it did not work well. F. Baoling et al. [5] complete effective image segmentation in bad light conditions, but the image of similar color is not segmentation. Using the depth information of Kinect can separate objects from environments, and the segmentation is not affected by the lighting effect and can improve the accuracy of detection. Chen. Zihao [6] use the deep information in partition, but his algorithm for segmentation of edge effect is not well.

This paper presents the double threshold method, the object model of the depth image is extracted, the method has the advantages that not only the depth information is not affected by the illumination and the surrounding environment, but also can obtain better segmentation of edge effect.

2 ACQUISITION OF KINECT DEPTH IMAGE

Figure 1 shows the overall structure of Kinect. Kinect have three cameras, the middle is a RGB camera, used to obtain color images of 640 × 480, and on both sides there are two depth sensors, the left one is the infrared transmitter, the right one is the infrared receiver, used to obtain the relative position of detecting objects. Before processing the data of depth, the depth image is to obtain firstly from the measured model.

The specific steps to obtain depth image is as follows:

1. Installing and checking the Kinect peripheral somatosensory is connected.
2. Initializing the Kinect, and set the depth of data stream into a usable state.
3. Using the depth of frame function to extract depth data.
4. Releasing the depth data frame.
5. Releasing the Kinect device object.

Figure 1. Hardware structure of Kinect.

3 THE DATA STRUCTURE OF KINECT DEPTH INFORMATION

Since Kinect is somatosensory game equipment, the access depth data of Kinect is not all the depth of information. The depth image data of Kinect contain two formats, and both use two bytes to store the depth value of one pixel.

1. Expressing depth value only: The lower 12 bits of the pixel depth value represent a depth value and higher 4 bits are unused.
2. Expressing depth value and player ID: The depth values of pixel are saved in the higher 13 bits, and the user number is saved in the lower 3 bits. As shown in Figure 2, each depth frame of pixel occupies 2 bytes (i.e. 16 bits): the depth value occupies the top 13 bits, and the game index occupies the back 3 bits. Therefore, in order to obtain the actual distance of pixel to Kinect, the shift operation is needed.

4 THRESHOLD METHOD

The depth of image commonly used pixel gray level to distinguish. When distance is closer to Kinect and the gray level is brighter, by contraries the gray level is darker. Therefore, the selection of an appropriate threshold can divide the pixel gray level of the object and background into two groups: all gray values greater than the threshold point are called as the object point; otherwise, they are called as background. Based on this idea, the threshold method in image processing can be expressed as the form of Eq.1:

$$\text{Im}\,ag(m,n) = \begin{cases} 0, Gray(m,n) \leq Thr\_Val \\ 1, Gray(m,n) > Thr\_Val \end{cases} \quad (1)$$

where Thr_Val is the selection of threshold, Gray (m, n) represents the gray level of pixels, and the pixel of segmentation target is marked as 1, background pixels are labeled as 0, Imag (m, n) is the new image obtained by after threshold treatment. When the Thr_Val depends only on the Gray (m, n), called the global threshold. The method based on image pixel scan and compared with the given threshold, then the pixel is marked as object or background. Optimal threshold segmentation method can effectively and accurately find the optimal threshold point.

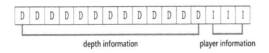

Figure 2. The depth frame structure of Kinect.

It is supposed that the gray level image of the object and background both have a normal probability density distribution, distribution functions are expressed respectively as $\alpha(x)$ and $\beta(x)$, and means are expressed respectively as Ω and $\mho$, and variances are expressed respectively as Γ^2 and Υ^2. The proportion of object in image area is K, and the background proportion is $1-K$. Based on these data, the probability density function of the total image gray:

$$K\alpha(x) + (1-K)\beta(x) \quad (2)$$

Setting the threshold Thr_Val, all the gray level less than Thr_Val is called the object point and all the gray level more than Thr_Val is known as the background point. The probability of error in putting the background point as object point is $f(Thr\_Val)$, and the probability of error in putting the object point as background point is $g(Thr\_Val)$, therefore:

$$f(Thr\_Val) = \int_{-\infty}^{Thr\_Val} \beta(x)dx \quad (3)$$

$$g(Thr\_Val) = \int_{Thr\_Val}^{-\infty} \alpha(x)dx = 1 - \int_{-\infty}^{Thr\_Val} \alpha(x)dx \quad (4)$$

The total error probability is:

$$(1-K)f(Thr\_Val) + Kg(Thr\_Val)$$
$$= K[1 - \int_{-\infty}^{Thr\_Val} \alpha(x)dx] + (1-K)f(Thr\_Val) \quad (5)$$

Obviously, Thr_Val is the best threshold when Eq.5 reaches the minimum value. It is needed that the derivative of Eq.5 equal to 0. Because

$$\frac{d}{dThr\_Val} \int_{-\infty}^{Thr\_Val} \alpha(x)dx = \alpha(Thr\_Val) \quad (6)$$

$$\frac{d}{dThr\_Val} \int_{-\infty}^{Thr\_Val} \beta(x)dx = \beta(Thr\_Val) \quad (7)$$

So, we can get

$$(1-K)\beta(Thr\_Val) - K\alpha(Thr\_Val) = 0 \quad (8)$$

That is

$$K\alpha(Thr\_Val) = (1-K)\beta(Thr\_Val) \quad (9)$$

Since

$$\alpha(Thr\_Val) = \frac{1}{\sqrt{2\pi}\Gamma} e^{-\frac{(Thr\_Val - \Omega)^2}{2\Gamma^2}} \quad (10)$$

$$\beta(Thr\_Val) = \frac{1}{\sqrt{2\pi}\Upsilon} e^{-\frac{(Thr\_Val - \mho)^2}{2\Upsilon^2}} \qquad (11)$$

So, when put Eq.10 and Eq.11 into Eq.9, and Eq.9 is simplified as

$$\frac{K}{\Upsilon} e^{-\frac{(Thr\_Val - \Omega)^2}{2\Gamma^2}} = \frac{1-K}{K} e^{-\frac{(Thr\_Val - \mho)^2}{2\Upsilon^2}} \qquad (12)$$

Further, we get

$$\frac{K\Upsilon}{(1-K)\Gamma} = e^{\frac{(Thr\_Val - \Omega)^2}{2\Gamma^2} - \frac{(Thr\_Val - \mho)^2}{2\Upsilon^2}} \qquad (13)$$

Natural logarithm of both sides and removes the denominator,

$$\Upsilon^2 (Thr\_Val - \Omega)^2 - \Gamma^2 (Thr\_Val - \mho)^2$$
$$= 2\Gamma^2 \Upsilon^2 \ln \frac{\Upsilon K}{\Gamma(1-K)} \qquad (14)$$

It is supposed that $K = \frac{1}{2}$, $\Upsilon = \theta$, then

$$\ln \frac{\Upsilon K}{\Gamma(1-K)} = 0 \qquad (15)$$

So, *Thr_Val* should meet

$$Thr\_Val = \frac{\Omega + \mho}{2} \qquad (16)$$

That is the best threshold value of image.

5 DUAL THRESHOLD METHOD

In fact, the processed image is often more complex, a single threshold cannot give good segmentation results. For example, if some parts of the object image and background are both brighter than other parts, the single threshold where can make a part of object and background images accurately separated, in another parts, may take the background as object segmentation. In order to solve this problem, there are two solutions: one is the gray level corrections of image pixel are corrected, and then use the single threshold; another method is to separate the image into different regions, and set different thresholds in each block respectively. If an image contains objects or background only, threshold will not be found in the image.

The double threshold method of depth image uses depth data of the object model as the threshold, rather than gray value. According to depth

level, the depth threshold reaches a partition of the set of pixels, each subset forms a corresponding with the real scene area, each region have the same attribute, and adjacent regions do not have this same property. When the threshold is determined, if the threshold is set too high, it will be regarded as model background, and if the threshold is set too low, the model point's collection will not be full. Multi threshold method can be considered to overcome the difficulty, according to the actual demand, the paper adopt two threshold.

Assume the image function image (x, y), for each depth values of the image segmentation, two threshold can be determined, t1 and t2, t1 > t2, when the pixel depth d is greater than or equal to t1 and less than or equal to t2, image (x, y) = 0, or image (x, y) = d.

$$image(x,y) = \begin{cases} 0((d \geq t1)or(d \leq t2)) \\ d((d < t1)and(d > t2)) \end{cases} \quad (t1 > t2)$$

$$(17)$$

The specific process algorithm is as follows:

1. start the application.
2. start the Kinect to obtain the depth image.
3. read depth frame and use shift operation to obtain depth data.
4. set the threshold of t1 and t2, t1 > t2;
   for (i = the first pixel to end)
   if (depth data of i pixel satisfies
       t1<=i.depth<=t2)
   {
       keep i.depth;
   }
   else
   {
       i.depth=0;
   }
5. display the selected depth image.
6. release depth frame.
7. release Kinect.
8. end.

6 EXPERIMENTAL RESULTS AND ANALYSIS

Running environment of this paper is the windows 7 operating system, processor Intel (R) Core (TM) i5-2450M CPU@2.50GHz 2.50GHz (RAM) 4.00GB memory. First, We use Kinect to get color images and depth image, as shown in Figure 3, Figure 3(a) is a color image, Figure 3(b) is the corresponding depth image.

If the best threshold value segmentation is used, segmentation image was obtained as shown in Figure 4. From the results of Figure 4, the optimal

(a) color image   (b) depth image

Figure 3. The color image and depth image of Kinect.

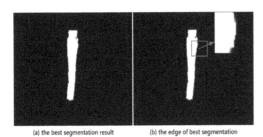

(a) the best segmentation result   (b) the edge of best segmentation

Figure 4. The optimal threshold segmentation.

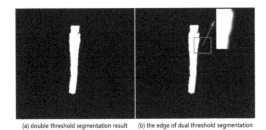

(a) double threshold segmentation result   (b) the edge of dual threshold segmentation

Figure 5. Double threshold segmentation.

segmentation method can complete the segmentation of the target object, as shown in Figure 4(a), but there are many saw tooth in the edge of segmentation image, and are not smooth, as shown in Figure 4(b).

Dual depth threshold method is used and t1 and t2 are selected such as t1 = 1300 mm, t2 = 900 mm to process in the depth map. Using this method, the obtained segmentation effect is shown in Figure 5. As you can see from Figure 5, the double threshold segmentation effect is, not only effective, as shown in Figure 5(a), but also effectively solves the problem of edge saw tooth, as shown in Figure 5(b).

ACKNOWLEDGEMENTS

This work is supported by the Guangdong Province Education Information Technology Special Topic (12JXN048). I would like to thank all the students of Game13 in Guangdong Polytechnic and especially acknowledge the contributions made by Prof. Zhang Jian and as well as contributions made by my students.

REFERENCES

[1] W. Bin, Level Set Methods for Image Segmentation, Xi'an Electronic and Science University, 2010.
[2] Yang C.J., Yang X. Abdominal CT image segmentation based on graph cuts and fast level set. CT Theory and Applications, 2011, 20(3):291–300.
[3] Bu Fuqing. Image Recognition and Classification Based on Artificial Neural Networks. Chengdu University of Technology, 2010.
[4] L. Zhiguo, L. Yan, X. Xin, "Research on Fast 3D Hand Motion Tracking System," Journal of Computer Research and Development, Vol.49 No.7, pp. 1398–1407, 2012.
[5] F. Baoling, W. Ming, D. Yingdi, "Hand Gesture Segmentation Based on Skin Color Detection Technology," Computer Technology and Development, pp. 105–108, 2008.
[6] Chen. Zihao, "Hand Detection and Tracking Based on Depth Maps," MS thesis. South China University of Technology, 2012.

Future Information Engineering and Manufacturing Science – Zheng (Ed)
© *2015 Taylor & Francis Group, London, ISBN 978-1-138-02644-5*

The characteristic and countermeasure research of physical training for civil aviators

Xian Kun Chang
Shanghai University of Engineering Science, Shanghai, China

ABSTRACT: Based on the methods of document literature, field visit, logical analysis, this paper studies the characteristic of physical training for civil aviators in China. Through this analysis, bonding the existing problems in present physical training, the author puts forward reference proposals in order to improve physical training for civil aviators.

Keywords: civil aviator; physical training; characteristic; proposal

1 INTRODUCTION

At present, the Chinese aviation industry has stood among the aviation world, and kept striding forward to the aim of being great-power aviation country. Accompanying with increasing opening up international routes, the equipped long-distance flight ability of the civil aviators is a necessary. Objectively speaking, this necessity requests the aviators to have toned body, to be able to fulfill kinds of long time, hard intensity and high standard air transport tasks under special conditions of cockpit in the air, and to stand up to the physical and mental fatigue from long-distance flight.

Under such circumstance, it appears so important that how to put forward rationalization proposals for special physical training, which should be more scientific and effective, of aviators.

2 BASIC FEATURES OF PHYSICAL TRAINING FOR CIVIL AVIATORS

2.1 *Bright industrial characteristic of physical training for civil aviators*

At present, flights are arranged quite tight in both domestic and abroad and the journey is farther than before. Abnormal situations happen from time to time, which requires aviators to highly concentrate their attention in flight and make prompt and right choices to the special and unexpected situations. Lacking of abilities of being sensitiveness, quick response and calmly handling problems will easily fall into a mess when emergency occurred, which sometimes would lead to even severe flight accidents because of such improper manipulation. Therefore, aviators are not only required to possess

good psychological quality, but also to have excellent physical quality. This is why content structures of physical training for civil aviators should be established from flight practice. All the contents should start from forming flight skills, improving flight quality, guarantying flight security and prolonging flight lifetime, and then develop special physical training programs for civil aviators.

For example, in the gymnastic programs of the physical education for aviators, it is better to choose technical movements related with rolling, overturning and rotating from high bar, parallel bars as well as mat tumbling, etc. Such kind of movements could improve vestibule stamina of the aviators. In the whole special physical training programs, almost all the contents permeate characteristic of flight skills and qualities, full of bright and rich flight features.

2.2 *High relevance between physical training effect and flight skill formation of aviators*

Through years study and deep research on physical education and training practice of aviators, it turns out that the aviators with good records on physical education practice get excellent flight skills. The two points are positively correlated. That is to say, practice effects directly influence the improvement of flight skills and the enhancement of the flight ability. Such as aviators with weak vestibule stamina, will easily get airsickness symptoms during flying like dizzy, headache, nausea, vomiting, pallor, and then get flight illusions. These symptoms affect aviators' ability of judgment, maneuvering, and reaction speed and so on, and then put air safety in danger and influence the fulfillment of the air transport tasks.

2.3 *Mandatory of carrying out the physical training for aviators*

Aviators undertake missions and sacred duties of the country. Driven by such special professional features, aviators dedicate themselves into the physical training with obvious initiatives and enthusiasm. The mandatory of physical training could not be weakened at any time. And the whole carrying out process of physical education should be strict and solemn. The physical training requests aviators to take part in, no matter if they are willing to or not, and does not shift due to their different hobbies.

2.4 *It is durable and lifelong for aviators to get physical training*

Due to flight professional characteristic, there is a special request for physical quality of aviators, that is, the physical education of aviators should be life-long. Fit physical quality and good stamina could never be done once and forever. It could only be gained and kept through special physical training of all the time. The life-long physical education for aviators is embodied in the two fields. For one thing, the influence of special physical education is life-long for each aviator; for another, as aviators, they are obligated to join physical training in order to keep their good special flight ability.

3 PROBLEMS EXISTING IN PHYSICAL TRAINING FOR AVIATORS

3.1 *It is incomplete of relevant departments to train*

Through interviewing civil aviators, a conclusion could be drawn that organization management of daily physical training and competition for civil aviators is weak and not on a regular basis, not to mention establishing system and developing it into scale.

3.2 *A part of civil aviators still have vulnerable awareness to get self-training*

Through interviewing, a part civil of aviators shows no interest in the special physical exercises and the programs. They do not have strong sense of urgency to be sturdy and do not consciously have more exercises besides the physical training classes.

3.3 *There are improper training programs and disappointing workout*

Currently, it is an increasing situation that young aviators are losing their abilities of balance and flight stamina, getting obesity and lumbar muscle degeneration. Aviators at school are dramatically declining their abilities of physical exercise, controlling, flight stamina and balance. It is obvious that civil aviators are declining in their physical quality. The more important thing is that their choice methods on the special physical exercise programs and training methods are incorrect. The chosen physical training programs could not fully reflect aviators' body needs. That's why the physical exercise does not work well.

4 ADVISEMENT

4.1 *Shift concept*

It is time to move on from the idea that aviators' special physical quality is indispensible to the brand new idea that civil excellent aviators should be versatile talents at present. Therefore, aviators must master high-tech theoretical knowledge, excellent flight skills and strong stamina for aviation.

4.2 *Take part in aviators' special physical exercises as much as possible*

In current piloting field for civil aviation, there are more and more flights for transporting passengers and goods. Duration of flight is becoming longer and across time zone flight are more often. Therefore, for aviators, abilities of good resisting flying fatigue and anti-biorhythm disorders are to be required in order to adapt night navigation. What's more, aviators work long hours in low pressure, anoxia, noise, vibration, bump, etc. This demands higher qualities of resisting airsickness, anti-anoxia and anti-low pressure for them. However, there are ways to strengthen these abilities. Aviators could use Rotation of the ladder, Fixed roller, Ground spinning operation to train their vestibule stamina so as to enhance the ability of anti-airsickness. They could use imitative exercise in high sky on the machine to strengthen their balance when bodies are unstable. They could also do aerobic exercises, such as long-distance race, ball games, to develop their flight stamina.

4.3 *Enhance the studying on scientific research and promote scientific progress of aviators' physical training*

More efforts should be put on researches of aviators' physique. More applications should be explored, practiced for theories and methods of aviators' physical fitness training. Research centers should be established for endeavoring guiding physical training of talents for civil aviation.

In need of the development of civil aviation, matched research works should be actively carried out. This works include assessment indicator index, assessment criterion, training methods and measures for aviators' physical exercise.

4.4 Complete training facilities, e.g. training field, equipment

Relevant units should often check and maintain equipment facilities, field for special physical training in order to make sure they are in good condition.

REFERENCES

[1] Tian Mai Jiu. Theories of Sport Training. Beijing: Beijing sports university press, 2000.

[2] Xiang Yu. Theory and Practice of Aviation Sports. Cheng du: Sichuan University Press, 2006.9.

[3] Huang Bo. Effects of Physical Exercises on Pilots Adaptation of the Aviation Environment. Journal of Chengdu Sport University, 2000, 26(1):60–62.

[4] Zhou Bao Hui, Chen Hua Wei, Guo Hong Bo. Investigation and retrospection of the current state of civil aviation physical education in China. Journal of Physical Education, 2009, 16(8):72–73.

[5] Sun Xue Chuan. On Modern Military Physical Fitness Training. Sports institute proceedings in the people's liberation army, 2001, 20(2).

[6] Hao Yu, Zhou Bao Hui, Ma Bing. The civil aviation college student pilots' aviation sports curriculum optimization Settings. Shangdong Physical Education Institute, 2009, 25(3):90–92.

Future Information Engineering and Manufacturing Science – Zheng (Ed)
© 2015 Taylor & Francis Group, London, ISBN 978-1-138-02644-5

Performance research of China's listed commercial banks based on factor analysis

S.X. He & W.J. He

Three Gorges University, Yichang, Hubei, China

ABSTRACT: Financial firms play a crucial role in the economic development, and commercial banks are important parts in the financial system. Commercial banks' competitiveness is reflected on their business performance. Therefore, measuring the performance of commercial banks has become an urgent task. In this paper, through the establishment of index system, using the factor analysis method of SPSS19.0 software with 11 listed commercial banks' 7 financial indices in 2013, we obtain the 11 listed commercial banks' operating performance ranking in 2013. Finally, we get the evaluation based on the analysis summary of China's commercial banks, aiming to provide reference for investors and bank management decision-making level, at the same time, promoting the steady development of financial market in China.

Keywords: commercial bank; factor analysis; performance evaluation

1 INTRODUCTION

As China's economic system reform deepens gradually, as well as the change of external environment such as China's accession to the WTO, China's commercial banks are facing unprecedented competitive pressure. Improving the management method of commercial banks, building effective performance evaluation system and improving the performance of commercial banks are the key problem for the sustainable development of China's financial sector and are also the basic requirement to the challenge of foreign banks.

As the biggest developing country in the world, China has conducted financial system reform for more than 20 years, and has achieved high grades with great contributions to economic development. At the same time, we realize that although China's banking industry is developing rapidly, it also faces many problems. For example, management efficiency is low, lacking of core competitiveness and non-performing loans ratio is high. These problems seriously weakened the commercial banks in China within the international market competitiveness[1]. With the opening of domestic financial market in China, the situation which state-owned commercial banks to unify the whole country has ceased, and the competition between banks is becoming more and more fierce. Moreover, foreign banks whose strengths are relatively strong also begin to enter the Chinese market, accelerating the pace of the banking market competition. In addition,

it also has brought unprecedented opportunities and challenges to China's banking industry. In this case, researching on the problem of China's commercial bank performance and finding out a reasonable way of performance evaluation are of great significance.

2 THE EMPIRICAL ANALYSIS OF THE PERFORMANCE EVALUATION OF CHINA'S LISTED COMMERCIAL BANKS

2.1 *Selection of samples and indexes*

Considering data availability, original sample data of this paper comes from the annual report data of listed commercial banks on financial websites, in newspapers as well as the "China financial yearbook". 11 listed commercial banks in the A-share are finally selected as samples. They are five large commercial banks: Industrial and Commercial Bank of China, China Construction Bank, Agricultural Bank of China, Bank of China, and Bank of Communications. Four joint-stock commercial banks: China Citic Bank, China Merchants Bank, Min Sheng Bank, and Societe Generale. Two city commercial banks: Bank of Beijing and Bank of Nanjing. Chapter 4 of the "Commercial banking law of the People's Republic of China" describes: "commercial banks set security, liquidity and profitability as the management principle[2]". This paper considers these three principles comprehensively when selecting indicators of commercial banks and chooses the average total assets return rate (X1), net profit margin (X2), provision for coverage (X3), capital adequacy ratio (X4), core capital adequacy ratio (X5), non-performing loan ratio (X6) and liquidity ratio (X7) as the seven commercial banks' business indicators.

2.2 Factor analysis

The basic idea of factor analysis is that according to the size of correlation to group the original variables, making the correlation between variables within the same group is higher, and the correlation between variables in different groups is lower[3]. The factor analysis model is described as follows[4]:

1. x = (x1, x2,..., xp)' is observable random vector, and the mean vector E(x) = 0, covariance matrix cov(x) = Σ. Moreover, covariance matrices Σ and related matrix R are equal;

2. F = (F1, F2,..., Fm)' (m < p) is immeasurable variables, the mean vector E(F) = 0, covariance matrix cov(F) = I, namely each component of vector F is independent;

3. ε = (ε1, ε2, ... εp)' is independent of F, and E (ε) = 0, the covariance matrix of ε Σε is a diagonal matrix, namely each component of ε is independent. The following model is called the factor model:

$$X1 = a11F1 + a12F2 + ... + a1\,mFm + ε1$$
$$X2 = a21F1 + a22F2 + ... + a2\,mFm + ε2$$

....

$$Xp = ap1F1 + ap2F2 + ... + ap3Fm + εp$$

The matrix form of the model is: X = AF + ε. Among them, F is the public factor of X, matrix A is the loading matrix of factor, and ε is the special factor for X[5].

KOM and Bartlett's sphericity test are provided by the SPSS method, checking whether it is suitable for factor analysis. It is generally believed that if KOM's value is less than 0.5, meaning that it is unfavorable for factor analysis[6], It is feasible to carry out factor analysis through the KMO and Bartlett's Test as showed in Table 1. In the table, KOM = 0.876 > 0.7 and Bartlett sphere Test P = 0.00 < 0.05.

We can get correlation coefficient matrix's characteristic value, the variance contribution rate and

Table 1. KMO and Bartlett's.

Sampling enough degrees of Kaiser-Meyer-Olkin measurements	0.876
Bartlett's sphericity test	
The approximate chi-square	314.278
df	28
Sig	0.000

Table 2. The total variance of explanation.

The initial eigenvalue			Rotate the sum of squares loaded		
Summation	Variance %	Cumulation %	Summation	Variance %	Cumulation %
2.743	39.187	39.187	2.402	34.316	34.316
1.910	27.281	66.468	1.826	26.080	60.396
1.241	17.731	84.199	1.666	22.803	83.199
0.658	9.395	93.594	1.053	11.855	95.054
0.322	4.603	98.198			
0.124	1.770	99.968			
0.002	0.032	100.00			

Table 3. The factor loading matrix.

Factor/index	F1	F2	F3	F4
Average total assets return rate (X1)	0.309	−0.123	0.740	0.156
Operating net margin (X2)	−0.013	0.156	0.932	−0.118
Coverage for provision (X3)	−0.172	0.915	0.114	−0.157
Capital adequacy (X4)	0.968	−0.121	0.025	0.114
Core capital adequacy ratio (X5)	0.937	−0.141	0.292	0.001
Non-performing loan ratio (X6)	−0.070	0.950	0.027	0.180
Liquidity ratio (X7)	0.182	0.016	−0.156	0.978

the cumulative contribution rate of the sample through the factor analysis of SPSS[7]. The results are shown in Table 2. We select four factors that the cumulated variance contribution ratio of the four factors is 95.054%. Meanwhile, using the maximum variance method to round factor of 25 orthogonal rotations, we can find that factor rotation doesn't change model performance for the data fitting but changes the eigenvalues of factor (Table 2). The rotating load matrix is shown in Table 3.

The index is divided into four types of common factors. According to the table above, capital adequacy X4 and core capital adequacy ratio X5 are big load on factor F1, which is called the capital adequacy factor; provision for coverage X3 and non-performing loan ratio X6 are high load on factor F2, which are called the loan safety factor; Average total assets return rate X1 and operating net margin X2 load on factor F3 are very high, which are called the profitability factor; Liquidity ratio X7 loading on factor F4 is very big, which is called the liquidity factor. It can be obtained by factor loading matrix showed in Table 4.

$F1 = 0.309 \, X1 - 0.013 \, X2 + \cdots + 0.182 \, X7$

$F2 = 0.123 \, X1 + 0.156 \, X2 + \cdots + 0.016 \, X7$

$F3 = 0.740 \, X1 + 0.932 \, X2 + \cdots + (-0.156 \, X7)$

$F4 = 0.156 \, X1 + (-0.118 \, X2) + \cdots + 0.978 \, X7$

Using factor score function of SPSS19.0 to rate four main factors respectively. Factor score coefficient is shown in Table 4.

2.3 The construction of performance evaluation model of China's listed commercial bank

After naming factor, preliminary model of the performance evaluation of commercial banks is determined according to the variance contribution rate of each factor and the total explained variance ratio:

$$FN = (34.316 * F1 + 26.080 * F2 + 22.803 * F3 + 11.855 * F4)/95.054$$

FN stands for the total score of the performance of commercial Banks (N = 1, 2, 3 ... 11). It can be obtained by Table 4:

$Fac1 = 0.320 \, X1 + 0.252 \, X2 + 0.320 \, X3 + 0.046 \, X4 - 0.012 \, X5 + 0.126 \, X6 + 0.243 \, X7$

$Fac2 = 0.201 \, X1 + 0.153 \, X2 + 0.112 \, X3 + 0.403 \, X4 + 0.320 \, X5 - 0.274 \, X6 - 0.332 \, X7$

$Fac3 = 0.335 \, X1 - 0.234 \, X2 + 0.118 \, X3 - 0.148 \, X4 + 0.095 \, X5 + 0.267 \, X6 + 0.132 \, X7$

$Fac4 = -0.092 \, X1 + 0.176 \, X2 - 0.125 \, X3 + 0.083 \, X4 + 0.876 \, X5 + 0.095 \, X6 + 0.257 \, X7$

Substituting these four formula into the preliminary evaluation model which contains four main

Table 4. The factor score matrix.

Index	Component			
	1	2	3	4
X1	0.320	0.201	0.335	−0.092
X2	0.252	0.153	−0.234	0.176
X3	0.301	0.112	0.118	−0.125
X4	0.046	0.403	−0.148	0.083
X5	−0.012	0.032	0.095	0.876
X6	0.126	−0.274	0.267	0.095
X7	0.243	−0.332	0.132	0.257

Table 5. The selected samples' factor score and sorting of commercial bank.

Name of bank	Fac1	Fac2	Fac3	Fac4	F	Sorting
Industrial and commercial bank	0.839	−0.709	1.360	−1.112	−0.008	5
Construction bank	0.900	−0.852	1.322	−1.296	−0.337	8
Agricultural bank	0.532	−1.622	−0.373	−0.454	−0.706	11
Bank of China	0.101	−0.927	0.502	−0.433	−0.056	7
Bank of communications	0.509	−1.001	0.643	−1.074	−0.430	10
China citic bank	−0.805	−0.287	0.622	1.183	0.170	4
China merchants bank	−0.339	0.573	−0.024	−0.534	−0.046	6
Min sheng bank	−0.641	0.256	−0.343	−1.043	−0.383	9
Societe generale	−0.272	1.778	0.655	−0.229	0.532	2
Bank of Beijing	0.426	0.718	0.516	−0.676	0.335	3
Bank of Nanjing	0.442	−0.237	0.869	−0.329	0.641	1

factors, we can get the comprehensive evaluation model of 11 listed commercial banks in China:

$$F = 0.240\ X1 + 0.099\ X2 + 0.152\ X3 + 0.102\ X4 + 0.136\ X5 + 0.046\ X6 + 0.060\ X7$$

We can get total score of the performance of commercial banks after calculation, and rank them according to the order from high to low (Table 5).

3 CONCLUSIONS

We can get the following conclusions through the factor analysis of SPSS software by comparing China's 11 A-share listed commercial banks' operating performance in 2013.

3.1 *Overall analysis of ranking results*

From comprehensive ranking in 2013, the best business performance is the Bank of Nanjing, and the worst is the Agricultural Bank. Among the large commercial banks, the most prominent is the Industrial and Commercial Bank and the worst is the Agricultural Bank. In addition, Bank of China and China Construction Bank have comparable level. Among Joint-stock commercial banks, the best is Societe Generale, and the worst is the Min Sheng Bank, China Citic Bank and China Merchants Bank's performance is medium. Among the two city commercial bank, Nanjing Bank's performance evaluation is very high and it also possesses the best performance of all the selected banks. On the whole, the city commercial Banks operating performance is best, followed by the joint-stock commercial banks, and the large commercial Banks are at last. But there is a big gap between the corresponding single rank and the comprehensive rank of listed commercial banks, meaning that listed commercial banks' unbalanced development in various factors, which affects the overall performance ranking.

3.2 *According to each factor ranking to analyze the listed commercial banks' performance*

1. On the capital adequacy factor, Industrial and Commercial Bank, Construction Bank, Agricultural Bank, and Bank of Communications are significantly ahead of other listed commercial banks, Bank of Nanjing and Bank of Beijing are next to these banks, showing that large commercial banks do well in terms of capital adequacy.
2. On the loan security factor, Societe Generale, Bank of Beijing, China Merchants Bank, Min Sheng Bank respectively rank in the top four, but Agricultural Bank and Bank of Communications come last and the second from bottom. The Agricultural Bank also ranks last in the profitability factor, which leads to its disadvantage position in the total performance evaluation in 2013.
3. On the profitability factor, Industrial and Commercial bank, Construction Bank and Bank of Nanjing are among the best of listed commercial banks, implying these banks' profitability is stronger than others. On the contrary, Industrial and Commercial Bank and China Construction Bank don't have advantages on the loan security.
4. On the liquidity factor, China Citic Bank ranks first, followed by Societe Generale and Bank of Nanjing. However, Industrial and Commercial Bank, Construction Bank and Bank of Communications perform poorly. This illustrates the large state-owned commercial banks' liquidity is poorer.

As a whole, our country's state-owned commercial bank's performance is poorer than other joint-stock banks, mainly because the capital structure of joint-stock commercial bank is more reasonable than the state-owned commercial banks and the operating mechanism is more flexible. So they possess stronger ability of business development. State-owned banks of our country have dual nature goal. On one hand, they hope the investor can maximize, on the other hand, they expect to promote the development of macro economy and maintain social fairness and stability through controlling bank[8]. However the general commercial banks' goal is to maximize their own rights and interests.

This paper selects seven operation indices of China's 11 listed commercial banks in 2013 to analyze their performance by using factor analysis of SPSS software, reflecting different degrees of differences on the capital adequacy, loan security, profitability and liquidity among China's large state-owned commercial banks, joint-stock commercial banks and city commercial banks. This makes investors not only pay attention to the comprehensive ranking of the various commercial banks but also analyze each factor variables. This paper also helps in understanding the cause of comprehensive evaluation results in many ways in order to assist investors to make right decisions. In the meantime, the bank managements should also analyze all aspects, only in this way can they improve the competitiveness of the bank.

REFERENCES

[1] Guanjun Xia. "Performance valuation of China's commercial banks". Strategy and management, 2004(2):145–149.

[2] Xiaoqun He. "Multivariate statistical analysis". Renmin university of China publishing house, 2008:192–196.

[3] Yu Lu, Xiaofeng Zhang. "Principal component factor analysis method in the application of the performance evaluation of listed banks". Special zone economy, 2005(10):99–101.

[4] Ji Lin, Yufeng Ji. "Business performance analysis of China's listed commercial banks based on factor analysis". Economist, 2009(10):8–10.

[5] Ning Wang. "China's listed commercial banks performance analysis-Based on factor analysis and cluster analysis". The financial times, 2012(3):95–97.

[6] Chen Huang. "Commercial banks evaluation research". Beijing: The people's bank of China research institute, 2002:79–81.

[7] Liqiu Xin. "An empirical analysis of the performance level of China's commercial banks". Contemporary economic, 2013(4):122–123.

[8] Kai Wei, Xuwei Hu. "Research of China's commercial banks' performance evaluation system". Business and management, 2013(2):91–94.

Future Information Engineering and Manufacturing Science – Zheng (Ed)
© *2015 Taylor & Francis Group, London, ISBN 978-1-138-02644-5*

Constructive research of Moodle-based mobile network teaching system

Min Yao

Library of North China Institute of Science and Technology, Sanhe, Hebei Province, China

ABSTRACT: In order to adapt to the development of ubiquitous network and mobile teaching, we need to expand the mobile teaching functions on the web teaching platform and integrate it into the network teaching environment to enable students to self-study according to their needs anytime, anywhere. Based on the contrast of current mobile web teaching system implement mode, this paper design a framework of supporting mobile teaching on Moodle platform, and introduce the key technologies of realizing mobile teaching expanding on Moodle platform.

Keywords: mobile network teaching system; Moodle platform

1   INTRODUCTION

With the development of network ubiquity, successful application of 3G/4G network, popularity of mobile terminals like smart phone or Pad etc, Mobile learning has became a hot topic in the research of educational technology and related fields as a branch of Electronic learning (E-learning)[1]. Mobile learning (M-learning) has a lot of learning characteristics such as mobility, high efficiency, ubiquity, sharing, interactivity and personalization etc. As a brand new learning mode, M-learning has immeasurable application potential in school education and training fields. Up to now, M-learning is still in the primary stage of development in not only theoretical research but also practical application. It is worth studying deeply that how to apply current mature network technology, mobile communication technology and open source course management system to set up high efficient and practical mobile learning management system to implement open, autonomous, personalized learning anytime and anywhere and dig the potential of mobile teaching application.

2   MOBILE NETWORK TEACHING SYSTEM IMPLEMENT MODE

As a new teaching assistance system, mobile network learning system is built for those learners whose learning space position and time are fluid and self-learning requests increase, which consists of mobile learning terminals, internet, mobile communication network, resource platform and support distant network learning and mobile learning[2]. From the view of technology, there are mainly three constructive modes of current mobile network learning system: Information-service-based mobile learning system, C/S-based mobile learning system, B/S-based mobile learning system.

2.1   *Information-service-based mobile learning system*

There are three types of technology are commonly used in developing short message mobile learning system by using short message gateway, short message cluster-sender and the third party short message group send service of mobile communication operators[3]. Information such as notice, arrangement, syllabus, after-class exercises, course and tests could be released to learners through the short message interactive platform provided by the operators, through which learners could master related teaching information timely and arrange learning time and schedule reasonably. Though this technology mode implements simply, fast and conveniently, it couldn't adapt to the complex learning environment due to limiting to simple text information transmission and lacking of the degree of interactivity.

2.2   *C/S-based mobile learning system*

C/S-based mobile learning system developed mobile application program in mobile phone terminal, learners run clients terminal program to interact information with services terminal by mobile devices. Though, system interactivity is good and function is powerful under this technology mode, it's very difficult to combine the content presentation with technology development due to the diversification of mobile terminal devices, complication of the operation system—common operation system are Android, Windows Phone, IOS, Symbian, BlackBerry, the system construction is complex as well.

2.3 B/S-based mobile learning system

Combining the mobile-device-based WAP which provide effective support for mobile terminal devices with the internet-based Web technology, B/S-based mobile learning system succeed in using mobile terminals to visit the rich teaching resources on the network teaching platform anytime and anywhere[4]. For its system development, it's only need to interactive with wireless application environment. Mainly using WML language to write WAP website, it support to add scripting language to implement interactive operation, release and run via WEB servers such as IIS, Apache etc.

B/S-based mobile learning system is suitable for the design and development for primary mobile learning system because the definition of its content representation is more mature to provide rich interactivity function and technology implement is easy relatively and system construction is convenient. Therefore, from the view of expanding the mobile function on the basis of network teaching platform, this paper construct B/S-based mobile network learning system, increase support to mobile devices with keeping PC terminal services to implement mobile teaching.

3 THE OVERALL FRAMEWORK OF MOODLE-BASED MOBILE TEACHING SYSTEM

Based on Web2.0, Moodle (Modula Object-oriented Development Learning Environment) is a set of full open source network teaching platform, which present not only most of modules request by online learning but also typical applications such as blogs, Wikis, RSS etc. Meanwhile, it could support a variety of forms such as one-to-one, one-to-many, synchronous and asynchronous between teachers and students or between learners and learners to communicate interactively[5]. Moodle provides an open platform for the majority of teachers to design, manage, implement and evaluate their own courses and teaching in the Web2.0 environment[6]. Duo to above advantages of Moodle, based on the original Moodle network curriculum management, we choose to use WLAN/GSM wireless communication system technology as an accessed way of mobile terminal, expand the mobile service function to build mobile network learning system.

Moodle-based mobile network learning system should have the following characterics:

1. To support a variety of handheld terminal devices (Smart phone, Pocket PC, I-pad, multimedia terminal, etc.) under different operating systems to login system through a variety of wireless communication network with keeping original PC terminal services function and provide convenient, fast and high-efficient mobile teaching services like PC connect to internet, To identify terminal devices used by users and authenticate identity.
2. To enable all kinds of teaching contents and activities provided by Moodle network teaching management system to use in different handheld devices without changing system construction of the original Moodle platform and affecting the original function implement and on the basis of normal use; to ensure the friendliness, convenience, accuracy and consistency when user look through the content.

Therefore, the expansion of supporting mobile learning is conduct to Moodle network teaching platform, its overall framework design as shown in Figure 1. To meet the demand of mobile learning, expanded Moodle platform will increase "User and erminal identification module", "Mobile terminal content self-adaptive module", "Mobile terminal device information database", "Mobile teaching function logical circuits module", "Mobile teaching resource management module".

"User and terminal device identification module" will confirm that if the users accessing

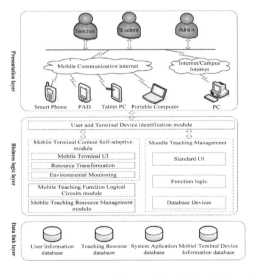

Figure 1. The overall framework of Moodle-based mobile teaching system.

Moodle is mobile communication network users or the Internet/Campus network users through the identification of the client explorer after users login the mobile network learning system through different networks and devices, at the same time, it also confirm the identity of the users and vest the corresponding privileges. If it is from the HTTP explorer, it should be PC users, transfer to current Moodle network teaching management system; if it's from a WAP explorer, it should be mobile terminal users, transfer to "Mobile terminals self-adaptation module". "Mobile terminals content self-adaptive module", including three parts: environmental monitoring, resource transformation and mobile terminal interface generation, provide exploration contents meeting its terminal devices to mobile terminal users. "Mobile teaching function logic module" implement the course management provide by Moodle network teaching management system and mobility application of teaching serves functions, meanwhile, "mobile teaching resource management module" complete the query and management for underlying layer database query and provide the request data for upper layer mobile teaching application.

4 KEY TECHNOLOGIES TO IMPLEMENT MOODLE PLATFORM MOBILE EXPANSION

The key technology in the design of Moodle network teaching platform mobile expansion are User and terminal devices identification and Mobile terminal self-adaptive.

4.1 *User and terminal devices identification*

Users and terminal devices identification is used to implement type identification of accessing users and mobile terminal devices. It is set on the client's explorer based on PHP technology. When a user access Moodle, it will check if the gateway contains Text/vnd.wap.vml MIME type first by STROPS function first. If yes, it will support WIML page to look through. The code is as follows:

```
$accpetStr=$_REQUEST['HTTP_ACCEPT']
if (stristr($accpetStr, 'xhtml')){
    define('WAPTYPE','wap2');
}else {
define('WAPTYPE','wap1');
header("Content-Type: text/vnd.wap.vml");}
```

And then confirm which user's explorer or mobile devices the accessing users come from by checking the HTTP-USER-AGENT.

```
function getBrowType(){
    if(strpos($_SERVER["HTTP_USER_
AGENT"],"MSIE 8.0"))
```

```
browType="Internet Explorer 8.0";
else
    if(strpos($_SERVER["HTTP_USER_
AGENT"],"Opera"))
    browType="Opera";
    else
    browType=$_SERVER["HTTP_USER_
AGENT"];
        Return $browType;}
    $_browType=getBrowType();
echo($_browType);
```

Accordingly you can confirm the type of user agent, then combine with mobile terminal devices information database, you can identify related information such as mobile terminal devices type and software and hardware performance and get ready for the generation of mobile terminal content self-adaptive.

4.2 *Mobile terminal content self-adaptive module*

This module, mainly consist of "mobile terminal interface generation" and "resources transition and recombination" two components, is used to transform learning contents provided by Moodle network teaching management system into exploring contents which are suitable for specific mobile phone terminal devices. For the convenience of mobile terminal devices explore, mobile terminal interface generated components using column priority technology, transform the layout of the page content display from multiple columns into a single column to adapt the contents are explored on the small screen of mobile terminal devices. According to the different device types and users' preferences, this module transform XML document into Web pages suit specific devices to explore through using XSLT when mobile terminal users access Moodle.

This module will provide appropriate learning resources according to user terminal devices type and accessing speed. First, it will transform some larger quantity, good quality, rich media resources into the same content but smaller quantity to save to the learning resources database. And then, recombine the recourses and interface generate component through "mobile teaching resources management module" to index the learning resource suit for the user devices characteristic, to generate user interfaces, return to the terminal user.

5 CONCLUSION

Though Moodle network teaching offers a good support for network teaching according to the characteristics such as full-open, EOU (Ease of use), modularity, scalability, supporting a variety of teaching methods etc, it is necessary to introduce mobile learning and set up mobile learning environment for learners. Only by which could adopt the urgent situation of network generalization to implement anytime, anywhere, open, personalized learning indeed. Through expanding mobile services on Moodle platform, this paper set up mobile network teaching system supporting different mobile devices access Moodle network teaching system by wireless communication network, to expand course teaching from curricular to extracurricular; It also combine the class teaching, network teaching and mobile teaching to meet teachers and students' need to get teaching resources conveniently, fast, and high efficiently and self learning anytime and anywhere.

ACKNOWLEDGEMENTS

This work was financially supported by the Fundamental Research Funds for the Central Universities (3142014099).

REFERENCES

[1] Jang, Xin. Chun. & Chen, Zhang. Qi. 2009 Research on the Application of Mobile Learning in Network Teaching Platform. *Modern Education Technology* 19(01): 109–112.

[2] Wei, Hong. wei. & Qiu, Jia-ji. 2009 Research and Practice of the Mobile learning Theory. *Computer Engineering and Science* 31(09): 135–137.

[3] Hou, Wei. & Zhu, Dong. Ming. 2010 Research on the Practice of Developing Technologies on Mobile learning. *Modern Education Technology* 20(01): 115–119.

[4] Li, Qian. & Gao, Ge. 2008 A Literature Review on Application Pattern of Mobile learning. *Modern Education Technology* 18(10): 64–68.

[5] Zen, Zong. Gen. 2010 *Moodle Network Course Platform*. Beijing: Beijing University Press.

[6] Chen, Yan, juan. 2010 Development Research of Network Course Based on Moodle. *Computer Era* 18(5): 64–66.

Future Information Engineering and Manufacturing Science – Zheng (Ed)
© *2015 Taylor & Francis Group, London, ISBN 978-1-138-02644-5*

Research on ethics games between the MNEs and Chinese government

Jian Feng Zhu

Economics and Management School, Wuhan University, Hubei, China

ABSTRACT: The MNEs are inevitably confronted with various discrepancies and even conflicts during their production and operation in China because of the profit-pursuing nature of capital. In order to solve these conflicts, ethics games between the MNEs and Chinese government are proposed in this paper. It can be deduced that there are four main factors accounting for the probability of cooperation, (1) the MNEs' payoff when strategy set 1 (cooperation & honoring commitment) is adopted; (2) the MNEs' payoff when strategy set 2 (cooperation & dishonoring commitment) is employed; (3) probability that China has a supervising and administrating government; (4) punishment on the MNEs by Chinese government. To be specific, the higher probability of a law-based government, the stronger punishment on illegal behaviors of the MNEs, and thus more payoff is yielded to the MNEs which choose strategy set 1 (cooperation & honoring commitment), but less to those which adopt strategy set 2 (cooperation & dishonoring commitment). Chinese regulatory government can stimulate the MNEs to honor their commitments that they should shoulder their social responsibility in China and manage their behaviors within the framework of business ethics and codes of ethics. Those findings will provide references for decision-makers, the government and other departments involved.

Keywords: MNEs; ethics games; regulatory government; ethics-restrained government

1 INTRODUCTION

With the growing economic globalization, more and more multinational enterprises (MNEs) have engaged themselves in production and operation in China. The MNEs are inevitably confronted with various discrepancies and even conflicts during their production and operation in China because of the profit-pursuing nature of capital, though they bring about abundant capitals, innovative technologies and advanced managerial experiences and lend a helping hand in optimizing and upgrading Chinese industrial structure. On the other hand, Chinese relevant laws and regulations on supervision for the MNEs are imperfect at present, leaving some space to many MNEs to take advantage of. The profit–pursuing nature plays a decisive role in their practices that the MNEs maximize their economic profits regardless of the moral and ethical principles. In fact, Chinese government requires that the MNEs should take their corresponding social responsibilities, and violate neither business ethics nor the legitimate rights and interests of Chinese government, labors, consumers and other stakeholders while exploiting cheap labor, relatively relaxed investment climate and huge market potential in China. Within such context, games on business ethics between MNEs and Chinese government come into being.

2 ETHICS GAME MODEL AND ASSUMPTION

To explain the games, some assumptions must be introduced, namely,

1. That the MNEs and Chinese government in the ethics games are absolutely rational.
2. That uncertainties remain in both parties, in other words, the MNEs have no idea whether China has a regulatory government based on law or an ethics-restrained one. It is a common knowledge that the probability of Chinese government belongs to the former type is α.
3. That in case of a regulatory state based on law, China metes out punishment m on the MNEs for their immoral practices against business ethics in China, and thus the Chinese government gets the payoff M.
4. That the probability that China has a regulatory government based on law and is willing to cooperate with the MNEs is p, while probability that Chinese government is ethics-restrained and agrees to cooperate with the MNEs is q. The probability that the MNEs self-manage their behaviors and take no action to violate business ethics is δ (despite the Chinese government pattern they are faced with).

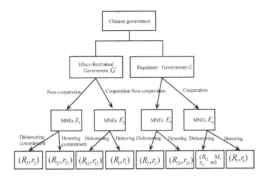

Figure 1. Ethics game tree between the MNEs and Chinese government.

The ethics game tree between the MNEs and Chinese government is shown as Figure 1.

Where (R_1, r_1) indicates payoffs both players earned respectively when the government adopts cooperation strategy and the MNEs keep their promises. (R_2, r_2) signifies payoffs produced respectively when the government decides not to cooperate and the MNEs break their promises. (R_{21}, r_{21}) presents payoffs when the government chooses not to cooperate but the MNEs honor their commitments. (R_{12}, r_{12}) shows payoffs when the government chooses to cooperate but the MNEs fail their commitments.

3 SOLUTION OF ETHICS GAMES

3.1 Decision of the MNEs

In this game, the MNEs take a later step after observing whether Chinese government makes her decision to cooperate or not. According to Bayesian theory, judging from their observation that Chinese government adopts cooperation strategy, the MNEs can calculate the probability that China is a regulatory state based on law

$$P(G \mid (G'E_2, GE_4)) = \frac{P(G)P(G'E_2, GE_4) \mid G)}{P(G'E_2, GE_4)}$$
$$= \frac{\alpha p}{\alpha p + (1-\alpha)q} \qquad (1)$$

However, the MNEs compute the probability that China has an ethics-restrained government restrained by ethics from the same observation,

$$P(G' \mid (G'E_2, GE_4)) = \frac{P(G')P(G'E_2, GE_4) \mid G')}{P(G'E_2, GE_4)}$$
$$= \frac{(1-\alpha)p}{\alpha p + (1-\alpha)q} \qquad (2)$$

At the same time, the MNEs either keep or break their promises on condition that Chinese government decides to cooperate, and the expected revenue of the MNEs which choose the former is given by,

$$P(CP \mid (G'E_2, GE_4)) = \frac{\alpha p r_1 + (1-\alpha)q r_1}{\alpha p + (1-\alpha)q} \qquad (3)$$

Otherwise, the MNEs, which adopt the latter strategy, bring in expected revenue of,

$$P(NCP \mid (G'E_2, GE_4)) = \frac{\alpha p(r_{12} - m) + (1-\alpha)q r_{12}}{\alpha p + (1-\alpha)q} \qquad (4)$$

Assume that the MNEs earn the same expected revenue irrespective of their promises. In such case, we get the equation

$$\frac{\alpha p r_1 + (1-\alpha)q r_1}{\alpha p + (1-\alpha)q} = \frac{\alpha p(r_{12} - m) + (1-\alpha)q r_{12}}{\alpha p + (1-\alpha)q} \qquad (5)$$

3.2 Decision of Chinese government

If China is administered and regulated by law and consents to engage in cooperation with the MNEs, the probability is

$$p = \frac{(1-\alpha)q(r_{12} - r_1)}{\alpha[r_1 - (r_{12} - m)]} \qquad (6)$$

And when keeping their commitments in the ethics games, the MNEs can get expected revenue of

$$E(CP) = \delta R_1 + (1-\delta)(R_{12} + M) \qquad (7)$$

However, if the MNEs fail them, the expected revenue of the MNEs is

$$E(NCP) = \delta R_{21} + (1-\delta)R_2 \qquad (8)$$

Assume that the expected revenue of the MNEs is in the same amount regardless of their commitments in the game, it can be expressed by

$$\delta R_1 + (1-\delta)(R_{12} + M) = \delta R_{21} + (1-\delta)R_2 \qquad (9)$$

Based on above computation, the probability of the MNEs restricting their behaviors and in no violation of business ethics is

$$\delta = \frac{(R_2 - R_{12} - M)}{R_1 + R_2 - R_{12} - R_{21} - M} \qquad (10)$$

By the same token, the expected revenue can be computed on condition that faced with an ethics-restrained government, the MNEs honor their commitments,

$$E(CP) = \delta R_1 + (1-\delta)R_{12} \qquad (11)$$

On the contrary, the MNEs, failing to honor their commitments, generate expected revenue of

$$E(NCP) = \delta R_{21} + (1-\delta)R_2 \qquad (12)$$

Provided that the expected revenue is equivalent whether the MNEs honor or dishonor their commitments, the MNEs will earn expected revenue of

$$\delta R_1 + (1-\delta)R_{12} = \delta R_{21} + (1-\delta)R_2 \qquad (13)$$

Assumption that the MNEs restrict their behaviors and don't engage in any violation against business ethics is followed with a probability of

$$\delta = \frac{(R_2 - R_{12})}{R_1 + R_2 - R_{12} - R_{21}} \qquad (14)$$

Given the above, when Chinese government is a supervising and administrating one based on law, both parties has two options respectively and then there are four possible strategy sets, and corresponding expected revenue Chinese government yields is sorted

$$R_1 > R_{12} + M > R_{21} > R_2 \qquad (15)$$

The reason that strategy set 2 (Cooperation & dishonoring commitment) brings higher payoff to the government compared with strategy set 3 (Non-cooperation & Honoring commitment) is that to a regulatory government based on law, China, in accordance with Chinese and international laws and regulations, is entitle to impose strict punishment on the MNEs when they fail to perform their social responsibilities or infringe Chinese business ethics and consequently gains an increase in her own payoff.

By contrast, with a ethics-restrained Chinese government, the corresponding expected revenue China generates is followed an order of

$$R_1 > R_{21} > R_{12} > R_2 \qquad (16)$$

In case of an ethical-restrained state, Chinese laws and regulations are not of much restraint to the MNEs, which, driven by their profit-pursuing nature, will neglect to fulfill their corporate social responsibilities in China for purpose of their own economic profits. Actually, the MNEs would take

their responsibilities only under the huge pressure from Chinese media and public opinion with an aim to maintain their corporation reputation. In such case, honoring their commitments is of great importance to the MNEs and exerts self-control effect on their behavior, which wields direct and positive influence on Chinese government part. Hence, strategy set 3 (Non-cooperation & Honoring commitment) dominates strategy set 2 (Cooperation & dishonoring commitment).

The expected revenue the MNEs earn correspondingly is followed an order of

$$r_{12} > r_2 > r_1 > r_{12} - m > r_{21} \qquad (17)$$

The MNEs' practices that they honor their commitments, namely, taking corporate social responsibilities in China, give rise to an increase in their costs and expenses and consequently a reduction on the expected revenue. However, the expected revenue the MNEs will earn is the highest of the four when the strategy set 2 (Cooperation & dishonoring commitment) is adopted. Instead, a regulatory government suggests the expected revenue will suffer a loss when the MNEs are against Chinese ethics and regulations and thus are subject to statutory misconducts.

3.3 *Equilibrium of ethics games*

In case that probability δ that the MNEs decide to fulfill their commitments meet equation $\delta R_1 + (1-\delta)R_{12} = \delta R_{21} + (1-\delta)R_2$ and $\delta R_1 + (1-\delta)(R_{12} + M) = \delta R_{21} + (1-\delta)R_2$. A comparison of above naturally turns out $\delta R_1 + (1-\delta)(R_{12} + M) > \delta R_{21} + (1-\delta)R_2$. It is revealed that the expected revenue is higher when the MNEs tend to cooperate than that when they don't on condition that China has a regulatory government based on law, expressed by $E(CP) > E(NCP)$.

Provided that China has a law-based supervising and administrating government and commits to cooperation with probability that obeys $p = 1$, it can be got

$$q = \frac{a(r_1 - (r_{12} - m))}{(1-a)(r_{12} - r_1)} \qquad (18)$$

Based on above analysis, the mixed-strategy at equilibrium of the ethics game between the MNEs and China within the context of economic globalization is given by,

$$\left\{ \delta = \frac{R_2 - R_{12} - M}{R_1 + R_2 - R_{12} - R_{21} - M}, \ p = 1, \ q = \frac{a(r_1 - (r_{12} - m))}{(1-a)(r_{12} - r_1)} \right\} \qquad (19)$$

It can be seen from the mixed-strategy at equilibrium that Chinese government insists on cooperation provided that the probability that the MNEs keep their commitments is no less than $(R_2 - R_{12} - M)/(R_1 + R_2 - R_{12} - R_{21} - M)$, with a regulatory state on Chinese part. On the other hand, probability of Chinese government deciding to cooperate can be given by $(a(r_1 - (r_{12} - m)))/((1-a)(r_{12} - r_1))$ with the same probability of the MNEs and an ethics-restricted government. Since $(a(r_1 - (r_{12} - m)))/((1-a)(r_{12} - r_1)) < 1$, China is more likely to engage in cooperation when administered and supervised by law than that the other type.

Differentiate (19) with respect to R_1 and R_2 respectively, and get

$$\begin{cases} \dfrac{\partial \delta}{\partial R_1} = \dfrac{R_2 - R_{12} - M}{(R_1 + R_2 - R_{12} - R_{21} - M)^2} < 0 \\ \dfrac{\partial \delta}{\partial R_2} = \dfrac{R_1 - R_{21} - M}{(R_1 + R_2 - R_{12} - R_{21} - M)^2} > 0 \end{cases} \quad (20)$$

$$\begin{cases} \dfrac{\partial \delta}{\partial R_1} = \dfrac{R_2 - R_{12}}{(R_1 + R_2 - R_{12} - R_{21} - M)^2} < 0 \\ \dfrac{\partial \delta}{\partial R_2} = \dfrac{R_1 - R_{21}}{(R_1 + R_2 - R_{12} - R_{21} - M)^2} > 0 \end{cases} \quad (21)$$

It can thus be seen that the following two inequations can both met in ignorance of the governing pattern of Chinese government.

$$\frac{\partial \delta}{\partial R_1} < 0, \quad \frac{\partial \delta}{\partial R_2} > 0$$

Differentiate the equation (19), with respect to a, m, r_1, r_2, respectively and get

$$\frac{\partial q}{\partial a} = \frac{r_1 - (r_{12} - m)}{(r_{12} - r_1)(1-a)^2} > 0 \quad (22)$$

$$\frac{\partial q}{\partial m} = \frac{a}{(r_{12} - r_1)(1-a)} > 0 \quad (23)$$

$$\frac{\partial q}{\partial r_1} = \frac{am}{(r_{12} - r_1)^2(1-a)} > 0 \quad (24)$$

$$\frac{\partial q}{\partial r_{12}} = \frac{am}{(r_{12} - r_1)^2(1-a)} < 0 \quad (25)$$

From above analysis, it can be deduced that there are two main factors contributing to the MNEs' decision on honoring commitments despite the pattern of Chinese government. Factor one is the government's payoff R_1 when strategy set 1 (Cooperation & honoring commitment) is

adopted, and the other one is the government's payoff R_2 as strategy set 4 (Non-cooperation & Dishonoring commitment) is employed. In case that R_1 increases or R_2 decreases, either way, probability δ of the MNEs honoring their commitments gets higher. The choice that Chinese government makes to cooperate suggests that the government and the MNEs share the social responsibility that should be taken by the MNEs alone. It cuts down the expected revenue R_1 of Chinese government. Cooperation between the government and the MNEs is way much beneficial to the latter and turns out a stimulus to their commitments. Conversely, assume that China pursue non-cooperation strategy, the expected revenue on China's part gains an increase as a result of a reduction of costs of the MNEs' social responsibility.

The profit-pursing nature drives the MNEs to maximize their economic profits and decline to shoulder their corporate social responsibility in China. In the event that it happens, Chinese social media and public opinion will give full play, under pressure of which, the MNEs, as a consequence, will still take their social responsibility for sake of their reputation. A law-based regulatory state will, pursuant to the law, make strict punishment on the MNEs' dishonoring commitment, which as a result will promote the MNEs to undertake their social responsibility when engaging in production and operational activities with a higher probability δ of commitments fulfillment.

4 CONCLUSION

A study on the ethics game between the MNEs and Chinese government reveals that there are four main factors accounting for the probability of cooperation, (1) the MNEs' payoff when strategy set 1 (cooperation & honoring commitment) is adopted; (2) the MNEs' payoff when strategy set 2 (cooperation & dishonoring commitment) is employed; (3) probability that Chinese government is administered and supervised based on law; (4) punishment on the MNEs by Chinese government. To be specific, the higher probability of a law-based regulatory government, the stronger punishment on illegal behaviors of the MNEs, and thus more payoff is yielded to the MNEs which choose strategy set 1 (cooperation & honoring commitment), but less to those which adopt strategy set 2 (cooperation & dishonoring commitment). The Chinese regulatory government can stimulate the MNEs to honor their commitments that they should shoulder their social responsibility in China and manage their behaviors within the framework of business ethics and codes of ethics. Those findings will provide references for decision-makers, the government and other departments involved.

REFERENCES

[1] Banai M. Sama L.M. 2000. Ethical dilemmas in MNC's international staffing policies: A conceptual framework. *Journal of Business Ethics* 3(25):345–355.

[2] Carasco E.F, Singh J.B. 2003. The content and focus of the codes of ethics of the world's largest transnational corporations. *Business and Society Review* 108(1): 71–94.

[3] Jackson K.T. 1997. Globalizing corporate ethics programs: Perils and prospects. *Journal of Business Ethics* 16(12–13): 1227–1235.

Future Information Engineering and Manufacturing Science – Zheng (Ed)
© *2015 Taylor & Francis Group, London, ISBN 978-1-138-02644-5*

Energy Balanced Self-adaptive Intelligent Water Drops routing algorithm

C. Xu
Shanghai Institute of Technology, Shanghai, China

S.Z. Ni
Shanghai Jiaotong University, Shanghai, China

H.G. Shao
Shanghai Institute of Technology, Shanghai, China
Shanghai Jiaotong University, Shanghai, China

ABSTRACT: We propose an Energy Balanced Self-adaptive Intelligent Water Drops (EBSIWD) routing algorithm. IWD algorithm has shortcomings in dealing with energy, end-to-end delivery delay and routing loops. Our algorithm makes up for these. In this paper, we make the following improvements on the basis of the IWD algorithm: (i) We improve the update mechanism of soil and HUD. The calculation formulas add the energy, end-to-end delivery delay and hops; (ii) We take advantage of IWD packets' life-time effectively avoid routing loops. The simulation results show that the proposed algorithm is better to balance the node energy, avoid the local optimal solution, enhance the network lifetime and have excellent performance than the traditional algorithm.

Keywords: energy balance; IWD; self-adaptive

1 INTRODUCTION

Ad-hoc is a Multi-hop Temporary autonomous system that be composed of a set of mobile terminals with wireless transceiver. This network can utilize wireless connections to constitute any network topology. It is very suitable to establish a temporary communication, such as military communication, environmental monitoring, medical care and disaster relief, etc. The primary goal of Ad-hoc network is to provide high quality of service and efficiently utilize the resource of network bandwidth. The main task of the current network routing protocol is to look for the low communication delay path between the source node and the destination node. Meanwhile, they considered how to improve the utilization rate of the whole network, avoid to produce traffic and to balance network traffic. But a few protocol notices energy consumption of nodes.

In 2007, Intelligent Water Drops Algorithm was first put forward and managed to solve the TSP problem. Later, intelligent water droplets algorithm was used to solve the problem of MKP, in a queen and robot path planning problems. This paper introduces an intelligent water drops optimization methods to solve the problem of energy-balanced in Ad-hoc network. It simulated the natural water system which forms river way through interaction between water and surrounding environment,

iterative calculation and optimization eventually results. It doesn't need to know the entire network topology and the link, only interactive calculation with the surrounding nodes is enough, and transmit information about the attributes of the intelligent water drops. So this protocol has less network traffic overhead, adapt to the network dynamic change and support redundant routing.

This protocol selects transmission route on the basis of the node energy utilization, so as to choose a more suitable path to transmit data packets. It not only improves the performance of the protocol, but also prolongs the network lifetime.

This paper is organized as follows: Section 2 describes the related work. Section 3 introduces the basic model of the EBSIWD. Computational experiment and results analysis in Sectconclusions are presented in Section 4. Finally, we conclude our work in Section 5.

2 RELATED WORK

2.1 *Energy aware*

Energy is the most important resource in the nodes. At present, many routing algorithm try to reduce the energy difference between nodes by sensing nodes available energy, so as to balance the node energy distribution and prolong network lifetime. SER routing protocol presented in [5] can sense nodes' energy, balance energy consumption, reduce routing overhead, and eliminate time slot between the nodes. In [3], this paper proposes ARAMA which senses node's energy information according to the routing request packets. Then, routes evaluate the corresponding path and update routing information, balance the node energy resources. In [1], on the basis of ARAMA routing algorithm, this paper joins the ant colony optimization method. It solves the problem of the path control by an adaptive energy awareness structure. In our paper, we combine node's energy, transmission delay and transmission distance as a standard of routing choose, thus to improve the performance of the network.

2.2 Decision path

In [1], they take the pheromones as the basis of route choice. According to the node energy and load distribution, they amend coefficient of pheromone update and heuristic function. They use forward ants to find propagation path. The nodes utilize the local pheromone table to update the propagation path. These tables update their pheromones by backward ants. Our paper introduces the IWD optimization to choose the optimal path. When Ad-hoc network is the period of route discovery, nodes judge whether their soil is equal to initial value. If not, there is an effective path between nodes. Otherwise, the nodes transmit packets in the light of the heuristic information probability formula. So the nodes both find an effective path and reduce their energy consumption.

2.3 Loop resolution

In the methods of eliminate routing loop that perform ant colony algorithm in [4], forward ants and backward ants are utilized to update routing tables. However, these methods can aggravate the amount of node transmitting data, resulting in redundant energy consumption of nodes. In this paper, we consider a IWD packet lifetime to solve the routing loop in Ad-hoc. Then intermediate node only forwards IWD packets. Since this process is a single transmission, its energy consumption is a half of ACO. We set a critical path in IWD packets. This path includes a IWD through all the nodes.

3 EBSIWD SYSTEM MODEL

3.1 The rule of IWD algorithm

In nature, the flow of water and river channel form the relationship between action and reaction. In reality, because of the influence of the obstacle, river flowing path is often bending twist. After some researches, it has been found that considering the distance from the source to the destination and the presence of obstacles, water drops tend to build up the most optimal river. A liquid drop generally has two attributes: a certain speed and a certain amount of the soil. The faster water drops, the larger its kinetic energy. The soil will move from a higher flow velocity place to a lower place. When the velocity slows, the soil under the action of gravity will be deposited. Therefore, we can conclude the relationship between water drops and soil satisfies the following three rules in [6]:

Rule 1: The drops of faster flow rate carry more soil than the slower ones.

Rule 2: The drops in the less soil path get more velocity than which in the more soil path.

Rule 3: The drops have greater probability to choose less soil paths.

3.2 The EBSIWD algorithm process

Step 1: In the stage of routing discovery, all of the source node random distribute in the whole network. The source nodes send periodically IWD packets to the destination nodes. We set a lifetime parameter in IWD packet. Its value is show the number of hop. The structure of IWD packet is illustrated in Figure 1.

S_ID is the source node address. D_ID is the destination node address. $SOIL$ is the value of the IWD packet's soil. VEL is the value of the IWD packet's velocity.

If the IWD packets' lifetime is zero, the intermediate nodes discard the packets. Then this node determine whether its address in the critical path of IWD packets. If this address already exists, this node deletes the critical path of IWD packets and forwards the IWD packets. When the nodes receive the same IWD packet at the same time, they discard the packet, which have less lifetime. This process is illustrated in Figure 2, in which the $path[\,]$ is the array of the critical path address.

Step 2: The relay nodes choose the forwarding nodes according to the soil anycast mechanism.

S_ID
D_ID
SOIL
VEL
path[]
lifetime

Figure 1. The structure of the IWD packet.

path[1,2,3,4]

path[1,2,3,4,1]

path[1]

Figure 2. The process of delete the cirical path.

We define an InitSoil which is the initial value of soil between nodes. There is not an effective path between nodes when their soil is not equal to InitSoil. The nodes forward the IWD packets by Equation 1.

$$P^{IWD}(i,j) = \begin{cases} \dfrac{f(soil(i,j))}{\sum_{k \in V(i)} f(soil(i,j))} & soil(i,j) = InitSoil \\ 1 & soil(i,j) \neq InitSoil \end{cases} \tag{1}$$

The set V(i) denotes the group of nodes that the IWD should not visit to satisfy the constraints of the problem. Each source node generates IWDs package and initializes. The initial $soil(i, j)=InitSoil$. The initial IWD packets' velocity is InitVel.

The constant ε_s is a small positive number to prevent singularity.

In this paper, we introduce three parameters to balance the network energy and optimize IWD algorithm: $D(i, j)$ said delay, $E(i, j)$ said energy, $C(i, j)$ said that the number hop from node i to node j. The new probability formula is defined as follows Equation 3:

$$f(soil(i,j)) = \frac{1}{g(soil(i,j))^{\alpha} \cdot D(i,j)^{\beta} \cdot E(i,j)^{\gamma} \cdot C(i,j)^{\theta}} \tag{2}$$

where α, β, γ and θ are the weight of each parameter. The function g ($soil$ (i, j)) is used to shift the $soil$ (i, j) and is described in Equation 4.

$$g(soil(i,j)) = \begin{cases} soil(i,j) & if \min(soil(i,j)) \geq 0 \\ soil(i,j) - \min(soil(i,j)) & else \end{cases} \tag{3}$$

Step 3: We use the Equation 5 to update the IWD packets' velocity. In [5], Hussein presents the basic of IWD algorithm, includes the basic formula. Since IWD packets are transmitted from node i to node j, their velocity will be changed according to the soil of edge. Their velocity update by the following Equation 4:

$$vel^{IWD}(t+1) = vel^{IWD}(t) + \frac{a_v}{b_v + c_v \cdot soil^2(i,j)} \tag{4}$$

The less amount of soil, the faster the increased velocity, and vice versa. a_v, b_v, c_v are the velocity of updating parameters, to ensure velocity in the original basis with the same amplitude increase. If speed increases too fast, IWDs fall into local optimal solution. If speed increases slowly, IWDs need to spend more time to get the optimal solution.

Step 4: The amount of soil increment can be calculated according to the Equation 5 in [4].

$$\Delta soil(i,j) = \frac{a_s}{b_s + c_s \cdot time^2(i,j;vel^{IWD}(t+1))} \tag{5}$$

where, a_s, b_s, c_s and are pre-defined positive parameters for the IWD algorithm.

In [2], [6], they not use the soil to determine paths, but use a heuristic function. It is called to $HUD(i, j)$. $HUD(i, j)$ has to be defined for a given problem to measure the undesirability of an IWD to move from node i to node j. The past IWD algorithm just use the simple parameters to form $HUD(i, j)$, such as the number of hop. For making the algorithm more adapted to the network, we consider the multiple factors to form $HUD(i, j)$, such as node energy, end-to-end delay, and the hop. The Equation 6 as follow:

$$HUD(i,j) = D(i,j)^{\lambda_D} + E(i,j)^{\lambda_E} + C(i,j)^{\lambda_C} \tag{6}$$

where λ_D, λ_E and λ_C respectively mean in the weight of end-to-end delay ($D(i, j)$), node energy ($E(i, j)$) and hop count($C(i, j)$). The greater value of $HUD(i, j)$ is, the worse path nodes choose. Meanwhile, the time spent by the IWD packet move from node i to node j is given by Equation 7.

$$time(i,j;vel^{IWD}(t+1)) = \frac{HUD(i,j)}{\max(\varepsilon_v, vel^{IWD}(t+1))} \tag{7}$$

Step 5: The node judge whether IWD packets reach the destination nodes. If it returns yes, the algorithm comes to an end. If it returns no, the nodes continue to forward IWD packets to the destination nodes by Step 1.

Step 6: In [2], an objective or quality function is needed to evaluate the fitness value of the solutions. We need to use the iterative equation to find the optimal solution. So the network needs to repeat Step 1 to 5. The node utilizes IWD packets to collect the iterative information. T^{IWD} is a solution founded by an IWD. The best solution T^{IB} can be calculated by Equation 8.

239

$$T^{IB} = \arg \max_{\forall T^{IWD}} q(T^{IWD}) \qquad (8)$$

We use the Equation 6 to update the *soil(i, j)*. It may result in a local optimum solution. In order to increase the opportunities of finding the global optimum solution, we use the Equation 9 to update the *soil(i, j)*.

$$soil(i,j) = (1 + \rho_{IWD})soil(i,j)$$
$$- \rho_{IWD} \frac{1}{(N_{IB} - 1)} soil_{IB}^{IWD} \quad \forall(i,j) \in T^{IB}$$
$$(9)$$

where $soil_{IB}^{IWD}$ is the soil of the current iteration best IWD when it reaches the destination. N_{IB} is the number of nodes belong to the best solution T^{IB}. ρ_{IWD} is the global soil updating parameter.

4 COMPUTATIONAL EXPERIMENTS AND RESULTS ANALYSIS

In this section, we utilise the NS2 to simulate EBSIWD algorithm. Meanwhile, we take this result to compare with AODV and IWD.

1. The number of nodes is set to 100. The network scale is 500 m × 500 m. Pause time gradually increased from 0 s to 100 s. The simulation results are as follows:

Figure 3 reports the simulation of end-to-end delay among three kinds of routing protocol. These results indicate that the EBSIWD exhibits higher quality than the IWD. The computational complexity of IWD protocol is larger than AODV protocol. In this paper, we optimize

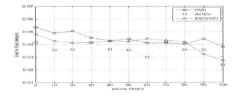

Figure 3. The simulation of end-to-end delay.

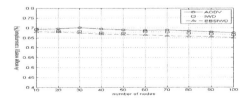

Figure 4. Perfomance with static phenomenon.

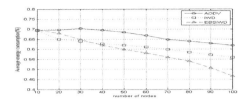

Figure 5. Performance with mobile phenomenon.

the IWD protocol eliminating the routing loop. However, our formulas are more complicated than the IWD ones. Therefore, the nodes will extend the computation time. The simulation results show that the performance of EBSIWD protocol processing end-to-end delay is better.

2. The node number increase from 10 to 100. Y-axis is the average energy consumption of the all network, expressed as a percentage. The simulation results are as follows:

The Figure 4 simulates a static network where the nodes are randomly deployed with a static phenomenon. The Figure 5 simulates a mobile network where the nodes are randomly deployed with a mobile phenomenon. In the static model, the network transmission path remains stable. The average energy consumption of network will not have a large floating. The Figure 4 illustrates the average energy consumption of EBSIWD is lower than the IWD. In the mobile model, the average energy consumption of network has a large floating according to the nodes randomly move. Comparing with results from Figure 5, the route optimization will be more and more effective along with the increase in the number of nodes. When the node number is 10, the optimal path is the same as the general path. Optimizations are rather useless. When the node number is 100, the EBSIWD can choose energy path, balance the network energy and reduce the network energy consumption.

5 CONCLUSION

In this paper, EBSIWD algorithm has been proposed. It solves some problems in the original IWD algorithm, such as the local optimal solution, the routing loop and the redutant energy consumption. It uses the soil anycast mechanism to erase the network storm. The node effectively use lifetime and critical path to avoid routing loop. This algorithm minimizes end-to-end delay and maximizes energy savings, contributes to expend the lifetime of the Ad-hoc network.

REFERENCES

[1] De Rango, F. & Tropea, M. 2009. Energy Saving and Load Balancing in Wireless Ad Hoc Networks through Ant-based Routing. *International Symposium on Performance Evaluation of Computer and Telecommunication Systems 2009, SPECTS 2009, Part of the 2009 Summer Simulation Multiconference, SummerSim 2009, 13-16 July 2009.* Florida: International Precious Metals Institute.

[2] Hoang, D.C. & Kumar, R. & Panda, S.K. 2012. Optimal data aggregation tree in wireless sensor networks based on intelligent water drops algorithm. *IET Wireless Sensor Systems* 2(3): 282–292.

[3] Hussein, O.H., Saadawi, T.N. & Lee, M.J. 2005. Probability routing algorithm for mobile ad hoc networks' resources management. *IEEE Journal on Selected Areas in Communications* 23(12): 2248–2259.

[4] Jiang, H., Wang, M.R., Liu, M. & Yan, J.W. 2012. A quantum-inspired ant-based routing algorithm for WSNs. *Proceedings of the 2012 IEEE 16th International Conference on Computer Supported Cooperative Work in Design, CSCWD 2012, 23–25 May 2012.* Piscataway: IEEE Computer Society.

[5] Kamruzzaman, S.M., Kim, E., Jeong, D.G. & Jeon, W.S. 2012. Energy-aware routing protocol for cognitive radio ad hoc networks. *IET Communications* 6(14): 2159–2168.

[6] Shah-Hosseini, H. 2007. Problem solving by intelligent water drops. *2007 IEEE Congress on Evolutionary Computation, CEC 2007, 25–28 September 2007.* Piscataway: Inst. of Elec. and Elec. Eng. Computer Society.

Future Information Engineering and Manufacturing Science – Zheng (Ed)
© *2015 Taylor & Francis Group, London, ISBN 978-1-138-02644-5*

Cellular automaton based three-lane traffic simulation model for traffic incident

W.H. Yang
Southeast University, Jiangsu, China

ABSTRACT: Not only does the traffic incident endanger the safety of vehicles and pedestrians, it also puts down road capacity and vehicle speed, affecting city's economic construction and the operation-efficiency of social development. Therefore, it is of great importance to research the impact of traffic incidents. The establishment of traffic flow simulation model to analyze the evolution process would contribute to the understanding of the traits and evolution law of traffic flow under the incident conditions, which lays the theoretical basis for the traffic incident guidance. Previous studies mainly focused on the two-lane area. However, with the transportation development and the growing number of the multi-lane roads, three-lane cellular automaton has more practical significance. Based on the traditional two-lane cellular automaton model, this paper proposes a three-lane model for traffic incident which is applied for incident simulation. The results show that the model performs well on the simulation.

Keywords: transportation engineering; cellular automaton; traffic flow simulation; three-lane model; traffic incident

1   INTRODUCTION

Traffic incident is the aperiodic event occurring in one or more lanes in the road that results in lane(s) or part of lane(s) closed for some reasons (Zhang et al. 2008). Traffic incident can lead to capacity decrease in the affected segment of the road and formation of traffic bottleneck (Hu et al. 2007). The key to releasing accurate incident information and constituting proper traffic guidance plan is to predict the disturbed scope and duration of an incident precisely (Ju 2011). Therefore, it needs to explore the evolution of traffic incident and traffic flow. As the data collection of traffic flow in incident conditions has difficulty in collecting with high cost, study can be taken by means of computer simulation methods.

Cellular automaton is a dynamical system evolving in accordance to certain local rules in the discrete time dimension, which is a completely discrete model. As the traffic elements are essentially disconnected, studying the traffic issues with a cellular automaton model has its unique advantages. NaSch model is the earliest cellular automaton model for traffic flow simulation (Nagel et al. 1992). Basing on the NaSch model, many researchers developed models simulating more realistic traffic phenomenon by modifying the rules and increasing the dimensions of methods. Evolved symmetrical lane changing rules out

of vehicles' lane changing behaviors, the Symmetrical Two-Lane Cellular Automaton (STCA) model was developed (Chowdhury et al. 1997). According to researches on lane changing behaviors in crowded segments, the STCA model was improved and can correspond to driving characteristics better under the congested situation (Wang 2010).

Current studies on cellular automaton based traffic flow model are mostly focused on two-lane areas. This is due to the simplicity of the model of two-lane. The lane changing cases can be determined easily and the lane changing rules are easier to develop. Besides, the studies about double lane cellular automaton are relatively mature, which make it easier when applied to practical problems. However, with the increasing urban transportation development and traffic flow, new roads are often built with three lanes or more, leading to the more complex lane changing behavior. Thus the two-lane cellular automaton is limited in the practical application.

This paper takes three-lane cellular automaton model of traffic flow for study for the following reasons. First, three-lane model is representative that can simulate drivers' choices in the multi-lane case involving multi-select lane changing, overtaking, etc. Meanwhile, three-lane cellular automaton is the most simple and clear model among multi-lane cellular automata. With the principal rules of

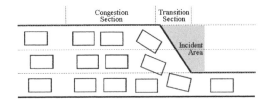

Figure 1. More complicated lane changing behavior in three-lane road.

STCA retained, new rules can be developed for the three-lane model with the actual situations.

Based on the traditional two-lane cellular automaton model and with the researches and statistics on driving characteristics of vehicles, this paper proposes a three-lane model for traffic incident. And it is used for specific incident simulation in MATLAB 7.0 to test its performance.

2 RULES OF THE CELLULAR AUTOMATON BASED THREE-LANE TRAFFIC SIMULATION MODEL

2.1 Following rules of traffic flow

According to the following model, when running without changing lanes, the speed of the vehicle is determined by the speed of vehicle in front and the distance between the two vehicles (Wang 2010).

$$\ddot{x}_{n+1}(t+T) = \frac{\alpha}{x_n(t) - x_{n+1}(t)}[\dot{x}_n(t) - \dot{x}_{n+1}(t)] \quad (1)$$

$\ddot{x}_{n+1}(t+T)$—The acceleration when following.
α—Parameter.
$x_{n+1}(t)$—The distance of the front vehicle.
$\dot{x}_n(t)$—The speed of the vehicle.

Given the length of passenger car and the necessary space between cars, 7.5 m is taken for the unit cell length. The distance is expressed as integers. And the unit time for the model is 2 seconds. Based on researches, statistics and the speed limitation for urban road according to relevant laws and regulations, the speed of most traveling vehicles in urban road is about 40 km/h and the maximum speed is limited within 60 km/h. In this case the speed for vehicle can be expressed as 0 to 4 cells per unit time, which can be abbreviated as [0,4].

When a vehicle is traveling without changing lane, the basic rules in mathematical expression are as follows.

a. Accelerate when one of the following conditions are satisfied: $d \geq 3$ and $V_2 > V_1$; $d \geq 8$. The new speed is $\min(V_1 + 1, V_{max})$.

b. Decelerate when one of the following conditions are satisfied: $d \leq 6$ and $V_2 < V_1$; $d < 2V_1$. The new speed is $\max(V_1 - 1, 0)$.
c. Brake when $d = 0$ and $V_2 < V_1$.
d. When not meet any of the above conditions, the vehicle traveling remains the same speed.

2.2 Probabilistic lane changing rules

Usually there are two reasons for vehicles to change lane: one is seeking acceleration, the other is forced by incident. Because different drivers change lane with different probability, in the same circumstances, the choices of different drivers are usually different.

According to statistics obtained from fieldwork, this paper develops a probabilistic lane changing model in looser judging situation, in order to mimic the different choices from different drivers. Different from models with harsh lane changing rules, probabilistic lane changing model can reflects drivers' subjectivity when they change lanes. Besides, because of the introduction of probability, the result of each time is not completely the same, which shows the variability of road conditions.

For the acceleration lane changing, assuming that when the accelerated conditions cannot be achieved in its lane but can be met in its adjacent lane, the vehicle change to the adjacent lane at a probability of 0.35.

For accidents, assuming the scope of lane changing is limited within a certain range of the accident, and in this range, the drivers may be aware of the accident by some means. So they are more likely to changing lane away from the accident area instead of closing. In this scope, the drivers will change lane away from the accident area as long as meeting the speed requirements.

2.3 Assumption for incident

In the case of incidents, the model assumptions are as follows:

a. Except for the accident blocked lane(s), the traffic flow of the lane(s) remains the same during the accident.
b. Vehicles outside the scope of accident comply with the following rules.
c. In the same lane, a vehicle's speed is determined only by the speed of the front vehicle, the distance between the two vehicles and the availability of adjacent lane(s).
d. When the vehicle meets lane changing requirements, it will change lane with a certain probability, which is obtained by statistics and to mimic the driver's lane changing choice.
e. Every vehicle is passenger car.

3 SIMULATION FOR TRAFFIC INCIDENT

3.1 *Basic settings of the model*

The cellular automaton based three-lane model with the rules above is used to simulate real road traffic incident, constructing traffic incident scene for simulation experiments.

Set the segment length for simulation of 3×200 cells. As unit cell length is 7.5 m, the length of segment for simulation is 1.5 km, three lanes. The unit time is 2 s and the simulation is of 1800 units, so the simulation duration is 3600s, namely 1h. A cell has only two states: occupied by a car or not. Vehicles are generated according to Poisson distribution by cellular automaton. Set the starting location of the selected segment of a signal intersection, then the initial speed of vehicles is zero. According to researches and statistics on urban roads, the average observed maximum speed of the vehicle is lower than 55 km/h. Set the speed interval for vehicles is [0,4]. Maximum speed $V_{max} = 4cell/2s = 4 \times 7.5m/2s = 15m/s = 54km/s$. Due to the short distance (1.5 km) of the selected segment, the impact of random slow effect is not taken into consideration. Mark the lane nearest the median strip as Lane One and the other lanes as Lane Two and Lane Three.

Combined with observation and analysis on a video about a traffic incident occurring in urban road in China, the basic conditions of the incident are as follows. The incident is a head-on collision which occurred after the start of the simulation for 10 min in the Lane One and Lane Two. The entire process lasted for 17 min and the site was 1.0 km away from the beginning of the simulation section. The incident resulted in one cell occupied in each lane of the Lane One and Lane Two. The distance that the drivers can be affected by the incident when they made choices to change lane is about 75 m, 10 unit cells. As the probabilistic lane changing rules set above and statistics about the incident, when conditions meets lane changing requirements, drivers will change lanes with a certain probability at 0.37 and 0.21.

3.2 *Simulation results*

By programming in MATLAB 7.0, the spatiotemporal patterns of the three lanes are as the Figure 2, Figure 3 and Figure 4 shows.

In the three figures, the time and distance are plotted as coordinates in an x-y plane. And the position of the point in the figure represents the distance from the starting point at a moment. Lane One and Lane Two are shut down because of the traffic incident, which happens at the point (1000, 600) in the figure. Lane Three is the only

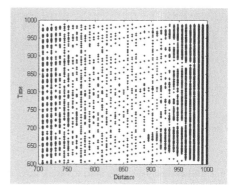

Figure 2.   Spatiotemporal pattern of Lane One.

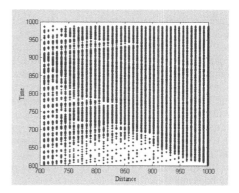

Figure 3.   Spatiotemporal pattern of Lane Two.

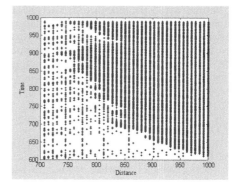

Figure 4.   Spatiotemporal pattern of Lane Three.

open lane. The spatiotemporal patterns show the distribution and the running situation of the passenger cars in the shut lanes and the open lane. In the figures, the dense areas represent the greater density of the cars, which can lead to

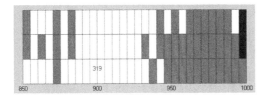

Figure 5. Cell pattern when the congestion starts to take shape.

queues and congestion. The sparse areas show the smaller density and in this condition, the traffic flow is smoother and the cars travels more freely. It can be seen from Figure 2 and Figure 3, when the Lane One and Lane Two are closed, queuing occurs immediately in the two lanes, and spread to the upstream. Cars from the Lane One need to first change to Lane Two and then change to Lane Three to leave the incident scope. Coupled with these cars, the congestion in the Lane Two is significant. The car queue is also longer than queue in Lane One. Almost the same time, the open lane (Lane Three) is affected by the adjacent shut lanes. The plot becomes dense immediately and the cars start queuing. For the cars from the two shut lanes have to change to the one open lane, the queue in Lane Three is quite long and just a little shorter than the queue in Lane Two. The simulation patterns basically accord with the characteristics of traffic flow in actual situation.

In the Figure 5, cells are represented by boxes, the black boxes represents the incident area and the gray boxes represents the cars. As the unit time is 2s, 319 times equals the time of 638s after the simulation starts and 38s after the incident happens. The pattern of the cells is basically accord with the video at the same time after the incident happens. Thus, the model can be used into the simulation of the incident which causes the closing of Lane One and Lane Two.

4 CONCLUSION

With the researches and statistics on driving characteristics of vehicles in urban roads, this paper proposes a three-lane traffic simulation model for traffic incident and develops probabilistic lane changing rules and following rules for three-lane roads based on the traditional two-lane cellular automaton model. The three-lane model is used to incident simulation in MATLAB 7.0. The incident results in the closing of Lane One and Lane Two. The simulation results show that the model performed well in describing the evolution of

traffic flow and the lane changing behaviors of the vehicles.

As the vehicle type is limited to passenger car in this paper, in future studies, it needs to develop more researches including more large car types. And in order to make the model accord with the driver's reaction in actual situation better, it also needs to improve the parameters by more data analysis. Thus the model would be more realistic and lead to more reliable and accurate simulation results.

ACKNOWLEDGEMENTS

This work is supported by the National High Technology Research and Development Program of China (Project No.2012AA112304) and the Student Research Training Program. I also thank Professor S.Y. Chen and Doctor K.J. Zhao for their kindness and constructive suggestions, and C. Liu, a student from Southeast University who devoted time to taking part in the simulation.

REFERENCES

[1] Chowdhury, D. & Wolf, D.E. & Schreckenberg. M. 1997. Physica A. 235–417.
[2] Hu, X.J. & Wang, W. & Lu, J. 2007. Determination Impact Area of Traffic Incident. *Journal of Southeast University (Natural Science Edition)*; 5:038.
[3] Ju, J.P. 2011. Study on the Impact of Variable Message Sings (VMS) on Properties of Traffic Flow Based on Cellular Automata Model. Beijing: Beijing Jiaotong University.
[4] Nagel, K. & Schreckenberg, M. 1992. A cellular automata model for freeway traffic. *J de Physique I*; 2:2221–2229.
[5] Nassab, K. & Schreckenberg, M. & Boulmakoul, A. et al. 2006. Effect of the lane reduction in the cellular automata models applied to the two-lane traffic. *Physica A: Statistical Mechanics and its Applications*; 369(2):841–852.
[6] Sheng, P. & Zhao, S.L. & Wang, J.F. et al. 2010. Study of Temporary Traffic Bottleneck Based on Cellular Automaton Model. *Acta Physica Sinica*; 59(6): 3831–3840.
[7] Wang, Y.M. 2010. Study of Traffic Congestion's Simulation Based on Cellular Automaton Model. *Journal of System Simulation*: 2010(9):2149–2154.
[8] Zhang, H.J. & Zhang, J. & Yang, X.G. 2008. Analysis and Simulation of Traffic State on Expressway during Incident. *Journal of Traffic and Transportation Engineering*; 8(2):116–122.
[9] Zhang, J. & Li, X. & Li, C. et al. 2010. Simulations of Traffic Flow Induced by the Reduction of Lanes under Different Management Measures with Cellular Automata Model. *Third International Conference on Transportation Engineering (ICTE)*.

Future Information Engineering and Manufacturing Science – Zheng (Ed)
© 2015 Taylor & Francis Group, London, ISBN 978-1-138-02644-5

Research on physical education personnel training mode in the perspective of market economy

Chuan-Qin Liu & Youming Zhang
College of Sports Science, Mudanjiang Normal University, Mudanjiang, Heilongjiang, P.R. China

ABSTRACT: By using the method of market economy and logic analysis, the paper describes the relation between the cultivation of Physical Education Talents and the talent demand from the market economics perspective under the system of market economy, pointing out the present situation of higher physical education talents training in China: oneness of training target, professional settings is less, the pattern of college physical education is old, curriculum development is lagging behind, teaching contents lack of innovation, teacher structure of PE teachers in colleges and universities is not reasonable. On the basis of this, the paper puts forward talent cultivation mode to meet the demand of market economy including: renewing education idea, broadening physical education target, constructing the reasonable curriculum structure, optimizing curriculum system, strengthening the construction of teachers and other reform measures.

Keywords: market economy; sports training; physical education

1 INTRODUCTION

Along with the development of the market economy, Sports education is influenced greatly by Planned economics system for a long time. In order to deepen reform of physical education, we should have a correct understanding of the market economy demand for sports talents. Thus, to better serve socialist modernization construction is an essential task.

2 MARKET ECONOMY

Market economy is the economy based on market as resource allocation of basic way and the dominant means, it is the resource allocation by way of an objective or objective demand, when all the commodity economy developed to mass production stage. Its core is the use of supply and demand and the law of value, through the free competition, to optimize the allocation of resources, and to achieve maximum economic benefits.

3 THE DEMANDING OF SPORTS TALENTED UNDER MARKET ECONOMY

The relationship between Sports education and market economy is related to the cultivation of sports talents. Social demand will be the dynamic mechanism of education structure and education content in the process of sports talent training. It directly affects the education content and education structure adjustment and reform.

In today's market economy, Sports fully shows great economic function shadowed in the planned economy period. Economy combines with sports, commerce, tourism, and entertainment sports become an indispensable part for economic development and prosperous enterprise. It is also an important part of enterprise culture. Enterprises can gain huge economic benefits from sponsored sports development to the establishment of sports, from sports publicity to Sports industry.

With the development of market economy and the improvement of people's living standards, people began to pursue a higher level of cultural and sports life. A lot of sports venue operators emerge, including table tennis, tennis, squash, bowling, golf, and skating. As the sports venues are built, and basketball and football club are set up, sports become an industry. With the rise of the sports industry, it puts forward the demand of sports professional talents.

Because of the rise of Sports industry and social sports groups and urban and rural masses fitness movement, society needs a large number of professional sports talents with technical and theoretical knowledge and the guidance of sports management cadres. In the face of such situation, it is necessary to adjust sports talents training target, to adapt to the market economy under the conditions of demand for emerging sports talents.

4 THE STATUS QUO OF HIGHER PHYSICAL EDUCATION PERSONNEL TRAINING IN OUR COUNTRY

4.1 Generic training objectives

For a long time, due to the effect of the Soviet Union education mode, China's higher sports education on the education goal setting is relatively narrow, merely emphasis on the sports skill teaching and master, mainly for the school sports, sports training specialized personnel.

The school physical education is one-sided to think the education target is the final purpose.

4.2 Professional settings is less

Professional Settings cannot meet the demand for sports talents under the market economy. Broad professional Settings is the important factor of sports talent cultivation, and the cultivation of sports talents should be closely linked with the demanding of social sports. If we cannot cultivate a large number of high level talents, it is impossible to adapt to social competition for sports talent demand.

4.3 Sports education pattern in colleges and universities is antiquated

For a long time, colleges' sports teaching is always influenced by the traditional athletics teaching mode. Motor skills teaching becomes the sports teaching content, also becomes the main index for the evaluation of teachers' teaching level.

4.4 Curriculum development lagging behind

Colleges and universities sports teaching has not established a scientific system. So, we should set up physical education curriculum system to satisfy the needs of the social economic development, social progress of sports discipline characteristic.

4.5 Sports teaching content lacked of innovation

The core content of sports teaching in colleges and universities has long been occupied by track and field and basketball, football, volleyball which repeated the middle school sports teaching content.

4.6 College PE teachers' personnel structure is not reasonable

First, the age structure of teachers directly affects the adding, flowing, retirement and promotion of teachers and other series of problems. The structure of teachers reflects the vitality of school teaching and scientific research, the trend of the rise and the fall of a school business level. Second, people who have master degree or doctor degree are less. Our teacher's theoretical knowledge is not solid, teaching and scientific research ability is not strong. These greatly influence the teaching quality and research level of university sports.

5 MEETING THE NEEDS OF THE MARKET ECONOMY AND REFORMING SPORTS TALENTS TRAINING MODE

Under the condition of market economy, our country sports education reform is imperative. To adapt to the market economy and to deepen the reform of professional sports education, we should not only emphasize the role of market economy, but also follow the principle of education. The object of education is people, not contents, the purpose of education is to cultivate talents who adapt to the social development in the sports professionals. Educational activities rules are different from the economic field. Therefore, the choice of the ways of physical education resource configuration and the reform of education system, the characteristics of physical education itself must be considered.

5.1 Updating education ideas

Higher education must adapt to the market economy system, to develop comprehensive quality talents as the main target. Transformation of education idea and renewing education concept are the forerunner. First, It is quality benefit idea, to change the idea of ignoring quality for a long time, from paying attention to expanding the denotation increase the quantity to pay attention to the letter and improve the quality of talent development and teaching efficiency, to put the quality consciousness and quality education combined closely. Secondly, it is to should know the idea, according to the social progress and the trend of the development of higher education reform, strengthen the awareness of the mission of colleges and universities facing the society, for the future, focus on the socialization of physical exercise, diversification development direction, to take the initiative to adapt to the needs of the development of regional and local construction, efforts to expand the social services, the enclosed education to open education.

In order to better meet the needs of the development of social sports in our country in many aspects, according to the requirement of the implementation of the "national fitness plan", we should combine social sports and the national fitness movement into the teaching system, and based on this, it is needed to establish a new teaching system adapted to the needs of society and the development of sports culture. Thirdly, distinctive concept, it is inevitable to have competition among universities in market economy, personality. And characteristic is one of the weapons to win in competition.

Personality is the basis of survival. The characteristic is the essence of competitiveness and vitality. Fourthly, the concept of talent quality, social progress requires improving talent quality and developing personality. The market economy environment requires a more flexible adaptability and more personalized competition ability. With rapid development of science and technology, talents should have stronger thinking and innovation ability, and have the comprehensive multidisciplinary knowledge base and operation ability of modern technology. With the increasing openness of international exchanges, talents should have a more in-depth understanding of the world culture and sports. On the cultivation of talents, therefore, we should pay attention to comprehensive quality education, namely, pay attention to scientific thinking, scientific method, practice skill, innovation ability, and the cultivation of the humanities knowledge and ideological and moral quality.

5.2 *Broadening the sports education training goal*

According to current basic situation, the cultivate aim of physical education in our country, obviously, can't keep up with market trends, lagging behind the rapid development of social economy.

The current sports education professional training objectives should be expanded from the teachers type to multiple type conversion, making the sports education to present the multi-function. Thus, it will become the teaching and training, fitness, management, and other sports talents training center and base, so, the students will become sports talents in the knowledge structure and specific ability to. At present, according to the demand of social development, it is essential to open some emerging disciplines appropriately, such as sports judicial talents, martial arts and security, etc. These new subject will objectively enhance the vitality of running a school. It will make the professional education to have the characteristics of diversity and open mode, the cultivation of sports talents will be from the simple to the pluripotent, quality, composite, thus we can make the sports education run into the track of benign development.

5.3 *Building a reasonable course structure for optimizing curriculum system*

Optimizing the curriculum structure system, further adjusting and perfecting the curriculum structure setting, are the main way to attain the goal of education and training objectives.

Physical education curriculum should not only consider social needs and the needs of the development of disciplines, but also to the needs of students' personality development.

In the optimization of course structure, we must pay attention to strengthening Elementary Education, because the change of sports course content in Elementary Education will change the sports education professional course reform in colleges[2]. We should focus on broadening students' knowledge, cultivating students' creativity and social adaptation ability and teaching ability. Setting the Curriculum should meet the requirement of training target, reducing the required courses, increasing the elective course, getting rid of the mode of competitive sports, increasing theory curriculum, stressing the content of skills classes, entertainment and fitness class, improving scientific and practical in physical education curriculum system.

6 EPILOGUE

To sum up, the establishment of the socialist market economic system requires us to do anything initiatively to meet the needs of the market economy. Only in this way can we survive in the spring tide of market economy. Sports education adapted to the development of socialist market economy is to cultivate sports talents who are needed in the socialist market economy under the conditions. Training of all kinds of top sports specialized personnel, not only to meet and adapt to the needs of the development of society in quantity, quality, level, professional and specifications, economy, but also plays a decisive role in the sports education's own development.

REFERENCES

[1] Wangyiwu. Market economics. Beijing: tsinghua university press, 2009, 12 Season.
[2] Jiliu. Our country elementary education sports curriculum reform of university sports education specialized curriculum reform revelation. Journal of Beijing sport university. 2004. 27(6):799–801.
[3] Under the state council. The central committee of the communist party of China's education reform and development compendium. 1993.

Future Information Engineering and Manufacturing Science – Zheng (Ed)
© *2015 Taylor & Francis Group, London, ISBN 978-1-138-02644-5*

Research on teaching method of Unified Modeling Language

Wei Quan, Yaohong Xue, Jing Tao Fan, Cheng Han, Zhengang Jiang & Huamin Yang
School of Computer Science and Technology, Changchun University of Science and Technology, Changchun, China

ABSTRACT: UML is a standard modeling language widely used in software development, and UML course is an important professional course for students of computer science and other related professions. This paper analyses the characteristics and the present situation of the UML course. And a staged teaching method combining with case-based and task-driven methods is presented to improve the teaching quality.

Keywords: UML; staged teaching method; case-based; task-driven

1 INTRODUCTION

Object-oriented technology is an important branch of software engineering. Analysis and design methods of object-oriented systems are the mainstream. UML (Unified Modeling Language) provides support for visual modeling of all phases in object-oriented software development process. UML is the most popular Modeling Language, and it has become a standard of software modeling today.

In recent years, many colleges and universities at home and abroad offer UML course for students of computer science and other related professions. During the course, students could learn the basic concepts of UML and the analysis and design methods of object-oriented system. Students are supposed to have the ability of analyzing, designing and modeling at the end of the course. The UML course involves many subjects such as software engineering, object-oriented technologies, etc, and the contents are abstract and difficult for college students because of their lacking in programming experiences. Many students said "they can't learn UML well for it is too difficult, and they can't do system analyzing and modeling using UML either." In order to improve the teaching quality of UML course, lots of teaching methods are studied by many colleagues, such as the case-based teaching method (Zheng & Tian. 2008, Yi & Ma. 2007), the task-driven teaching method (Shen & Sun & Zhang. 2007) and the teaching method of combination of these twos (Zhang & Li & Hao. 2012). However, the teaching effect is still not good. It is because that students usually can't understand object-oriented concepts and ideas actually for lacking in programming experiences, which are

basises of UML. They can't learn knowledge points well at the beginning of UML course, so any teaching methods, such as case-based method, task-driven method, etc, don't work.

This paper analyzes and summarizes the characteristics of UML course and the present situation of course teaching firstly. And then a staged teaching method combining with case-based methods and task-driven methods are presented according to the author's practical teaching experiences, in order to improve the teaching effect and the students' abilities of system analyzing and designing based on UML.

2 CHARACTERISTICS AND PRESENT SITUATION OF UML COURSE

2.1 *Characteristics of UML course*

UML is a well-defined, easy to express, powerful and widespread used modeling language. It can effectively support not only the analysis and design of object-oriented systems, but also the whole processes of software development. The features of UML course can be summarized as follows.

2.1.1 *A large number of incompact knowledge points*

There are ten model diagrams in UML, which are: use case diagram, class diagram, object diagram, package diagram, sequence diagram, collaboration diagram, activity diagram, state diagram, component diagram and configuration diagram. Each kind of diagrams is used to describe different aspects of the system. The system could be described completely using these ten kinds of diagrams reasonably. The main content of UML course is to teaching these ten kinds of model diagrams, so that students can skillfully use them for object-oriented system analysis and design. Each kind of these model diagrams has its unique concepts, figures and modeling methods. So, there are a large number of incompact knowledge points in the whole classes.

2.1.2 *Abstract contents required more related knowledge and higher practicality*

The contents of UML course are analyzing, designing and modeling of object-oriented system. These are abstract and difficult to be understood for students. Besides, the course covers all phases of the software development, and involves knowledges from object-oriented, software engineering, and other related subjects. UML course is a highly practical course. So it will affect course learning seriously that lacking of software development experiences.

2.2 *Present situation of UML course*

In the process of UML teaching, the current method is to teach the knowledge points one by one, then let the students practice in the experimental classes. Some methods like the case-based or the task-driven teaching method are used to promote the students' understanding and mastering of knowledge.

In the author's school—Changchun University of science and technology, UML is a specialized elective course. The juniors of computer science and technology, network engineering, software engineering and other related professions can choose it to learn. Before they learn UML, they usually have already studied the software engineering, and an object-oriented programming language (such as c++), etc. However, the teaching effect of UML curriculum is still not well. The reason can be summarized as below:

2.2.1 *The understanding of object-oriented concepts and ideology not well and lacking in programming experiences*

Before learning UML class, Students have already studied the software engineering and an object-oriented programming language. But a lot of students can't understand the basic concepts and ideology of object-oriented well, and they lack practical programming experiences in object-oriented software development. Whereas, UML is base on these courses and knowledges, as well as the reference books. Therefore, if students can't understand in class, they are not able to read reference books after class either. As time goes by, knowledges that students can't understand become

more and more. As a result they can't learn UML well.

2.2.2 *Effect of the existing teaching methods not ideal*

Case-based teaching method is beneficial to promote students' understanding of the abstract UML knowledges; Task-driven teaching method is able to mobilize students' initiative and enthusiasm of UML learning; The two methods play positive roles in UML teaching. However, in the early stages in the UML class, many students cannot understand the knowledge points, and even lose the interest. In this case, teaching methods like case-based method, task-driven method, etc, don't work. The teaching effect is still not good.

2.2.3 *Students do not pay enough attention to UML course*

Although the UML is very useful for the actual software development. But, students have little work experience, and don't know the actual meaning and importance of UML. In addition, UML is an elective course, and it is abstract and hard to understand. So, many students prefer to spend time on learning a programming language rather than UML. Students do not pay enough attention to UML course, and they study just to pass the exam, or even give up learning.

3 A STAGED TEACHING METHOD COMBINING WITH CASE-BASED AND TASK-DRIVEN METHODS

3.1 *The staged teaching method*

Staged teaching method stems from human cognitive law. This method has characteristics such as: from the shallower to the deeper, from the outside to the inside, from the concrete to the abstract, and from simple to complex. The staged teaching method is a staged cognitive activity which is divided into different stages under the guidance of teachers (Pang. 2003). And suitable teaching tasks are arranged in each stage. So, students could understand new knowledge gradually, and this will help them to truly master the knowledge.

3.2 *Case-based and task-driven methods*

The case-based teaching method is presented to solve the problem that the knowledge of curriculum is abstract and difficult to be understood. In order to improve the teaching effect, abstract knowledge is explained under the guidance of a series of practical cases. In this way, students can understand these knowledge well.

The task-driven teaching method takes tasks as the main line, students as the main body. The whole teaching process is dominated by the teacher. During the process, the teacher explains task firstly. Then, the task is divided into subtasks which are assigned to students. Students is asked to complete the task by themselves. Finally, that what they have done will be demonstrated and examined by the teacher and other students. This method can efficiently arouse the students' studying interesting. They could think positively and participate actively in teaching activities. In addition, teachers can obtain feedbacks about how students learn, and they can adjust teaching contents correspondingly.

3.3 A staged teaching method combining with case-based and task-driven methods

A staged teaching method combining with case-based and task-driven methods for UML course is presented in this paper. The model of our methods is show as Figure 1.

Firstly, teaching contents are organized according to the software development sequence. It is divided into four parts: use case modeling, static modeling, dynamic modeling, system modeling. The use case models are established to describe the system requirements. And, in use case modeling part, the use case diagram, activity diagram are taught. In the phase of system analysis and design, static models and dynamic models are established to describe both the static structure and the dynamic behavior of the system. Class diagrams, object diagrams, package diagrams are taught in the static modeling part, and sequence diagrams, collaboration diagrams, state diagrams and activity diagrams are taught in the dynamic modeling part. As to the system modeling part, component diagrams and deployment diagrams are taught, which can describe the architecture of software and hardware of the system.

Secondly, The staged teaching method combining with case-based and task-driven methods is applied on teaching each UML model diagrams. It is because that UML course contains a large amount of incompact knowledge and abstract

contents, and it requires more related knowledge and higher practicality, so that students can't learn UML well. Whereas, as a modeling language, UML is just to use its concepts and symbols to describe the software model actually. So, if in the early teaching period of each model, concepts and symbols are explained and models are use to describe something which has nothing to do with the software development. In this way, students only need to remember some concepts and symbols, the studying difficulty will be greatly reduced. Therefore, when teaching each model, the teaching process is divided into two stages according to the students' cognitive law. They are primary stage and advanced stage, which are discussed following.

3.3.1 Primary stage

This stage focuses on explaining the basic concepts and symbols. The target of this stage is to enable students to use the UML model to describe certain things, but not the software model. This stage should involve object-oriented knowledge as little as possible in order to reduce the learning difficulty and arouse the students' study interesting. In this stage, case-based teaching method is used, in order to facilitate understanding of knowledge points.

For example, when teaching the use case diagram, cases, which students are familiar with, such as library management system, are selected,. Firstly, the teacher describes requirements of the system by natural language, and then, shows students how to establish use-case diagrams. Namely: the use case diagram is regarded as a tool with which requirements described by the natural language are "translated" into the use case diagrams. In this way, the basic concepts and symbols of use case diagram are explained easily. Students can understand these knowledge well, and could be able to read and draw use case diagrams.

3.3.2 Advanced stage

When coming to advanced stage, students have been able to use the UML model to describe things. Therefore, the target of this stage is teaching students how to analyze, design system and how to modeling in the software development phase. Case-based methods and task-driven methods are used in this stage, and object-oriented knowledge and ideology should be involved in cases.

We still take teaching of the use case diagram as the example here. First of all, the use case diagram modeling method is explained through a case. For example, a software system is take as an example, the teacher explains how to recognize actors and use cases and how to determine the relationship among them. This software system can be the system used in the primary stage, and it should be analyzed further and be refined; the system can

Figure 1.   Model of the staged teaching method combining with case-based and task-driven methods.

also be a new software system, which students must be familiar with. Secondly, the analysis of a similar system could be arranged as homework. This task is divided into some subtasks, and the students are divided into groups to complete these tasks. Finally, these homework will be examined and discussed in class.

In addition, in the advanced stage, there should be a case throughout every part of the course. In every part when teaching each model diagram, the case is used. In this way, when finishing teaching the ten models, a complete method of how to analysis and modeling with UML has shown to students.

4 CONCLUSION

Continuous improvement of teaching methods and teaching quality, is the eternal pursuit of the educator. This paper summarizes characteristics and the present teaching situation of UML course, and presents a staged teaching method combining with case-based and task-driven methods, in which the teaching activity is divided into two stages, according to student's cognitive law. This method inherits the advantages of staged method, case-based method and task-driven method. And it can effectively improve the quality and the teaching effect of UML course.

REFERENCES

[1] Zheng Yu & Tian Wei. 2008. Research on "Unified Modeling Language UML" Case Teaching. *China Electric Power Education,* 2008, 23: 86–87.
[2] Yi Yang & Ma Feiteng. 2007. A Integrity Specification by OO Visual Modeling of UML—a Case Based Study in Computing Education. *Acta Scientiarum Naturalium University Sunyatseni,* 2007, 46(2): 74–78.
[3] Shen Weizheng & Sun Hongmin & Zhang Xihai. 2007. Exploration and Practice of the Task-driven Teaching Mode in the Teaching of Computer Course. *Journal of Northeast Agricultural University (Social Science Edition),* 2007(1): 89–90.
[4] Zhang Liping & Li Song & Hao Xiaohong. 2012. Research on Teaching UML Combining with Case-based and Task-driven Methods. *Journal of Heilongjiang College of Education,* 2012.3, 31(3): 50–52.
[5] Pang Weiguo. 2003. *Autonomous Learning—Learning and Teaching Theory and Strategy.* Shanghai: Publisher of East China Normal University.

Future Information Engineering and Manufacturing Science – Zheng (Ed)
© 2015 Taylor & Francis Group, London, ISBN 978-1-138-02644-5

Research on the reputation evaluation system of online group buying based on consumer preference—taking catering category as an example

L. Zhong, J.C. Xiang & L.H. Xu
School of Business, Hunan Normal University, Changsha, Hunan, China

ABSTRACT: Based on the reputation management of online group buying in China nowadays, this paper analyzes the catering enterprises' reputation evaluation indexes which influence customer's purchase decision in online group buying from the perspective of consumer preference. Matrix of catering enterprises' relatively important reputation indexes is determined by the AHP method in this paper and the weight of reputation evaluation indexes are calculated by the eigenvector method used in making multi-objective decision. The enterprises' reputation score that meet consumer's different preference is worked out and the operational processes of online group buying reputation evaluation system is analyzed in this paper. This paper provides reference for consumers in making a purchase decision and also provides the theoretical guidance for online group buying in improving the reputation management system in China.

Keywords: Catering enterprise; Eigenvector method; Consumer preference; Reputation evaluation indexes; Online group buying

1 INTRODUCTION

Online group buying is a type of e-commerce model, which is increasingly becoming a fixed demand of consumers in China. The accumulative transactions of online group buying market in 2013 reached 53.289 billion yuan and it increased by 18.404 billion yuan comparing with 34.885 billion in 2012, the transactions increased by 52.8%.

Under the e-commerce environment, online group buying websites permit anonymous interactions in order to enhance consumption experience. With the continuous extension of enterprise resources, There is a dramatically increasing partners number for consumers to choose in online group buying. So it makes reputation management to be more complex and difficult. At present, Most of the big domestic online group buying websites only use a single general reputation score as a reference for consumers to make a purchase decision except Meituan has made a preliminary classification of enterprises' reputation score on the main category. As a result, consumers cannot quickly filter and search the enterprise which satisfy their preference. As is known to all, the enterprise will have different comprehensive reputation when it faces the raters who have different preference. If we only use a single general reputation score, it would

make the high quality enterprises' core competitiveness on certain aspect such as price, consumption environment, service quality unable to get a fair treatment and the enterprise cannot be filtered by consumers who have these preference. That Consumer preference can't be met and the high quality enterprises' core competitiveness can't get a fair treatment make the current reputation evaluation system of online group buying is not able to adapt to its development. So it is necessary to explore a new reputation evaluation system which fits for the development of online group buying in China.

2 BACKGROUND ON REPUTATION MANAGEMENT SYSTEMS

Reputation is a cross-disciplinary concept, incorporating ideas from economic, marketing, social psychology, and decision science. From Leuthesser (1988) and Bromley (1993) in marketing and Emler (1990) in social psychology we know that social entities, people and organizations actively monitor their own reputation.

An important aspect of reputation is the meaning of reputation information. A study by Resnick and Zeckhauser (2000) concluded, "A reputation system collects, distributes and aggregates feedback about participants' past behavior. That these systems help people decide whom to trust, encourage trustworthy behavior and deter participation by those who are unskilled or dishonest. They help each other to examine how to build trust naturally in long-term relationships."

In terms of reputation management system design and improvement, Dellarocas (2000) identify two types of reputation system fraud problems: reputation feedback unfair and sellers' discrimination. In his study, he suggested we should use controlled anonymity and cluster filtering to eliminate or reduce the negative effects of the two kinds of fraud. Zacharia et al. (2000) took advantage of collaborative filtering technology to realize the reputation management in e-commerce market. Yang Qi-feng et al. (2011) established a "participants' basic characteristic, biological characteristics and credibility" information genetic model and measurement model of B2B online payment system based on the theory of order parameter. At the application of online group buying, Zhao Peiqing et al. (2010) proposed ant algorithm to optimize the online group buying reputation evaluation model in view of the reputation problems.

The goal of this paper is to explore and solve the following related problems when constructing a reputation evaluation system of online group buying such as the construction of catering enterprise's reputation evaluation indexes, the calculation of enterprise's comprehensive reputation score and the reputation evaluation process analysis of online group buying.

3 CATERING ENTERPRISE'S REPUTATION EVALUATION INDEX ANALYSIS AND CONSTRUCTING AFFECTING CONSUMER SATISFACTION IN ONLINE GROUP BUYING

Currently, there are a lot of segmentation categories in online group buying such as cosmetics, movie, hotel, food, etc. According to the consumption forms, these categories can be roughly classified into physical and service class. Considering the physical goods via online group buying is similar to traditional e-commerce and the reputation research on it is mature, so in this paper, we take this typical category catering service as the research object.

Combined with the characteristics of catering online group buying, Referring to the reputation evaluation research results related to the catering industry, this paper summarizes three dimensions which may affect consumer satisfaction such as the product, service and environment.

3.1 Dimension 1—product related

In the current catering online group buying, other consumers' rating scores on taste, quality, quantity, etc, which will be a reference index of the consumer purchase decision. Pan Guang-jie et al. (2007) stated, "The products that catering enterprises provide include tangible diet and the service. Food quality directly affect the quality of food and various sensory attribute index of the food also affect consumer satisfaction." Many other indexes affect online group buying consumer satisfaction including invisible consumption, the match degree of goods with online publicity (Pan Ting, 2012). Participating in online group buying, consumers buy goods at a lower price, but it is at the expense of his personal preferences (Tokuro Matsuo et al., 2002). So this paper extracts taste, quality, quantity, price, invisible consumption, product consistency as the segmentation indexes of product related dimension.

3.2 Dimension 2—service related

Service quality is the overall experience of consumer's feeling (Carman et al., 1990). An important aspect of service quality is the division of dimension, which includes not only the result quality, but also the process of service delivery (Sasser, 1978). Brady and Cronin (2001) defined three dimensions of service quality. In their study, each dimension was further divided into three dimensions, that is the interactive quality, physical environment quality and the results quality. The interactive quality includes the interaction between the consumers and the staff in the process of service delivery, staff attitude, behavior and professional ability. The result quality includes the output end of the service, the waiting time of customer, tangible results and the attitude towards the service result. Yu Yin-yan et al. (2009) state," the personalized service and complaint handling service are the factors that have impact on consumer satisfaction in B2C e-commerce." So, referring to the above research and combining with the characteristics of catering online group buying, this paper extracts the convenience of booking, equality of treatment between consumers who purchase the service via online group buying and not, waiting time, the degree of personalized demand be met, consumer complaint handling and invoice issued as the subdivision indexes of service related dimension.

3.3 Dimension 3—environment related

In addition to the advantage of price, the goal that consumers choose catering online group buying is just that they decide to have dinner with friends or families to enhance communication. So the

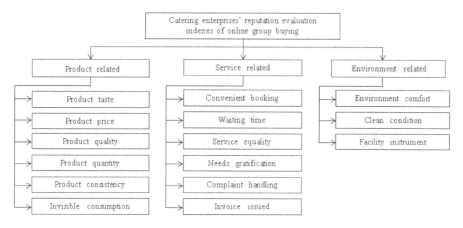

Figure 1. Catering enterprises' reputation evaluation indexes of online group buying.

comfortable degree of the consumption environment, the clean condition will also affect consumer satisfaction. Rust and Oliver (1994) propose the physical environment that consumer experience service should be included in the measurement of service quality. In their study, the service environment include the external environment of the service delivery, the atmosphere, the physical facilities, etc. So, this paper extracts the environmental comfort, clean condition, facilities as the subdivision indexes of environment related dimension.

Through analysis on catering enterprise's reputation evaluation indexes which influence consumer satisfaction in online group buying above, this paper constructs a reputation evaluation index system which consists of three dimensions and fifteen indexes, as shown in figure 1.

4 ENTERPRISE'S COMPREHENSIVE REPUTATION SCORE MODEL OF ONLINE GROUP BUYING BASED ON CONSUMER PREFERENCE

In terms of the choice of catering enterprises, different consumers may have different preference, even the same consumer will have different preference in different cases. As a result, the enterprise gets different reputation score when it provides the same service to consumers who have different preference. Designing the reputation management system is bound to face two critical problems: one is how users evaluate their trading partners' reputation and the second is how to get all the reputation scores feedback gather for comprehensive reputation score (Malaga, 2001). In addition to the two problems above, enterprises reputation management system based on consumer preference have to consider adopting the appropriate method to give

the reputation evaluation indexes different weights (Jiang Wen-yu et al., 2007). Weights reflect three factors: the importance of the target to policymakers, the different degree between the target attribute value and the reliability of the target attribute value (Zhang Wen-zhu et al., 2010). Different consumers have different preference for reputation indexes of the trading partners, so the weights are introduced to solve the multi-objective decision.

4.1 The calculation of enterprise's comprehensive reputation score

Likert five-point attitude scale is applied in this paper to get enterprise's reputation score of online group buying, which is further detailed below: "very good" add 5 points, "better" add 4 points, "medium" add 3 points, "poor" add 2 points, "very poor" add 1 point. The formula of calculating reputation score is used the one that put forward by Malaga in 2001. This formula can solve several problems that many sites suffer from such as use of a single general reputation score, equations that do not accurately reflect reputation, there is no ability to filter or search by reputation score, etc. (Malaga, 2001). The modified formula is as follows:

$$R_{it} = \frac{\sum_{k=1}^{t-1} R_{ik} \cdot \omega(R_{kt}) + s}{t-1} \quad (1)$$

where:
R_{it} is an individual's i reputation at transaction t,
R_{ik} is an individual's i reputation rating for a specific transaction,
R_{kt} is a rater's reputation score at transaction k,
ω is a constant used to determine how much weight to give a rater's reputation,
s is a starting reputation score.

257

As a rater's reputation score is higher, the evaluation of he makes is more credibly.

To calculate ω: Assuming that the upper limit of a rater's reputation score is M and the rater's reputation is R_{kt}, through standardizing R_{kt}, we will get a value R_{kt}/M within the range of 0~1, the value R_{kt}/M is the weight of the rater.

An individual's i reputation at transaction t is R_{it}, thus, the individual's comprehensive reputation R_t at transaction t is

$$R_t = \sum_{i=1}^{n} R_{it} \cdot \omega_i \qquad (2)$$

where:

$\sum_{i=1}^{n} \omega_i = 1$ and $\omega_i > 0$ ($n = 1, 2, \ldots$), n is for the number of reputation evaluation indexes.

4.2 The calculation of enterprise's comprehensive reputation score based on consumer preference

Different consumers have different preference for the trading partners' reputation indexes, the consumers who have different preference will give different weight to the reputation indexes. Eigenvector is an important and effective method to measure weight, it calculates the largest eigenvalue of matrix through a comparison property matrix. The corresponding eigenvectors is the requested weigh vector.

The comparison property matrix can be obtained by AHP method, due to the differences in the raters' preference, they are not clear that each reputation index should add how much weight at the start, but they can obtain the relative importance of every index through comparing it to the others. The thought of AHP method is that the raters compare every index to other indexes, then get the relative importance of every index. The radio a_{ij} is equal to the importance of i reputation index divided by the importance of j reputation index for the consumer, Furthermore, a_{ij} is approximately equal to the radio of ω_i divided by ω_j (see Equation (3)).

$$A = \begin{pmatrix} a_{11} & a_{12} & \cdots & a_{1n} \\ a_{21} & a_{22} & \cdots & a_{2n} \\ \cdots & \cdots & \cdots & \cdots \\ a_{n1} & a_{n2} & \cdots & a_{nn} \end{pmatrix}$$

$$= \begin{pmatrix} 1 & \omega_1/\omega_2 & \cdots & \omega_1/\omega_n \\ \omega_2/\omega_1 & 1 & \cdots & \omega_2/\omega_n \\ \cdots & \cdots & \cdots & \cdots \\ \omega_n/\omega_1 & \omega_n/\omega & \cdots & 1 \end{pmatrix} \qquad (3)$$

where:

A is a comparison property matrix,

a_{ij} is the radio of the i reputation index's importance divided by j reputation index's importance for the consumer,

ω_i is the weight of i reputation index.

Assuming that $w = (\omega_1, \omega_2, \ldots, \omega_n)^T$, according to Equation (3), we will get $A\omega = n\omega$, that is:

$$(A - nI)\omega = 0 \qquad (4)$$

I is the unit matrix in Equation (4). According to the eigenvector method, if the estimate of matrix A is accurately, Equation (4) would be strictly equal to 0. If the estimate exists errors and cannot accurately make Equation (4) equal to 0, the small perturbation of elements in matrix A would mean the small perturbation of the eigenvalue. So there will be:

$$Aw = \lambda_{\max} w \qquad (5)$$

$\lambda_{\max}$ is the largest eigenvalue of matrix A. $w = (\omega_1, \omega_2, \ldots, \omega_n)^T$ can be calculated according to Equation (5). Then the enterprise's comprehensive reputation $R_t = \sum_{i=1}^{n} R_{it} \cdot \omega_i$ at transaction t (see Equation (2)) also can be calculated. In this way, we get enterprise's reputation conditions R_t based on consumer preference. It provides reference for consumer in making a purchase decision.

4.3 Examples of calculus

Take the catering enterprise in Meituan as an example. The following reputation evaluation indexes are using in the website such as the taste, service and environment. The rating system that enterprises to consumers has not been established in Meituan currently. So, it gives every consumer the same weight to rate, that is $\omega(R_{kt}) = 1$ in Equation (1). Suppose a catering enterprise got three rater's rating in the past, and the feedback matrix of three rates is as following:

$$R = \begin{pmatrix} 3 & 4 & 3 \\ 4 & 5 & 4 \\ 4 & 4 & 4 \end{pmatrix} \qquad (6)$$

In Equation (6), every row in the matrix R represents a rater's evaluation to the enterprise, and score represent the raters rating for taste, service and environment index. For example, the first row in the matrix represents the first rater gave the index taste 3 points, the service 4 points, the environment 3 points. Assuming that R_1 is the enterprise's taste score, R_2 is the service score, R_3 is the environment score. According to Equation (1), we will get $R_1 = 11/3$; $R_2 = 13/3$; $R_3 = 11/3$.

Now, consumer A plans to trade with the enterprise, according to his preference, he think the taste is more important than the service 2 times, taste is more important than the environment 2 times and service is equal to environment. According to Equation (3), The matrix A that represent consumer A's preference is detailed below:

$$A = \begin{pmatrix} 1 & 2 & 2 \\ 1/2 & 1 & 1 \\ 1/2 & 1 & 1 \end{pmatrix} \tag{7}$$

Using eigenvector method, according to Equation (4), we will have

$$\begin{vmatrix} 1-\lambda & 2 & 2 \\ 1/2 & 1-\lambda & 1 \\ 1/2 & 1 & 1-\lambda \end{vmatrix} = 0 \tag{8}$$

From Equation (8), we get $\lambda = 3$, then, according to Equation (5), we have

$$\begin{pmatrix} -2 & 2 & 2 \\ 1/2 & -2 & 1 \\ 1/2 & 1 & -2 \end{pmatrix} \begin{pmatrix} \omega_1 \\ \omega_2 \\ \omega_3 \end{pmatrix} = 0 \tag{9}$$

$$\omega_1 + \omega_2 + \omega_3 = 1 \tag{10}$$

According to Equation (9) and Equation (10), we get $W^T = (\omega_1, \omega_2, \omega_3) = (0.5, 0.25, 0.25)$. The consumer A give the enterprise's reputation evaluation index (taste, service, environment) weight is $(0.5, 0.25, 0.25)$.

According to Equation (2), get the enterprise's comprehensive reputation R_A

$$\begin{aligned} R_A &= (R_1, R_2, R_3)W^T \\ &= (11/3, 13/3, 11/3) \cdot (0.5, 0.25, 0.25)^T \\ &\approx 3.83 \end{aligned} \tag{11}$$

In this way, we get enterprise's reputation condition R_A which reflects consumer A's preference. It provides reference for consumer A in making a purchase decision.

5 ONLINE GROUP BUYING REPUTATION EVALUATION PROCESS

In the early stage of the purchase, consumers log in some online group buying websites and subsequently make sure the consumption business circle and category firstly, then according to their preference for the reputation indexes in that category, determine the ratio of two indicators, then the system established the relative important matrix by AHP method. Through the processing of reputation evaluation algorithm, consumers get the enterprise's comprehensive reputation. They filter and search the trading partner through reputation, and refer to the reputation condition to finish the deal. After consumption, online group buying websites take certain incentives to encourage consumers to feedback. Lastly, consumer rate the partner's indexes according to satisfaction to each indexes, and this provides the later consumer a reference for making a trading decision.

The subjective feeling of the consumer will affect their evaluation to the enterprise. Because of the existing of information asymmetry, the reputation evaluation system needs enterprises to reverse the feedback of consumer behavior to prevent "bad review" phenomenon. Enterprise rate consumer's reputation by using a single general reputation score, and the comprehensive reputation score of consumers can be calculated by Equation (1). Online group buying reputation evaluation process is shown in figure 2.

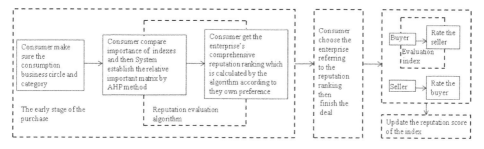

Figure 2. Online group buying reputation evaluation process.

6 CONCLUSION

This paper outlines the need for reputation management systems in online group buying markets. It extracts fifteen indexes as catering enterprise's reputation evaluation indexes in online group buying. According to consumer preference, it designs an algorithm to calculate enterprise's comprehensive reputation in online group buying.

The calculating of comprehensive reputation needs the rater to feedback. While most Chinese online group buying websites have no incentive to encourage the rater to feedback, the appropriate incentives are worth discussing.

Considering the practicality, the appropriate number of evaluation indexes should be determined by the consumer preference and behavior. The more the evaluation indexes are, the heavier the evaluation workload for consumer are.

The intensifying competition makes online group buying websites crazily loot offline business resources, as a result, the enterprise which has poor reputation can transfer to other platforms easily, so, how to realize the cross-platform reputation management is worth exploring.

ACKNOWLEDGEMENT

This research was supported by the research project of Science and Technology Department of Hunan Province under Grant 2013GK3131.

REFERENCES

[1] Bromley, D.B. 1993. *Reputation, Image, and Impression Management*. West Sussex, England: John Wiley & Sons.

[2] Carman, James M. 1990. Consumer perceptions of service quality: An assessment of the SERVQUAL dimensions. *Journal of Retailing*, 66(1):33–55.

[3] Dellaroca, C. 2000. Immunizing Online Reputation Reporting Systems Against Unfair Ratings and Discriminatory Behavior. *Proceedings of the 2nd ACM Conference on Electronic Commerce*. Minneapolis, USA, 150–157.

[4] Emler, N. 1990. A Social Psychology of Reputation. In W. Stroebe and M. Hewstone(eds.), *European Review of Sociology*. New York: Academic Press, 173–237.

[5] Jiang, Wen-yu. & Mi, Zhong-chun. 2007. Multi-dimensional Graded Model of Reputation Management System in E-commerce. *Chinese Journal of Management*, 4(3):302–305.

[6] Lingtuanwang, 2014. *The national online group buying market statistics report in December 2013*. http://zixun.lingtuan.com/article-34380-1.html.

[7] Leuthesser, L. 1988. "Defining, Measuring and Managing Brand Equity." In *Conference Summary Report No. 88–104*. Cambridge, MA: Marketing Science Institute.

[8] Li, Jing. & Zhen Yong-ji. 2010. Service Quality Factors of Influence on Customer Satisfaction—Centered on The Regulating Effect of Sex and Relationship Length. *Journal of Yanbian University (Social Science Edition)*, 43(6).

[9] Michael, K. et al. 2001. Some New Thoughts on Conceptualizing Perceived Service Quality: A Hierarchical Approach. *The Journal of Marketing*, 65(3):34–49.

[10] Malaga, R.A. 2001. Web-Based Reputation Management Systems: Problems and Suggested Solutions. *Electronic Commerce Research*, (1):403–417.

[11] Pan, Guang-jie. et al. 2007. Exploring on Improving Customer Satisfaction in Catering Industry. *Market Modernization*, 36(19):208–209.

[12] Pan, Ting. 2012. Research on Influencing Factors of Customer Satisfaction in Catering Online Groupbuying. Shandong: Shandong University.

[13] Resnick, P. et al. 2000. Reputation System. *Communications of the ACM*, 43(12):45–58.

[14] Tokuro, Matsuo. & Takayuki, Ito. 2002. A Decision Support System for Group Buying based on Buyers' Preferences in Electronic Commerce. *Proceedings of the 11th World Wide Web Conference*, 84–89.

[15] Earl, W. Sasser. 1978. *Management of Service Operations: Text and Cases*, Allyn & Bacon Press.

[16] Yu, Yin-yan. & Yang, Shan-yuan. 2009. B2C E-commerce Customer Satisfaction Survey Research. *Statistics and Decision*, 25(20):187–188.

[17] Yang, Qi-feng. et al. 2011. Measure Model for Information Gene of B2B Online Payment Evolutionary System. *Journal of Wuhan University of Technology*, 33(5):689–693.

[18] Zacharia, G. et al. 2000. Collaborative Reputation Mechanisms for Electronic Market Places. *Decision Support Systems*, 29(4):371–388.

[19] Zhao, Pei-qing. 2010. *Research on The Credit Evaluation of Group Shopping Trades Based on C2B Model*. Hebei: Yanshan University.

[20] Zhang, Wen-zhu. et al. 2010. Effective Strategy for Terminal Reconfiguration under The Framework of IEEE 1990.4. *Journal of Xidian University*, 37(3):594–601.

Future Information Engineering and Manufacturing Science – Zheng (Ed)
© *2015 Taylor & Francis Group, London, ISBN 978-1-138-02644-5*

A novel contourlet-based No-Reference Image Quality Assessment metric

Jie Wang & Yaning Wang
Beijing University of Posts and Telecommunication, Beijing, China

ABSTRACT: No-Reference (NR) Image Quality Assessment (IQA) is of fundamental importance to numerous image processing applications. Whose goal is to evaluate the image quality without a reference image and without knowing the distortion present in the image. To solve this problem, we extract a set of statistical features from a computed image contourlet representation to learn a new no-reference image quality assessment model. In addition, we attach more attention on the distortion of image salient regions considering characteristic of human visual system. Our results indicate that these features are sensitive to the presence and severity of image distortion. Operating within a 2-stage framework of distortion classification followed by quality assessment, a distortion classification and quality prediction model is trained by a Support Vector Machine (SVM). The result of Experiment on LIVE database shows that the statistical performance of our algorithm is comparable to the state-of-the-art NR/FR IQA algorithms.

Keywords: Image Quality Assessment; No-Reference; Natural Scene Statistic; Contourlet Domain

1 INTRODUCTION

With tremendous growth in the use of digital images for representing and communicating information, the research of Image Quality Assessment (IQA) is of much significance in numerous applications. The most reliable way for IQA is subjective evaluation; however, it's expensive and time consuming. Hence, an objective image quality method consistent with the subjective perception is in demand.

Objective quality metrics can be divided into reference [7, 16] and no-reference [1–4, 8, 14–16] methods, depending on whether the method require a reference image for comparison. In most practical cases, a reference image is not available. So we propose a new no-reference IQA metric in this work, which does not demand that any original image be present or even exist.

Existing NR-IQA algorithms can be broadly classified as (1) Natural Scene Statistic (NSS) based approaches [1–3, 15] and (2) Training-based approaches [4, 9]. NSS based approaches are based on the hypothesis that natural scenes possess certain statistical properties after the long time evolution, and those properties will be disrupted by the presence of distortions. Many statistical models have been built under spatial [2, 3], wavelet [1] or cosine [15] domain. In this paper, we study the efficacy of utilizing a powerful image descriptor, the contourlet transform, to construct a no-reference

IQA model. Thus an improved NSS model is established under contourlet domain.

The rest of the paper is organized as follows. In Section 2 we review previous work in NR IQA and introduce the contourlet transform. In Section 3 we describe the features extracted from images to be trained on the dataset. In Section 4 we describe the classification and regression process used to predict image quality. Experimental results and a thorough analysis of our results are presented in Section 4. Finally, Section 5 concludes with a summary of our work.

2 RELATED WORK

2.1 *No-Reference Image Quality Assessment*

Recently, many NR-IQA algorithms has been proposed. Tang et al. [9] proposed a Learning-Based Blind Image Quality (LBIQ) measure that combines and incorporates numerous low-level image quality features stemming from natural image measures and texture statistics with a regression algorithm. Li et al. [4] developed a NR method based on General Regression Neural Network (GRNN), which assess image quality by approximating a function relationship between features and subjective mean opinion scores using GRNN. Ye and Doermann [14] proposed a block based NR IQA method which it use raw-image-patches extracted from a set of unlabeled images to learn a dictionary in an unsupervised manner. Then learns mappings from the quantization feature space to image quality scores using support vector machine.

NSS based approaches are based on the hypothesis that natural scenes possess certain statistical properties which will be affected by the presence of distortion. Different NSS-based NR algorithms often have different features extracted and different training methods applied. Anish Mittal et al. [2] proposed a NSS-based NR IQA algorithm called BRISQUE which operates in the spatial domain. The features used derive from the empirical distribution of locally normalized luminance and products of locally normalized luminance under spatial domain makes it have a low computational complexity and well suited for real time applications. BIQI [1], another state-of-art NSS-based NR-IQA algorithm, which build a 2-stageframework for blind image quality indices. It extracts a set of 18 features in wavelet domain from each image. Given a distorted image, the algorithm estimates the presence of a set of distortions by probability in the image. Then, the quality of the image is then expressed as a probability-weighted summation. Recently, Lixiong Liu et al. [12] found that the curvelet transform has rich information of scale and orientation and can change its own window size to extract the scale information without image scaling, making it well suited for extracting features highly relevant to natural image quality. Operating within a 2-stage framework introduced by Moorthy et al. [1], features are used to classify the distorted image into one of finite set of possible distortions. Then evaluate the distortion-specific quality of the image by the same set of features. Our work is inspired by this paper.

2.2 Contourlet transform

The contourlet transform is a new extension to wavelet transform in two dimensions using nonseparable and directional filter banks, Which is proposed by Do and Vetterli [13]. With rich set of basis images oriented at varying directions in multiple scales, Contourlet transform can effectively capture the smooth contours, which are the dominant features in natural images, with only a small number of coefficients. It combines the multi-scale and multi-direction decomposition together, employs Laplacian Pyramid (LP) to capture the point discontinuities and then uses Directional Filter Banks (DFB) to link these point discontinuities into linear structures. Figure 1 shows the framework of contourlet decomposition which applies the LP and DFB in succession to generate subbands with different frequencies and directions. It shows great advantages for constructing a sparse expansion for typical

images because of its intrinsic characteristics, such as good spatial localization, good direction selectivity, anisotropy. In essence, it first uses a wavelet-like transform for edge detection, and then a local directional transform for contour segment detection. Thus the contourlet transform provides a sparse representation for two-dimensional piecewise smooth signals that resemble images. The resulting image expansion is a directional multi-resolution analysis framework composed of contour segments, which makes it overcome the challenges of wavelet and curvelet transform. An example of the contourlet decomposition of the "caps.bmp" image is shown in Figure 2.

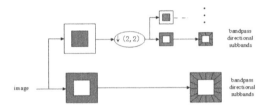

Figure 1. Contourlet filter bank. First, a multi-scale decomposition into octave bands by the LP is computed, and then a DFB is applied to each bandpass channel.

Figure 2. Contourlet decomposition of caps.bmp in three scales with 8 subbands in each scale.

3 EXTRACTED FEATURES

Feature selection is of fundamental importance to the creation of NR IQA algorithms, the features should be consistent to the degree of image distortion and be independent of image content. The NSS features used in [1–3, 8–9, 15–16] achieve this goal and deliver good image quality predictive power. Low level image features like pixel or high level image features like semantic features can be deployed to model construction. Here, we employ intermediate-level image features extracted from the contourlet image transform. Our work based on the hypothesis that these mid-level features capture characteristics arising in low-level NSS models in a localized way and consequently capture perceptual image distortions in a content-independent way.

3.1 Contourlet NSS

From the LIVE database, we chose a natural image and its distortion versions of 5 different distortion types to conduct experiment.

For an image, we process four scalar LP decomposition and 8 directional DFB decomposition for each scale. Because the contourlet coefficients at finer scales represent the high frequency components of an image and image distortion often affect them as a way that blur attenuates the high frequencies while noise may increase them. To model the marginal distribution of the finest scale coefficients with larger amplitude more effectively, we calculate he empirical Probability Distribution Function (PDF) of the logarithm (base 10) of the magnitude of the contourlet coefficients at scale 4.

$$h(x) = \mathrm{pdf}(\log_{10}(|\theta_j|)) \tag{1}$$

where θ_j is a set of contourlet coefficients of a random subband at scale 4.

Figure 3 plots $h(x)$ for the images we selected.

From the curve of original image we can see that a peak appears near the zero coefficient and the asymmetry and the tail behavior in $h(x)$ can be easily found, which indicate the sparsity of the contourlet coefficients. While the distortions change the shape of the curve. We use an Asymmetric Generalized Gaussian Distribution Model (AGGD) to fit the curve of $h(x)$.

$$f(x; \alpha, \mu, \sigma_l, \sigma_r) =$$
$$\begin{cases} \dfrac{\alpha}{2(\beta_l + \beta_r) + \Gamma(1/\alpha)} \exp\left(-\left(\dfrac{x-\mu}{\beta_l}\right)\right), x < 0 \\[3mm] \dfrac{\alpha}{2(\beta_l + \beta_r) + \Gamma(1/\alpha)} \exp\left(-\left(\dfrac{x-\mu}{\beta_r}\right)\right), x \geq 0 \end{cases} \tag{2}$$

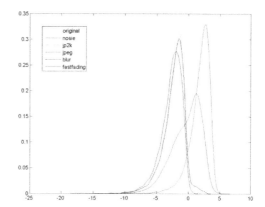

Figure 3. PDF of the logarithm of magnitude of the contourlet coefficients of a subband at finest scale 4 for image.

where

$$\beta_l = \sigma_l \sqrt{\dfrac{\Gamma\left(\dfrac{1}{\alpha}\right)}{\Gamma\left(\dfrac{3}{\alpha}\right)}} \tag{3}$$

$$\beta_r = \sigma_r \sqrt{\dfrac{\Gamma\left(\dfrac{1}{\alpha}\right)}{\Gamma\left(\dfrac{3}{\alpha}\right)}} \tag{4}$$

The shape parameter α controls the "shape" of the distribution and (σ_l, σ_r) are scale parameter that controls the spread on each side of the mode, respectively. As a tool deployed to model skewed heavy-tailed distributions of image texture, the parameters of AGGD $(\alpha, \mu, \beta_l, \beta_r)$ are estimated using the moment-matching based approach proposed in [11]. We use the fitting parameters $(\alpha, \mu, \beta_l, \beta_r)$ as the features.

As is shown in Figure 3, the presence of many kinds of distortion changes the distribution of the pristine image. In the case of additive white noise, the noise like some kind of contour and increase the high frequency component, so the peak shifts to the right. While the situation is reversed for blur, the blurriness destroy the contour with a way of attenuate the high frequency component, hence the peak shifts to the left. From Figure 3, we can see the distortions affect the contourlet coefficients' distribution characteristics remarkably, which means these features correlate well with human judgments of quality.

3.2 Orientation energy distribution

As a directional multi-resolution analysis transform, contourlet has rich orientation information on images and their distortions. Whereas image distortion can modify the orientation energy distribution in an unnatural manner [1,8], and these changes can be perceived by cortical neurons sensitively.

Through 3 scalar LP decomposition followed by 8 directional DFB decomposition, we get 24 subbands for each image. We use the mean magnitude of the coefficient in each orientation matrix to capture the orientation energy, and in consideration of simplicity, we use combination likes (2–7), (3–6), (1–8), (4–5) in the same scale, so we select 12 orientation metrics as features to capture orientation energy.

$$E = E(|\theta_1|) \tag{5}$$

where $E(x)$ is the mean of x and θ_1 is a set of coefficients of the orientation matrix with orientation index is 1 ($l = 1, 2, …, 12$).

Figure 4 plots the orientation energy distribution for each image.

As is shown in Fig. 5, we can observe that white noise changes the distribution of energy by the way of increasing all the magnitude randomly and

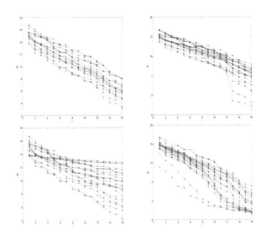

Figure 5. The statistic of the distribution of contourlet coefficients in salient regions of different distorted types. (a) original image, (b) JPEG2000 compressed image, (c) white noise image, (d) Gaussian blur image.

weakening the difference between them. Which can be explained that the white noise add more contours to the original image generally, but homogenize the orientation energy distribution at the same time. By contrast, Blur weakens the magnitude generally, while make the peak which can be treated as the cardinal of the image more outstanding. Blur weaken the contour of the image, but sharp the cardinal peaks should account for this change. What's more, the more white noise and more blur makes the change more distinct.

As has found by previous studies, image distortion will destroy the anisotropy of the natural image. We treat the stem plot as a histogram, and use the μ and cv to capture this variation.

$$cv = \sigma/\mu \tag{6}$$

where μ and σ are the mean and standard deviation of the orientation energies.

3.3 Salient region distribution

The distribution of contourlet coefficients show some characteristics like AGGD, which reflects the structure correlation in the contourlet domain and the similarity between the scales. While the image quality is closely relevant to the structure of the image, the variation of the NSS of contourlet coefficients will definitely reveal the degree of image distortion. For example, blur and jpeg2000 weaken the high frequency component, and make coefficients of the finer scale subbands draw close to zero. By contrast, white noise increase the coefficients of the contourlet. Hence, these distortions

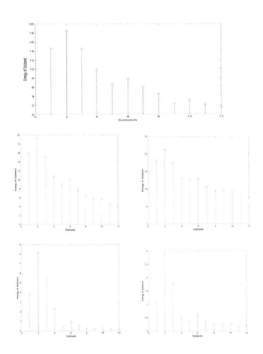

Figure 4. Orientation energy distribution of the caps.bmp image and its distortion versions. (a) Original image, (b) noise 1 distorted image with DMOS = 33.54, (c) noise 2 distorted image with DMOS = 41.56, (d) blur 1 distorted image with DMOS = 60.83, (e) blur 2 distorted image with DMOS = 69.15.

will change the statistics characteristic of image in contourlet domain.

However, to attach more importance to distortions in salient regions of image according to Human Visual System (HVS), we select the texture region of image to extract features. We process a 4 scalar 8 directional contourlet transform just like above, then analyze the subbands in the finest scale namely the fourth scale firstly. We extract the coefficients whose absolute value is greater than the mean of all the each subband coefficient as the texture region, then use the logarithmic mean of the texture region as the mean of the subband. As to the coarser scale, the salient region can be available by interpolation scale by scale. After studying the distribution of this value of all the subbands, we find that it present a linear distribution when the image is pristine. By contrast, we do the same analyze in the images of distorted images which has been selected in LIVE image database. As is shown in Figure 5, distortion will destroy the linear characteristic in natural image.

In Figure 5, I (1, 2, …, 16) is the index of the subbands. From Fig. 6 we can see, the mean values of the JPEG2000 image drop, but white noise increases the mean value generally, and blur reduces the mean value and makes it more disperse. To capture this change in texture region, we calculate the logarithm of magnitude of the contourlet coefficients of each subband as follows.

$$T_i = mean(\log_{10}(|C_i|)) \tag{7}$$

where C_i is a set of the coefficients in the texture region of each subband indexed by i.

Thus, so far we defined a series of features (see Table 1 which are extracted under contourlet domain and described how image distortion affect these features.

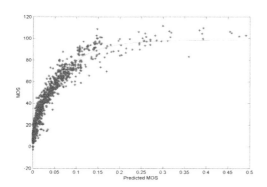

Figure 6. The scatter diagram of predicted MOS and MOS.

Table 1. Features extracted in contourlet domain.

Feature vector	Feature description
$f_1 - f_4$	AGGD model fitting parameters of the histogram of logarithm of magnitude of contourlet coefficients
$f_5 - f_6$	Description of the orientation energy distribution
$f_7 - f_{22}$	Description of distribution of the mean of logarithm of magnitude in texture region

4 LEARNING IMAGE QUALITY EVALUATION

While each feature is motivated by the NNS properties of image distortions in contourlet domain and carefully designed, we cannot expect each individual feature would work all well across all the distortion type. Hence we utilize a 2-stage framework [1] to establish a mapping from the extracted features to image quality evaluation, which firstly uses a series of features to classify images into distortion categories with probability, then evaluate the degree of each kind of distortion in the image and finally give the predictive score with a linearly combination.

In our work, we employ the latest LIBSVM-3.17 package [5] to perform our experiment. We use the LIVE image quality assessment database [10]. Firstly, we randomly select 20 reference images and their associated distorted versions for training and the rest 14 reference images and their associated distorted versions for testing. In stage 1 of distortion type classification, we utilize the C-SVC with the Radial Basis Function (RBF) kernel to train the classifier, which calculate each probability p_i {$i = 1, …, 5$} that the distortion type of the distorted image belongs to. The second stage evaluates the image quality along each of these distortions. We firstly train regression models for each category of distorted images which based on the v-SVR with RBF kernel. Then we obtain n regression models to yield an n-dimensional quality prediction vector q_i {$i = 1, …, 5$}. Thus the quality of the image can be expressed as a probability-weighted summation.

$$Q = \sum p_i * q_i \tag{8}$$

where $i = 1, …, 5$.

5 EXPERIMENTAL RESULTS

We validate our algorithm on the laboratory for LIVE database [10]. This database contains 29

Table 2.   Median Root-Mean-Squared-Error (RMSE) between algorithms and MOS.

	JP2K	JPEG	NOISE	BLUR	FF	ALL
PNSR	8.2834	9.8265	3.2476	10.0246	8.1582	10.2831
BIQI	13.8621	17.2308	5.4573	9.5932	17.3543	16.3227
BRISQUE	9.1386	10.2791	3.8152	8.2106	9.5738	8.9632
Proposed	7.5205	10.0326	4.7935	4.6833	10.2451	8.8401

Table 3.   Median Pearson Linear Correlation Coefficient (LCC) between algorithms and MOS.

	JP2K	JPEG	NOISE	BLUR	FF	ALL
PNSR	0.8265	0.8617	0.9548	0.7724	0.8479	0.8242
BIQI	0.8642	0.7571	0.9536	0.8925	0.7345	0.7526
BRISQUE	0.9394	0.9255	0.9674	0.9596	0.9021	0.9271
Proposed	0.9281	0.9017	0.8933	0.9689	0.8932	0.9022

Table 4.   Median Spearman's Rank Ordered Correlation Coefficient (SROCC) between algorithms and MOS.

	JP2K	JPEG	NOISE	BLUR	FF	ALL
PNSR	0.8796	0.8563	0.9676	0.8174	0.8865	0.8178
BIQI	0.8415	0.7814	0.9721	0.9284	0.7561	0.7540
BRISQUE	0.9367	0.9215	0.9719	0.9512	0.8895	0.9365
Proposed	0.9210	0.9207	0.9277	0.9655	0.8512	0.9197

high-resolution 24 bits/pixel RGB color images as reference and corresponding five types of distorted images: 175 JPEG, 169 JPEG2000, 145 White Noisy (WN), 145 Gaussian Blurred (Gblur) and 145 Fast-Fading (FF) Rayleigh channel noisy images at a range of quality levels. The Difference Mean Opinion Scores (DMOS) of the image is provided to describe the subjective quality of the degraded image. To demonstrate that our algorithm is independent of content and specific distortion severity we randomly divided our overall database into the training set (80% of the overall database) and the testing set (20% of the overall database).

Over such train-test sets, we conduct 1000 times experiments with random permutations of the LIVE database. We get the result of the median classification accuracy was 84.3652% (mean = 84.3652%, std. Dev. = 2.9804%). For the performance of the quality estimate, we measure three indices to evaluate, including Root Mean Square Error (RMSE), Pearson Linear Correlation Coefficient (PLCC), and Spearman Rank Order Correlation Coefficient (ROCC). Figure 6 shows the scatter diagram of the predicted MOS value and MOS value. Tables 2, 3, 4 shows the median value of above indices over 1000 trails in Figure 6. For comparison, we also list the median values for the Peak Signal to Noise Ratio (PNSR) metric (a FR-IQA metric), the BIQI metric, and the BRISQUE metric.

From the experimental results, we can see that our proposed algorithm performs better comparing to PSNR and BIQI, and achieves the performance close to the state-of-the-art NR-IQA algorithm BRISQUE. Because the features employed in our algorithm based on contourlet coefficients, it works quite well for JPEG compression and blur distortion which affect the contours distinctly. However the performance for JPEG2000 and FF distortion is less impressive. Future work will involve improving the SVR measure of quality for JEPG2000 and FF distortion images.

6   CONCLUSION

We proposed a new NR-IQA methods that based on contourlet domain. It achieve an improved performance to do image quality assessment without any reference by capturing the NSS characteristics in an image. The image is decomposed by contourlet transform into multi-scale and multi-directional subbands. We extract a set of features to capture the image structural information, especially attach more importance in the salient region. The final objective quality score is calculated upon a 2-stage framework, BIQI.

Our trails demonstrate our algorithm correlates well with human perception, and is promising to apply in the real-time applications.

REFERENCES

[1] A.K. Moorthy and A.C. Bovik 2010. A two-step framework for constructing blind image quality indices. IEEE Signal Process vol.17, no. 5: pp. 513–516.

[2] A.K. Moorthy and A.C. Bovik 2011a. Blind image quality assessment: From natural scene statistics to perceptual quality. IEEE Trans. Image Process vol. 20, no. 12: pp. 3350–3364.

[3] Anish Mittal and A.K. Moorthy and A.C. Bovik 2011b. Blind/Referenceless Image Spatial Quality Evaluator. Signal, systems and computers: pp. 723–727.

[4] C. Li, A.C. Bovik, and X.Wu 2011. Blind image quality assessment using a general regression neural network. IEEE Trans. Neural Netw vol. 22, no. 5: pp. 793–799.

[5] C. Chang, C. Lin 2001. LIBSVM: a library for support vector machines. http://www.csie.ntu.edu.tw/cjlin/libsvm.

[6] D.J. Field 1987. Relations between the statistics of natural images and the response properties of cortical cells. J. Opt. Soc. Am. A: Opt. Image Sci vol. 4, no. 12: pp. 2379–2394.

[7] H.R. Sheikh, A.C. Bovik, and G. de Veciana 2005. An information fidelity criterion for image quality assessment using natural scene statistics. IEEE Trans. Image Process vol. 14, no. 12: pp. 2117–2128.

[8] H.R. Sheikh, A.C. Bovik, L. Cormack 2005. No-reference quality assessment using natural scene statistics: JPEG2000. IEEE Trans. Image Process vol. 14, no. 11: pp. 1918–1927.

[9] H. Tang, N. Joshi, and A. Kapoor. 2011. Learning a blind measure of perceptual image quality. Proc. IEEE Conf. Comput. Vis. Pattern Recog: pp. 305–312.

[10] H.R. Sheikh, Z. Wang, L. Cormack, A.C. Bovik LIVE Image Quality Assessment Database Release 2 (online). http://live.ece.utexas.edu/research/quality.

[11] K. Sharifi and A. Leon-Garcia 1995. Estimation of shape parameter for generalized Gaussian distributions in subband decompositions of video. IEEE Trans. Circuits Syst. Video Technol vol. 5, no. 1: pp. 52–56.

[12] Lixiong Liu, Hongping Dong, Hua Huang and Alan C. Bovik 2014. No-reference image quality assessment in curvelet domain Signal Processing: Image Communication: pp. 494–505.

[13] M.N. Do, M. Vetterli The contourlet transform: an efficient directional multiresolution image representation. IEEE Transactions on Image Processing vol. 14, no. 12: pp. 2091–2106.

[14] P. Ye and D. Doermann 2011. No-reference image quality assessment based on visual codebook. Proc. IEEE Int. Conf. Image Process: pp. 3089–3092.

[15] T. Brandao, M.P. Queluz 2008. No-reference image quality assessment based on DCT-domain statistics. Signal Process vol. 88, no. 4: pp. 822–833.

[16] W. Lu, K. Zeng, D. Tao, Y. Yuan, and X. Gao. No-reference image quality assessment in contourlet domain. Neurocomputing vol. 73, no. 4–6: pp. 784–794.

[17] Z. Wang, A.C. Bovik, H.R. Sheikh, and E.P. Simoncelli. 2004. Image quality assessment: From error visibility to structural similarity. IEEE Trans. Image Process vol. 13, no. 4: pp. 600–612.

Future Information Engineering and Manufacturing Science – Zheng (Ed)
© 2015 Taylor & Francis Group, London, ISBN 978-1-138-02644-5

Blind image quality assessment based on saliency extraction and contourlet transformation

Yaning Wang & Jie Wang
Beijing University of Posts and Telecommunication, Beijing, China

ABSTRACT: In this paper, we present a blind image quality assessment based on saliency extraction and Contourlet transformation (BIQSC). The BIQSC first uses a method to extract the salient regions from the image and gives more emphasis on the salient regions. By combining Contourlet transform with a version of the hidden Markov mode—non-Gaussianty, the marginal distributions of neighbor coefficients in the Contourlet domain are modeled. With the Contourlet transform, the marginal histogram of coefficients in each subband can be well fitted by Asymmetric Generalized Gaussian Distribution (AGGD) after divisive normalization transforming. The parameters extracted from the Asymmetric Generalized Gaussian Distribution (AGGD) will be selected as the feature parameters. Experiments shows that the proposed metric has good consistency with the human subjective perception.

Keywords: Contourlet Transformation; Saliency; BIQSC; SROCC; PLCC

1 INTRODUCTION

In recent years, the need to accurate and reliable objective Image Quality Assessment (IQA) metrics has become more pressing with the advent of new applications and services [1]. Objective image quality assessment can be divided into three categories [2]: Full-Reference (FR), Reduce-Reference (RR) and No-Reference (NR). Both FR and RR are essentially image similarity or fidelity measurement since they need the original undistorted image as a reference. However, in many practical applications, an image quality assessment system does not have access to the reference images. Therefore, the Blind Image Assessment (BIQA) which can evaluate image quality without using any reference image information, has become the focus of the research.

Now the majority of image quality assessment are based on the Natural Scene Statistics (NSS), and these methods have good performance. But these methods didn't consider that when we watch an image we put more attention to the salient regions and the distortion of the non-salient regions has little influence of the quality of the whole image. So before assessing the image quality we extract the salient regions to make improvement. In our framework, the method of extracting the salient regions can be regardless of the type of the distortion and it can work when the image is seriously distorted.

The Contourlet transform is a new two-dimensional extension of the wavelet transform using multi-scale and directional filter banks. The Contourlet expansion is composed of basis images oriented at various directions in multiple scales, with flexible aspect ratios. Given this rich set of basis images, the Contourlet transform effectively captures smooth contours that are the dominant feature in natural images [7]. So we decide to use the Contourlet transform to assess the quality of the image.

The rest of the paper describes the model in detail and is organized as follows. In Section 2, previous work on saliency detection and Contourlet transform is briefly reviewed. In Section 3, we present framework of the proposed quality evaluation metrics. And we conduct the thorough experiments on the Laboratory for Image and Video Engineering (LIVE) database II and the CSIQ database. Section 4 concludes this paper.

2 RELATED WORK

2.1 Saliency extraction

Visual saliency is the perceptual quality that makes an object, person, or pixel stand out relative to its neighbors and thus capture our attention [3]. Detection of visual salient image regions is useful for applications like object segmentation, adaptive compression, and object recognition. Saliency estimation methods can broadly be classified as: top-down method and bottom-up method. The top-down model is used for searching some specific targets. But the bottom-up method can be used to extract the salient regions for any image. The bottom-up method uses one or more features of intensity, color, and orientation to determine contrast of image regions relative to their surroundings [4].

In this paper, we used a bottom-up method. This method uses a Spectral Residual (SR) approach to construct the corresponding saliency map in spatial domain [5].

Given an image I(x), we use the SR to calculate the saliency map S(x), and the size of S(x) is 64×64. With a bilinear interpolation, we get a new saliency map S'(x) with the same size as input image. We separate I(x) into N different $B \times B$ image patches $P = [P_1, P_2, ..., P_N]$. Then calculate the saliency s_{pi} of each patch P_i,

$$s_{pi} = \sum_{x \in pi} s'(x), i = 1, 2, 3, ..., N \tag{1}$$

Local descriptors $H = [h_1, h_2, ..., h_M]$ of the input image I(x) are chosen from P, where $s_{pi} >$ threshold

$$\{h_i\} = \{p_j \mid s_{pi} > threshold\}, i = 1, 2, ..., M; \\ j = 1, 2, ..., N; \tag{2}$$

In the field of object detection, the selection of threshold is a trade-off problem between false alarm and neglect of objects. In [12], Hou and Zhang set threshold $E(S(x)) \times 3$ to detect proto-objects in a saliency map, where $E(S(x))$ is the average intensity of the saliency map. In this case, the threshold $> E(S(x))$.

$$\{h_i\} = \left\{ p_j \mid s_{pj} > \frac{1}{N} \sum_k^N s_k \right\}, i = 1, 2, ..., M; \\ j = 1, 2, ..., N; \tag{3}$$

Then we normalize the chosen patches by subtracting the mean and dividing by the standard deviation of its elements. Figure 1 shows that the SR method can find salient regions correctly when the image is serious distorted.

2.2 Contourlet transform and non-Gaussianity

Contourlet transform has features of multi-resolution representation, localized analysis and direction-sensitive [6]. Before discussing the behaviors of the Contourlet coefficients, we define some important Contourlet coefficient relationships as depicted in Figure 2. For each Contourlet coefficient X, we define its eight adjacent coefficients in the same subband as its neighbors. Next, the coeffi-

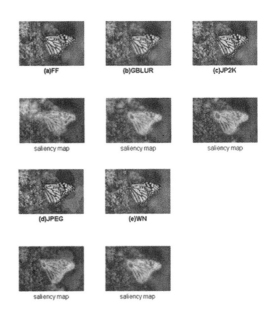

Figure 1. Local feature extracted from image with five distortions. (a) FF (fastfading), (b) GBLUR (Gaussian-blurring), (c) JP2K, (d) JPEG, (e) WN (white Gaussian noise).

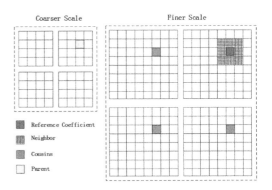

Figure 2. The contourlet coefficients relationships.

cient in the same spatial location of the coarse scale is defined as its parent, and those in the same spatial location of the finer scale are its children. Note that each child has one parent and each parent has four children. We also define coefficients at the same scale and spatial location but in different directions as cousin. Combing the relationships across scales, space, and directions, we refer to the collective set of parent, neighbors, and cousins of each coefficient X as its generalized neighborhood [17].

The subband marginal distributions of natural images in the Contourlet domain are highly non-Gaussian but conditionally Gaussian conditioned [7]. This NG has been applied to various image

270

processing problems, such as noise removal and compression.

Once the image is distorted, the marginal distribution will deviate from the original one. Figure 3 shows the marginal distribution of the reference image and the distorted images with the five types of distortions. From the Figure 3 we can see the marginal distribution of the distortions is not a zero mode, so we can use the mean value μ of the marginal distribution as a feature. And we can also see the distortions change the marginal distribution of the image. So we can use the coefficients that can describe the marginal distribution of the image as the features. The products of neighboring coefficients are well-modeled as following a zero mode Asymmetric Generalized Gaussian Distribution (AGGD) [8]:

$$f\left(x; \gamma, \sigma_l^2, \sigma_r^2\right) =$$

$$\begin{cases} \dfrac{\gamma}{(\beta_l + \beta_r)\Gamma\left(\dfrac{1}{\gamma}\right)} exp\left(-\left(\dfrac{-x}{\beta_l}\right)^{\gamma}\right) & x < 0 \\\\ \dfrac{\gamma}{(\beta_l + \beta_r)\Gamma\left(\dfrac{1}{\gamma}\right)} exp\left(-\left(\dfrac{-x}{\beta_r}\right)^{\gamma}\right) & x > 0 \end{cases} \quad (4)$$

where

$$\beta_l = \sigma_l \sqrt{\dfrac{\Gamma\left(\dfrac{1}{\gamma}\right)}{\Gamma\left(\dfrac{3}{\gamma}\right)}} \quad (5)$$

$$\beta_r = \sigma_r \sqrt{\dfrac{\Gamma\left(\dfrac{1}{\gamma}\right)}{\Gamma\left(\dfrac{3}{\gamma}\right)}} \quad (6)$$

The shape parameter γ control the 'shape' of the distribution while σ_l^2 and σ_r^2 are scale parameters that control the spread on each side of the mode, respectively. So we use the parameters of $(u, \gamma, \sigma_l^2, \sigma_\gamma^2)$ as the NG features. Those parameters can be efficiently estimated using the moment-matching based approach.

For the Contourlet transform, we use the method proposed by Cohen and Daubechies. It is the 9–7 biorthogonal filters for the multi-scale decomposition stage and the CD filters for the multidirectional decomposition stage. The number of scale is three. Then all these scales are divided

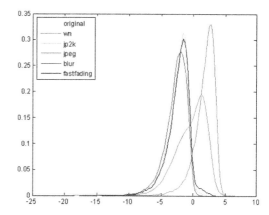

Figure 3. The marginal distribution of original image and five type distortion images.

into four directional subbands, respectively. For each subband, we use 13 neighboring coefficients, including 9 from the same subband, 1 from the parent band, and 3 from the same spatial location in the other orientation bands at the same scale. Total of 48 features are extracted for a reference image.

3 EXPERIMENT AND RESULTS

In this section, we will implement the proposed algorithm. To examine the performance of our metric, we use the LIVE database II [9]. There is five datasets of 982 subject-rated images with five types of distortions at different distortions levels in the LIVE database II. The distortion types include JPEG2000 compression, JPEG compression, Gaussian blur, white noise contamination and fast fading. The Figure 4 shows the framework of our metric. First we extract the salient regions from the distorted image, and cut the image to two patches: the salient region and non-salient region. Second we conduct the Contourlet transform and extract the features of the NG of the salient region and non-salient region. Third we use the NSS extracted from the natural images from the LIVE database II to assessing the distortion levels of the salient and non-salient regions and give the score. Finally we use the equation (7) to get the quality of the distorted image.

$$Q = \alpha Q_s + \beta Q_{ns} \quad (7)$$

In the equation (7), the Q_s is the salient region's quality which is calculated by comparing with the NSS of natural image in Contourlet domain. And the Q_{ns} is the non-salient region's quality. We used the weighting parameters α and β to get the quality of distorted image. The sum of the weighting parameters α and β is equal to 1.

VQEG provides the comparison criterion among the metrics [10]. So we use the two criterions to assess the performance of our metric. The Spearman Rank Ordered Correlation Coefficient (SROCC) is showed in the Table 1. The Pearson linear correlation coefficient is showed in the Table 2. In the two tables, we also tabulated the performance of three full-reference indices: Peak–Signal-to-Noise Ratio (PSNR), Structural Similarity Index (SSIM) [11] and Multi-Scale Structural Similarity Index (MS-SSIM) [12]. The PSNK is always used as the benchmark. The SSIM and the MS-SSIM have good performance. At the same time, we also tabulated the performance of some no-reference indices: CBQI [13], LBIQ [14] and BLIINDS-II [15]. From the Table 1 and Table 2, we can know that the performance of the BIQSC has a certain degree of improvement.

To verify that the BIQSC is not just work well on the LIVE database II, we also test the BIQSC on the CSIQ database [16]. We select the 4 distortion types from the CSIQ database. The 4 distortion types include JPEG2000 compression, JPEG compression, Gaussian blur and white noise contamination. In the Table 3 and Table 4, we used the SROCC and PLCC to test its performance. From the Table 3 and Table 4, we can see the performance of the BIQSC on the CSIQ has a relatively small decline comparing with the performance on the LIVE database II. So the BIQSC performs well in terms of correlation with human perception of quality and the performance does not depend on the database.

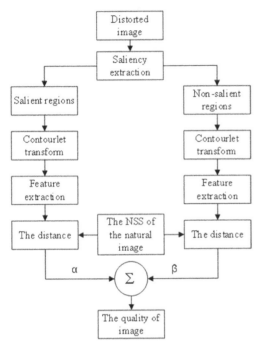

Figure 4. The framework of the proposed metric.

Table 3. The spearman rank ordered correlation coefficient on the CSIQ database.

	JP2K	JPEG	WN	BLUR	ALL
BIQSC	0.8567	0.8771	0.8741	0.9259	0.8835

Table 1. The spearman rank ordered correlation coefficient on the LIVE IQA database.

	JP2K	JPEG	WN	BLUR	FF	ALL
PSNR	0.8646	0.8831	0.9410	0.7515	0.8736	0.8636
SSIM	0.9389	0.9466	0.9635	0.9046	0.9393	0.9129
MS-SSIM	0.9627	0.9785	0.9773	0.9542	0.9386	0.9535
CBIQ	0.8935	0.9418	0.9582	0.9324	0.8727	0.8954
LBIQ	0.9040	0.9291	0.9702	0.8983	0.8222	0.9063
BLIIDS-II	0.9323	0.9331	0.9463	0.8912	0.8519	0.9124
BIQSC	0.9192	0.9235	0.9425	0.9150	0.9050	0.9210

Table 2. The pearson linear correlation coefficient on the LIVE IQA database.

	JP2K	JPEG	WN	BLUR	FF	ALL
PSNR	0.8762	0.9029	0.9173	0.7801	0.8795	0.8592
SSIM	0.9405	0.9462	0.9824	0.9004	0.9514	0.9066
MS-SSIM	0.9746	0.9793	0.9883	0.9645	0.9488	0.9511
CBIQ	0.8898	0.9454	0.9533	0.9338	0.8951	0.8955
LBIQ	0.9103	0.9345	0.9761	0.9104	0.8382	0.9087
BLIIDS-II	0.9386	0.9462	0.9635	0.8994	0.8790	0.9164
BIQSC	0.9137	0.9092	0.9250	0.9490	0.9074	0.9208

Table 4. The pearson linear correlation coefficient on the CSIQ database.

	JP2K	JPEG	WN	BLUR	ALL
BIQSC	0.8691	0.8956	0.8861	0.9229	0.8934

4 CONCLUSION

In this paper, we first point out the value of extracting salient regions, and we use a method to extract the salient regions regardless of the distortions. To adjust the HVS, we consider the salient region and the non-salient region respectively, and we give the more emphasis to the salient regions. Second we used the method of Contourlet transform to assess the distortion levels of images. From the Tables 1–4, we can get the conclusion that the BIQAC can perform well.

REFERENCES

[1] Z. Wang and A.C. Bovik. 2006. Modern Image Quality Assessment, Morgan & Clavpool Publishers, Mar.

[2] Wang, Z., et al. 2004. Image quality Assessment: From error visibility to structural similarity. Image Processing, IEEE Transactions on.13(4): p. 600–612.

[3] Niebur, E and C. Koch. 1998. Computational architectures for attention. The attentive brain: p.163–186.

[4] Xiaodi, H. and Z. liqing. 2007. Saliency Detection: A Spectral Residual Approach. In Computer Vision and Pattern Recognition. CVPR'07. IEEE Conference on. 2007.

[5] Zhang Hong. And Feng Ren, Saliency-based feature learning for no-reference image quality assessment. IEEE International Conference on Green Computing and Communications and IEEE Internet of Things and IEEE Cyber, Physical and Social Computing.

[6] M.N. Do and M. Vetterli. 2005. The Contourlet transform: an efficient directional multiresolution image representation, IEEE Transactions Image on Processing, 14(12): 2091–2106.

[7] D.D. –Y. Po and M.N. Do. 2006. Directional multiscale modeling of image using the Contourlet transform, IEEE Trans. On Image Processing, 15(6):1610–1620.

[8] Lasmar, N. E Stition, Y. and Berthoumien, Y. 2009. Multisacale Skewed heavy tailed model fot texture analysis, in IEEE Internat'l Conf Image Process, 2281–2284.

[9] H.R. Sheikh, Z. Wang, A.C. Bovik, Cormack, Image and video quality assessment research at LIVE, http://live.ece.utexas.edu/reasearch/quality/.

[10] VQEG. Validation of reduced-reference and no-reference objective models for standard definition television, phase i, 2009. Available Online: http://www.vqeg.org/.

[11] Z. Wang, A.C. Bovik, H.R. Sheikh, and E.P. Simoncelli. Apr. 2004. Image quality assessment: From error visibility to structural similarity, IEEE Trans. Image Process, vol. 13, no. 4, p. 600–612.

[12] Z. Wang, E.P. Simoncelli, and A.C. Bovik. 2003. Multiscale structural similarity for image quality assessment, in Proc. Asilomar Conf. Signals, Syst. Comput., vol. 2. pp. 1398–1402.

[13] P. Ye and D. Jul. 2011 Doermann, No-reference image quality assessment using visual codebooks, in Proc. IEEE Int. Conf. Image Process, pp. 3129–3138.

[14] H. Tang, N. Joshi, and A. Kapoor. Jun. 2011. Learning a blind measure of perceptual image quality, in Proc. Int. Conf. Comput. Vis. Pattern Recognit. pp. 305–312.

[15] M. Saad, A.C. Bovik, and C. Charrier. Aug. 2012. Blind image quality assessment: A natural scene statistics approach in the DCT domain, IEEE Trans. Image Process, vol. 21, no. 8, pp. 3339–3352.

[16] E.C. Larson and D.M. Chandler. March 2010. Most apparent distortion: full-reference image quality assessment and the role of strategy, Journal of Electronic Imaging, 19 (1). PDF bibtex.

[17] Xu, Wang and Gangyi Jiang. 2009. Reduced Reference Image Quality Assessment Based on Contourlet Domain and Natural Image Statistics. 2009 Fifth International Conference on Image and Graphics.

Future Information Engineering and Manufacturing Science – Zheng (Ed)
© *2015 Taylor & Francis Group, London, ISBN 978-1-138-02644-5*

Based on fuzzy FMEA and FMECA of aircraft fuel system compare analysis

Ruili Zhang, Yanming Yang, Yang Gao & Yue Teng
Naval Aeronautical Engineering Institute Qingdao Branch, Qingdao, China

ABSTRACT: The paper organized the failure of an aircraft fuel system parts in recent years, established fuzzy FMEA and failure modes, and effects and criticality analysis, based on the severity of impact and probability of occurrence faults, sorting to each fault impact of the fuel system, according to the possible dangers, the maintenance department taken corrective measures or repair strategies to improve equipment reliability.

Keywords: failure mode; fuzzy theory; risk priority; criticality analysis

1 INTRODUCTION

Currently, the complex system maintenance and support mode change significantly in the world, to carry out intelligent maintenance prediction, diagnosis and risk analysis has become an important goal of quality management, Attention to equipment failure modes and effects analysis in the use phase, it has an important impact for the design of effective intelligent maintenance program, reduce the maintenance staff and reform the maintenance mode, while providing meaningful reference to production and product improvement and new product development.

Failure mode and effects analysis is able to analyze and identify a variety of potential failure modes of a product in using process, determine their priority, calculate the risk priority value through systematic analysis of all possible failure modes, causes and consequences. According to this value to determine whether there is a need to improve or determine the degree of improvement priorities, and improve system reliability.

With the growth and enhance troubleshooting capabilities, maintenance staff can not limited only by the available data and resources to diagnose faulty components, and we should focus on anti-failure detection monitoring, from the current priorities and division of labor point of view, basic maintenance staff focus on the discovery and analysis of faults, and isolate the faulty process and methods, maintenance technical management attention to failure time, specific content guidance, management of repair process, equipment office focus on maintenance mode and resources if has a reasonable allocation, industrial production

sector is study of macro failure, failure modes and reliability based on basic level maintenance experience.

2 FMEA BASED ON FUZZY THEORY

The system is based on the component, therefore, system failure is caused mainly by a component failure, study and analyze the fault of the individual components and its failure modes is a basic research of system failure, using FMEA risk analysis methods, which can identify system failures mode and effect, reduce the quality risk, eliminate defects, but when analysis a system, there are many complex and uncertain factors inevitably, it is difficult to quantitatively describe the primary and secondary relationship between different failure modes of system components, using fuzzy mathematical method to deal with uncertain information, quantitative analysis of the degree of impact of various risk factors on the system, according to the fuzzy sort of risk priority value of each risk factor, which can control factors of key components fuel systems to reduce the use of risk.

2.1 Traditional FMEA method

Explanation of statistical data is as the followings: Put specific checks and aircraft change season into a class, mechanical day and preparatory maintenance prepare date statistics combined together, due to the current filled card does not regulate, we sort out many similar failures data as one class, such as combined aircraft body and airfoil tank relief valve, combined fuel oil discharge switch and electric switch leakage into one category, combined fuel level control and signal failure of pressure controller while plus pressure into one category, related issues combined statistics. Failure mode of the aircraft fuel systems and impact analysis shown in Table 1.

The traditional method uses the RPN to sort for the quality risk, through the severity of the failure mode, the probability and detection to assessment risk, which ranges from [1, 10], with quantitative indicators to determine high risk of failure modes and key factors.

Where: RPN = SOD.

Obviously, three factors given different weights will produce different risk priority number, because factors itself has a ambiguity and uncertainty, the traditional expert evaluation method to determine the weight of each factors can not reflect the exact real impact extent of evaluation factors objectively, so we consider obfuscate to the value of three factors.

2.2 Fuzzy risk priority of FMEA based on fuzzy average weight

Suppose there are n fault mode Fi (I = 1, ..., n), needs assessment, evaluation team has m expert Ej (J = 1, ..., m). Several risk models fuzzy rating of risk factors O, S, D are as follows:

$\widetilde{R}_{ij}^{o} = (R_{ijL}^{o}, R_{ijM}^{o}, R_{ijU}^{o}), \widetilde{R}_{ij}^{S} = (R_{ijL}^{S}, R_{ijM}^{S}, R_{ijU}^{S})$. And $\widetilde{R}_{ij}^{D} = (R_{ijL}^{D}, R_{ijM}^{D}, R_{ijU}^{D})$. (L, M, and U are the risk factor levels).

M ExpertS (EJ) provides fuzzy weight of three risk factors of FMEA:

$$\widetilde{w}_{j}^{o} = (w_{jL}^{o}, w_{jM}^{o}, w_{jU}^{o}), \quad \widetilde{w}_{j}^{s} = (w_{jL}^{s}, w_{jM}^{s}, w_{jU}^{s}),$$
$$\widetilde{w}_{j}^{D} = (w_{jL}^{D}, w_{jM}^{D}, w_{jU}^{D}) \tag{1}$$

KJ (J = 1, ..., M) is a relatively important weight of the experts, and satisfy $\Sigma_{j=1}^{m} kj = 1$, $K_j > 0$ (J = 1, ..., M).

fuzzy weight of n fuzzy numbers mean:

$$\widetilde{y}_G = f_G(\widetilde{x}_1, ..., \widetilde{x}_n; \widetilde{w}_1, ..., \widetilde{w}_n) = \prod_{i=1}^{n} (\widetilde{x}_i)^{\frac{\widetilde{w}_i}{\sum_{j=1}^{n} \widetilde{w}_j}}, \tag{2}$$

$\widetilde{x}i$: N positive fuzzy numbers to be weighted; $\widetilde{w}_i$ is respective weights, i = (1, ..., n).

based on subjective opinion of experts to calculate overall incidence of the various failure modes, severity, detection of degree level:

$$\widetilde{R}_i^O = \sum_{j=1}^{m} k_j \widetilde{R}_{ij}^O, \quad \widetilde{R}_i^s = \sum_{j=1}^{m} k_j \widetilde{R}_{ij}^s,$$
$$\widetilde{R}_i^D = \sum_{j=1}^{m} k_j \widetilde{R}_{ij}^D, \quad I = 1, ..., n \tag{3}$$

The overall fuzzy weight of O, S, Dis:

$$\widetilde{w}^O = \sum_{j=1}^{m} k_j \widetilde{w}_j^o, \quad \widetilde{w}^s = \sum_{j=1}^{m} k_j \widetilde{w}_j^s, \quad \widetilde{w}^D = \sum_{j=1}^{m} k_j \widetilde{w}_j^D \tag{4}$$

Table 1. Fuel system FMEA table.

No	Component	Symptom	Failure mode	Cause	Failure affects	Basic maintenance measures
1	Fuselage (wing) fuel tank relief valve	When refueling, "relief vent" does not light	Light does not shine	Internal circuit failure	Local influence	Replacement parts
2	Fuel electric switch	Electric switch indicator rod junction leakage oil	Oil spill	Sealing gasket aging	Affect performance	Replacement parts
3	Fuel gauge indicator	Former group, the total group instruction swing	Swing. Jitter.	Oil indicator of internal bridge fault	Damage	Replacement parts
4	Switch	Conversion switch set to "left" wing when the indicator is incorrect	Indicating abnormal	Internal mechanical failure	Affect performance	Replacement parts
5	Gas turbine starter	Not turn the engine when starting	Function failure	Starter low efficiency	Function failure	Replacement parts
6	Fuel level controller	Lights are not bright when refueling	Light does not shine	Fuel oil controller damage	Local influence	Replacement parts

Table 2. Risk factors fuzzy weights.

Failure mode	The frequency	Severity	Detection	Priority
1 (tank does not light)	(5.2, 6.2, 7)	(3.2, 4.2, 5)	(4.2, 5.2, 6.2)	5
2 (junction leakage)	(5.55, 6.55, 7.3)	(3.8, 4.2, 5.8)	(6.2, 7.2, 8.2)	2
3 (fuel gauge indicates jitter)	(7.05, 8.2, 8.6)	(6.05, 7.2, 8)	(4.2, 5.2, 6.1)	4
4 (switch indicates abnormal)	(3.2, 3 .. 7, 4.7)	(5.8, 6.2, 7)	(4.8, 5.8, 6.7)	3
5 (starter function failure)	(2.2, 4 .. 7, 4.1)	(5.2, 4.7, 4.2)	(2.2, 5.7, 4.0)	1
6 (level controller does not light)	(3.15, 3.5, 2.7)	(4.2, 3.3, 4.7)	(6.2, 3.5, 5.7)	6
Importance weights	(4.2, 6.2, 8.2)	(5.7, 7.7, 8.8)	(1.2, 3.2, 5.6)	

Fuzzy risk priority of failure modes are as follows:

$$FRPN_i = \left(\tilde{R}_i^o\right)^{\frac{\tilde{w}^o}{\tilde{w}^o + \tilde{w}^s + \tilde{w}^D}} \times \left(\tilde{R}_i^s\right)^{\frac{\tilde{w}^s}{\tilde{w}^o + \tilde{w}^s + \tilde{w}^D}}$$

$$\times \left(\tilde{R}_i^D\right)^{\frac{\tilde{w}^D}{\tilde{w}^o + \tilde{w}^s + \tilde{w}^D}}, \quad I = 1, ..., n \quad (5)$$

Therefore, we select four different professional and technical experts involved in the fuel system of FMEA assessment, first, given different weights based on experience and knowledge of the experts, relative weights are: 0.3 (E1), 2.5 (E2), 3.5 (E3), 0.1 (E1). According to Table 1, the failure mode of the fuel system have five kinds, according to the frequency of fault occurrence, the degree of severity and the detection to optimize 5 failure mode, according to the level of blurring listed in Table 1 and the above equation, we can obtain the frequency of the each failure mode, severity and the Overall fuzzy evaluation information degree of detected difficulty and overall evaluation degree of the importance information, according to the formula we calculate the risk of failure priority level, results are shown in Table 2.

We can obtain the risk priority class of failure modes from Table 2.

FRPN5> FRPN2> FRPN4> FRPN3> FRPN1> FRPN6.

3 FMECA METHOD

FMECA is failure modes effects and criticality analysis, for analysis failure, causes and effects of product, so that we can modified the various stages of production and the feasibility study, provide a meaningful reference for new product development or evaluation. Using this method, we need to determine the probability and severity rating categories of failure.

3.1 Determine severity type and probability level of failure

Severity category is classified for the most serious potential consequences caused by the failure, at first, by severity it is classified into four categories:

When the failure mode can not be explicitly expressed in the four categories, according to the extent of the loss that we will represent an approximation of the classification, according to the severity of the failure, the system components can be divided into four categories as follows:

Class I (disaster): May cause death or system damage all;

Class II (fatal): serious injury, equipment damage or task termination;

Class III (critical): mild injury, equipment damage or performance degradation;

Class IV (mild): does not cause injury or damage to the equipment, but if left unattended, lockouts, which may lead to equipment failure.

It is related to not only the damage level but also the probability of the failure mode occurrence in using if the system is damaged, the probability of failure is divided into five grades as follows:

Level A (frequent): 0.2 <P ≤ 1.0
Level B (sometimes): 0.1 <P ≤ 0.2
Level C (occasional): 0.01 <P ≤ 0.1
Level D (rare): 0.001 <P ≤ 0.01
Level E (basically does not happen): 0 <P ≤ 0.001

For six kinds of failure modes, failure probability and severity grading are shown in Table 3.

We use the FMECA analysis for the importance of each failure mode and the probability of occurrence to determine the fault law of the various components of the fuel system, failure mode, provides useful information for maintenance

Table 3. Failure probability and severity level of 6 failure modes.

No	Failure mode	Severities	Failure probability level
1	Light does not shine	IV	C
2	Oil spill	III	C
3	Swing. Jitter.	III	D
4	Indicating abnormal	IV	D
5	Function failure	II	E
6	Light does not shine	IV	B

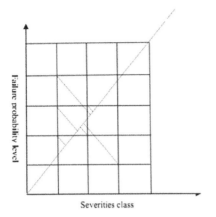

Figure 1. Harmfulness matrix analysis of failure modes.

plans and troubleshooting, and take preventive measures.

3.2 *Hazard analysis*

The Hazard analysis is performed by the combined effects of the occurrence probability and harm's degree of system failure mode, it is classified into quantitative and qualitative matrix method, using the former method when there is incomplete data, when there is sufficient data to use the latter, the paper collected failure data from 2009 to 2011, the data are more full, therefore we use qualitative analysis.

Judgment method: the method of dangers is: the distribution point of failure mode marked diagonal (dashed line OP in Fig), the distance from the intersection of the vertical and diagonal to the origin as a measure of the dangers of failure modes; the longer the distance, the harm is greater. As the qualitative analysis, we require the point in the same box marked have same dangers (such as (2) and (6) have the same dangers, (1), (3) and (5)). From the dangers of the matrix, (2) has maximum harm to the system, that is leaking fuel failure of electric switch, it is aging seals located in the landing gear compartment, it is mechanical engineering failures; (6) fuel level controller failure, it is the oil level controller fault, it is mechanical damage, (5) the gas turbine starter is a II fault severity level, but because of its level of probability failure is small, it is E so not particularly high harmfulness. (2), (6) and (1) probability failure is large relatively. (3) Quantity indicator for the ad hoc is internal bridge failures, it is professional failure, and probability rating is D.

The dangers sort order of fuel system faulty is as follows:

2> 6> 5> 3> 1> 4.

4 CONCLUSION

This paper analyzed the fuzzy FMEA and FMECA of a certain type of aircraft fuel system fault conditions, we get priority hazardous sort of six kinds of failure mode and dangers sort in the fuel system, we put it as the only fault in the system, in practical, if the fault is undetectable, we should be further analyzed to determine the impact of other failures related, design special control measures, research departments can update FMEA report timely, provide empirical data for a new round of equipment development, along the fault we can predict the remaining life of components accurately, so as to promote the realization of the equipment precision support gradually.

REFERENCES

[1] He Guofang, Xu Haibao reliability of data collection and of Beijing: National Defence Industry Press, 1995.
[2] Ho Ming-jian editor aviation engine reliability, maintainability fault diagnosis Beijing: Aviation Industry Press, 1998.
[3] Zhang Haibo, JIA Ya-zhou, Zhou Guangwen, Pei Yonghong, Chen Yun CNC system failure modes, effects and criticality analysis. Mechanical Engineering, 2004 (6).
[4] Wang Shaoping reliability engineering Beijing: Beijing Aerospace University Press, 2000.
[5] Jiang Xing Wei, Song Zheng Ji, Wang Xiaochen reliability engineering technology [M] Harbin: Harbin Institute of Technology Press, 2005.
[6] GJB/Z 1391–2006 Failure Modes, Effects and Criticality Analysis Guide [S] Beijing: China Standard Press, 2006.
[7] Li Tianmei, Qiu Jing, Liu Kuan-chun test failure rate based verification tests fault sample allocation scheme Aviation Journal, 2009, 30 (9): 1661–1665.
[8] MIL - H DBK -. 470 A Designing and Developing Maintainable Product and Systems USA, 1997.
[9] GJB/Z1391 -. 2006 Failure Modes, Effects and Criticality Analysis Guide General Armament Department of Military Standard Publishing, 2006.
[10] National Military Standard failure modes, effects and criticality analysis programs, fault tree analysis Implementation Guide. National Defense Ministry military standard published in 1992.
[11] Associated with a complex system FMEA Fuzzy Context. Zhejiang University, 2013 (5).
[12] Jun-Feng Zhou, Xu Leping, bamboo Kenford so on. Failure mode and effect analysis method in the ship's main propulsion system. Boat Ocean Engineering 200635684–87.
[13] Deng Haifei, Lv Yanmei and so on. UAV parachute recovery system failure modes, effects and criticality analysis Equipment Environmental Engineering 2012.06: 90
[14] Yang, Duan Weisiang. Simon Noel front landing gear failure modes.

Future Information Engineering and Manufacturing Science – Zheng (Ed)
© 2015 Taylor & Francis Group, London, ISBN 978-1-138-02644-5

The economic and social development issues on school abolishing and merging

Jingjing Zhang, Tao Wang & Li Ling
Shaanxi Normal University, Xi'an, Shaanxi, China

ABSTRACT: The school abolishing and merging policy is good to improving the efficiency and quality of education in primary and secondary schools. And it has more and more influence on economic and social development. It is providing not only a number of development opportunities but also some new challenges. We analyzed the relationship among the school abolishing and merging, finance, taxation, urbanization, industrial structure and other social development issues and collected a lot of data. Then we given some suggestion to resolve these problems.

Keywords: The school abolishing and merging; Optimize the allocation of resources; Economic development

1 INTRODUCTION

In recent years, the discussion of the perspective in public on School Abolishing and Merging has been once again thrown into the spotlight resulted from the repeated occurrence in safety accident of school bus. There are various opinions on this topic after the implementation of School Abolishing and Merging policies. People who in favor of the policy think it benefits the resource combination and optimizes the efficiency of the resources allocation. Opponents hold the opinion that rural children study in far away school will not only bring about safety problems but also lead to the family economic burden heavier. What is more, it does not significantly mean the improvement of big-class education quality. As the old saying goes, a slight move in one part may affect the situation as a whole. The widespread controversial question in education field has closely connected with economic development. School Abolishing and Merging policies have brought profound influence to the development of local economy and society, which not only has promoted the urbanization process and the development of the service industry, but also contribute to revolution. This requires the government to capture this opportunity with a keen eye, reasonable to operate the implementation of the policy and make the comprehensive development of local economic.

2 THE DEFINITION AND CURRENT SITUATION OF LAYOUT ADJUSTMENT OF RURAL ELEMENTARY AND MIDDLE SCHOOLS

"School Abolishing and Merging" refers to the integration of rural education resources for the purpose of optimizing the allocation of resources in China, comprehensively improving the efficiency and quality of education in terms of improving the efficiency of educational investment and education quality in primary and middle schools and accelerating the sustainable and healthy development of the rural basic education. In particular, that is to largely abolish the native rural primary and middle schools so that to bring students together in a small part school in town.

During 14 years from 1997 to 2010, 371470 primary schools have been reduced across the country, which contains the decrease of 302099 rural primary schools accounting for 81.3 per cent of the total loss of the national elementary school. Excessive School Abolishing and Merging will cause the problem of going to school far and expensively and lead to the increase in dropout rate and the "desertification" of rural civilization. In "Post-age of School Abolishing and Merging", the rural education should follow the principle of the entrance to the nearest school, equity priority, and adapting with rural demand. In addition, the guarantee of rural education inputs, doing a good job in necessary teaching school are significant as well. Don not let small-scale schools become synonym for low education level. [1]

3 SOME CLOSELY RELATED ECONOMIC PROBLEMS ASSOCIATED WITH SCHOOL ABOLISHING AND MERGING

3.1 The change of fiscal and tax policies accelerate the practice of school abolishing and merging

Fiscal function is beyond the will of human being, it means that the inherent functions and responsibilities of finance which is determined by the nature of fiscal finances. The performance of financial function is expressed through the distribution and interaction between each component of the national economy and obtains a certain effect. The country's fiscal revenue refers to the funds that the country raises through certain forms and channels, such as the revenue from printing money or bonds, tax revenue, and so on. In general, the government's fiscal income includes profit income, liability revenue, tax revenue and other income. Tax revenue is the most important and the most common form of fiscal revenue, which occupies the dominant role in fiscal revenue and it is the most important source of revenue.

In order to further reduce the burden on farmers and regulate the charges in rural fees, central government explicitly put forward to reform the current system of rural taxes and fees, and pilot in several provinces and cities step by step which starts from 2001. Its main contents can be summarized as: "three cancel, two adjustment and one reform". "Three cancel" refers to call off the administrative fees and government-managed funds or capital raising which are specifically imposed taxes from farmers in terms of the raised fund of the highly centralized plan; butchering tax and compulsory employees. "Two adjustment" refers to adjust the current policy on agricultural tax and taxes on special agricultural products. "Reform" refers to reform the existing CunDiLiu levying method.

Before the reform of rural taxes and fees, the source of rural primary and middle school education investment mainly comes from the villages and towns, namely, education surcharges, education funding and financial allocation, in which the first two ways are the main funding channel of rural compulsory education. After the reform of taxes and fees, with the abolition of the agricultural surtax and rural financing, the township financial revenue is greatly reduced. Despite the country promulgated the corresponding adjustment policy, the shortage of funds is still serious. Take the four cities of Luochuan, YiChuan, Fu county and huanglong in Shaanxi Province for example, the county finance income in 4 April 2004 followed by 57.15 million yuan, 13.2 million yuan, 28.13 million yuan and 6.81 million yuan. Only the teachers' salaries spending this year in 4 county are 46.49 million yuan, 26.06 million yuan, 32.43 million yuan and 17.72 million yuan. In addition to the Luochuan county, it still has multiplied deficit in rest 3 county after all of the local finance income used for teachers' salaries. [2]

Enormous financial, therefore, compel the government to choose school abolishing and merging, that is to say, by compressing teaching points to expand the scale of school, achieve economies of scale, improve the efficiency of resource utilization, ultimately to ease the mounting financial pressure.

3.2 The mutual relationship between urbanization and school abolishing and merging

Since the reform and open policy, the urbanization in China has rapidly developed. More and more small towns are emerging, urbanization rate has increased substantially and the population of permanent residents in cities and towns has a remarkable growth. This implies the inevitable tendency of urbanization. By the end of 2000, the small town has reached 20300, increased 4 times compared with 978 (2173). The municipal town has occupied 46.7 per cent of total villages and towns, in which 14 provinces account for more than 50 per cent of the municipal town. [3] By the end of 1999, the household number in municipal town has reached 25.79 million, accounting for 11.6 per cent of the persons in rural homes across the country and the population number is 91.13 million, accounted for 10.2% of the rural population. [4]

Therefore, under the condition of accelerating urbanization process and the decline in rural school-age population, nowadays the newly-born population in most of the administrative villages has not enough to start a complete primary school. Consequently, it is urgent to break the administrative organizational system of school instruction in rural primary and middle schools and to build primary school in traffic convenience, trade developed administration and to set and develop primary and middle school education in the central town of the county, in other words, according to population size and transfer planning to layout school is a historically inevitable trend. [5]

At the same time, the policy of school abolishing and merging makes more and more people from remote mountainous area come into towns that promote the further development of local urbanization because they may choose to work in the city to take care of the children or some people will settle down in the city.

3.3 The effects on the local industrial structure

Compared with the original school, 68% of parents believe that the spending on their children's education expenses has been increased. Besides, 34% of the parents will feel have difficulty in supplying the cost in children's accommodation and living expenses. At the same time, after school abolishing and emerging, 40% of parents have ever rent a house near children's attending school, and 22% of parents chose to buy a house nearby in order to facilitate their children to school. [6]

These data suggest that the policy of school abolishing and merging will absolutely facilitate the development of he third industry near the central school in town. With a large number of rural populations are gathering to urban, it is inevitable has requirement of development of the corresponding infrastructure or service industry. Maslow-Need-hierarchy theory has divided the human needs into five categories, in proper order from a lower level to higher level are physiological needs, security needs, social communication needs, esteem needs and self-actualization need. Physiological needs are the most basic requirement of human to maintain their own survival, including hunger, thirst, clothing, shelter and transportation, and so on. Therefore, regardless of the level of residents' consumption, physiological needs and security needs is imperative, otherwise people will appear survival obstacles. So, the local restaurants, small shops, supermarkets, barber shops, bath center, recreation centers, cinemas will been gradually established along with the development of demand, the mainly service-oriented third industry will been continuously developed along with the development of the consumption.

Meanwhile, in order to make the school attendance more convenient, there is a increasing phenomenon of renting or buying a house near urban central school which will lead to the reduce in rural labor force in return and finally affect the agricultural production which will make the farmers to increase the labor intensity, add the burden of work even cause the decrease of the decrease in soil fertility, what is more, will make serious damage to land resources.

3.4 The great amount of original school idleness and resources wasting

Due to the unrealistic evacuation and vulgar management, there is a huge waste of the abolishing school. In 2011, the school building construction area in national common elementary schools is 56913.n square meters, and in middle school is 455.4631 million square meters. The total area of primary and middle school buildings is 1.0245942 billion square meters. The above data give us the implementation that the amount of giving up school building is close to 400,000 since the implementation of school abolishing and merging. If calculate the each average primary school of a area of 50000 square meters, only the land of the schoolhouse will reach 20 billion square meters, namely 2 million hectares. [7]

The school building will appear damaged even divide the school resources privately with the timing going due to the lack of management resulted from the unclearly defined property rights after the no longer existence of teaching organization. Many rural schools turn into the venue for chickens, pigs, or wood processing plants, even sold by removal of the primary school to repay debt in the village. On the one hand, it is the waste original education resources.

4 THE SUGGESTIONS TO THE DEVELOPMENT OF ECONOMY SOCIETY UNDER THE BACKGROUND OF SCHOOL ABOLISHING AND EMERGING

4.1 Supporting the development of township enterprises and resettlement of surplus labor force

After the implementation, a part of parents choose accompany studying or settle down in the city in order to take care of their children and let their children receive better education. But for the masses of long engagement in agricultural production. After living in the city, maybe they are unable to get the source of income result in the lack of certain professional skills. For society, it is also a serious waste of human resources.

In this case, the government should play a positive role, making guidance, encouraging the development of township enterprises as well as to provide professional training with idle human resources. To speed up the development of township enterprises, the government must make one batch of agricultural products processing enterprises bigger and stronger. At the same time, we should in the light of the actual circumstances, closely rely on resource advantage, develop characteristic agricultural product processing industry, especially in the intensive processing, make the existing industrial chain shorter, and continuously extending industrial chain, hinging around the advantage industry, leading industry, the characteristic industry, focus on "characteristics", letting the feature agricultural product processing industry which possess real advantages on local resource bigger and stronger. Furthermore, to strengthen the training of idle human resources so as to make resources to realize the reasonable allocation and promote the better development.

Practice proves that the township enterprises play a unique and irreplaceable role in promoting agricultural the development of agriculture and rural economy and make contributions to the transition of rural labor force and increasing farmers' income. Therefore, in order to realize the reasonable use of social resources and the development of county economy, the development of township enterprises must be placed in an important agenda.

4.2 Promoting rural productivity through adding the investment in agricultural science and technology

Due to the rapidly pace of school abolishing and merging and urbanization, more and more migrant workers go to the city, which cause the lack of rural labor force phenomenon. In the situation of lower investment of mechanization and technicalization, it adds uncertainty to agriculture. The agricultural mechanization dependency on financial investment currently in China is only 10.44 per cent, which tell us that we need to strengthen overall fiscal support to the development of agricultural mechanization. [8]

Under this background, how to guarantee the sustainable supply of our nation's agricultural products is the most urgent challenge which we faced. Pig prices in 2012 directly reflect the problem, the price of pigs associated with feed, feed is associated with corn. China imported 300–4 million tons of corn last year, so the final target is refers to the agricultural production. Therefore, we should promote the development of agriculture by developing agricultural mechanization. There will be some part of people who still living in the countryside after most of farmers selected to work. At that point, we can choose contracting in order to concentrating the land or using agricultural machinery production to achieve economies of scale.

Besides, the measures of establishing and improving the in agricultural extension system in counties and townships can guarantee that there is various severs in all sorts of agricultural extension line, such as methane, plant protection, gardening and so on, which ensure that farmers can get personnel guidance and services when they do agricultural production.

In education aspect, the government should put forward the technology training among farmers in order to make them become excellent cultivation experts or fruitcake expert or green thumb. The last training is mainly to stay in the village of farmers, migrant work for their services. This training should focus on how to improve the ability and level of science and technology.

4.3 Making fully use of idle school resources for the sake of optimizing the allocation of resources

The town central school campus presented a thriving scene but the school which has been abolished appear be a cold and cheerless sense. Although the government had to explored before utilizing these resources, because of the lack of policy or unclear property right of, some idle buildings lack of maintenance then has turn into a dangerous house. This requires that when dealing with these resources, we should follow the guideline of government-oriented, social participation and reasonable utilization.

1. Following education priority principle, transforming the original school into a host preschool education or adult education institutions, and so on. Or adjusting as the base of towns public welfare education institutions such as rural elementary and middle schools, vocational education and training center, community education, villages and so as to fully improve the utilization rate of idle buildings.
2. Used as rural culture and welfare service, transferring the idle buildings into village-level technology school, village-level cultural compound and community neighborhood center, and so on. Furthermore, carrying out the training of rural labor transfer, practical technologies and putting the community education into the key position. These measures will improve peasants' science and technology culture level and people's lives standards, then ultimately progress the prosperity of rural economy.

ACKNOWLEDGEMENT

This article is the graduate student cultivating innovation fund achievement in Shaanxi Normal university, item number: 2013CXB021, project name: The Economic and Social Development Issues On School Abolishing and Merging.

REFERENCES

[1] Guohua Zeng, Chao Lv. Research on adjustment of rural middle and primary school for ten years. Exploration and Free Views. 2013-02.

[2] Xianzuo Fan. The background, purpose, methods, results, problems and countermeasure on the adjustment of rural elementary and middle school in our country based on the investigation and analysis of the midwest 6 provinces, 38 counties, 177 villages and towns. Academic journal of CCNU. 2008-07'(4).

[3] NBS. The development report of China. China Statistics Press. 2011:340.

[4] Zude Xian. Construction of small towns and rural labor transfer. China Statistics Press. 2001:7.

[5] Fan Yi, Xuping Zeng. The problems and counter-measures on the adjustment in rural primary and middle school. Expert Viewpoint. 2009-05.

[6] Qiuxiang Lan. The dynamic and future direction on the adjustment in rural primary and middle school. Education and management. 2013-06-15.

[7] Li Wang. The thinking of school abolishing and merging and file management of educational engineering. Managing forums. 2012-11: 346.

[8] Huajun Zhu, Zhihong Tian, Lujia HAN, Maohua Wang. The research on the dependency of how agricultural mechanization development depends on financial investment. Nuclear Science and Techniques. 2007, (3).

Future Information Engineering and Manufacturing Science – Zheng (Ed)
© 2015 Taylor & Francis Group, London, ISBN 978-1-138-02644-5

Traffic flow prediction based on artificial life and RBF neural network

Xianwen Zhu & Jing Zhang
International College, Huanghuai University, Zhumadian, China

ABSTRACT: The prediction of traffic flow is very key importance for the traffic plan, traffic management, traffic control and safety. RBF neural network is a kind of feed-forward neural network, which is composed of the different layers includes the input layer, hidden layer and output layer. The hybrid method of artificial life and RBF neural network is presented to predict the traffic flow. The experimental data are used to testify the traffic flow prediction ability of the hybrid method of artificial life and RBF neural network. It can be seen that the traffic flow prediction results of the hybrid method of artificial life and RBF neural network are better than those of RBF neural network and BP neural network.

Keywords: Traffic flow; Artificial life; RBF neural network; Prediction method

1 INTRODUCTION

The prediction of traffic flow is very key importance for the traffic plan, traffic management, traffic control and safety. Recently, the experts present all kinds of prediction models. The artificial neural network is a supervised training technique, which has good paralleling process ability. RBF neural network is a kind of feed-forward neural network, which is composed of the different layers includes the input layer, hidden layer and output layer. The artificial life is the important research method for the traditional biology and the ecology, and artificial life does not emulate the life form of carbohydrate. The research of artificial life contributes to indicate the most essential feature that life needs and basic regulation of life evolution. The hybrid method of artificial life and RBF neural network is presented to predict the traffic flow. The experimental data are used to testify the traffic flow prediction ability of the hybrid method of artificial life and RBF neural network. It can be seen that the traffic flow prediction results of the hybrid method of artificial life and RBF neural network are better than those of RBF neural network and BP neural network.

2 RBF NEURAL NETWORK

RBF neural network is a kind of feed-forward neural network, which is composed of the different layers includes the input layer, hidden layer and output layer [1–3]. RBF neural network has the global approximation and convergence, which makes it the primary choice for the traffic flow prediction.

As shown in Figure 1, the structure of three layers includes the input layer, the hidden layer and the output layer. Among which the input layer collects the input information and creates the input vector. The hidden nodes which apply non-linear transformation to the input vector are employed in the hidden layer. The final responses are given in the output layer.

The hidden node is defined by the Gaussian function, which is expressed as in the followings:

$$h_i = \exp\left(-\frac{1}{2\sigma_i^2}\sqrt{\sum_{i=1}^{m}(x_i - c_i)^2}\right) \tag{1}$$

where c_i is the center of the ith RBF hidden unit and x_i is the width of ith RBF hidden unit.

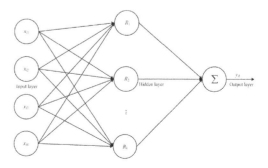

Figure 1. The structure of RBF neural network.

3 THE THEORY OF ARTIFICIAL LIFE

The behavior of artificial life is a certain system of basic life characteristic that we usually think of. The basic life characteristic generally includes own multiply, evolution and self-organization. The creed of artificial life is that the life do not exist in the single material, but exists in the combination of material, the life behavior appears in the interaction of great quantities of simple material from top to bottom [4, 5].

The artificial life is the important research method for the traditional biology and the ecology, artificial life do not emulate the life form of carbohydrate, but regardless the forms of the lives all have the basic characteristic together.

This means that the research of artificial life contributes to indicate the most essential feature that life needs and basic regulation of life evolution. The life form of accurate control which is easy to establish can speed the process of studying the life essence.

4 EXPERIMENTAL ANALYSIS FOR TRAFFIC FLOW PREDICTION BASED ON ARTIFICIAL LIFE AND RBF NEURAL NETWORK

The experimental data which are shown in Figure 2 are used to testify the traffic flow prediction ability of the hybrid method of artificial life and RBF neural network. The hybrid model of artificial life and RBF neural network is established, and RBF neural network and BP neural network are applied to compare with the hybrid method of artificial life and RBF neural network. The testing results of the hybrid method of artificial life and RBF neural network are shown in Figure 3, the testing results of RBF neural network are shown in Figure 4 and the testing results of BP neural network are shown in Figure 5. The comparison error among the three

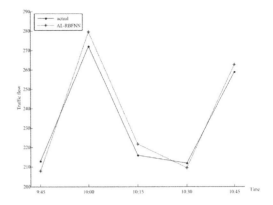

Figure 3. The testing results of the hybrid method of artificial life and RBF neural network.

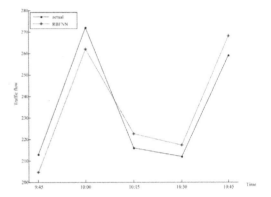

Figure 4. The testing results of RBF neural network.

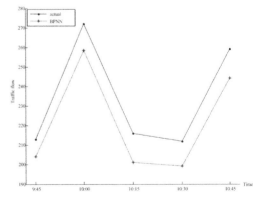

Figure 5. The testing results of BP neural network.

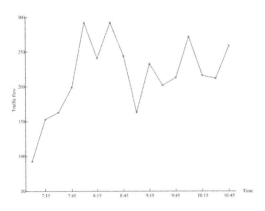

Figure 2. The experimental data.

methods is shown in Figure 6. It can be seen that the traffic flow prediction results of the hybrid method of artificial life and RBF neural network are better than those of RBF neural network and BP neural network.

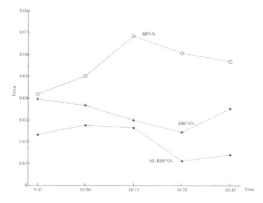

Figure 6. The comparison error among the three methods.

5 CONCLUSION

The hybrid method of artificial life and RBF neural network is presented to predict the traffic flow. The research of artificial life contributes to indicate the most essential feature that life needs and basic regulation of life evolution. The experimental data are used to testify the traffic flow prediction ability of the hybrid method of artificial life and RBF neural network. The experimental results indicate that the traffic flow prediction results of the hybrid method of artificial life and RBF neural network are better than those of RBF neural network and BP neural network.

REFERENCES

[1] Chiung-Hsin Tsai, H.-T. Han-Tung Chuang, "Dead-zone compensation based on constrained RBF neural network", Journal of the Franklin Institute, 2004, vol.341, no.4, pp. 361–374.

[2] Antreas Afantitis, Georgia Melagraki, Kalliopi Makridima, Alex Alexandridis, Haralambos Sarimveis, Olga Iglessi-Markopoulou, "Prediction of high weight polymers glass transition temperature using RBF neural networks", Journal of Molecular Structure: THEOCHEM, 2005, vol.716, no.1–3, pp. 193–198.

[3] Gh. A. Montazer, Reza Sabzevari, Fatemeh Ghorbani, "Three-phase strategy for the OSD learning method in RBF neural networks", Neurocomputing, 2009, vol.72, no.7–9, pp. 1797–1802.

[4] R.C. Dommarco, P.C. Bastias, C.A. Rubin, G.T. Hahn, "The influence of material build up around artificial defects on rolling contact fatigue life and failure mechanism", Wear, 2006, vol.260, no.11–12, pp. 1317–1323.

[5] M.M. Olsen, N. Siegelmann-Danieli, H.T. Siegelmann, "Robust artificial life via artificial programmed death", Artificial Intelligence, 2008, vol.172, no.6–7, pp. 884–898.

Future Information Engineering and Manufacturing Science – Zheng (Ed)
© 2015 Taylor & Francis Group, London, ISBN 978-1-138-02644-5

Optimization for concurrent WLAN Offload/Anti-offload and 3GPP HO processes for 5G wireless communication systems

Linlin Feng, Zhizhong Zhang, Haonan Hu & Fang Cheng
Key Laboratory on Communication Networks and Testing Technology,
Chongqing University of Posts and Telecommunications, Chongqing, China

ABSTRACT: The increasing growth of data traffic and the popularity of the intelligent terminals lead to the fact that 4G network cannot meet the demand in terms of capacity, speed, and the spectrum. Thereby the fifth generation (5G) mobile communication network comes into being. According to the state-of-the-art research, we propose a potential 5G cellular architecture, and discuss some potential optimization methods for concurrent WLAN Offload/Anti-offload and 3GPP HO Processes which the intra-cell HO case was focused.

Keywords: 5G; WLAN Offload/Anti-offload; optimization

1   INTRODUCTION

The third generation (3G) mobile services have been available for years all over the world. In December 2013, China officially issued fourth-generation mobile network licenses to the country's three telecom operators. The licenses are all based on Time Division—Long Term Evolution standard. 4G wireless networks shall support data rates of up to 1 Gbps for low mobility and hot spot coverage, such as nomadic/local wireless access, and up to 100 Mbps for high mobility and wide area coverage, such as mobile access. Long-Term Evolution (LTE) and its extension, LTE-Advanced systems, as practical 4G systems, have recently been deployed or soon will be deployed around the globe. However, the explosive growth of mobile smart devices and mobile data traffic, severe shortage of spectrum resource, diversity of mobile services and communication scenarios make the fourth generation mobile communication system faced with enormous technical challenges [1][2]. On the other hand, 4G networks have just about reached the theoretical limit on the data rate with current technologies and therefore are not sufficient to accommodate the above challenges. It is extremely urgent and important that we start the research on the general technology of 5G mobile communication system.

In November 2012, European Union launched the world's first large-scale international research projects 5G—METIS (Mobile and Wireless Communications Enablers for the Twenty-Twenty (2020) Information Society); In February 2013, China officially founded IMT-2020 (5G) working group, a platform for 5G project update of requirement, frequency, technology and standard. The main goals focused by IMT-2020 working group are as follow: a) 1000-fold traffic growth, and significant improvement in the throughput per unit area; b) 100-fold growth of the number of connecting devices; c) A peak rate of 10 Gbps; d) Lower latency and higher reliability; e) Higher spectral efficiency; f) significant improvement in energy efficiency. Especially, IMT-2020 working group focus on the evolution of the system architecture to simplify and coordinate heterogeneous networks and provide collaborative optimization.

Heterogeneity is one of the important characteristics of 5G mobile networks [3]. By way of integrating macro, pico, femto, relay base station and Wi-Fi AP, the network shall be deployed flexibly with low cost and provide broadband access for users to enhance the spectrum efficiency per unit area. Another, the network capacity and coverage shall be improved through deploying heterogeneous network in a random way. 5G mobile networks must be deployed in high density [4]. The low-power base station (e.g., pico, femto, relay base station and Wi-Fi AP) shall eliminate blind areas only with macro and improve network capacity for hotspot area while macro is the base stone of the network. Meanwhile, optimization for concurrent WLAN Offload/Anti-offload and 3GPP HO Processes, seriously affecting the quality of communication for users, is a problem that must be addressed for 5G mobile networks.

As proceedings for its SI (Study Item), the RAN#62 newly approved Rel-12 WI (Work Item)

"WLAN 3GPP radio interworking" aims to specify enhanced mechanism, which enables NW to control UE access network selection/traffic routing between 3GPP and WLAN cell more efficiently [5]. With either eANDSF or enhanced RAN policy/rule, UE can be more dynamical and flexible at traffic steering between WLAN and 3GPP RATs (e.g., more frequent WLAN offload/anti-offload occurs). As the WLAN offload/anti-offload procedure may occur in parallel to the intra-3GPP HO procedure, this may imply some potential optimization issues, which we shall elaborate on in this contribution.

The remainder of this article is organized as follows. We first propose a potential 5G mobile architecture. We analyse the scenarios, different UE mobility behaviours and performance inefficiency. Some potential optimization methods for concurrent WLAN Offload/Anti-offload and 3GPP HO Processes which the intra-cell HO case was focused are highlighted. Finally, conclusions are drawn.

2 A POTENTIAL 5G MOBILE ARCHITECTURE

Future steps for 5G cellular networks based on the existing network architecture will be the further flattening extension and enhancement. 5G cellular networks were planned to increase network capacity, transmission rate through integrating cellular networks with different coverage in high dense deployment to increase network capacity and transmission rate, thereby improving spectrum efficiency per unit area and QoS (Quality of Service).

The proposed 5G architecture is illustrated in Figure 1. 5G cellular networks consist of three fields: core network, radio access network and UE. In terms of core network, we can conclude that: a) with IT virtualization technology the mobile core network equipments would be migrated to a standardized server, namely signaling plane integration. Business functions will be virtualized as "control" and "forward" capabilities, which can be deployed

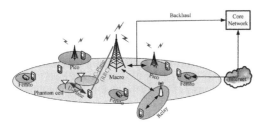

Figure 1. A proposed 5G heterogeneous wireless cellular architecture.

flexibly and instantiated according to the demand; b) a separation of control and forward of gateways in user plane for core network, namely user plane decoupling; centralized controller will be responsible for topology-awareness, routing decision and so on; the forwarding devices base on dedicated hardware will implement high-speed data forwarding. As for radio access network, we adopt: a) C-RAN, namely the base station simplified as RRU (Radio Remote Unit), centralized processing baseband pool and clouding baseband processing; b) flat RAN, namely further enhancement of the base station with user plane functions of core network sinking to the base station [6].

3 OPTIMIZATION FOR CONCURRENT WLAN OFFLOAD/ANTI-OFFLOAD AND GPP HO PROCESSES

With tighter integration between WLAN and 3GPP systems in Rel-12, NW/UE would actually be able to maintain two separate radio connections more frequently: one within 3GPP cells which follows the 3GPP specified mobility procedures alone and the other within WLAN cell which follows the IEEE specified mobility procedures as well as 3GPP specified radio interworking procedures. Above mobility processes can be tangled together up to various 3GPP/WLAN cell deployment scenarios. Currently, it is purely UE internal implementation that solves above "WLAN/3GPP mobility tangling", so that different UE mobility behaviours and performance inefficiency can be observed from time to time. In this contribution, we shall continue shedding thoughts on the improvement in those regards.

For simplicity, we shall still focus on the intra-cell HO case, while the relevant issues and principles can also be applied for intra-UMTS and inter-RAT HO cases as well.

3.1 *A. Intermediate AP offload*

As illustrated in Figure 2 (scenario I): UE (not in offload status) gets ready to perform normal HO procedure at cell border, where it has already entered WLAN coverage (e.g., AP2). As the source cell 1's RF condition becomes rather bad, hence UE tries to perform offload based on the criteria/rules obtained from source cell 1 ahead. If AP2 is a suitable offload target, then it is possible for UE to perform offload with AP2 before HO procedure is finished. Then after a short while, UE finishes HO towards the target cell 2, so should update its offload criteria/rules obtained from target cell 2. However, unluckily AP2 is not suitable or highest prioritised offload target per cell 2's offload criteria/rule, hence UE has to perform anti-offload/re-offload again. From above process, we can observe some offload ping-pong (we may name it as intermediate AP offload), which shall bring unnecessary offload relevant signalling and service interruption (WLAN data transfer suspended during anti-offload/re-offload). The potential preventive method is as follows:

Once UE realizes itself to be ready for HO process, it requires the target cell to convey its offload criteria/rule to the source cell, so that the source cell can further filter out the unsuitable AP2 seen from its target perspective, hence UE shall not perform intermediate offload towards AP2 during HO process. This method is representative for NW based solution, and can have multiple variants.

For the above method, the issue of intermediate AP offload can be solved softly without impacting offloading performance much, and the schematic procedure is highlighted in Figure 3 below.

It is worth noting that in case of HO failure, UE shall fallback to the previous offload criteria/rules obtained from source cell 1. The above "WLAN Offload Criteria/Rules Coordination" takes effect only during HO process, and UE shall update its offload criteria/rules from target cell 2 after HO process is finished.

3.2 B. Deferred AP offload

As illustrated in Figure 4 (scenario II): UE (in offload status with AP1) gets ready to perform normal HO procedure at cell border, where it has entered WLAN coverage (e.g., AP2). As the source AP1's RF condition becomes rather bad, UE should perform anti-offload/re-offload from AP1. Based on the criteria/rules obtained from source cell 1, AP2 is not suitable offload target, so it is impossible for UE to perform offload with AP2 before HO process is finished. Then after a short while, UE finishes HO towards the target cell 2, so should update its offload criteria/rules obtained from target cell 2. Now AP2 becomes suitable/highest prioritized offload target per cell 2's offload criteria/rules, then UE shall perform offload with AP2 at a relative later timing point. From above process, we can observe some offload status interruption or transmission latency (we may name it as deferred AP offload), which may degrade user's experiences such as throughput rate or challenge target cell's admission control due to larger 3GPP traffic load after anti-offload. The potential preventive method is as follows:

Once UE realizes itself to be ready for HO process, it requires the target cell to convey its offload criteria/rules to the source cell, so that UE can be allowed to perform offload with suitable AP2 seen from its target perspective, hence UE can perform offload towards AP2 even before HO process is finished. This method is representative for NW based solution, and can also have multiple variants.

For the above method, the issue of deferred AP offload can be solved safely under NW's control, and the schematic procedure is highlighted in Figure 5 below.

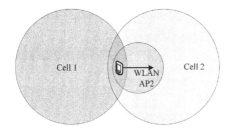

Figure 2.   Offload process → HO process → Anti-offload/Re-offload process.

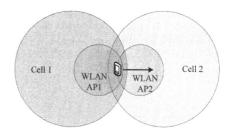

Figure 4.   Anti-offload process → HO process → Re-offload process.

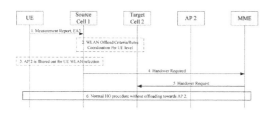

Figure 3.   NW based solution for intermediate AP offload.

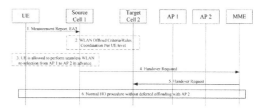

Figure 5.   NW based solution for deferred AP offload.

3.3 C. Intermediate AP anti-offload

As illustrated in Figure 6 (scenario III): UE (in offload status with AP1) gets ready to perform normal HO procedure at cell border, where UE is always under AP1's coverage across the cell border. In case AP1 is not suitable from target cell 2 perspective, then UE shall perform anti-offload after HO procedure for sure; In case AP1 is still suitable/high prioritized from target cell 2 perspective, it is supposed that UE's WLAN offload status is not interrupted by its HO process in parallel, namely no anti-offload with Cell 1 → offload with Cell 2 again (we may name it as intermediate AP anti-offload), so that it can maintain user's experiences such as throughput rate and not challenge target cell's admission control during HO process.

3.4 D. Deferred AP anti-offload

As illustrated in Figure 7 (scenario IV): UE (in offload status with AP1) gets ready to perform normal HO procedure at cell border, where it is about to leave AP1's coverage soon. As the source AP1's RF condition becomes rather bad, hence UE tries to perform anti-offload from AP1, meanwhile as the source cell 1's RF condition becomes worse as well, UE tries to perform HO from Cell 1 towards Cell 2. We can observe two different mobility processes, up to the exact timing differences between the anti-offload and HO processes.

Process Alt 1: UE shall finish anti-offload process in Cell 1 before HO process is finished, then UE finishes HO towards Cell 2 with aggregated total traffic load after anti-offload.

Process Alt 2: UE shall finish HO process towards Cell 2 with original 3GPP traffic load before anti-offload is finished, then UE finishes anti-offload in Cell 2 with original WLAN traffic load.

There is no much difference for the final UE status, but it is assumed that if UE's WLAN anti-offload occurs after HO is finished, UE can perform anti-offload directly in the target cell (we may name it as deferred AP anti-offload). The associated benefits are: some intermediate signalling and eNB processing effort for the intermediate anti-offload with source Cell 1 can be avoided; it shall also less challenge target cell's admission control due to smaller 3GPP traffic load during HO process. The potential method to achieve Process Alt 2 is as follows:

Once UE in offload status realizes itself to be ready for HO process (e.g., detecting event A3 but cannot find any other suitable target WLAN AP), NW shall indicate UE whether to hold on its WLAN anti-offload process, then after UE finishes its HO process towards certain target cell, UE shall start anti-offload process directly in the target cell. This method is representative for NW based solution, and can have multiple variants.

For above method, the intention of deferred AP anti-offload can be achieved per NW's need without eliminating the possibility of Process Alt 1, and the schematic procedure is highlighted in Figure 8.

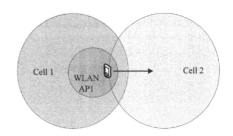

Figure 7. Direct Anti-offload in the target cell.

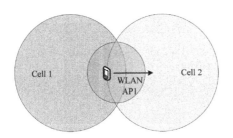

Figure 6. Stay in the same WLAN AP coverage during HO process.

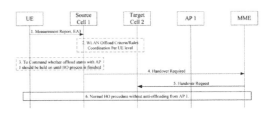

Figure 8. NW based solution for deferred AP anti-offload.

4 CONCLUSIONS

In this article, a new heterogeneous 5G cellular architecture have been proposed. We have also illustrated four typical scenarios where WLAN offload/anti-offload process may happen concurrently with HO process, and described manifold performance deficiencies to occur. In order to overcome those deficiencies, we proposed optimized handling in standardized means.

ACKNOWLEDGMENT

The work was supported by the National High Technology Research and Development Program of China (863 Program) (No.2014AA01A706) and the Project Supported by Program for Innovation Team Building at Institutions of High Education in Chongqing (2013).

REFERENCES

[1] Cisco Visual Networking Index: Global Mobile Data Traffic Forecast Update, 2013–2018, white paper, 2014.2.

[2] J. Zander and P. Mähönen, Riding the Data Tsunami in the Cloud: Myths and Challenges in Future Wireless access, IEEE Commun. Mag., vol. 51, no. 3, Mar. 2013, pp. 145–51.

[3] J. Andrews, The Seven Ways HetNets Are a Paradigm Shift, IEEE Commun. Mag., vol. 51, no. 3, Mar. 2013, pp. 136–44.

[4] Y. Kishiyama et al., Future Steps of LTE-A: Evolution towards Integration of Local Area and Wide Area Systems, IEEE Wireless Commun., vol. 20, no. 1, Feb. 2013, pp. 12–18.

[5] TR 37.834 v12.0.0, Study on Wireless Local Area Network (WLAN)—3GPP radio interworking.

[6] C-RAN: The Road Towards Green RAN, China Mobile Research Inst., Beijing, China, white paper, v. 2.5, Oct. 2011.

Future Information Engineering and Manufacturing Science – Zheng (Ed)
© 2015 Taylor & Francis Group, London, ISBN 978-1-138-02644-5

The inspiration of Marx's scientific and technological innovation theory on cultivating talents of computer science and technology by universities

Weili Zhao
Northwest University of China, Xi'an, Shaanxi, China

ABSTRACT: As computer science and technology progresses at an extremely rapid speed today, it has been essential for universities to cultivate high level talents on computer science and technology to fulfill the national talent pool, and to promote scientific and technological innovation, and social development. This paper holds the view that the Marx's scientific and technological innovation theory will guide universities in developing innovative talents on computer science and technology by providing them a theoretical foundation and practical basis.

Keywords: Marx's scientific and technological innovation theory; cultivation of talents on computer science and technology; philosophy training

1 INTRODUCTION

Nowadays, science and technology innovation has been becoming an outstanding feature of national prosperity and progress, what is more, accelerated development of science and technology propels the transformation of the whole human society. Science and technology of computer as one of the most advanced innovation in the 20th century, has been raised sharply with the instantaneous changes. In the meantime, cultivation of talents on computer science and technology has been to a new stage. Science and technology innovation activities require innovative consciousness should be combined with creative thought and practice and cultivation of talents on innovation activities is an inevitable requirement for the development of the time.

In terms of cultivation of talents on science and technology innovation ability for computer, at first, the most important is the cultivation of innovative consciousness and creative thought. The talent on science and technology of computer with ability of creative thoughts is dominant people who lead the development of science and technology in future society. The key point of cultivation of talents on computer science and technology is that whether the cultivation of innovative consciousness and creative thought are properly consistent with the target of innovation talents training. The cultivation concept of talents on computer science and technology is aim at stimulating the innovative consciousness and then further to cultivate and shape of innovation ability and personality though the way of practice on the basis of formation of creative ideas. Second, innovation activities on science and technology are needed to meet the demand of social development, for which society needs are the power of science and technology innovation. Third, the improvement of practice ability depends on the provision guarantees of consciousness and thought. Without the formation of creative thought, practical ability of innovation will not be performed. The Marx's theories of scientific and technological innovation meet the requirements of cultivation of talents on science and technology of computer, which provides the basis of theory and practice for cultivation of talents on science and technology of computer.

2 MARX'S SCIENTIFIC AND TECHNOLOGICAL INNOVATION THEORY

2.1 *The outstanding of nature of scientific and technological innovation in Marxist*

The Marx's scientific and technological innovation theory that originate from the thought to the understanding of theory and practice between human and nature, are an understanding to the historical and logical relationship between innovation of science & technology and development of human beings. In Marx's opinion, the scientific and technological innovation is regarded as a special social practice of human activities, which is an external expression for human exploration and innovation to the nature and adaptability for the changing time, and which is also a re-recognition for the objective world by people. Scientific and technological innovation as a special human productive activity different from general activities contains construction and application of knowledge subject to human beings. This kind of scientific and technological activity is generated to be a force by external macro social factors and internal micro epistemology, which is achieved dialectical unity between manual labor activity and mental labor activity and also between individual labor activity and social collective labor activity. In Marx's view, scientific and technological innovation is senior labor activity. *Engels* pointed out that the significance of theoretical thinking on science and technology innovation. He once said that, *"If a nation wants to stand at the forefront of science, it must have the theoretical thinking all the time."* [1] *(Engels, 1971)* In consequence, the understanding of nature of science and technology in Marxist is that to be the understanding of conversion process from human mind to social productivity.

2.2 *Scientific and technological innovation team in the views of Marxist*

With respect to the original innovation in the fields of the science and technology, it is hard to achieve development and improvement of the science and technology relied on personal thinking and practice, especially in the large-scale development of science and technology of computer, the spirit of teamwork plays an important role in the process of scientific and technological innovation. Marx had a profound understanding of teamwork in his concept on scientific and technological thinking. He once said that, *"Cooperation not only helps to improve individual productivity, but also to create a kind of productivity which must be a collective force itself."* [1] *(Marx, 2004)* in the view of Marxist, it is said that the activity of scientific and technological innovation is a group of joint activity and its innovation achievement is the wisdom of individual and collective cooperation.

2.3 *The outstanding of subject development of scientific and technological innovation on Marxist*

Marxist believed that all the activities of human being is to own existence and development, and the basic way of human development is to explore and reform their own potentials. Thus, human being has played the decisive role in the development of society. The basic essence of social development is the development and progress of people. Only own development and progress can promote social progress. [2] Thinking and activity of people create

the reform of nature and society and serves human itself in the same time for satisfying the increasing needs of human survival and development. Human is the most important key factor in the activity of scientific and technological innovation. Concept of people-oriented should be insisted on the cultivation of talents of scientific and technological innovation. Only the development of main body for innovation is regarded as the highest value orientation, can the reform of science and technology be truly achieved progress of society.

2.4 *Driving force of scientific and technological innovation in the views of Marxist—social demand*

In Marx's opinion, science and technology is not only a historical phenomenon in human society but also it is an important social productivity promoted historical progress. Society needs to foster the birth of science and technology, which is the activity for the productive development and applied research according to related target and plan based on social demands, especially market demand. Social demand provides research direction and topic for innovation and development of science and technology. All the scientific discovery and technology inventions are centered on the social demand and human survival development demand in relation to objective issues of actual demand. [3] Marx believed that the growth of customer demand was faster than the speed of production; in this case, people must abandon the traditional manufacturing with low-efficiency. The internal demand of human for discretionary body and mind and external social demand are the fundamental power to propel the activity of science and technology innovation, and it plays a key role for a long time. The innovation achievements of science and technology will be accompanied by the need of society, and it will get promotion and application to the market in the short term. The spiritual wealth and material wealth created are to encourage human continuously explore in the way of development of science and technology.

3 THE CULTIVATION OF TALENTS FOR COMPUTER IN THE VIEWS OF MARX'S THOUGHT ON SCIENTIFIC AND TECHNOLOGICAL INNOVATION

Computer science technology has been flourishing in today's information era. The cultivation of talents on computer science and technology is an inevitable demand of the times. If talents of computer science technology intend to follow the pave of development of the world and rank in the former of the world, they must make an explosive breakthrough on the present basis. As the important training bases, colleges and universities have responsibilities for the correct guides on the view of methodology and understanding to the talents of computer science technology. At the present, Marx's scientific and technological innovation ideas for the development of the cultivation of innovative talents of science and technology still has a long-term vision and distinct social value, which paves a sustainable development way for the cultivation of talents of computer science technology.

3.1 *Lay on solid foundation, pay more attention to educational ideals*

Knowledge and experience are the foundation of innovation. Laying solid theory foundation is the precondition of science and technology education during the cultivation of computer talents in colleges and universities. Education concept of science and technological innovation aims at cultivating the talents who will put action to the faith of innovation. The concept of education is meant to guide student to take initiative to pursue their own development, which is a kind of process of active exploration for perfecting themselves. Purpose of concept education is for stimulating student's innovation consciousness and giving them some enlightenment of innovation ideal. It is historically proved that philosophy views and scientific methods have a direct impact on the formation of cognitive process and scientific theory. *Lenin believed that the consciousness of people not only reflected the objective world, but also created the objective world.* [4] Innovation consciousness dominates the attitude and behavior of computer talents for innovation activity. Meanwhile, it is also an important factor for exciting and stimulating the potential of leaner.

How to shape the habit of innovation thinking for the talents of computer science and technology? First, it is necessary to further stimulate innovation consciousness on the basis of exiting consciousness for the talent of science and technology in order to cultivate stronger consciousness on the science and technology of computer. Innovative thinking is also named the creative thinking of human. Supported by solid theory, innovative thinking provides a series of way and method to solve problems or explain issues related. Innovative thinking, with strong pioneering and creativity is a way of thinking which is beyond human normal ways and is a way for exploring the unknown. Innovative thinking, established on the basis of connection of perceptual knowledge and rational practice, is the ways of creative recognition and understanding for the thing, which is the exploration of brain potential, the promotion of wisdom and combination of divergent thinking and convergent thinking. Second, colleges and universities

should actively encourage students to dare to challenge authority and to break prejudice.

Only concept first, can get rid of constraint from the traditional education, can actively stimulate leaner's curiosity and desire for knowledge. In this case, it can give full play to leaner's innovative potential. Third, concept education is an asymptotic process of cultivation of innovative thinking and innovative consciousness rather than a result. Only in this way, cultivation of innovative thinking should truly be promoted in the future.

3.2 *Strengthen personality shaped and cultivate spirit of teamwork*

As main point of the Marx's scientific and technological thought, the development of the people will be eventually promoted the development of society through the way of scientific and technological innovation. People-oriented must be regarded as a basic principle of cultivation of talents of computer science and technology. In traditional cramming education mode, students receive the knowledge passively leading to curiosity and desire for the knowledge of students is suppressed. In the process of cultivating talents of computer science and technology, it is necessary to abandon traditional educational thinking and should empower students for a role in teaching subject in order to cultivate student's creative personality. Scientific and technological talents with creative personality should have some kind of qualities, such as strong willpower, optimistic spirit, independent learning attitude and strong sense of responsibility. These individual personalities achieved by personal emotion and willpower play a regulatory role in actor's behavior and psychology and are important internal personality of scientific and technological talents who play a key role in the interactive of innovation. At the same time, scientific and technological innovation team can improve individual productivity on a large scale in Marx's view. Computer science and technology innovation is a complex social work, so it achieved in the need of collaboration with each other. Only the individual and collective work with each other, think and explore together, can the achievement continuously exceed in the process of the scientific and technological innovation. In normal teaching practice, we should strive to cultivate the spirit of teamwork to computer talents in order to help them better function in new times for the requirement of scientific and technological innovation.

3.3 *Change evaluation mode and raise practical ability*

With the professional theoretical foundation, skilled technology application ability and innovative spirit of constant striving for excellence are requirements for cultivating talents of science and technology in colleges. The major of computer, as a subject standing at forefront of science, is widely used in all fields of areas; however, innovation is only eternal theme in development of computer science and technology. Currently, the appraisal of computer course is mainly still using traditional pattern paper test. As we know, talent cultivation of computer science and technology should focus on the training of application development and practical operation ability. Traditional paper test is not consistent with the requirements of talent cultivation of computer science and technology. Therefore, evaluation method focus on student's comprehensive abilities should be changed from the traditional paper test to various ways of assessment, including experimental performance, program design or technological thesis. In accordance with the rules of informational society, scientific innovation and technological innovation must have achieved dialectical unity and manual labor and mental labor, individual and collective labor can complement each other in the fields of scientific and technological innovation of computer. Marx indicated that education of scientific and technological innovation thoughts had helped bring about promotion for production of scientific and technological innovation, moreover, the major force and basic solution of human liberation and development rely on the existence of people.

3.4 Adjust the profession direction setting to meet the needs of social development

Colleges and universities are a basement for cultivating and providing qualified personnel to the society. Talent cultivation should be in line with social demand in order that the talent can provide a powerful guarantee in the practice of scientific and technological innovation. Whether education works, it has a direct effect on the big subject of the nation building and scientific and technological innovation. Currently, part of the first line of university teachers are lack of social practice experience, and their research and teaching are only staying in the school, so lack of awareness of innovation consciousness and social demand. What's the worse, teaching material and profession setting are difficult to keep up with the pace of social development, thus the whole school teaching can't meet market demand.

According to the talent cultivation of Marx's thought on scientific and technological innovation, at first, colleges and universities should establish a teacher team with development vision standing in subject frontier field. They are the major force on talent cultivation of computer science and technology in colleges and universities. [6] Second, the setting direction of computer subject should be made significant adjustments and should establish new curriculum system to meet the social demand.

Third, the teaching mode of university-industry cooperation has not been formed in colleges and universities in China. Lack of sustainability and comprehensiveness, the cooperation relationship is always unstable when participating in enterprise or national project. In this way, it is difficult for colleges and universities to gain a long-term technique support and necessary fund, and some long-term tracking research are unable to achieve. In a word, the most effective way to solve disadvantage of university-industry cooperation is that colleges and universities actively promote independent research and innovation and ensure more adoption of research result. Meanwhile, colleges and universities can invite some professionals in enterprise to instruct students directly in order to cultivate talents of computer science and technology with market skills and technological skills. The social demand is considered a fundamental power by Marxist for promoting activities of scientific and technological innovation.

4 SUMMARY

About scientific and technological innovation is an important part of the Marxist theory; it is an important modern value for cultivating talents of computer science and technology innovation in universities and colleges in China. Computer science and technology involved in all aspects of whole social and living production, has been becoming an important indicator of national competitiveness.

Talent cultivation of scientific and technological innovation in computer subject needs to advanced scientific views for instructing. With respect to the scientific and technological innovation in Marxist provides a good reference for cultivating talent of innovation in our colleges and universities. It still has a long way to go to develop the cultivating mode for talent of scientific and technological innovation in computer subject by trail and error.

REFERENCES

[1] Karl Heinrich Marx. 2004. *Das Kapital (first volumes)*. Beijing: Beijing people press.
[2] Rongfang Zhang. 2009. Science and technology innovation ideas of Marx Engels. Southeast University.
[3] Panpan Ma. 2012. The Exploration of Enhancing the Innovative Scientific and Technological Talents Quality of Marxism. Soochow University.
[4] Gang Wu. 2007. *Expounding the Thoughts of Scientific And Technological Innovation that Marx and Engels Put Forward*. Central China Normal University.
[5] Xiufang Xu, Yongzheng Tang. 2011. *Research and exploration of cultivating innovative talents in local colleges and universities computer*. China Electric Power Education.

Future Information Engineering and Manufacturing Science – Zheng (Ed)
© 2015 Taylor & Francis Group, London, ISBN 978-1-138-02644-5

Deep exploration test in Taiwan Straits applying OBS

M.Q. Sun, Y.X. Xu, P. Fu & H.L. Huang
Xiamen Seismic Survey Research Center of Xiamen, China

ABSTRACT: In 2012, the first deep exploration test was conducted in Taiwan Straits applying Ocean Bottom Seismograph (OBS) of Iggacas made in China and SEDIS V made by GeoPro in Germany. The paper reviews and analyzes the deep exploration test in Taiwan Straits, describes the main instrumental performance of Iggcas self-float OBS, and briefly introduces the measure line layout, instrument retrieval and data logging. Then implications of operation at sea and data processing are concluded. IGGCAS OBS made in China is smaller in size, easy to operate and less energy-consuming with longer logging time. Another advantage is that the high-capacity flash memory substantially reduces the energy consumption and the noise. Water depth of Taiwan Straits is too shallow and environmental noises from things like waves may affect the data receiving of the instrument. On the other hand, flaws exist in the designing of IGGCAS. There are problems with the seismometer with both positive and negative polarities. Moreover, the hollow base of the instrument and its high center of gravity easily result in serious trembling and high failure rate of the seismometer, and need be more improved.

Keywords: Taiwan Straits; OBS; Ocean deep exploration

1 INTRODUCTION

The deep ocean seismic exploration in developed countries like the U.S., Japan and Germany started in the 1960s, and has gone through three stages, namely sonobuoy, common midpoint stacking double boat extended profiling and OBS. Since the 1980s, nearly ten teams have conducted seismic exploration in South China Sea and have yielded fruitful results. In 2010 and 2011, several research institutions cooperated and acquired data of two deep seismic exploration profiles in Bohai Gulf. Research into the deep structure of eastern sea of Taiwan was also conducted more than once. Though the shallow layer data of Taiwan Straits is relatively complete, the deep seismic structure exploration of Taiwan Straits is still a gap to be filled up.

2 THE DEEP EXPLORATION TEST IN TAIWAN STRAITS

The deep exploration test in Taiwan Straits in the July of 2012 was part of Project ATSEE cooperated by Fujian Seismological Bureau and Taiwan. Sea tests were organized by Fujian Seismological Bureau and co-conducted by Second Oceanographic Institution of State Oceanic Administration and Institute of Geology and Geophysics of Chinese Academy of Sciences. The method adopted was land shooting and joint receiving of land and

sea. Four measure lines were laid out in Taiwan Straits (Fig. 1).

2.1 Equipment

The equipment in the test includes sixteen IGGCAS OBS from Chinese Academy of Sciences, among which eight are of seven channels and the other

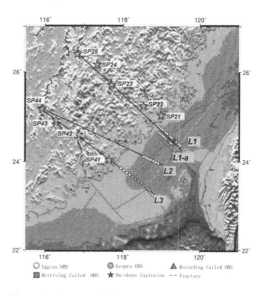

Figure 1. Measure line layout in Taiwan Straits.

eight are of four channels, two OBM, five newly developed small-sized OBS, as well as ten SEDIS V OBS by GeoPro in Germany and three TT801 deck machineries by iXsea in France.

Compared with the foreign products, IGGCAS OBS is the same in regards of designing principle, the seismometer and the logger, but it is different in regards of its exterior structure, floating system, power system and data reading. The designing of domestic OBS fits in the trends of international technology development and employs new techniques such as rechargeable lithium cells, Bluetooth input parameter data reading and GPS communication. For example, in the deck acoustic response unit, domestic OBS employs GPS wireless data transmission module while German and French OBS employ radio transmitting and receiving technology.

2.2 Measure line layout at sea

OBS were distributed along four measure lines in the test (fig. 1): L1: 11 OBS (Iggcas) with dot pitch of 5 km and measure line length of 72 km; L2: 7 OBS (Iggcas) with dot pitch of 5 km and measure line length of 30 km; L3: 12 OBS (4 Iggcas and 8 Geopro) with dot pitch of 10 km and measure line length of 95 km; L1-a: CAS experimental measure line, 4 OBS, 5 MOBS and 2 OBM with dot pitch of 5 km and measure line length of 41 km.

2.3 Instrument retrieving and data logging

31 OBS were distributed in the test and instrument retrieving rate was 97% (100% for Geopro; one IGGCAS lost, 95%), data retrieving rate was 94% (no logging for one of the IGGCAS OBS).

In the test, in shot points SP21 and SP22, nine OBS showed seismic signals in L1; in SP41, only one OBS showed seismic signals in L3, which means that totally there were only 19 effective source-receiver data. The shortest offset distance was 34.81 km, and the longest, 136.41 km.

Figure 2 is SP22 seismic phase comparison between the nine OBS that showed seismic signals in L1 and the recording of Quanzhou station. The seismic phase is clear. IGGCAS OBS showed serious trembling which caused that the secondary seismic phase unrecognizable by the velocity seismometer.

L3 employed both IGGCAS and SEDIS-4 OBS and only received signal from the point No. 1 (offset distance of 36.8 km) and no signal from the other nine (none from any of the frequency bands). It is inferred that there is thick unconsolidated sediment at the bottom of the sea which absorbs most of the seismic wave. Strong interference was also observed from the logging which may result from stream impacts and ship motors and so on.

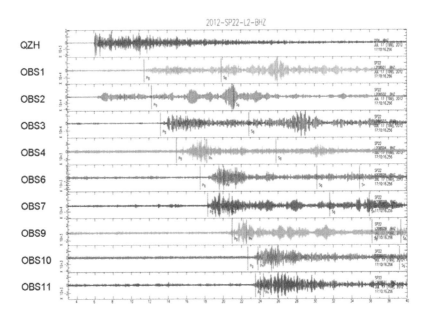

Figure 2.   SP22 seismic phase comparison between the nine OBS that showed seismic signals in L1 and the recording of Quanzhou station.

300

3 CONCLUSIONS

3.1 *Text and indenting*

Applying OBS to the deep ocean earth crust research is of high risk. If one OBS fails to be retrieved, it means huge loss of money and material, not to mention the loss of the valuable data that are urgently needed. According to the operation practices and data processing in this test, the following implications can be concluded.

1. IGGCAS OBS made in China is smaller in size, easy to operate and less energy-consuming with longer logging time. Another advantage is that the high-capacity flash memory substantially reduces the energy consumption and the noise.
2. The shooting time recorded by navigation is not accurate enough with big errors, while the time recorded by the deck is highly accurate, so the deck plays a crucial role in rectifying the shooting time. A seismometer should be placed on the deck to record the shooting time.
3. Sufficient information should be collected before going to the sea. Submarine topography features of each OBS launching site should be fully studied to avoid placing OBS in sites of complicated topography. Hence it is necessary to collect multi-beam data and sediment data near each measure line and evaluate whether to launch OBS according to the water-depth variation scope and sediment conditions of the measure lines.
4. Water depth of Taiwan Straits is too shallow and environmental noises from things like waves may affect the data receiving of the instrument. On the other hand, flaws exist in the designing of IGGCAS. There are problems with the seismometer with both positive and negative polarities. Moreover, the hollow base of the instrument and its high center of gravity easily result in serious trembling and high failure rate of the seismometer, and need be more improve.

REFERENCES

[1] Ruan, A.G. et al. 2009. Wide aperture seismic sounding in the margin seas of China. *South China Journal of Seismology* 29(2): 10–18.
[2] Ruan, A.G. et al. 2010. The experiment of broad band I-4C type OBS in the Southwest India ridge. *Chinese Journal of* Geophysics 53(4):1015–1018.
[3] Li, J. et al. 2010. Development of broadband Ocean Bottom Seismograph (OBS). Acta Seismologica Sinica, 32(5): 610–618.

Future Information Engineering and Manufacturing Science – Zheng (Ed)
© 2015 Taylor & Francis Group, London, ISBN 978-1-138-02644-5

School games in the United Kingdom for teenagers

Guohua Ding
Department of Physical Education, Shanghai University of Engineering and Science, Shanghai, China

ABSTRACT: By arranging and analyzing the relevant literature, the article systematically introduces the United Kingdom ongoing school games as in the guiding ideology, contribution, organization and management of the competition as well as formats, and then forwards four suggestions for the operation of Sunshine Sports Program in China: (1) expands the range of benefits of Sunshine Sports Program especially for students with disabilities, (2) forms a functional system of the linkage at school, community and family level, (3) gradually establishes the "government guides, social groups leads" Sun Sports Program administration pattern through socialization and market-oriented operation, (4) positively develops and innovates game system and formats with local characteristics.

Keywords: School Games in United Kingdom; primary and secondary school of China; Sunshine Sports Program

1 INTRODUCTION

Since January 2011, the British government has carried out curriculum reform by experts from various fields, where physical education is one of the most important areas. Michael Gove, UK Secretary of State for Education, encouraged schools at all levels to hold more competitive activities in and out of schools when was talking about this reform and said that the competitive sports activities would be included in syllabus. British Prime Minister David Cameron also said that "We has begun to hold annual school games and are currently rewriting the syllabus by adding competitive sports in the physical education curriculum and we also help schools and local sports clubs work together, which are the part of the government's £100 million education strategy for teenagers."

Over £128 m of Lottery and Government funding is being invested to support the School Games. This includes: up to £35.5 m Lottery funding from Sport England between 2010–2015; £28.4 m exchequer funding from the Department of Health and Sport England to fund 450 School Games Organisers (SGOs) and Change 4 Life Clubs; £65 m exchequer funding from the DOE to release a PE teacher for one day a week in all secondary schools.

In China, in order to thoroughly implement the policy named "Guidance On enhancing youth sports activities and strengthening teenagers' health" issued by the CPC Central Committee and the State Council in 2007, local governments have come out successively with a series of policies and reform measures in recent years, which contribute to the wide development of Sunshine Sports Program and the fulfillment of "Health First" guiding ideology and are great help to the good effect of "one hour a day sports activities" on campus for primary and secondary school students nationwide and the forming of the good situation of young students' actively participating in physical exercises. However, there are also many problems such as inadequate publicity, exercise time and coverage of students (Xia, YUAN Hong, ZHOU Xiao-min, 2011), lack of interesting and innovative activities and void of evaluation and feedback system in administrative departments of education for supervising schools at all levels (SHEN Ling, 2008).

Therefore, it is necessary for us to learn from the foreign school games in order to promote the development of China's primary and secondary schools Sunshine Sports Program scientifically and institutionally.

2 ADVANTAGES OF SAINSBURY'S SCHOOL GAMES

Compared with the previous school games system, Sainsbury's School Games has the following two advantages: First, the number of students' participation in this program is greatly increased. Differences in various regions, schools and students are taken in full account in the design process.

The project calls for more students to participate in intramural and inter-school competitions including students with disabilities. Competitions at local and national level would be asked to encourage students who are good at other courses like photography to take part in the process of the entire game in order to develop their interests and skills.

The second is novelty. There not only have traditional events like basketball, football and sailing, but also have local events like boccia, orienteering and rounders. The project has also designed some events for disable students like table cricket and wheelchair basketball. Both traditional and local events have made some adjustments in the rules in order to enable all students at different level to participate in the games.

3 MISSIONS AND OBJECTIVES OF SAINSBURY'S SCHOOL GAMES

The Sainsbury's School Games is a unique opportunity to motivate and inspire millions of young people across the country to take part in more competitive school sport. One of the key outcomes of the School Games is that young people are helped to transit from school sport into lifelong participation.

By using Sainsbury's School Games, schools can achieve the following objectives.

Build on existing practices to provide more opportunities for those currently not engaged in competitive sport, and provide an appropriate pathway for all young people to achieve their potential.

Maximize the impact of competitive school sport to develop student's personal skills and contributing to school life.

Supporting the achievement of school priorities and outcomes.

At same time students can achieve the following objectives:

Creating positive aspirations and attitudes to learn.

Developing leadership skills and expressing citizenship through volunteer roles such as officials and team managers.

Through four levels of competitive sport, young people have the opportunity to get involved, have fun, challenge themselves and progress.

4 ORGANIZATIONS AND MANAGEMENT OF SAINSBURY'S SCHOOL GAMES

4.1 *The administrative organizations and staffs*

The Games are made up of four levels of activity: competition in schools, between schools, at county/area level and a national finals event. Each level has a different organizational structure, manager team and partners. Table 1 provides an overview of all levels, the manager team as well as assisting resources.

4.2 *Arrangement and implementation of the project*

First, the intra-school competition (Level 1) is the foundation of the project and most of the students have their first competition experience at this level, that is the reason why this level should accommodate all students The Department for Education

Table 1. The school games network.

Level	People	Support and guidance provided by the YST
Level 1	• Teacher Release posts, • School PE departments, • School Sports Organizing Committee	• Schools Guide including NGB guidance and sports formats • School Games website • YST Development Manager • School Games roadshows • YST conference • Networking and Training
Level 2	• Schools, PE departments, • School Games Organizers, • Young Officials and volunteers	• Schools Manual including NGB guidance and sports formats • School Games website • YST Development Manager • YST conference and Training • School Games roadshows
Level 3	• Local Organizing Committee With 50% workforce for School Games festival to be young volunteers • County Sports Partnership	• Lottery Funding • LOC guidance • NGB guidance and sports formats • School Games website • Festival branding
Level 4	NATIONAL ORGANISING COMMITTEE	

has made funding available to every secondary school in England to enable a PE teacher to be released one day a week to provide support for PE and school sport, both within their own school and across their family of feeder primary schools. A school holds games around a year with the sports formats designed for its level. It holds the School Games Day in the end of the year when school staffs, parents, local residents would be asked to participate in the event. The Games is arranged and managed by the School Sport Organizing Committees which are made up of students who join together to lead the planning and delivery of school sport clubs and intra-school competition programmes in their school. They influence and shape the school sport offer for their peers through deciding on the types of competitions to be held, their structure and where and when they take place, making it more attractive and accessible for all young people.

Next is Inter-school Competitions (Level 2) which is held in schools in a relatively small area (equivalent to a county or city in China) with 3–13 years students participated in this level. Individuals and teams are selected to represent their schools in local inter-school competitions, many of which will consist of leagues. The School Games Organizers (SGOs) funded by the Department for Culture, Media and Sport and the Department of Health are non-government organizations to promote, develop and organize inter-school competitions. There are 450 School Games Organizers nationally which generally are host schools across the country. They have a common set of tasks as detailed below:

Competition and Talent. Responsibility for encouraging as many schools as possible to take part in School Games; work with School Games LOC to ensure clear links between Level 1, 2 and 3 (Mallika Kanyal, Linda Cooper, 2010).

Volunteering and Coaching. Develop the workforce of teachers, non teaching staff, parents and volunteers and young leaders; develop a workforce of Active Lifestyle Coaches to engage least active young people; provide access to specialist coaches, training and equipment—lunchtime/after school programs; support young people to access quality coaching, higher level competition and move into clubs.

Participation and Engagement. Sustain/grow the network of Change4Life sports clubs to support progression of levels 1 and 2; ensure provision for young disabled people to access School Games and stage local events; ensure development of a cultural component to the School Games.

Monitoring and Evaluation. Ensure quality assurance and local impact measurement mechanisms are in place.

Then comes to county/area sport festivals (Level 3), they are held in schools at county (equivalent to China's province) level. Each county/area will host a multi-sport showcase event as a culmination of the year-round programme of competition between schools that is offered at level 2. Local Organizing Committees (LOCs) chaired by a Head Teacher and supported by the County Sports Partnership are responsible for the planning and managing of the Level 3 competitions.

Local Organizing Committees can help to achieve the following objectives:

Delivering bespoke training and providing local advice and guidance to SGOs.

Driving the establishment and implementation of more local competitive opportunities for young people.

Sustaining young people's participation beyond their involvement in competition through the development of school club activities.

Implementing Project Ability to encourage schools to develop meaningful high profile sport competitions for young disabled people.

5 CASE STUDY-CASTLECROFT PRIMARY SCHOOL

The Castlecroft Primary School, which is a suburb of Wolver Hampton, West Midlands, located on the edge of the city, WSW of the city centre, has its advantages to deliver Level 1 competitions because it is a local sports college union school and has been accepted into the Get Set group of schools preparing for the London Olympics 2012.

When organizing Sainsbury's School Games Day in 2012, Castlecroft Primary school adopted a number of measures to attract more students to be involved:

Afternoon of activity in July made up of track athletics and traditional sports based on the 1950s (sack races and tug of war) (Stuart J. Fair clough, Zoe H. Butcher, Gareth Stratton, 2007).

Open to all parents and governors of the school.

Every young person competed in at least one event for the chance to win points for their house team. The overall winning team then received the cup.

The event was fully inclusive, enabling our young people with disabilities to compete against their able-bodied peers in a variety of events such as the sprints and relay races.

Castlecroft Primary School held an opening ceremony led by its school samba band around the field with the whole school involved in a flag bearing procession. Young people were allocated a country's flag and had spent time in lessons learning about that country. They also ended the event with a closing ceremony with a parade and music and also had an Olympic torch relay race where young people had to compete in a relay holding their torches to celebrate the 2012.

When talking about how to run their Level 1 event next year, the Castlecroft primary School gave the tips as follows:

Plan early.

Allow enough time for the procession to take place and to get the young people seated.

All faculties should participate.

Keep the activities that are meaningful to your school and young people as well as trying new things.

Ensure there is a mixture of fun activities as well as more competitive ones to enable all young people to represent their house and participate.

6 ENLIGHTENMENT TO THE OPERATION OF SUNSHINE SPORTS PROGRAM IN CHINA

6.1 *It should expand the range of benefits of sunshine sports program especially for students with disabilities*

The organizers should expand the range of benefits of Sunshine Sports Program through multilevel events such as intra-school, inter-school, locally as well as nationally and a variety of accompanying activities such as sports and fitness essay and photography competition, animation collection exhibition, sports reporter competitions and cheerleaders show. At the same time, more attention and consideration should be taken into students with disabilities. Also, more opportunities should be given to them to compete alongside non-disabled athletes.

6.2 *Forming a functional system of the linkage at school, community and family level*

School leads, community supports and the family backs. In the development of Sunshine Sports Program, we must fight for the community's support and guarantee of the family, or the social environment of the Sunshine Sports Program can not be

a radical improved and the effect of the Sunshine Sports Program will be greatly affected because the parents will deprive of children's activity time with the excuse of bad influence on study.

6.3 *Gradually establishes the "government guides, social groups leads" Sun Sports Program administration pattern through socialization and market-oriented operation*

The governments at all level should play the role of system and policy making and supervising when the daily operation function should be handed over to a number of social groups such as the sports federations, all walks of sponsors and primary and secondary schools at all levels.

6.4 *Positively develops and innovates game system and formats with local characteristic*

The event of this project should be designed according to the different locations and different seasons. Revising the rules, the playground's layout and competition system based on the formal sports formats are encouraged, so that more young people can join the Games and have enthusiasm for fitness.

REFERENCES

[1] Xia, Yuan Hong, Zhou Xiao-min, 2011. Influencing factor of operation mechanism of "sunny sports" at primary and secondary schools. Journal of Shandong Institute of Physical Education and Sports 27(2):83–87.

[2] Shen Ling, 2008. Analysis and Proposal for Current Situation of Sunlight Sports Carries out in Basic Education Stage. Journal of Xi'an Physical Education University 25(6):122–124.

[3] Mallika Kanyal, Linda Cooper, 2010. Young children's perceptions of their school experience: a comparative study between England and India. Procedia—Social and Behavioral Sciences 2(2): 3605–3613.

[4] Stuart J. Fair clough, Zoe H. Butcher, Gareth Stratton, 2007. Whole-day and segmented-day physical activity variability of northwest England school children. Preventive Medicine 44(5):421–425.

Future Information Engineering and Manufacturing Science – Zheng (Ed)
© 2015 Taylor & Francis Group, London, ISBN 978-1-138-02644-5

Researched on software evolution process modeling

Zuo Jiang
School of Software, Yunnan University, Kunming, China
School of Mathematics and Computer Science, Yunnan Nationality University, Kunming, Yunnan, China
Key Laboratory in Software Engineering of Yunnan Province, Kunming, China

Zhuo-jia Jiang
China Telecom Corporation Ltd., Yunnan Branch, Kunming, Yunnan, China

Zhi-tao Fu
School of Mathematics and Computer Science, Yunnan Nationality University, Kunming, Yunnan, China

ABSTRACT: An Object-Oriented Software Evolution Process Meta-Model (OO-EPMM), formal OCL constraint of meta-model and a software evolution process model are presented in this paper. OO-EPMM can not only represent software development process, but also represent software evolution.

Keywords: MetaModel; Software process; Software evolution process; OCL

1 INTRODUCTION

A software process is a set of activities, methods, practices, and transformations that people use to develop and maintain software and associated products (project plans, design documents, code, test cases, user manuals, and so on) [1]. Process is a collection of activities, activity is the collection of tasks and the task is an operation converting input into output operation. Software evolution has become an important characteristic in the software life cycle [2]. A software evolution process is a set of interrelated software processes under which the corresponding software is evolving [3]. A well-managed software evolution process will lead to high quality and efficient evolution of software systems on time and under budget. Software evolution process models describe the software process activities, roles, work products, resources and other information which is used to analyze and implement the evolution process. Activities, roles and work products are the most basic elements in all these process elements [4]. The relationship among Activities, Roles and Work products are depicted in Figure 1.

Based on the extended Petri Net, Hoare Logic and Backus-Naur Form, a formal Evolution Process Meta-Model (EPMM) and the corresponding descriptions to support software evolution were proposed in [3]. The Software & Systems Process Engineering Meta-Model 2.0 (SPEM2.0) developing by the Object Management Group (OMG)

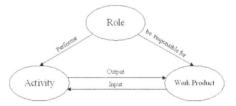

Figure 1. SPEM 2.0 meta-model architecture.

is defined as a meta-model as well as a UML 2 Profile, which serves as a guide for understanding the semantics of this meta-model as well as its direct application for all method and process modeling activities [5].

In order to make the EPMM meta-model supporting MDA framework, Inter operating between models and getting UML CASE tools support, SPEM2.0 architecture are referenced in this paper. By using rich process elements provided in SPEM2.0, a complete objecting EPMM meta-model (OO-EPMM) is presented.

2 OO-EPMM ARCHITECTURE

Object Oriented modeling for enterprise-level concurrent engineering provides a fundamental framework for shifting from structured type traditional methods to an object technology [6]. Object

Oriented Evolution Process Meta-Model (OO-EPMM) using object technology is used to define software evolution process which have double features of software process and software evolution. Therefore, OO-EPMM can not only represent software development process, but also represent software evolution. The goal of OO-EPMM is designed to represent a large range of method entities and different process styles or lifecycle models. OO-EPMM focuses on optional and flexible software development information process structures and method entity contents that are used to processes model. Moreover, OO-EPMM provides reference classes, by which some behavior models from the third parties such as Petri nets, Activity diagrams, State machine etc. can be modeled. The OO-EPMM is structured into six meta-model packages as depicted in Figure 2.

There are six meta-model packages in OO-EPMM meta-model Architecture and the following capabilities are provided by these packages:

1. **Core**: The Core package includes the classes which are the basic classes of all other packages' classes. That is to say, all the common classes defining the core of OO-EPMM have been placed here.
2. **EProcessStructure**: EProcessStructure package is the basis of all process models. A breakdown and nested activities are its core data structure which maintain references to Role classes which perform activities and work product classes which are input or output parameters of activities.
3. **EntityContent**: The EntityContent package is a development knowledge base independent of any specific processes and development projects. It includes concepts and reusable method content elements of software development methods, techniques to define lifecycle.
4. **EProcessBehavior**: EProcessBehavior package extends structures models, which can be represented by any behavior models such as Petri nets, Activity diagrams, State machine etc.
5. **EProcessWithEntity**: EProcessWithEntity package integrate structures and the instances of EntityContent meta-model package concepts. Basic entity contents EntityContent packages are placed into lifecycle model comprising, such as phases and milestones. Some entity content, such as Tasks, Roles, and Work Products are applied in the specific parts of the process.
6. **EProcessLibrary**: In EProcessLibrary package, some concepts are used to design and manage reusable maintainable, flexible, large scale, and configurable libraries of entity contents and processes are provided. Extensibility and variability mechanisms about method entity contents and process are provided in the package. A more flexible process will be created by using the mechanisms.

In OO-EPMM, there are some key concepts in Process structure package and EntityContent package. The elements in EntityContent package mainly express method entity concepts such as task definition, role definition, work product definition, step, guidance etc. In process structure, these elements such as process, activity, task use, role use, work product use are defined. Among these elements, the breakdown structures of activities are most important content. In addition, task use, role use, work product use express respectively reference to task definition, role definition, work product definition.

3 MAIN CLASSES AND CORRESPONDING ABSTRACT SYNTAX

Overall, by using UML 2 package mechanism, OO-EPMM realizes gradual extensions of the modeling capabilities from lower levels to high levels. Usually, these classes included in lower level packages are simply, and those classes in higher level package extend the modeling capability by the UML 2 package merge mechanism with additional properties and relationships to realize more complex process modeling. Length of be confined to, we can't touch on all the classes in every packages here. We only

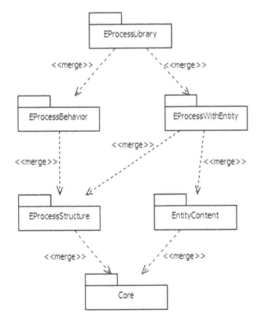

Figure 2. OO-EPMM architecture.

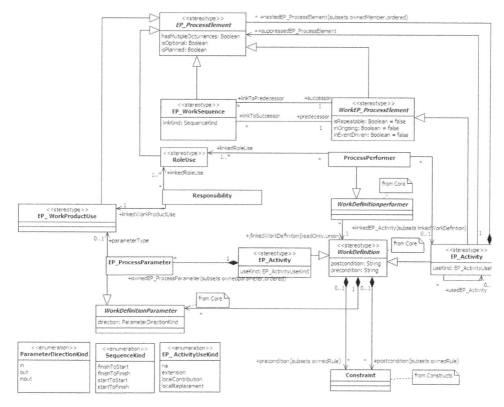

Figure 3. Main classes and associations defined in the Process Structure package.

present main classes and associations of process structure package and corresponding abstract syntax in Figure 3.

Because UML class diagrams syntactic added is not sufficient to semantically define the metamodel, OMG OCL (Object Constrains Language) are used to add formal constraints which limit the possible instantiations and thus the valid process models [8]. Some constraints are as follows:

[C1] A Processformer must be responsible for all the products carried out by activities of which he perform, and reciprocally.
context ProcessPerformer
inv: let productsActivities: Set{WorkProduct} =
self. linkedEP_Activity → select(a:
EP_ProcessElement |a.oclIsTypeOf (EP_
WorkProductUse)).
self. linkedRoleUse. linkedWorkProductUse
→ asSet()
in productsActivities
[C2] ProcessPerformer prohibit the instanciation:
context ProcessPerformer
inv: self.allInstances() → size() = 0

[C3] EP_WorkSequence has four Sequence Kinds, and these Sequence Kinds respectively four values.
context EP_WorkSequence inv:
if SequenceKind = finishToStart then
LinkKind = 0
endif
if SequenceKind = finishToFinish then
LinkKind = 1
endif
if SequenceKind = startToStart then
LinkKind = 2
endif
if SequenceKind = startToFinish then
LinkKind = 3
[C4] Enumeration Type ParameterDirectionKind has 3 common values.
context ParameterDirectionKind inv:
self.values → size() = 3
[C5] Association with A Responsibility is defined to have just single RoleUse and single EP_WorkProductUse.
context Responsibility inv:
self. linkedRoleUse → size() = 1
self. linkedWorkProductUse → size() = 1

309

4 SOFTWARE EVOLUTION PROCESS MODEL

Software evolution process is about a project management, which relates to the management of software development process, people and tools. Based on OO-EPMM, a software evolution process model can be expressed in formal way. In this paper, by using the Petri net, a software evolution process model was presented. A software evolution process model is an 8-tuple:

$$EPM = (S, T, F, V, A_F, \phi_T, \theta_S, M_0)$$

1. Let $N = (S, T, F)$ be a net (c.f [3]);
2. $V = \{x_1, x_2, ..., x_k, ...\}$, $x_i \in V$, V *is* a finite set of variables;
3. A_F: $F \rightarrow P_f (V)$, A_F is first-order predicate formula set, which is defined on the directed edge set F;
4. ϕ_T: $T \rightarrow P_t (V)$, ϕ_T is first-order predicate formula set, which is defined on the transition set T;
5. θ_S: $S \rightarrow P_s (V)$, θ_S is first-order predicate formula set, which is defined on the place set P;
6. $M_0 = \cap s \in M_0(s)$, M_0 is initial marking, $M_0(s)$ represents tokens set in place s.

5 CONCLUSION

In order to not only represent software development process, but also represent software evolution, firstly, an OO-EPMM architecture are presented; secondly, key concepts in meta-model package are discussed; thirdly, part abstract syntax of OO-EPMM is presented; finally, a software evolution process model is expressed in formal way in this paper. Our further work is to design concrete algorithms to get helpful guidance from OO-EPMM meta-models in software process implementation.

ACKNOWLEDGEMENTS

This work was financially supported by the National Science Foudation of China under Grant No. 61262024, 61379032, by the Science Foundation of Key Laboratory in Software Engineering under Grant No. 2012SE402, by the Science Foundation of Yunnan Province, China under Grant No. 2012FD005.

REFERENCES

[1] Paulk M C, Curtis B, Chrissis M B, et al. Capability maturity model, version 1.1. Software, IEEE, 1993, 10(4): 18–27.
[2] Oreizy P, Medvidovic N, Taylor R N. Architecture-based runtime software evolution//Proceedings of the 20th international conference on Software engineering. IEEE Computer Society, 1998: 177–186.
[3] Li, Tong. An approach to modelling software evolution processes. Springer, 2008.
[4] Paulk M. Capability maturity model for software. John Wiley & Sons, Inc., 1993.
[5] OMG, "Software & Systems Process Engineering Metamodel Specification (SPEM) Version 2.0", Tech. Rep., Object Management Group., 2008.
[6] Blaha M, Rumbaugh J. Object-oriented modeling and design with UML. Upper Saddle River: Pearson Education, 2005.
[7] Pan Wen-lin, Liu Da-xin. "Abstract Syntax of Object Role Modeling". 2010 International Conference on Software and Computing Technology (ICSCT2010), Kunming, China, 2010. IEEE: 25–29P.
[8] Combemale B, Crégut X, Caplain A, et al. Towards a Rigorous Process Modeling with SPEM //ICEIS (3). 2006: 530–533.

Future Information Engineering and Manufacturing Science – Zheng (Ed)
© 2015 Taylor & Francis Group, London, ISBN 978-1-138-02644-5

The effect and simulation of one-way traffic on urban transportation environment

Weixiong Zha & Tingting Zhu

Institute of Transportation and Economics—Humanities and Social Research Base of Jiangxi Province, East China Jiaotong University, Nanchang, Jiangxi, China

ABSTRACT: As a special and effective traffic organization, not only the one-way traffic can improve the vehicle's traveling speed, operating efficiency and intersection's traffic safety in the region, but also have a high-value on reducing the urban environmental pollution. Its special organizational model is destined to improve the road travel speed and capacity. However, the improvement of systematic operating efficiency can increase the ratio of energy utilization and reduce emission rate. Air pollutants caused by vehicles are mainly CO, NOx and HC compound. So this study focuses on the three exhausts' emission rate. We use the general function of TRANSYT7-F model to analyze the problem. Thus, it needs to be evaluated depending on the actual situation. This paper takes Bayi Square in Nanchang as an example, compared with the implementations of traffic measures forward and backward, analyzing its implementation's effect on low-carbon traffic.

Keywords: One-way Traffic; Low-carbon Traffic; Urban Transportation Environment; Pollutant Emission; Energy Consumption

1 INTRODUCTION

At present, implementing various urban traffic control policies are aimed at reducing traffic congestion and increasing road capacity. With more and more attention to urban environment, building low-carbon system is mainly to reduce transport energy consumption and exhaust emission and meet the travel demand so as to promote sustainable development in urban traffic under through organizing scientifically and optimizing transport resources legitimately. Thus, exhaust emission should be an important indicator in traffic environmental impact assessment, to evaluating the goals of satisfying travel demand and environmental sustainable development.

Liu Wei, et al. (2010) calculated the one-way traffic energy consumption and exhaust emission using vehicle emission and the energy consumption models and established an energy consumption discrimination model with the ratio method.

Zhao Tong, et al. (2010) proposed the heurist algorithm which is based on the idea of the Branch and Bound to solve the model of maximal OD demand in urban transport discrete network design problem under environment objective.

Yao Zhiliang, et al. (2010) introduced IVE model and presented the method to quantify the main parameters by taking Beijing city for example, compared with MOBILE 6 and IVE model.

February 2011, Tianjin, Chongqing, Shenzhen, Xiamen, Hangzhou, Nanchang, Guiyang, Baoding, Wuhan and Wuxi are selected to carry out low-carbon transport system construction.

2 ONE-WAY TRAFFIC

One-way traffic refers to the traffic of vehicles travelling in one direction on the road. As an effective urban transport organization, it makes the vehicles travelling in one direction, reducing the interference of vehicles in opposite direction, makes the vehicles in opposite direction round to the other roads, equilibrating the traffics. The increase of travel speed not only improves the efficiency of energy utilization, but also reduces the carbon emissions. If the entrance lanes are to one-way traffic instead, the parking frequency at intersection will be reduced and the air pollution caused by automobile exhausts will be improved.

3 ANALYSIS OF THE ONE-WAY TRAFFIC ON POLLUTANT EMISSION

The implementation of one-way traffic will change traffic flow in original road network, reduce the complexity of the intersection and improve road capacity and travel speed, but meanwhile, will increase the circuitous distance and change energy consumption and exhaust emission in region accordingly. Therefore, it is necessary to analyze and evaluate the effect of one-way traffic scheme.

We get the traffic flow and travel speed by survey data and calculate the exhaust emission by means of the model of vehicle emission to appraise the energy efficiency with the one-way traffic scheme applied pre and post.

Air pollutants caused by vehicles are mainly *CO*, *NOx* and *HC* compound. It's showed that about 90% *CO* and 50% *NOx* in urban pollution are from vehicles' exhaust. So this study focuses on the three exhausts emission rates.

According to the different simulation method, vehicle emission model is classed into average speed and driving conditions. The average speed model takes the average speed as characteristic parameters, gets the gross of contamination via modified emission factors and establishes a mathematical model.

3.1 Calculation method

Southeast University Wang Wei et al. study vehicles' emission factors are sensitive to the speed, and they proposed a modified method about exhaust emissions based on MOBILE5 model to be suitable for China traffic planning. The estimating formulas for emission factors of hybrid vehicles are seen for example Equation 1, 2 and 3 below:

$$R_{CO} = 7.344 + 318.445/v \tag{1}$$

$$R_{HC} = 0.825 + 33.875/v \tag{2}$$

$$R_{NOx} = 3.879 - 0.099v + 0.002v^2 \tag{3}$$

We assume that, the OD demand on road network is constant, vehicle emission factors are invariant in addition to speed and the delay in intersection is assigned to the road. So we can get the following function:

$$t_i(x_{ij}) = 1_j \cdot R_{ij} \tag{4}$$

$$T_i(x_{ij}) = x_{ij} \cdot l_j \cdot R_{ij} \tag{5}$$

$$t = \sum_i \sum_j x_{ij} \cdot l_j \cdot R_{ij} \tag{6}$$

$$t' = \sum_i \sum_j x'_{ij} \cdot l_j \cdot R'_{ij} \tag{7}$$

$$\rho = \frac{t' - t}{t} \tag{8}$$

Where i = the kinds of pollutants, j = the link, v_j = the average speed of link (m/s), R_{ij} = the pollutant emission rate (g/(pcu·m)), l_j = the length of link j. x_j and x'_j = the traffic flow on link j, R_{ij} and R_{ij}' = the i-th pollutant emission rate of exhaust on link j, t and t' = the total emissions within the system which the one-way traffic scheme applied pre and post. ρ = the rate of total emissions change.

If ρ is positive, it indicates that the implementation of the one-way traffic makes the pollutant emissions increase. Otherwise, the one-way traffic planning reduces the pollutant emissions, which has important positive implications for the construction of urban low-carbon traffic system.

4 EXAMPLE ANALYSIS

The paper takes Nanchang Bayi Square as an example. Bayi Square is located in the central area of old city in Nanchang, with heavy traffic and road in saturation state. Since the construction of Metro Line 1 takes up 50 percent of road, but the road cannot be taken the measure of occupying one lane and returning another one, so the microcirculation in region of Bayi Square is considered setting. Originally, the intra-regional roads are two-direction traffic. The microcirculation is shown in the following figure 1. □—□ represents the primary node of regional boundaries, (1)—(4) stands for sections, < → > is the changes in the number of lanes which the one-way traffic scheme applied pre and post.

According to 2012 data and the survey results in 2014, we got the main operating indicators of lanes in region which the one-way traffic scheme applied pre and post, see Table 1.

Then we can calculate the exhaust emissions, the results are seen in Table 2.

According to Table 2, total emissions dropped 323.0 (g/h), the rate of change is −0.201. It indicates the microcirculation in Bayi Square can reduce the

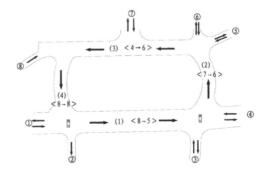

Figure 1. Caption of one-way traffic scheme.

Table 1. The main operating indicators of lanes in region.

Lane	L(km)	Traffic flow		Average speed	
		1	2	1	2
(1)	0.373	8100	4950	23	26
(2)	0.166	4984	4306	22	29
(3)	0.397	3685	5096	24	28
(4)	0.166	5700	5440	28	33

Table 2. The exhaust emissions.

Lane	R_{CO}		R_{HC}	
	1	2	1	2
(1)	21.19	19.59	2.30	2.13
(2)	21.82	18.33	2.37	1.99
(3)	20.61	18.72	2.34	2.04
(4)	18.82	16.99	2.04	1.85

Lane	R_{NOx}		Exhaust emissions	
	1	2	1	2
(1)	2.66	2.66	789.99	450.08
(2)	2.67	2.69	222.16	164.46
(3)	2.66	2.68	373.11	473.95
(4)	2.68	2.79	221.67	195.38
Total			1606.9	1283.9
The rate of change			−0.201	

pollutant emissions, have an important positive significance to the construction of transportation environment in Nanchang and be conducive to the sustainable development of urban traffic.

5 CONCLUSION

The one-way traffic not only can improve the vehicle's traveling speed, operating efficiency and intersection's traffic safety in the region and reduce delays in intersection, but also can reduce the urban environmental pollution. The dates in the paper are based on survey, but for planning one-way traffic scheme, we can assume the OD in region unchanged, simulate the scheme by use of TransCad and VISSIM, and then judge the pollutant emissions of the one-way traffic scheme applied pre and post.

REFERENCES

[1] Zhao Tong, Guo Tiande, Gao Ziyou 2005. An Optimal and Solution Algorithm for Maximal OD Travel Demand in Urban Transport Discrete Network Design Problem under Environment Objective. China Civil Engineering Journal 3(3): 119–124.
[2] Yao Zhiliang, et al. 2006. Application Study of IVE Vehicle Emission Model. Environment Science 10(27):1928–1933.
[3] Liu Wei, Hua Wenting, Fu Qingsong 2010. One Way Traffic Organization Scheme Evaluation Based on TransCad. Journal of Transport Information and Safety 5: 99–103.
[4] Jia Shunping, Mao Baohua 2010. Calculation and Analysis of Transportation Energy Consumption Level in China. Journal of Transportation Systems Engineering and Information Technology 2(1): 22–27.
[5] Liu Wei, Gao Jianjie, Fu Qingsong 2010. Discrimination Method of Urban One-way Traffic Plan in Energy Consumption and Emission. Journal of Transportation Engineering and Information 6(8): 27–42.
[6] Wu Bing, et al. 2007. Traffic Management and Control. Beijing: China Communication Press.

Future Information Engineering and Manufacturing Science – Zheng (Ed)
© 2015 Taylor & Francis Group, London, ISBN 978-1-138-02644-5

Simulation of Brushed DC Motor driving system based on fuzzy PID

Yaru Xu, Jingping Lu & Chuanhong Zhang
School of Mechanical Engineering, Guangxi University, Nanning, China

ABSTRACT: Brushed DC Motor drives the nut screw mechanism to realize forklift automatic clutch. Referring to the non-linearity, strong-coupled time-varying and lagging behind of the drive system, Fuzzy self-tuning PID algorithm is adopted and simulated with MATLAB. The result of simulation shows that the fuzzy self-tuning can change the PID parameters in the dynamic process, be able to realize stability in a shorter period of time, has better robustness and adaptability than the conventional PID. The algorithm has better control of separation rapidly and engagement smoothly for forklift clutch.

Keywords: Automatic clutch; Brushed DC Motor; Fuzzy self-tuning PID

1 INTRODUCTION

The forklift, as the small distance transporting machinery, needs to use clutch frequently when it starts, runs and works [1]. Compared with cars, the forklift clutch is used more frequently [2]. The clutch, which needs put the foot on the pedal, is replaced gradually with automation control mechanism, namely, the conventional clutch is assembled with an actuator driven by motor to realize forklift automatic clutch, thus lighting drivers' work and mental burden, reducing shifting jerks, improving ride comfort and avoiding differences due to different technique [3]. The actuator includes motor, drive mechanism and disengaging lever.

Screw transmission with reverse self-locking characteristic [3] can increase the torque transferred by motor and has wide selection of sizes, so the screw rod and screw nut mechanism is selected as the drive mechanism; stepper motor has high control precision, but the problem of out-of-step and under-capacity are solved by increasing the motor's size and the weight; Brushless DC Motor requires less maintenance, lasts longer and produces greater output power, but the control process is very complicated, and there is the problem of torque ripple and is easy to produce vibrations against precise control; the life of Brushed DC Motor is short, but it is simple to control, has good controlling capability [4], high startup and tune performance. Automatic Clutch System has a limited mounting space for the forklifts, adopting Brushed DC Motor can reduce manufacturing costs and the difficulty of arrangement, and it is integrated conveniently. Failing to address some uncertain information of the drive system for

the conventional PID causes it can't reach good propulsion performance; fuzzy control with strong robustness has no need for exact mathematic model of the controlled plant and is better fit to the problems of non-linearity, strong-coupled time-varying and lagging behind, but it has low steady accuracy and can't even define control objectives. Combining the conventional PID with fuzzy control applies to Brushed DC Motor drive system to have better control of separation rapidly and engagement smoothly for forklift clutch.

2 BUILD MATHEMATICAL MODEL OF ACTUATOR

To serve the liberation of drivers' left feet, replace the original clutch pedal with the actuator to realize power transmission of the Brushed DC Motor and the positive inversion of the motor will be transformed into the engagement and disengagement of the clutch.

The working principle: Brushed DC Motor 1 is driven by ECU and drives screw 4 rotating through coupling 2 and gear reducer 3, nut 5 converts rotating movement into linear movement and pushes the disengaging lever 6 moving forward and backward to realize the engagement and disengagement of the clutch7, as shown in Figure 1. The speed and state of the engagement and disengagement for clutch are determined by turning speed and turning direction of motor controlled by ECU, which stores operational experiences, thus guaranteeing the clutch can work under the optimal work condition to realize the target of rapid separation and smoothly engage for the clutch.

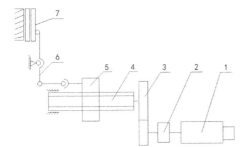

1. Brushed DC Motor 2. coupling 3. gear reducer

4. screw 5. nut 6. disengaging lever 7. clutch

Figure 1.   The actuator of Brushed DC Motor driving the clutch.

For the separately excited DC motor, the following equations can be written out [5]:

Electric network equilibrium equation:

$$L_a \frac{dI_a}{dt} + R_a I_a + E_a = U_a \qquad (1)$$

Mechanical equilibrium equation:

$$J \frac{d\omega}{dt} = T_a - T_L \qquad (2)$$

Electromotive force balance equation:

$$E_a = C_e \omega \qquad (3)$$

Torque balance equation:

$$T_a = C_c I_a \qquad (4)$$

where L_a is the motor inductance; I_a is the motor armature current; R_a is the motor resistance; E_a is the induced electromotive of armature winding; U_a is the terminal voltage of motor armature; C_e is the electromotive force constant, C_c is the electromagnetic torque constant, and both of them are controlled by structure parameters of motor, and torque coefficient $C_e = C_c = C = p \times M \times I_f$, I_f is the motor excitation current, p is the motor magnetic poles; M is the mutual inductance between field windings and armature winding; ω is the motor shaft output angular velocities; J is the total of inertia; T_a is the electromagnetic torque; T_L is the load torque.

Connect the above four equations and the transfer function can be obtained by using Laplace transform [6]:

$$G_1(s) = \frac{\omega_n^2/C}{s^2 + \varsigma \times \omega_n \times s + \omega_n^2} \qquad (5)$$

where ω_n is the natural oscillation frequency, $\omega_n = \frac{1}{\sqrt{T_a \times T_m}}$; ς is decay factor or damper rate, $\varsigma = \sqrt{\frac{T_m}{T_a}}; T_m = \frac{JR_a}{C^2}, T_a = \frac{L_a}{R_a}$.

Use Laplace transform on each side for $z_1 n_i(t) = z_2 n_o(t), z_1 N_i(s) = z_2 N_o(s)$, can be obtained, then transfer function of gear pair is:

$$G_2(s) = \frac{N_o(s)}{N_i(s)} = \frac{z_1}{z_2} = K \qquad (6)$$

where $n_i(t)$ is the input shaft speed, $n_o(t)$ is the output shaft speed, z_1 and z_2 are the teeth of the driving gear and the driven gear, respectively.

The transfer function of screw rod and screw nut mechanism is:

$$G_3(s) = \frac{X(s)}{\theta_2(s)} = P \qquad (7)$$

where P is the lead of screw.

The transfer function mathematical model of the drive system is:

$$G(s) = G_1(s)G_2(s)G_3(s) = \frac{KP\omega_n^2/C}{s^2 + \varsigma \times \omega_n \times s + \omega_n^2} \qquad (8)$$

3   DESIGN OF FUZZY SELF-TUNING PID CONTROLLER

3.1   The routine PID control

PID is a kind of linear controller, the control law is:

$$u(t) = K_P e(t) + K_I \int_0^t e(t) + K_D \frac{de(t)}{dt} \qquad (9)$$

The corresponding transfer function is:

$$G(s) = \frac{U(s)}{E(s)} = k_p \left(1 + \frac{1}{T_1 s} + T_D s\right) \qquad (10)$$

where $r(t)$ is the system given value; $u(t)$ is the controlled quantity; $e(t)$ is the systematic bias, $e(t) = r(t) - y(t)$, $y(t)$ is the actual output; K_P is the proportionality coefficient, used in response proportionately $e(t)$; K_I is the integral coefficient, used in clean the static deviation, $K_I = K_P/T_I$, T_I is the integration time constant; K_D is the derivative coefficient, used in react the variation speed of $e(t)$, $K_D = K_P T_D$, T_D is the differentiation time constant. K_P, K_I and K_D guarantee the stability, accuracy and rapidity of the system, respectively.

3.2 Fuzzy control

Fuzzy control is a kind of control mode to operate system work by adopting the control rules described by fuzzy mathematics linguistics, and the two key players are the fuzzy presentation of expertise and the fuzzy inference of on-line control.

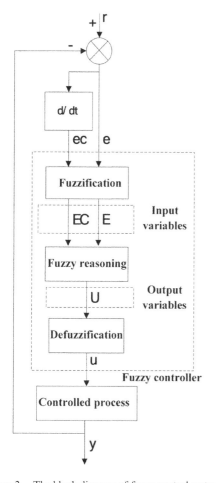

Figure 2. The block diagram of fuzzy control system.

As shown in Figure 2, after blurred of deviations e (the difference between the feedback value of the controlled objects and the target value) and deviating rate ec, the input lingual variables E and EC are obtained. The key is determining the rule of fuzzy control; and then the controlled quantity u after defuzzification performs the control action.

3.3 Design of fuzzy self-tuning PID controller

Fuzzy self-tuning PID can self-tune the PID parameters online with fuzzy reasoning [7].

As shown in Figure 3, use e and ec as inputs, look for the fuzzy relationship between the PID parameters and e, ec, measure constantly e and ec in motion, modify the PID parameters online with fuzzy reasoning to satisfy the e and ec with different times requirement for self-tuning of PID parameters [8]. The key steps of designing fuzzy self-tuning PID controller are as follows:

1. Open up the FIS Editor Window.
2. Define variables. Use e and ec as inputs, PID parameters correction ΔK_P, ΔK_I, ΔK_D as outputs [9]. Select the controller type is Mamdani under the File menu.
3. Choose the universe of input variables and output variables. Select E, EC = {−6, −4, −2, 0, 2, 4, 6} as the input parameter linguistic variables, e, ec = {NB, NM, NS, O, PS, PM, PB} as the linguistic variables fuzzy subsets [9]. Select the universe of ΔK_P, ΔK_I, ΔK_D are all [−3, 3].
4. Determine the membership functions. Select the membership functions of e, ec, ΔK_P, ΔK_I and ΔK_D are all trimf.

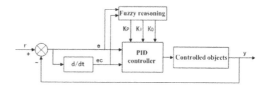

Figure 3. The structure of fuzzy self-tuning PID controller.

Table 1. Fuzzy control rules of ΔK_P, ΔK_I, ΔK_D.

e	Ec						
	NB	NM	NS	ZO	PS	PM	PB
NB	PB/NB/PS	PB/NB/PS	PM/NM/NB	PM/NM/NB	PS/NS/NB	ZO/ZO/NM	ZO/ZO/PS
NM	PB/NB/PS	PB/NB/PS	PM/NM/NB	PS/NS/NM	PS/NS/NM	ZO/ZO/NS	NS/ZO/ZO
NS	PM/NB/ZO	PM/NM/NS	PM/NS/NM	PS/NS/NM	ZO/ZO/NS	NS/PS/NS	NS/PS/ZO
ZO	PM/NM/ZO	PM/NM/NS	PS/NS/NS	ZO/ZO/NS	NS/PS/NS	NM/PM/NS	NM/PM/ZO
PS	PS/NM/ZO	PS/NS/ZO	ZO/ZO/ZO	NS/PS/ZO	NS/PS/ZO	NM/PM/ZO	NM/PB/ZO
PM	PS/ZO/PB	ZO/ZO/NS	NS/PS/PS	NM/PS/PS	NM/PM/PS	NM/PB/PS	NB/PB/PB
PB	ZO/ZO/PB	ZO/ZO/PM	NM/PS/PM	NM/PM/PM	NM/PM/PS	NB/PB/PS	NB/PB/PB

5. Design the control rules. According to the relationship between ΔK_P, ΔK_I, ΔK_D and e, ec, and the tuning fundamental principles [10], establish the table of fuzzy control rule, combined with superior technical knowledge and practical experience, as shown in Table 1 [8]. Use the rules under the Edit menu to realize the edit of fuzzy rules.
6. Establish the method of fuzzification, which is discretization of the accurate quantity, and defuzzification, which is centroid.

Up to this point, construct the fuzzy controller with a two-input (e, ec) and three-output (ΔK_P, ΔK_I, ΔK_D) based on the Table 1, and name BDCM-FUZZY_PID.fis.

4 SIMULATION RESULTS AND CORRESPONDING ANALYSIS OF MATLAB

In the Brushed DC Motor drive system described in the article, $P = 2$ mm, $K = 0.1$, $J = 0.91\,\mathrm{Kg \cdot cm^2}$, $L_a = 0.72\ H$, $R_a = 0.53\Omega$, $p = 2$, $M = 0.034\,\mathrm{mH}$, $I_f = 17.2\ MA$, then the transfer function of the drive system is:

$$G(s) = \frac{0.35}{s^2 + 0.73s + 2.07} \tag{11}$$

Firstly, read BDCMFUZZY_PID.fis, set sample time and simulation time, and discretize the transfer function of the system; then set the initial value of the system, input step signal; finally, perform computations of fuzzy reasoning, and then the output of fuzzy PID is obtained. In order to validate the performance of the fuzzy self-tuning PID controller, the simulation and comparison with the routine PID control are given, as shown in Figure 4.

Simulation results indicate that Fuzzy self-tuning PID with the feature of strong adaptability can make system have rapid speed response and strong robustness, improve greatly the adjusting precision and steady performance than the routine PID control with long transition time and poor stability.

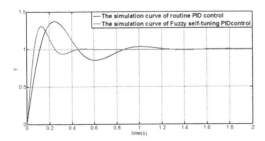

Figure 4.   The simulation curves.

5 CONCLUSION

Adopt Fuzzy self-tuning PID algorithm to control Brushed DC Motor driving the nut screw mechanism to realize forklift automatic clutch. Modify the PID parameters online with fuzzy inference reasoning by calculating current error and variety rate of error of the system based on the routine PID control. The method with better robustness and self-adaptation can be stabilized much faster than the routine PID control.

ACKNOWLEDGEMENTS

Project name and number: Project supported by the Key Laboratory of manufacturing system and advanced technology of Guangxi: coordinate control method and applied research for the comprehensive campaign of multiple spindle linkage system, Contract Number: 11-031-12S03.

REFERENCES

[1] Niu Wenbin, Ji Junrong, Huang Jizhong. Failure reason analysis and resolution of the machinery forklift truck clutch and release bearings. *Journal of information science and technology*, 2011 (18): 389–389.
[2] Zeng Wenzhang. Preliminary design research on forklift truck clutch manipulate. *Journal of forklift truck technology*, 2005 (3): 7–9.
[3] Liu Chengjun. Control strategy research and prototyping design on automatic clutch for mini cars. *Wu Han University of Technology*, 2009.
[4] Xie Xianping. Control strategy research on the engaging process of automobile automatic clutch control strategy research. *Harbin university of science and technology*, 2009.
[5] Wang Yanying, Wang Zhen, Guo Lihuan. Experiment study on the transfer function for DC motor. *Journal of experimental technology and management*, 2008, 25 (8): 38–40.
[6] http://wenku.baidu.com/link?url = KOHLCGGOv IXYCwZNlkk4l026vXUITYzhRY3_6uRLYIIUfz WMvyCSvpAFC5gACaFnG5xayLHIlqA2kO4fjx6-p0L9CCBeFKEuAgRIflUGbyL_
[7] Tian Hai, Li Jun. Simulation Study of Fuzzy Self-tuning PID Controller Based on Matlab. *Industrial control computer*, 2011, 24 (6): 24–25.
[8] Liu Jinkun. The advanced PID control and MATLAB simulation. *Electronic industry press*, 2004.
[9] Jiang W, Jiang X. Design of an Intelligent Temperature Control System Based on the Fuzzy Self-Tuning PID. *Procedia Engineering*, 2012, 43: 307–311.
[10] Wang An, Yang Qingqing, Yan Wenyu. Design and simulation on Fuzzy self-tuning PID controller. *Computer simulation*, 2012, 29 (12): 224–228.

Future Information Engineering and Manufacturing Science – Zheng (Ed)
© 2015 Taylor & Francis Group, London, ISBN 978-1-138-02644-5

Analysis of Chinese human capital investment strategy of scientific outlook on development

Xi Zhu
Postgraduate Student of Business School of Jiang Xi Normal University, Nanchang, China

Jiawen Huang
Vice President of Jiang Xi Normal University, Nanchang, China

ABSTRACT: With the approach of the knowledge economy era, a higher standard of human capital stock and quality is essential for a country that wants a better development. And the investment situation of human capital is very important for the development of the national economy and people's living standard. Using the method of statistical data analysis, this paper analyses the current education investment situation of China in recent years from the perspective of macro and micro, finds out some problems of human capital investment, and puts forward some suggestions for reference to promote the human capital investment strategy to combine with Scientific Outlook on Development, and lay good foundation for the sound and rapid development of Chinese economy.

Keywords: Human capital investment; Physical capital; Scientific Outlook on Development

1 INTRODUCTION

Human capital refers to the capital cost in terms of human health, education, training or other forms. The investment in human capital is the activities which will have impacts on future currency and the material income by increasing human resources, including increased investment in education and training, social security and health insurance.

With the advent of the era of knowledge economy, human capital is becoming increasingly important for a country's economic development. Under the guidance of the Scientific Outlook on Development, Chinese economic development pattern has gradually transferred from extensive to intensive, and the transformation of economic development mode has put forward higher requirements on knowledge, technology, and high-quality human capital. So increasing human capital and improving human capital investment structure has become an imperative. The founder of modern human capital, Schultz and Baker, demonstrated the problems of human capital investment systematically from the ability of workers. They believed that the investment in human capital plays an important role in promoting economic growth and improving labor income.

2 GETTING STARTED

2.1 *The analysis of human capital investment present situation and problems in China*

2.1.1 *The comparative analysis of education investment and material investment in China*

"Human Capital" which wrote by Nobel prize-winning economist, Garys Becker, analyzed the importance of the human capital for economic growth and the increase of national income. Besides, he demonstrated that the education investment is a kind of productive investment, which could have a huge impact on the development of economy. Combining the current reality, China has made its efforts in developing education career, as is shown in Table 1, the finance for total education investment in China is increasing year by year.

Although the total amount of education investment has an increasing trend, its relative amounts that compared with the material investment are still in a low level during the recent decades. But according to the Schultz's research findings, the future rate of return received by education investment are greater than that of material investment. In addition, according to surveys, an increase of 100 million of human capital investment can bring nearly 600 million GDP's increasing next year, while an increase of 100 million of material capital investment can only bring 200 million GDP's increasing. So the education investment of human capital is still in a low level in our country. And if the education investment of human capital is lagging behind the material investment in a long term, it would lead to bad consequences, which is that excellent technology and equipment would lack talents to make use of, then it may reduce the material capital investment efficiency.

Table 1. The financial education funds and material investment in our country in 1998–2012.

Year	State financial education funds (one hundred million Yuan)	Material investment (one hundred million Yuan)
1998	2032.5	28406.2
1999	2287.2	28954.7
2000	2562.6	32917.7
2001	3057.0	37213.5
2002	3491.4	43499.9
2003	3850.6	55566.6
2004	7242.6	70477.4
2005	5161.1	88773.6
2006	6348.4	109998.2
2007	8280.2	137232.9
2008	10449.6	172828.4
2009	12231.1	224598.8
2010	14670.1	251683.8
2011	18586.7	311485.1
2012	21984.0	374694.4

In a word, our county is paying more and more attention to human capital investment and the trend of investment in education in total national finance is gradually increasing. But compared with the material capital investment, it is still relatively backward.

2.1.2 *The regional human capital investment gap is too large between provinces in China*

In China, high-quality talents are habitual to go to the eastern coastal areas for their own career. Although facing this reality, the scale of human capital investment of eastern area is larger than the western area in China. Table 2, education funding of 2011, was selected as a good evidence, it can be seen that regional education investment is gradually widening. As we all know, the talents gap between regions have been very big. If the regional education investment gap is still significantly large, the quality and quantity of our country's regional talents would become larger and larger, which would eventually lead to that the regional economic development gap would become larger and larger.

2.1.3 *The state, enterprises and individuals all ignored vocational training*

Formal education and on-the-job training are the important ways of human capital investment. Upon completion of school education, on-the-job training is a good opportunity for persons to get a better improving. When analyzing the formation of human capital, Becker also emphasized the equal role of education and training. Besides, he analyzed the relationship between on-the-job training and one person's income level in the future.

Table 2. Education funding of different regions in 2011 (Unite: one hundred million Yuan).

ER*	EI*	CR*	EI	WR*	EI
AVE*	729.3	AVE	541.0	AVE	389.4
Beijing	627.7	Heilongjiang	385.9	Neimenggu	446.4
Tianjin	338.9	Jilin	354.3	Chongqing	383.2
Hebei	684.5	Shanxi	445.2	Sichuan	801.7
Shandong	1122.5	Henan	929.2	Guizhou	387.0
Liaoning	632.6	Anhui	646.1	Yunnan	565.3
Jiangsu	1176.8	Hubei	478.8	Xizang	80.7
Shanghai	584.4	Hunan	584.7	Shanxi	525.0
Zhejiang	873.3	Jiangxi	503.7	Gansu	312.9
Fujian	485.6			Qinghai	147.0
Guangdong	1359.2			Ningxia	114.7
Hainan	137.1			Xinjiang	418.3
				Guangxi	490.6

*ER refers to the eastern region of China.
CR refers to the central region of China.
WR refers to the western region of China.
EI refers to education investment.
AVE refers to the average value.

He mentioned that there is a difference between formal school education and vocational training. School education pays more attention to the improvement of book knowledge, and on-the-job training focuses on the application to work-related knowledge, skills and the ability enhancement on various aspects.

After these years of efforts, the enterprises have paid more attention to staff training on the job. According to National survey of more than 2100 companies of Enterprise Development Research Center of the State Council (2004), 69% of the enterprises have made the employee training plan. And Chinese enterprise training management status survey (2010) which carried out through network showed that 24% of the enterprises owned "a sound training system", 65% of the enterprises owned "basic training management", and only 10% of the enterprises are "training clueless" or "never trained management". Still the-job training of Chinese enterprises exist a number of issues. According to the State Council Development Center survey (2004), the proportion that enterprises spent on staff training in sales revenue accounted for 3%–5% or more is only 8.7%, and the proportion that enterprises which spent on staff training in sales revenue accounted for 0.5% or less is as high as 48.2%.

We can compare the situation with developed countries. According to statistics, American enterprises' annual funding for job training reached $ 210 billion per year, which has respectively exceed the funding of secondary education and higher education. Besides, UK costs about 33 billion pounds on in-service training per year. By comparing, no matter from the perspective of individuals, companies or states, the investment in vocational training is very few.

2.2 Guided by the scientific outlook on development, and set up the sustainable development of human capital investment strategy

With the changes of the world, the updating speed of technology turns faster, the degree of foreign trade is deepening, and there will be more and more opportunities for talents in market. All these have proposed the demands for quantity and quality of talents. So we should always adhere to the strategy of rejuvenating the country through science and education, and the strategy of reinvigorating the country through human resource development. Although the effect of the investment in human capital in a short term is difficult to see, our country must adhere to attach great importance to the human capital investment and follow the direction of Scientific Outlook on Development, which is a

important guarantee to carry out the new route in the new period and build a well-off society in an all-round way.

2.2.1 Increase the investment for education of finance, and accelerate the education system's revolution

Scientific Outlook on Development is a kind of coordinated development between economy and society. And human capital investment is a fundamental way to promote economic and social development. Therefore, the state should increase investment in human capital. Through the prior analysis, the total amount of investment of education funding in human capital is too small, which is far below the national proportion of investment for physical capital. However, with the emphasis on the accumulation of physical capital and the neglect of the human capital relatively, it is impossible to ensure sustained economic growth. Therefore, state must coordinate the relationship between investment in human capital and physical capital investment and promote each other, make our country's education budget proportion closer to the developed countries.

When the investment of education has been raised, it is also necessary for state to focus on the education funds investment structure and think over on the education system of our country to accelerate the education system's revolution. For college education, controlling the enrollment scale and strictly treating the approval of good professional settings all deserve to think over. For vocational education, we need to learn from other foreign countries and cultivate specialized talented person. Only in this way, can the effectiveness of the human capital investment turn more maximized.

2.2.2 Coordinate the gap in human capital between regions and optimize the human capital allocation

The basic requirements of Scientific Outlook on Development are comprehensive, coordinated and sustainable. Under the condition that the pace of economic development in the eastern region is much faster than in the central and western regions, there should be greater investment in human capital for central and western regions. So on capital investment, state needs to shift toward the central regions and rural areas, enhance economic growth with a more balanced portfolio policy and try to minimize unequal investment. Meanwhile, it should develop appropriate policies to attract talents to contribute to the development of central and western regions. Only by doing these, it can truly carry out the coordinated development in the Scientific Outlook on Development.

2.2.3 *Attach great impotence to the role of the training institutions and enhance the relation between training institutions and universities*

Facing the current situation of Chinese training market, we should learn from the experience of some developed countries and increase the government intervention appropriately. Government should provide some basic low-skilled workers with basic skill training, also they need to formulate relevant laws and policies to supervise the training investment. Besides, enterprises should combine actual needs with the production and make overall training plan. Government should also increase cooperation between enterprises and universities or training institutions, attach great impotence to the role of universities and vocational training institutions through various businesses, improve and perfect highly skilled personnel training system, and gradually form a modern enterprise training system.

2.2.4 *Human capital investment should pay more attention to people's health and well-being*

Human capital is composed of knowledge, technology, ability, health and other qualities, which are condensed upon workers and have economic value, and they are reflections of the labor quality (Yang Jianfang etc., 2006). In these factors, level of education and health are the two key factors of human capital ownership. In a sense, healthy body, good mood and great happiness can make the workers energetic, and then improve their work efficiency. Besides, the better the labor's health, the higher their happiness index is, the better the human capital investment efficiency is. So when making human capital investment strategy, state should pay attention to people's health and well-being at the same time and increase the investment of medical health care, social security and public infrastructure to create a good social environment and higher work efficiency service for the people's working.

REFERENCES

[1] Garys Becker (1975) Human Capital.
[2] Hun-Yao Tseng et al. (2005) Intellectual capital and corporate value in an emerging economy. Empirical Study of Taiwanese Manufacturers. R and D Management.
[3] Li Ling (2005) Analysis of Chinese enterprise investment in on-the-job training situation. Human Resource Exploitation of China.
[4] Matouschek et al. (2005) The role of human capital investments in the location decision of firms. Regional Science & Urban Economics.
[5] Mincer J. (1985) Investment in human capacity and personal income distribution. Journal of Political Economy.
[6] Sun Yanan (2012) Pushing effect research of human capital on economic growth. Journal of Technology Economy and Management Research.
[7] Tao Xiaolong et al. (2012) Effect of China's economic growth, human capital structure research. Journal of Social Science in Guang Xi Province.
[8] The bureau of the People's Republic of China (2013) Statistical Yearbook of China.
[9] Theodore Schultz (1981) Investment in Human Capital.
[10] Yang Jianfang et al. (2006) The formation of human capital and its influence on economic growth. Management World.

Future Information Engineering and Manufacturing Science – Zheng (Ed)
© 2015 Taylor & Francis Group, London, ISBN 978-1-138-02644-5

Robot odor localization based on evolutionary gradient algorithm under the Gaussian plume model

Zhiqi Zhao & Jiandong Fang

Jin Chuan Campus of Inner Mongolia University of Technology, Inner Mongolia, Hohhot, China

ABSTRACT: Inspired by the biological odor localization, this paper presents an active olfaction positioning technology used in mobile robots. An assignment method is used to search plume randomly in the design and a search algorithm is used to track the plume, the algorithm is called evolutionary gradient search algorithm. The experimental results show that this algorithm can guarantee mobile robot move along the plume from low concentration to high concentration. Plume Environment the design simulates is point source diffusion in the breeze environment, so the Gaussian plume model is used as Plume model which is widely used in the international; the model is suitable for point source diffusion. The design ideas were completed in Matlab, verify the correctness of the design ideas on robot to search and track the gas.

Keywords: Active Olfaction Positioning; Evolutionary Gradient Search; Assignment; Gaussian Plume

1 INTRODUCTION

With the development of robotics, robotics applications are more widely now. For example, abnormal temperature and humidity conditions, and operating under abnormal pressure, work under hazardous gases or dust, and radioactive environment, and so on [1]. Under the above conditions of work, relying solely on human strength is difficult to do theses work, and then requires the robot to replace humans to accomplish these tasks, so people aware of the degree of intelligence and ability of robots are increasingly high requirements [2]. Now the study of perception of robots focused on the tactile, visual and auditory have developed a number of single-function with visual, hearing even talking robot. However, it is relatively small for the robot olfactory aspects of application and research [3].

In long evolutionary process of animal, the sense of smell is one of the most primitive sensory function, it has been accompanied by the development of animal evolution, and it is the most important survival skills of many animals [4]. Inspired by the biological, this paper proposes a robot active olfaction algorithm, the algorithm is applied to a mobile robot, to make robot proactively track and locate odor source. The Gaussian plume model is a smoke plume model widely is used internationally, this design also used for the smoke source tracking and positioning algorithm evolutionary gradient search algorithm.

The design propose the mobile robot active olfaction positioning technology can be widely used in environmental monitoring [5], contraband goods inspection, hazardous gas tank leak detection and security in a large factory warehouse, and so on.

2 PLUME MODEL

Under the conditions of the indoor and breeze, the Gaussian plume model [6] is used as the study criteria. Gaussian plume model is suitable for point source diffusion; it was used in the early fifties and sixties, this model is based on statistical methods to examine the quality of the diffusion concentration distribution. Many standard adopted by the U.S. Environmental Protection Agency EPA are developed based on Gaussian model, and these standards are still being widely used. Because this model is in line with experimental simulation of environmental conditions, so the odor source localization can also be established on basic model.

In the three-dimensional spatial coordinates, wind along the axis and the average wind speed is constant, turbulence is isotropic and homogeneous, and then this area can use the Gaussian turbulence diffusion equation.

Gaussian model, also known as medium-density cloud continuous diffusion model or Gaussian diffusion model, its expression is (1) as follows:

$$c(x,y,z) = \frac{Q}{\pi \sigma_y \sigma_z u} \exp\left[-\frac{1}{2}\left\{\frac{y^2}{\sigma_y^2} + \frac{z^2}{\sigma_z^2}\right\}\right] \qquad (1)$$

In the formula c(x, y, z) represents the concentration of pollutants to the point (x, y, z) in the case of continuous emissions, the units is mg/m³.

Q represents a continuous material flow discharge, the unit is mg/s;

X represents the downwind distance, the unit is m;

Y represents a cross-wind distance, the unit is m;

Z represents the distance from the ground, the unit is m;

σ_x, σ_y, σ_z respectively represent the dispersion parameter of x-axis, y-axis and z-axis and have the relationship with downwind distance x, effective surface roughness Z0 and the atmospheric temperature.

Gaussian plume model is written as a block using the Matlab language. Internal program defaults the smoke source strength is 500 mg/s, wind speed is 1.5 m/s, the accuracy is 0.5 m and ground roughness is 0.5. Mapping plume concentration distribution at different solution concentration. The simulation results are shown in Figures 1–3.

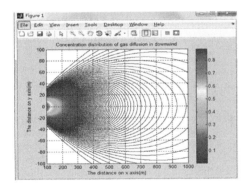

Figure 1.   The solution concentration is 500 mg/m³.

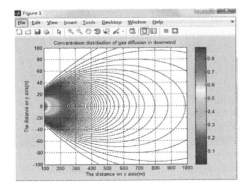

Figure 2.   The solution concentration is 250 mg/m³.

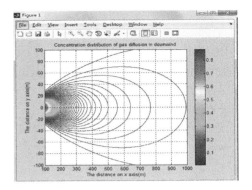

Figure 3.   The solution concentration is 100 mg/m³.

3   ODOR SOURCE TRACKING AND POSITIONING ALGORITHM

3.1   *Evolutionary gradient search algorithm*

Mobile robot [7] controlled by evolutionary gradient search algorithm, also called gradient search algorithm. The initial position of the mobile robot is O and its concentration is C (O). Compare the concentration between C (O) and eight points around O, the eight points are right in front of point A, right point of B, right rear point of C, front point of D, rear point of E, left front point of F, left point of G and the left rear point of H. After comparing the above-mentioned, the robot moves to the point where it has the highest concentration, and this cycle continues. Assume that the initial position of the mobile robot is O point, the coordinates is (tr, tc), the concentration of the point C (O) is C (tr, tc). First, the mobile robot moves a distance OA reaches point A, the concentration of point A is C (A), if C (A) > C (O), put the value of C (A) assigned to C (O); If C (A) < (O), to maintain C (O) = C (tr, tc). Then the mobile robot moves a distance AB reaches point B, the concentration of point B is C (B), If C (B) > C (O), put the value of C (B) assigned to C (O); If C (B) < (O), to maintain C (O) = C (tr, tc), accordance with the above procedure, successively compare the concentration between point of A and point of C, D, E, F, G and H, make the higher concentration of point assigned to C (O). After completion of the comparison, assuming that the point of G has the highest concentration, the next step will use the point of G as the search starting point of O, compare the concentration between G and the 8 point between G, make the higher concentration of point values assigned to C (O) and this cycle continues. Use this method of assignment search the plume randomly and in accordance with the gradient search algorithm to track the evolution of plume can guarantee the mobile robot move to the place where have the higher concentrations of plume. Therefore, the trajectory of the mobile robot will become increasingly close to the odor source. When Robot cannot find the point where have higher concentration than its own in the surrounding eight points, this shows that the robot reaches the vicinity of the odor source, robot will stop moving. The flow chart of evolutionary gradient search algorithm is shown in Figure 4.

3.2 Simulation of Matlab

1. The robot has the same initial position and concentration field different solution concentration

Internal program defaults smoke source strength is 500 mg/s, wind speed is 1.5 m/s, the accuracy is 0.5 m and the ground roughness is 0.5. The initial position of the robot (87.5, 375). Mapping robot odor source localization at different solution concentration. Search algorithm uses evolutionary gradient search algorithm.

The trajectories of robot are shown in black thick line. The Coordinate have the largest concentration in the concentration field is (0,0). Figures 5–7 show the analysis under the condition of robot has the same initial position and concentration field has different solution concentration.

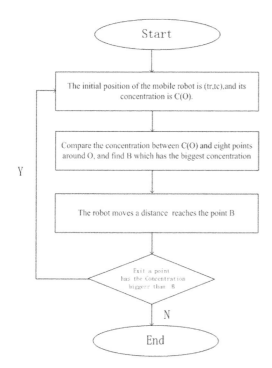

Figure 4. The flow chart of evolutionary gradient search algorithm.

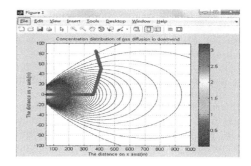

Figure 5. The solution concentration is 500 mg/m³.

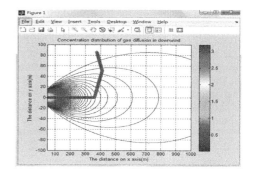

Figure 6. The solution concentration is 250 mg/m³.

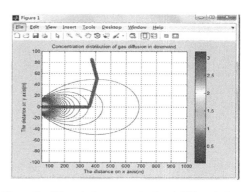

Figure 7. The solution concentration is 100 mg/m³.

The simulation showing that the plume have different diffusion distance in the X-axis and Y-axis, solution concentration of 500 mg/m³ corresponding to the maximum diffusion distance and solution concentration of 100 mg/m³ corresponding to the minimum diffusion distance. However, the robot in three figures are all start moving from the point of (87.5, 375) and finally navigate to the maximum concentration point (0,0), this proves that the robot find the odor source successfully. The above simulation results show that different solution concentration affect the diffusion distance of plume in the X-axis and Y-axis, but it does not affect the robot to locate the odor source in the condition of have the same starting point.

2. The robot has the different initial position and concentration field has same solution concentration

Internal program defaults smoke source strength is 500 mg/s, wind speed is 1.5 m/s, the accuracy is 0.5 m, ground roughness is 0.5 and the solution concentration is 165 mg/s. Mapping robot odor localization at different initial position. Evolutionary gradient search algorithm is used as search algorithm.

The trajectories of robot are shown in black thick line. The Coordinate have the largest concentration in the concentration field is (0,0). Figures 8–10 show the analysis in the condition of the robot has the different initial position and concentration field has same solution concentration, the robot successfully targeted to the point of maximum concentration, and the point is (0,0). The above simulation results show that the starting point of the robot does not affect the type of

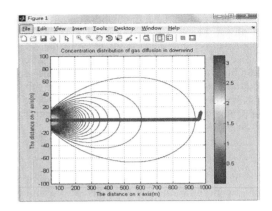

Figure 10. The initial position is (17, 980).

robot for positioning the odor source at the same concentration field.

4 CONCLUSION

This paper investigates a robot active olfaction technology. In the Gaussian plume model, using gradient evolutionary algorithms let the robot to find and track the plume and by experimental verification of this algorithm correctness. Under the condition of robot has the same initial position and concentration field has different solution concentration, and in the condition of concentration field has same concentration and robot has different initial position, the robot successfully targeted to the point of maximum concentration in both cases. In the future, the robot active olfaction application in practice will be studied more by us.

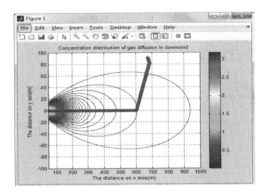

Figure 8. The initial position is (92.5, 675).

ACKNOWLEDGEMENTS

Inner Mongolia Science and Technology Project "Embedded intelligent control system of Multifunctional robot spraying plant protection" (20120304).

Inner Mongolia Science Foundation "Regulation algorithm and its application for resource-based industries intelligent optimization system for the position of spatial data" (2009MS0913).

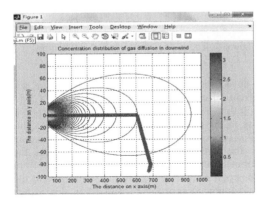

Figure 9. The initial position is (−92.5, 675).

REFERENCES

[1] A.J. Lilienthal, T. Duckett, Creating Gas Concentration Gridmaps with a Mobile Robot. Proceedings of the 2003 IEEE/RSJ International Conference on Intelligent Robots and Systems (IROS 2003). 2003:118–123.

[2] Zhang Jian, Xie Hui, Cao Xiedong, et al. Inner examination technology on pipeline and the development trend. Instrumentation Analysis Monitoring, 2011, 4(2): 39–41.

[3] Lino Marques, Anibal T. de Almeida. Electronic Nose-Based Odour Source Localization. IEEE. 2000:36–40.

[4] J S Elkinton, C Schal, T Ono, et al. Pheromone puff trajectory and upwind flight of male gypsy moths in a forest. Physiol Entomology. 1987, 2: 399–406.

[5] Frank W. Grasso, Jelle Atema. Integration of Flow and Chemical Sensing for Guidance of Autonomous Marine Robots in Turbulent Flows. Environmental Fluid Mechanics.2002, 2:95–114.

[6] T. Nakamoto, H. Ishida, T. Moriizumi. An Odor Compass for Localizing an Odor Source. Sensors and Actuators.1996, B 35–36:32–3.

[7] Qinghao Meng, Fei Li, Junwen Sun, et al. Multi-robot based odor source localization. RAS Newsletter—University of Waterloo. 2009, (7): 10–16.

Future Information Engineering and Manufacturing Science – Zheng (Ed)
© 2015 Taylor & Francis Group, London, ISBN 978-1-138-02644-5

The relationship between the perceptions of teachers' expectation, parents' expectation and academic performance of junior middle school students

Pinru Zhang

Shaanxi Normal University, Xi'an, China
Shaanxi Academy of Governance, Xi'an, China

ABSTRACT: The improvement of academic performance depends not only on students' own efforts, but also on the collaboration of teachers and parents. This paper, using quantitative research, through a questionnaire survey of 98 students from 4 middle schools, measured the perceptions of teachers' expectation and parents' expectation and academic performance of junior middle school students and analyzed the relationship between them. The results indicate that: (1) the students' perception of teachers' expectation has significant positive effect on their academic performance; (2) the perception of parents' expectation has significant mediating effect on their perception of the relationship between teachers' expectation and academic performance. Such research results will have reference significance for improving the academic performance of junior middle school students.

Keywords: Perceiving the teachers' expectation; perceiving the parents' expectation; academic performance

1 INTRODUCTION

1.1 *Conceptual definition*

The teachers' expectation may refer to some kind of the teachers' predictive cognition on the behavior consequences of the students based on the teachers' knowledge of the students. Domestically, there are mainly two sides of the results of the research on the teachers' expectation. The first side is the analysis of the cause antecedent of the teachers' expectation to explore what factors results in the teachers' different expectations. For example, the middle school teachers and primary school teachers were tested and the factors affecting the expectations of the middle school teachers and primary school teachers in China were analyzed in Song Guangwen and Wang Lijun (1998). The results indicated that basically, there were the same factors, that is, the tested results (mainly the academic performance and ability), character features and moral features, affecting the teachers to have high or low expectations on their students. In addition, appearance, gender, family conditions and etc., also played a certain role in the teachers' expectations, however, not as significant as the former three factors mentioned did. The students at Grade Two at junior middle school were tested and the relationship between the teachers' expectation, sense of self-value and target orientation was studied in Zheng Haiyan, Liu Xiaoming and Mo Lei (2004), which found out that gender and class have impacts on the teachers' expectation, sense of self-value and target orientation.

Combining the previous researches, this paper proposed the following assumptions:

Assumption 1: The teachers' expectation perceived by the students in the junior middle school has a positive impact on the academic performance.

Assumption 2: The parents' expectation perceived by the students in the junior middle school doesn't have a positive impact on the academic performance.

Assumption 3: The parents' expectation perceived by the students may play a role in mediating the relationship between the teachers' expectation and the academic performance.

This paper aimed to explore the impact of the parents' expectation and the teachers' expectation on the academic performance. To reasonably master the parents' expectation and the teachers' expectation may have guidance significance for increasing the academic performance of the students in junior middle school.

2 EMPIRICAL STUDY

2.1 *Subjects*

The subjects of this study came from the students sampled from four junior middle schools in Xi' An city. The tested students were all from general classes. The academic performances of those students were not significantly different. Totally 98 questionnaires were sent out and 96 reclaimed, with a callback rate of 98%. The conditions of the subjects were shown as Table 1.

2.2 *Scale*

The perception of the teachers' expectation: This study applied a maturity scale about the perception of the teachers' behaviors. Totally there were 12 questions with totally three dimensions, that is, learning support, emotion support and ability support. Of which, there were totally 11 questions for positive score and 1 question for negative score.

The perception of the parents' expectation: This study referred to the scale used by many scholars about the parent's expectation. There were totally 11 questions with totally five dimensions, that is, academic performance, future achievement, behavior performance, interpersonal communication and physical and mental health. Of which, there were totally 8 questions for positive score and 3 question for negative score.

Academic performance: The final test results, that is, the mean score of Chinese, Maths and English, were applied.

This study applied AMOS software and SPSS statistical software to conduct the following analysis of the data of the reclaimed questionnaires: Descriptive statistics analysis, reliability and validation analysis, correlation analysis and regression analysis.

Table 1. The descriptive statistics of the samples.

Variables	Statistical characteristics	Frequency (n = 96)	Percentage (%)
Gender	Male	42	43.8
	Female	54	56.3
Grade	Grade one at junior middle school	55	57.3
	Grade two at junior middle school	19	19.8
	Grade three at junior middle school	22	22.9

2.3 *Reliability analysis*

This paper conducted a reliability analysis of the variables in this study, as shown in Table 2. SPSS analysis results indicated that the α coefficients of the two variables: the teachers' expectation and the parents' expectation are above 0.6. This indicated that the variables in this study have general consistency and thus acceptable.

2.4 *Exploratory factor analysis*

This paper conducted exploratory factor analysis over the data acquired in this survey by using principal component analysis method. This paper followed a principle of eigenvalues equal to or above 1 for factor sampling. And factor loading less than 0.4 may not be shown. The exploratory factor analysis results are as shown in Table 3 and Table 4.

Table 2. Reliability analysis of scale (N = 96).

	Question item	Consistency coefficient	Consistency coefficients after removal of this question item
Parents' expectation	Academic performance 1	0.675	0.662
	Academic performance 2		0.642
	Future achievement 1		0.649
	Future achievement 2		0.692
	Behavior performance 1		0.645
	Behavior performance 2		0.640
	Interpersonal communication 1		0.649
	Interpersonal communication 2		0.663
	Interpersonal communication 3		0.658
	Physical and mental health 1		0.637
	Physical and mental health 2		0.647
Teachers' expectations	Learning support 1	0.840	0.828
	Learning support 2		0.834
	Learning support 3		0.822
	Learning support 4		0.823
	Learning support 5		0.827
	Learning support 6		0.828
	Ability support 1		0.826
	Ability support 2		0.826
	Ability support 3		0.821
	Ability support 4		0.821
	Emotion support 1		0.825
	Emotion support 2		0.847

2.5 Correlation analysis

In this study, Pearson correlation coefficient was applied to express the correlation among the independent variable, mediating variable and dependent variable in the study. The results are as shown in Table 5. It can be seen from the table that the teachers' expectation is in greater positive correlation to the academic performance; that the teachers' expectation is in greater positive correlation to the parents' expectation and that the parents' expectation is in no significant correlation to the academic performance. The correlation analysis results indicated that the variables in this study are inter-correlative at different levels. This is consistent with the research thoughts and assumptions of this paper.

Table 3. Parent's expectation factor analysis.

	Parent's expectation				
	1	2	3	4	5
Interpersonal communication 2	0.853				
Interpersonal communication 1	0.836				
Interpersonal communication 3	0.694				
Academic performance 1		0.883			
Academic performance 2		0.757			
Behavior performance 2			0.777		
Behavior performance 1			0.693		
Physical and mental health 2				0.876	
Physical and mental health 1				0.809	
Future achievement 2					0.906
Future achievement 1					0.783

2.6 Verifiable factor analysis and goodness of fit

In order to verify the structure validity of the scale, verifiable factor analysis of the measurement variables was conducted by using AMOS 7.0 analysis software. The analysis results indicated the x^2/df value less than 2, RMSEA values all less than 0.10, the values of IFI, TFI, CFI and GFI all more than 0.90. Thus, the indexes above meet the model fit valuation standards and can be considered as a good fit.

2.7 Regression analysis of mediating effect

The paper conducted regression analysis by using SPSS in which the teachers' expectation was an independent variable and the academic performance was dependent variable. The statistical results were as shown in Table 7. From which, it can be seen that this linear model was significant.

3 ENLIGHTENMENT AND DISCUSSION

These research conclusions have certain enlightenment significance for the management practice.

Table 4. Teachers' expectation factor analysis.

	Component		
	1	2	3
Learning support 2	0.736		
Learning support 1	0.720		
Learning support 3	0.704		
Learning support 6	0.671		
Learning support 4	0.612		
Learning support 5	0.563		
Ability support 2		0.898	
Ability support 4		0.797	
Ability support 1		0.759	
Ability support 3		0.652	
Emotion support 2			0.851
Emotion support 1			0.441

Table 5. Mean, standard deviation and correlation coefficient of variables (N = 96).

Variables	Mean	S.D.	Teachers' expectation	Parents' expectation	Academic performance
Teachers' expectation	4.00	0.67			
Parents' expectation	4.32	0.44	0.467**		
Academic performance	83.66	6.15	0.222*	0.13	

Table 6. Model goodness of fit.

Variables	χ^2/df	RMSEA	NFI	IFI	TLI	CFI	GFI	RFI	AGFI
Fit index	1.211	0.047	0.830	0.966	0.941	0.962	0.942	0.733	0.887

Table 7. Stepwise regression of parents' expectation and teachers' expectation.

Entry of model and variables

	Equation 1	Equation 2
Constant	74.492**	76.097**
Teachers' expectation	2.044*	
Parents' expectation		1.753
R^2	0.222	0.126
F	4.895*	1.517
df	(1,94)	(1,94)

Table 8. Multi-level regression of mediating effect.

Entry of model and variables	Academic performance		
	Equation 1	Equation 2	Equation3
Step 1: Constants	82.841**	84.176**	83.506**
Gender	2.468*	1.750	1.919
Grade	−1.832*	−1.961**	−1.916**
Step 2: Teachers' expectation (centralized)		1.482	1.886
Parents' expectation (centralized)		1.039	2.842
Step 3: Teachers' expectation by parents' expectation			2.427*
R^2	0.092	0.131	0.181
ΔR^2	0.092	0.039	0.05
F	4.734*	3.44*	3.985**
df	(2,93)	(4,91)	(5,90)

First, put the parents' expectation and the teachers' expectation into the model at the same time. The results validate a consensus: "The academic performance of the students is not always higher as the parents' expectation goes higher". There is no significant linear relationship between the academic performance and the parents' expectation. This also enlightens our parents to reasonably master the degree of the expectation on their children. Meanwhile, there is a significantly predictive relationship between the teachers' expectation and the academic performance. This is probably related to the students relatively trust the teachers and follow the teachers' words. And there is some psychological transmission mechanism. Therefore, the teachers shall have higher expectations on the students.

Second, from the impact of the parents' expectation on the teachers' expectation and the academic performance, when the teachers' expectation is relatively low, for example, when a student just studies in a new school and is not so familiar with the teachers or the classmates, the parents shall have a medium expectation and shall neither give too much pressure to the child nor have too high expectation on the child. However, the parents shall not be indulgent parents, either. When the teachers' expectation is relatively high and when the students are familiar with the teachers and the classmates and can also give full play to their abilities, the parents shall have very high expectation on their children.

Finally, this paper has some shortcomings that need to be improved, such as, there are too few samples. The mechanism of the expectation acting on the academic performance shall be discussed in depth and etc.

REFERENCES

[1] Liu Lihong & Chen Peng. Study on Effect Transmission Mechanism of Teachers' Expectation. Education Exploration, 1997 (2).

[2] Lu Juan, Lu Jian & Li Jian. Relationship between the Study of the Military Medical College Students and Their Parents' Concerns and Expectations on Their Study. China Journal of Health Psychology, 2010 (5).

[3] Pang Weiguo, Xu Xiaobo, Lin Lijia & Ren Youqun. Study on the Relationship between Family Socioeconomic Status and the Academic Performance of the Students of Middle School. Global Education Outlook, 2013 (2).

[4] Song Baozhong & Wang Pingchuan. Construction of Scientific Expectation of Parents according to the Education Value. Journal of Xi'an University of Arts and Science (Social Science Edition), 2005 (5).

[5] Song Guangzhi & Wang Lijun. Study on the Factors Affecting the Expectation of the Middle School and Primary School Teachers. Psychological Science, 1998 (2).

[6] Wang Li. Survey and Analysis of Anxiety Status of Middle School and Primary School Students and Their Parents' Expectations. China Journal of Healthy Psychology, 2010 (4).

[7] Wen Zhonglin, Zhang Lei & Hou Jietai. Mediated Moderator and Moderated Mediator. Journal of Psychology, 2006 (3).

[8] Zheng Haiyan, Liu Xiaoming & Mo Lei. Study on Relationships among the Teachers' Expectation Perceived by Junior Grade Two Students, the Self-value and Target Orientation of the Junior Grade Two Students. Psychological Development and Education, 2004 (3).

[9] Babad, E. Measuring and changing teachers' differential behavior as perceived by students and teachers. Journal of Educational Psychology, 1990 (4).

Future Information Engineering and Manufacturing Science – Zheng (Ed)
© 2015 Taylor & Francis Group, London, ISBN 978-1-138-02644-5

The optimizing strategy of capital structure of SMEs

Fan Hu
Shaanxi University of Technology, Hanzhong, Shaanxi, China

ABSTRACT: SMEs play an irreplaceable role in the implementation of the New-type urbanization strategy. Arrangement of capital structure is an important factor affecting the development of SMEs. For the promotion of sustainable development, it is quite significant to look into the Characteristics of capital structure of SMES, probe into improving the financing efficiency, and optimize macro and micro strategies of the capital structure, which is aimed at current unreasonable term and ownership of capital structure.

Keywords: SMEs; capital structure; macro strategy; micro strategy

1 INTRODUCTION

With the large-scale promotion of new urbanization strategy, small and medium-sized enterprises (SEMs) will play an irreplaceable role in promoting regional economic development, maintaining market competition, and ensuring full employment. But for a long time, difficulties in financing and unreasonable capital arrangement have become the main factors restricting the development of SMEs. The defects in the present capital structure of SMEs in China is mainly reflected in the capital shortage, imbalance between internal and external financing. One the one hand, a heavy dependence on the internal financing is restricted by the scale of financing, which led to the total shortage of capital; on the other hand, exogenous financing channel is so narrow that the enterprise mainly relies on short-term loans from banks, resulting in the unreasonable ownership capital structure and maturity capital structure. Therefore, the most urgent problems for the medium and small enterprises are to optimize the capital structure, as well as broaden the financing channels. This paper will explore the methods to optimize the capital structure of small and medium-sized enterprises from both macro and micro dimensions, hoping to promote the sustained and healthy development of small and medium enterprises.

2 CURRENT SITUATION OF CAPITAL STRUCTURE OF CHINA'S SMALL AND MEDIUM-SIZED ENTERPRISES

2.1 Capital shortage

As known to all, in the process of production and management of enterprises, adequate capital stock is not only the guarantee for maintaining normal operation, but is a necessary condition for technological innovation and the expansion of production scale. According to a World Bank study, 81% of China's small and medium-sized enterprises think that liquidity within a year cannot meet business needs, and 60% enterprises fail to obtain long-term loans. Due to the small scale, poor efficiency, and single financing channel, the financing environment for the medium and small enterprises in China is tight. For them, the main source of capital is endogenous financing. Endogenous financing is the accumulation of financing within the enterprise itself. The specific forms include capital funds, depreciation funds transformed into replacement investment and retained earnings into new investment. On one hand, although the endogenous financing can save the cost of capital, reduce tax expenditure, and help establish the ownership structure, it depends heavily on the size of the business and profitability; on the other hand, the majority of China's small and medium-sized enterprises have not set up modern enterprise system yet, and the lack of self-discipline, self accumulation mechanism and reasonable distribution system result in the corporate retained earnings failing to give priority for long-term development of enterprises. The earnings are mainly used in the interest of the investors' profit sharing, which directly leads to the shortage of capital in small and medium sized enterprises.

2.2 Capital ownership structure is not reasonable

Ownership structure of the capital is the value composition and proportion of different ownership capital relations. According to the pecking order

Table 1. Asset-liability ratio of China's SMEs and large enterprises 2007–2009.

Year	SMEs	SMEs board listed	Large enterprises
2007	53.7%	40.88%	68.54%
2008	52.98%	38.26%	64.42%
2009	51.5%	38.00%	67.79%

*From China Statistical Yearbook 2007–2009.

Table 2. Current liability ratio of China's SMEs and large enterprises 2007–2009.

Year	SMEs	Large enterprises
2007	85.69%	65.69%
2008	86.50%	66.79%
2009	85.34%	65.32%

*From China Statistical Yearbook 2007–2009.

theory of capital structure, firms prefer internal financing. When there is need for external financing, they should first consider the debt financing, before finally turning to equity financing. Financing structure of western developed countries is in line with pecking order theory of financing pattern. This is because when the endogenous financing is insufficient, if the enterprise's expected investment return rate is higher than the cost of capital, debt can create more value for shareholders using financial leverage, then realize the maximization of enterprise value, though it will also bring financial risk to the enterprise. In our country, on the contrary, endogenous financing is influenced by many factors. As mentioned previously, financing amount is limited, and the amount of capital enterprise is set up at the beginning of the equity capital financing. Although SMEs have created 60% of GDP, their loan limit set by the bank is less than 1/4 of commercial loans. Coupled with increased production costs, the SME financing deficit is up to 30%.

The rate of assets and liabilities is a comprehensive index to evaluate the level of corporate debt, also a measure of the company's ability of utilizing debt capital in business activities. As shown in table 1, the debt capital level of China's small and medium-sized enterprise is lower than that of the large enterprises of our country at the same period, which does not only restrict the enterprise financing scale, but also impedes the full use of financial leverage to enhance corporate value. The capital structure is not reasonable.

2.3 *The debt maturity structure is irrational*

According to the theory of debt maturity, if the debt maturity is not appropriate in collocation the governance effect of debt financing will be affected, and the financial security will be endangered; if the maturity structure of debt and assets match well with the term structure, enterprises will be able to reduce the agency cost of debt, so as to realize the control of risk and financial distress. In the overall debt capital, current liabilities is mainly used to maintain the daily business needs, but at the same time will improve the enterprise debt repayment pressure, increase the enterprise financial risk. In China, while the capital debt rate is, the

small and medium-sized enterprise liabilities rate ratio showed a high level. According to the data, 52% listing Corporations in Shenzhen SME Board has a long-term debt ratio of 0, the average is only 3%. The situation of long-term debt to support SMEs long-term construction is poor. There is no long-term loans for 60% small and medium-sized enterprises, so the enterprises are unable to make effective long-term development planning.

2.4 *Capital structure lacks financial flexibility*

Financial flexibility is the enterprise's ability to deal with unforeseen and uncertain issues in the future and its ability to grasp the investment opportunities, to use idle funds and the remaining debt, and to allocate financial resources. In theory, an enterprise can increase the financial flexibility through the equity capital financing. This is because the property debt capital cost is fixed and mandatory. Regardless of the operating results of enterprises, they must regularly pay the principal and interest. But in fact, the small and medium-sized enterprise always worry about the financial manager of elastic problems in capital structure decision because of their own limitations and imperfect capital markets, as well as the little room for capital structure adjustment. While the big enterprise can adjust by the internal financing and security debt, the small and medium-sized enterprise can only adjust the capital structure of external capital market through the issuance of equity. When in the face of new investment opportunities, SEMS can do nothing but issuing more stock rights, which further leads to the capital of the institution is not reasonable.

3 THE MACRO STRATEGY TO OPTIMIZE THE CAPITAL STRUCTURE OF SMALL AND MEDIUM ENTERPRISES

3.1 *Formulate relevant laws and policies, creating a relaxed financing environment for SMEs*

Capital shortage is the main problem that China's small and medium-sized enterprises face at present. In view of the irreplaceable role that SMEs play in the national, some of the world's developed countries and regions expanded the sources of funds for SMEs and implemented the relevant laws and regulations in order to support the development of SMEs. In addition to the "Anti-monopoly Law," "Anti-unfair competition Law," "Fair Trading Act", the federal government of USA also formulated the "Enterprise Law", "Small Business Investment Act", "Small Business Rewards Method", Japan and South Korea issued the "Basic Law of SMEs". Our government should further develop and perfect relevant laws and policies and improve the financing environment of SMEs from the macro policy level, to create favorable financing environment for the development of SMEs.

3.2 *Optimize financial services, improve the government support to SMEs*

Learn from the economically developed areas, establish specialized financing institutions for SMEs, and make financing policies accordingly. Take U.S as an example, on the one hand, American SME Bureau, the commercial banks, USA import and export banks have set up special financial institutions, taken specialized financial support measures for small business, and provide 90% credit guarantee for the business loans by SMEs. On the other hand, the Small Business Administration is in charge of the examination and approval of the establishment of the small business investment company. The investment company can receive up to $90 million of concessional financing from the federal government.

In China, first, we need to establish local and regional small and medium-sized financial institutions, which may help to understand the operating status, development prospects and credit level of local SMEs. Information sharing platform should be designed for SMEs, which may not only overcome the information asymmetry since incomplete information leads to high transaction cost and moral hazard, but also help small and medium-sized enterprises to establish risk consciousness. Second, turn to venture capital firms. Vc firm can be established under the guidance of the government using private capital. The government is responsible for formulating various technical innovation and development plans for SMEs, and granting appropriate special subsidies. SME venture capital is mainly invested to scientific and technological type of SMEs with high expected yield, so as to fund scientific research innovations and promote the optimization of capital structure.

Credit guarantee system is indispensable for solving the financing problems of small and medium-sized enterprises. In the whole world 48% countries and regions have established SME credit guarantee system. For instance, Korea Credit Guarantee Fund (KCGF) and Korea Technology Credit Guarantee Fund (KCGF) are to provide credit guarantee to all the SMEs with good reputation, SEMs of high technology and the high-risk enterprises. In addition, South Korea also has 14 local guarantee funds to provide credit guarantee service. In our country, the SME credit guarantee demand is unable to meet this situation. We need to draw lessons from the international experience. On the one hand, establish a "one body two wings four layer" model—the government as the main body, the commercial guarantee and inter-enterprise guarantee coexisting as the two wings. Establish government policy guidance, open various channels to raise guarantee fund. On the other hand, we must establish and perfect the laws and regulations to promote the healthy development of guarantee industry, strengthen the credit guarantee industry regulation and supervision. Increase policy support and financial support and perfect the supporting measures to further implement the "Provisions for Promoting SMEs". Set up special funds to support the construction of credit guarantee system for small and medium enterprises, and provide loss compensation and incentive mechanism for guarantee agencies who guarantee for small and medium-sized enterprises. Do regional reguarantee fund pilot as soon as possible. Establish and perfect the guarantee organization risk prevention, control and sharing mechanism.

An important way to promote the optimization of capital structure of small and medium-sized enterprises is creating and developing capital market, and increasing its support for small and medium enterprises. Central to the development of multi hierarchy capital market is the promotion of direct financing for SEMs and expansion of direct financing space. Direct financing and capital supply is to directly negotiate the loans or via issuance of shares. The capital supplier and demander directly transfer money in the capital market, which is with high transparency. As far as the direct absorption of financing situation is concerned with the small and medium-sized enterprises at present, first, although the "secondary market" has partly provided direct financing platform for small business, it cannot fully meet the demand of the majority of SEMs because of its high threshold, small size, particularly for SEMs in the western and other less developed regions. So we need to reduce the threshold as soon as possible to expand the scale and provide financing platform for the most high-quality small and medium enterprises. Second, the strict market access system issued by China leaves very little possibility of bond for SEMs, directly restricting the optimization of its capital structure. We can learn from the developed countries in the world to solve the problem. In 1970's, American introduced high-yield bonds to alleviate the problem of financing, making it possible for SMEs to enter the mainstream capital markets, so as to solve the problem of shortage of funds via direct financing. Therefore, we need to accelerate the establishment of a coordination mechanism for SEMs bond market, innovate product for bond market, broaden the financing channels for small and medium enterprises. Based on "the second board market", we need to speed up improving the financing mechanism of the "third board market", so as to provide strong financial support for small and medium enterprises that cannot be traded directly.

4 OPTIMIZE THE MICRO STRATEGY OF THE CAPITAL STRUCTURE FOR SEMs

To optimize the micro strategies of the capital structure of SEMs, the key is to follow the value maximization of the business objective. Start from the status quo of the enterprise itself, and strive to establish a modern enterprise system. Understand well the self accumulation capacity and operating risk of the enterprise, and adjust the capital structure accordingly.

4.1 Clarify the property rights of SEMs, establish governance structure specifications

A modern enterprise system and a clear property right SEMs, as well as effective corporate governance is an important guarantee of improving endogenous financing deficiency of SEMs. In China, most small and medium enterprises do not have clear property right and responsibility because they use family type management mode. Unavoidably, they focus on the pursuit of short-term interests and ignore the long-term development. The establishment of modern enterprise system can not only improve the governance structure between the owners, operators and producers of mutual supervision and mutual restriction, improve the management level, but more importantly, the self-restraint, self accumulation and reasonable mechanism of modern enterprise system has the potential to promote enterprise internal financing, to give full play to the advantages of internal financing, and to effectively promote the retained profits and depreciation fund transforming into needed funds for the enterprise. The scale of its own capital is thus expanded and the corporate capital structure improved.

4.2 Clarify the goals of financial management, adjust and optimize the capital structure

Generally speaking, the enterprise takes as its goal to maximize shareholder wealth. The value of the

company is thus maximized via maximization of the shareholder wealth. According to the MM theory, even if there are financial constraints and agency cost, the change of capital structure will greatly affect the value of the company. If lacking money, the enterprise must determine reasonable cost structure and ownership structure based on a good knowledge of its solvency and profitability and adjust the priority selection according to the actual situation. Given the preference for inner financing, the enterprise should not only make full use of financial leverage, but also fully consider the financial leverage risk. The enterprise should improve the manager's financial management ability and adjust the financing strategy of the enterprise according to profitability and management level in different periods, constantly optimize the capital structure with the aim of the maximization to the company value.

4.3 Build the concept of honesty, improve their credit level

The credit level of SEMs is an important factor restricting the development of their financing channels. Most financial institutions shall strictly examine the credit status of the financing subject for the control of their own risk. SEMs must establish perfect and standardize their credit system and improve the information transparency, meet their financial obligations, in order to obtain trust and support from the financial institutions. Credit management should be included into enterprise culture construction, made widely known, in order to expand financing channels.

REFERENCES

[1] Jing Xin, Wang Huacheng, Liu Junyan. Financial Management, Renmin University of China press, 2009.
[2] Financing of small and medium enterprises in Jiangsu province, Small and Medium Enterprises, 2010 (4), 19–20.
[3] Zhang Hailong, Li Bingxiang. The value of the company, capital structure and managerial entrenchment, Soft Science, 2012 (6), 111–114.
[4] Zhang Anling, Wang Yong, Wang Yingqi. Analysis of capital structure of the listed Corporation of SME board, Journal of Shanxi University of Finance and Economics, 2011 (2), 16–17.
[5] China Statistical Yearbook [R], 2007–2009.
[6] Lin Li. The optimization strategy of financing and capital structure for SEMs, Journal of Sichuan Normal University, 2013 (7), 74–78.

Future Information Engineering and Manufacturing Science – Zheng (Ed)
© 2015 Taylor & Francis Group, London, ISBN 978-1-138-02644-5

An enlighten of the development of micro-course for ideological and political education of higher education based on word frequency

Liangliang Wang

School of Humanities, Economics and Law, Northwestern Polytechnical University, China

ABSTRACT: The main purpose of this paper is to explore the development of micro-course for enlightenment and reference of ideological and political education in our higher education. Using the word frequency statistics method, it illustrated that in recent years the scholars had paid a great attention to this topic and pointed that the necessity of its research, which will promote the education of online learning. At the same time, the study maintained in our country that the teachers and learners have changed a lot with the development of information technology. Therefore, we should build our system of micro-course at once, which provides some ideas and reference to ideological and political education of China.

Keywords: The development of micro-course; Ideological and political education; Higher education; Word frequency; Enlighten

1 INTRODUCTION

Nowadays, human beings have entered into a all media era with the development of information technology. As the micro-blog, micro-message, micro-novel, micro-movie, micro-journal and other new media continue to emerge, massive information constantly had impacted on people's brain and each person always surrounded in a variety of information every day. When the micro era arrived, the form, diversity, individualization, flexibility of learning content became more and more import and learning is no longer confined within the classroom. Therefore, the emerge of micro-course so timely. Micro-course (microlecture and microless), from the literal meaning we can see that it can meet the needs of online' learners for its short and concise form, which focus on the relatively small content, and promotes a better learning experience. This is not only means the birth of a new mode of learning, but also a rise of new way of living in information age. The ideological and political education that cultivated, educated, led people should not depend on its traditional form such as classrooms, textbooks and other traditional media and it will change a lot in the modern society.

2 LITERATURE

The term micro-course is not used here to refer to microcontent for microlearning, but to actual instructional content that is formatted for online and mobile learning using a constructivist approach. Micro-course developed relatively early in abroad. In 1960, mini-course first was proposed by USA scholars. In 1998, MicroLESSONS project implemented by Singapore Ministry designed many courses over the field. Its main purpose is to train teachers ban built the mini-course, which is generally from 39 minutes to 1 hour, only focusing the teaching goal. More specifically, as described by (David, 2009) in the Chronicle of Higher Education. There are approximately 60 second presentations with a specific structure. They are not just brief (one minute) presentations: although Dr. McGrew (McGrew, 1993) had a success with one minute lectures at the University of Northern Iowa as did by Dr. Kee (Kee, 1995) at the University of Leeds.

As stated (Shea, 2009), these specific lectures are combined with specific activities designed to promote the epistemic engagement of the learner. Micro-course has undergone a relatively short and slow development process. In recent years, with the accelerating process of globalization, foreign micro-course research to all types of schools at all levels, including the higher education.These micro-curriculums and other teaching activities to promote the development of the learner's knowledge. Nevertheless, the response of the higher education community was mixed, with some positives (Loginquitas, 2009), and some negatives (Krajewski, 2009).

The interest surrounding the use of microlectures has continued to grow, even outside of the

United States, to the places such as Hong Kong University (2009). In the United States, the use of microlectures is even considered as a vital part of the Pandemic Response Plans. In addition, even scholars at schools like Princeton University (Brady, 2009), UNC's School of Government, (Cunningham, 2009), Humboldt State University, (HSU, 2011), University of West Florida (2009) and University of Illinois Urbana-Champaign (CITES Academic Technology Services, 2012) support the importance of an innovative teaching-learning approach for learners in the 21st century.

We note that the foreign experience to the development and extension line by line, the most typical of the Khan Aca. Studied on micro-course in China is relatively late, in recent years our scholars have made great exploration in theory and practice. In theory, as stated (Hu TS, 2010) first proposed the concept of micro-course in China. Moreover, Jiahou maintained that micro-course refers to a small course within 10 minute, a clear teaching objective, simple content and focused on a problem. However, Jingran hold that the micro-course is a five minutes that developed by teachers themselves from practical teaching to solve the difficult problems in the work for teachers. (Zhang JC, 2013).

The practice of micro-course launched in the southeast coastal provinces in China and some of them are excellent. For example, Guangdong Province Teacher Networks Classroom (GPTNWC), Fushun City Micro-Course Construction (FCMCC), Shenzhen City Collection and Online Exhibition (SCCOE) and Tianjin city air classroom project.

Through the analysis of the development of micro-course comparing with the domestic and international, we have found that (1) Micro-course had a profound influence on Chinese Education. (2) Micro-course serviced to the teaching, as a professional full-time teachers themselves development, playing an important role in teaching. (3) The micro-course applied of multimedia technology and built a platform in the exchange of information for their teaching. (4) In our country, the research on micro-course at present mainly in the middle and primary schools, served for the teaching of primary and middle schools teaching.

Admittedly, no one can deny the fact that micro-course had a profound impact on education and many scholars have studied a lot in past years. However, we want to know that how long it has impacted our education and how many studies have researched this topic and how does it impact a special course named ideological and political education in our higher education. Consequently, we carried out this study.

3 METHOD

Methods in this study attempts to use the word frequency statistics to display our country academic circles research concerns for the issues of micro-course. Word frequency analysis is a kind of statistical method that takes a advantage of the core keywords or subject in the literature of the frequency to determine the level and development trend of research. If a keyword or subject in the field of literature appears repeatedly, they can tell the theme of the keywords or subject word representation is a hot research topic in the field.

3.1 The journal source table of word frequency statistics of micro-course

We regarded the micro-course as a keyword search for some mainly and import database such as China National Knowledge Infrastructure (CNKI), the outstanding master and doctoral dissertations databases since 2010 (see Table 1). We input the article titles to Excel table, which are not included the journal coil, news, advertisement, notice, dynamic, forum, meeting minutes and then built 5 title databases by years.

3.2 The explanation of the reliability and validity of the research

Reliability refers to the stability of measurement data and consistency degree. We not only use text word frequency and keywords word frequency of the micro-course, but also search through a database and CAJviewer statistical software search tools. Furthermore, we reviewed the data manually for a relatively small number of items. Therefore, we believe that this study has a high reliability.

Validity refers to whether the data accurately and effectively explained the questions study put forward. Although we searched for keyword micro-course can already see the degree of concern of scholars on this issue is very high, the database search engine is strictly in accordance to the character sequence to search for. So, excepted for using the core keywords

Table 1. The journal source table of word frequency statistics of micro-course.

Classification of journals	The numbers of retrieval journals		Total
The core journals	164		
The noncore journals	Important journals	189	598
	The general journals	243	

*The classification of journal comes from the source of the tenth session committee of Chinese social science citation index in 2013.

micro-course, we also use other words such as micro-lecture, minicouse for the second retrieval, so that ensuring the validity of our study.

4 RESULT

4.1 *Micro-course text frequency statistical from 2010 to 2014*

The author found out the related papers about the micro-course studies and classified, merged and counts carefully this information corresponding to text frequency number. Detailed content is shown in Table 2.

In order to show the research status of micro-course more clearly, we draw the line graph, so that the reader can understand deeply the development trend of micro-course in China. Detail content is shown in Figure 1.

We can see there only a few of papers our Chinese scholars introduced and studied on this topic in 2010 (Figure 1). But from 2012 to 2014, there is a dramatically increase. There we must illustrate the word frequency number in 2014 is no less than that in 2013, the fact is not as shown as we see. That is because our data collection is till the June of 2014. According to the trend, we believe the degree that the number will be increased than 2013 in the half of this year.

4.2 *Micro-course word frequency statistics from 2010 to 2014*

Although the text word frequency statistics had also told that there was a high attention for the

micro-course, we want to take advantage core keyword word frequency again (see Table 3).

Based on the above-mentioned statistics, from 2010 to 2014, scholars gradually strengthen the understanding and research on micro-course, frequency results rose in a straight line. Micro-course problem gradually receives the attention of the scholars in China, we statistic all the related articles of word frequency of micro-course two times again to verify our results. The statistical results showed that it consistent with the trend of text frequency statistical results, also showed an increasing trend year by year.

Accordingly, its figure about this topic as follows:

4.3 *The school category of micro-course word frequency statistics*

In the research of micro-course, we have found that the research on micro-courses not only wrote by primary the curriculum reform and development of micro-course design, also have the attention from vocational colleges and higher education, especially from 2012 to 2013. See as Table 4.

In order to show the research status of micro-course school category more clearly, we draw the line graph, so that the reader can understand deeply the development trend of micro-course in China. Detail content is shown in Figure 3.

Table 2. Micro-course text word frequency statistics from 2010 to 2014.

Years		2010	2011	2012	2013	2014
Text frequency	Journals	1	5	40	355	198
	Master and doctoral theses	0	0	2	21	3
	Subtotal	1	5	42	376	201

Table 3. Micro-course word frequency statistics from 2010 to 2014.

Years		2010	2011	2012	2013	2014
Word frequency	Journals	1	4	31	315	184
	Master and doctoral theses	0	0	2	23	3
	Subtotal	1	4	33	338	187

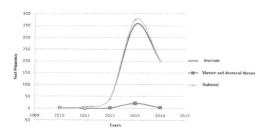

Figure 1. Micro-course text frequency statistic from 2010 to 2014.

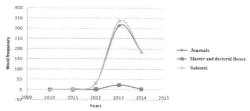

Figure 2. Micro-course word frequency statistical from 2010 to 2014.

Table 4. Micro-course school category word frequency statistics from 2010 to 2014.

Years		2010	2011	2012	2013	2014
Word frequency	Primary and middle schools	1	3	20	147	139
	Occupation colleges	0	1	4	51	14
	Universities	0	2	9	140	34

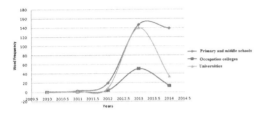

Figure 3. Micro-course school category word frequency statistics from 2010 to 2014.

5 DISCUSSIONS AND CONCLUSION

Based on the word frequency statistic, the results obtained in the course of this investigation clearly demonstrate that our scholars paid a great of attention to this topic and the micro-course effect our education and also brought an enlighten for ideological and political education.

5.1 *In the past few years, micro-course had a profound influence on Chinese Education*

The frequency statistic described in our findings is dynamic. It showed that in recent years the research on micro-course is obvious change, whether text frequency or word frequency, only from 1 article about the micro-course discussion in 2010 to 376 articles in 2013, which had explained the academic scholars, paid a great of attention to micro-course study. Investigated its reason, we not only had to think why in the past more than four years micro-course caused such a high degree of attention, but also further reflection on our practice.

5.2 *Micro-course is a major challenge to the ideological and political education in all media era*

When the network ideological and political education has been popular, when we actively concern and profound understanding the role of network even bring huge impact to the ideology of network on college students, why do not we think about how to use the network to the effect of ideological and political education. Micro-course is a combination course with multimedia. Because the micro-course has many advantages such as short and professional, very suitable to mobile learning way of life for students, whether in the public, school, even leisure time, we can be arbitrary learn the course of ideological and political education.

5.3 *Micro-course changes the learning method of ideological education*

Micro-course encourages students learn themselves, who can choose courses independently according to their own progress. They can also decide to pause or replay micro-course. It can focus on knowledge or learning difficulties and can use any recording mode to explain the content which greatly saves time the classroom to explain the concept of problem. A different classroom produced and more discussion and interaction and students and teachers are required to provide help when micro-course provides face to face mode of communication, which avoids other indirect factors' impact in the classroom teaching.

5.4 *Micro-course puts forward higher request to the ideological and political education teaching mode*

Although ideological and political education micro-course is a new thing, teachers in colleges and universities taught ideological and political theory course generally have higher education and more responsibility for the designation and development of micro-course. Therefore, they have to improve their abilities and adapt this special teaching. First of all, they should be more receptive to new things. Second, in addition to continuous research materials, teaching dynamic, they also learn to design micro class, new fabrication method of ideological and political education micro-course and make themselves better and faster to master this new form of teaching.

REFERENCES

[1] Shieh D. 2009. These lectures are gone in 60 seconds. *Chronicle of Higher Education* 55(26), A1, A13.
[2] McGrew LA. 1993. A 60-second course in Organic Chemistry. *Journal of Chemistry Education* 70 (7): 543–544.
[3] Kee TP. 1995. The one minute lecture. *Education in Chemistry* (32): 100–101.
[4] Shea P, Bidjerano T. 2009. Community of inquiry as a theoretical framework to foster epistemic engagement and cognitive presence in online education. *Computers and Education* (3): 543–553.

[5] Krajewski B. 2009. Lilliputians of Higher Education Invent Microlecture. http://brucekrajewski. wordpress.com/2009/03/03/lilliputians-of-higher-education/.

[6] Brady A. 2009. Microlectures in distance and online learning. https://blogs.princeton.edu/hrc/2009/03/ microlectures_in_distance_and_online_learning. html.

[7] Cunningham C. 2009. Micro-lectures: A cure for content bloat. *Teaching and Learning Support.* http://sogweb.sog.unc.edu/blogs/tls/?p = 483.

[8] Joan VD. 2011. Just-in-time teaching for critical topics and skills. http://www.humboldt.edu/celt/ tips/micro-lectures_just-intime_teaching_for_ critical_topics_and_skills.

[9] A microlectures workshop. CITESA cademic Technology Services, 2012. http://www.library.uiuc.edu/ blog/infolit/2012/07/ microlecture_workshop.html.

[10] Hu JT. 2013. Under the Environment of Mooc Micro-course Design-A Case Study of Computational Thinking micro-course as an example: 8–9.

[11] Zhang JC. 2012. Review Micro-course, *China Information Technology Education* (11):19–21.

Future Information Engineering and Manufacturing Science – Zheng (Ed)
© 2015 Taylor & Francis Group, London, ISBN 978-1-138-02644-5

Application of the VERICUT using in university course of mechanical manufacturing

Duanling Li & Xuejiao Wang
Beijing University of Post and Telecommunication, Beijing, China

Dongwei Ji
Beijing Phic Technology Co. Ltd., Beijing, China

ABSTRACT: This paper introduces how to use VERICUT which is a software based on the process simulation in university course of mechanical manufacturing, and explains how to use VERICUT to build simulation teaching environment, elements of the simulation environment, induction of simulation 5 + 1, the creation of MCTPA and so on, then cites a typical case of author's teaching work to explain the application of VERICUT in teaching. It explores a new teaching model for the national high technology university to cultivate innovative talents.

Keywords: simulation teaching environment; simulation 5 + 1; MCTPA

1 INTRODUCTION

Mechanical manufacturing is a professional basic course of mechanical manufacture and automation major. It plays an important role for students to get a solid professional foundation. However, to learn this course, merely simple theoretical learning is not enough. It is difficult to make students master the course enough to solve production problems in practice. The characteristics of the subjects determine this course needs much more practical, so as to rely too much on various styles of modern manufacturing equipment. For this purpose, a way which can make up for our weak school virtual network teaching is found by using the virtual simulation software VERICUT, then teaching problem of mechanical manufacturing practice link is improved.

VERICUT software is the NC machining simulation system developed by American CGTECH company. And it can process simulation of CNClathe, milling machine, machining center, line cutting machine and multi axis machine tools and other machining equipment. It also has function of optimizing NC program, shortening processing time, extending tool life, improving surface quality, checking the overcut and undercutting, preventing the machine collision and over stroke. The VERICUT software applies in the course of mechanical manufacturing process design, analogize the whole process, replaces the traditional way of cutting experiment has great prospects to enhance students' perceptual ability. Therefore, application of VERICUT in course of mechanical manufacturing is discussed by several years teaching experience.

2 ANALYSIS OF SIMULATING TEACHING ENVIRONMENT

Using the teaching environment VERICUT builds, we set out from the actual processing conditions. The whole manufacturing process system contains elements of machine tool, control system, cutting tool, fixture, blank and manual operation and so on. If it is automated manufacturing, it also has elements of NC system, NC program and so on, while the manual operation in the digital processing becomes artificial installed, operating tool setting.

Based on the analysis above, we can build teaching environment within the VERICUT. Then students can use the teaching environment for manual, or automatically manufacturing process experience. A particular analysis of processing results can be done after processing, and using measuring tool blank can measure and analyze the machining process of blank. For the part that can't be measured, through leading 3D design models of machining part into the workpiece, we can compare the coincidence degree of design model and machining blank to analyze the concrete over cut or residue by using automatic comparison function of the software. So the best processing technic plan and report of process can be acquired.

3 SIMULATION 5 + 1

We can find the mixed type, the enormous number by analyzing the teaching simulation environment elements, it is difficult for students to remember so many things. Therefore, in the process of teaching, we have summarized the related factor formula, called simulation 5 + 1.

"5" as the five basic conditions contain machine tool, cutting tool, system, program (automatic processing) and accessories.

"+1" is a manual operation, the common machining for artificial processing, automatic processing for artificial alignment tool setting.

The accessories are blank, fixture, design model.

The analysis of the process can make the students master how to use VERICUT for practical manufacturing environment, and not easy to forget.

4 BUILDING TEACHING ENVIRONMENT MCTPA

4.1 Machine tool

Software itself contains many machine tool library, VERICUT covers almost international manufacturing equipment, students can choose directly in it, also equipment model can be customized for special circumstances. Students can build topological structure, logical relationship same with the production of manufacturing equipment, and design geometry of each kinematic axis, motion limit, realize the zero distance with practical manufacturing equipment.

4.2 System

Software itself contains many system library, VERICUT covers almost international system, students can choose directly in it, also control system can be customized for special circumstances.

4.3 The tool

Through tool management system VERCUT carries we can design knife tool, cutter and other parts including turning, milling, boring cutter, molding cutter, special knives and other cutting tools.

4.4 Program

We can write processing code through the part of the program. If the manufacturing equipment is general machinery, this part does not need to write, students can complete it through MDI manual control.

4.5 Accessories

Forvice, three grasping such simple fixture can be directly modeled. For the complex professional fixture, tooling and other systems, can be modeled by other CAD software, then import in it.

5 APPLICATION EXAMPLES

According to many years teaching and factory working experience, a practical teaching process which uses VERICUT in mechanical manufacturing course is shown below.

5.1 Project drawings

As figure 1 demonstrates, analyze drawings for the parts, and finish machining manufacturing for the parts.

5.2 Analyze drawing for the parts

Guide students to analyze the characteristics, key points and difficult process size of the parts.

5.3 Draft preliminary process scheme

According to the requirements of the drawings for the parts and site conditions (blank, equipment etc.), process preliminary plan is rough turning tapered side—rough turning ball side—milling flat—drilling—the hole milling.

5.4 Simulation

Using VERICUT processes plan.

As figures 2–6 demonstrates, according to primary scheme, building all processes simulation environment within the VERICUT. Because the similar establishment process, only one example is cited here.

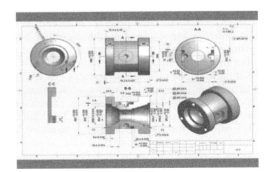

Figure 1.

Figure 2. Rough turning tapered side.

Figure 3. Rough turning ball side.

Figure 4. Milling flat (3-axis clamping 2 times).

Figure 5. 3-axis drilling (vertical assembly).

Figure 6. The hole milling of 4-axis.

5.4.1 Analysis of simulation elements

Machine tool: LATHE.MCH basic-3axes-vmill. MCH 4axis. MCH

System: fan21t.ctl fan15 m.ctl fan31im.ctl
Accessories: vice vblock three jaw chunk
Blank: 90 × 76.2 cylindrical blank

Program: ZUODUAN YOUDUAN milling flat ZK F03

Tool: turning tool. TLS M5_TOOL.TLS

Name	Number
C80 Turing tool	T01
4MM Cutter	T10
D22 Bit	T09
D8 Bit	T08
6MM Boring tool	T03
D20 Milling cutter	T07

+1: Artificial alignment process, programming zero location refers to each process diagram.

5.4.2 Building simulation environment: (turning the parts)

1. System of simulation 5 + 1: select system procedure specifies.

According to simulation 5 + 1 process, setting simulation processing as shown in figure 7, right click after controlling in the project tree region, then select the open command in the pop-up commands, popopen control system dialog box, find and open the FAN21IT numerical control system in the software installation directory database.

2. Machine tool of simulation 5 + 1: select machine tool procedure specifies.

As figure 8 demonstrates, find and select the machine tool in the project tree area, and then select open in the pop-up list of commands in the

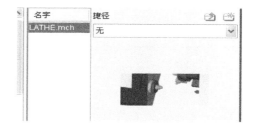

Figure 7. Select system.

Figure 8. Select machine tool.

pop-up dialog box, select and open the machine tool LATHE. Mch file shown in the figure in the pop-up dialog box of open the machine, then machine tools and CNC system is loaded successfully.

3. Machining tool of simulation 5 + 1: set machining tool.

As figure 9 demonstrates, select machining tools in the project tree side, and right click on it, then in the pop-up commands select the command open, find and open the tool under the corresponding path in the pop-up dialog box, can increase the cutting tool successfully.

4. Program of simulation 5 + 1: add NC program.

As figure 10 demonstrates, select NC program in the project tree region, and right click on it, and then select Add NC program file command in the popup command list, find the program ZUODUAN.TXT under the corresponding path in the pop-up dialog box, click the right button, determine.

5. Annex of simulation 5 + 1: add fixture, blank and so on.

As figure 11 demonstrates, select fixture, in the project tree, and right click on it, and then select add model file command in the command list popup, find fixture model under the corresponding path in the pop-up dialog box, add.

Figure 9.　Select machining tool.

Figure 10.　Select NC program file.

Figure 11.　Add annex.

Figure 12.　Artificial tool setting.

Figure 13.　Automatic comparision result.

6. +1 of simulation 5 + 1: artificial tool setting.

As figure 12 demonstrates, select the G-code bias in the project tree region, then set the original point according to the figure below.

7. Experience simulation.

When process configuration over, click button, simulate from the start to the end, you can experience the machining process.

5.5　*Analysis of the results and optimization*

Reasonable parts processing scheme and correct program are important tasks of the practical teaching link. We take advantage of the automatic comparison function by the software to analyze the machining quality and precision. As figure 13 demonstrates, after completing process simulation, select the menu command in the main menu, analysis > automatic comparison, pop-up automatic comparison dialog box, and then set the page in the dialog box, set the entity comparison method (compare way is two), click the compare button and then compare them, the results as shown in figure 13. By analyzing the graph, we can find the residual error in the orifice chamfer, then correct the program or find other solutions optimization by the related situation.

Conclusion: when the parts process at the left tend, the programming ignores processing orifice chamfering.

Solution: add chamfered measures in the process.

6 CONCLUSIONS

Practice teaching has proved that, the application of VERICUT in teaching the college courses of mechanical manufacturing has obtained the good teaching effect. By using virtual simulation training environment to analogy the whole process, the problem of lacking of equipment types and quantity, hard processing conditions and other problems are solved. It opens a new situation of educational development for the development of university personnel training and national manufacturing.

REFERENCE

[1] "VERICUT 7.2 NC machining simulation technology training course", Tsinghua University press, 2013.

Future Information Engineering and Manufacturing Science – Zheng (Ed)
© 2015 Taylor & Francis Group, London, ISBN 978-1-138-02644-5

Deficiency and countermeasure of ideological and political education in colleges

Caihong Zhang
Xi'an Physical Education University, Xi'an, Shaanxi, China

ABSTRACT: In order to improve the efficiency of ideological and political education in colleges, the work analyzed the deficiency of university students' ideological and political education in current higher education institutions by the method of analyzing current situation. Countermeasures of the existing problems are proposed. It is concluded according to the analysis that innovated teaching method, updated teaching content and personalized education should be considered in university students' ideological and political education. The main innovation point in this work is the comprehensive analysis on internal and external factors of the target issue using contradiction analysis.

Keywords: Ideological and Political Education; Colleges; College students; Deficiency; Countermeasure

1   INTRODUCTION

Students in colleges are the successors of Chinese socialism in the future, the great hope in the whole process of national development, and the future of this country. Therefore, the quality of university students rises and falls with the country development and national rejuvenation. In recent years China attached great importance to fostering talents, especially the ones in colleges and universities, with the education of comprehensive development in moral, intelligence, sports and aesthetics for the students as the direction [1]. However, dilemma appears in those students, including those who completed the education in higher education institutions and graduated successfully. Suicides and attacking incidents also occur frequently. Those problems challenge the ideological and political education in higher education institutions, requiring integrated analysis for practical scientific countermeasures.

2   CIRCUMSTANCES OF THE DEFICIENCY OF UNIVERSITY STUDENT'S IDEOLOGICAL AND POLITICAL EDUCATION

The main reason for this problem is the insufficient attention paid to ideological and political education in higher education institutions. The education can neither correct students' thoughts and behaviors as a function of FAQ, nor inspire them to consider their life value, resulting in deviation of their thoughts or behaviors. Currently, most of the colleges and universities don't put ideological and political education in an important position, and the undergraduates don't value this education in improving their quality either. Neither the teachers nor students have too much interest in ideological and political education. Another problem is the misbehavior of the undergraduates in higher education institutions because of their unsatisfactory social performance. Suicide, mental confusion and diseases are also the evidence of the necessity of ideological and political education.

2.1   Indication of disregarding the life in campus injury incidents

The injury incidents between students are very frequent in China's colleges and universities. Some of them even happened in the same dormitory, such as the poisoning case in Fudan University. Such injury incidents in campus are actually disregarding the life [2]. As well as suicides in campus, those incidents are resulted from lacking capability of handling emotion issues, high pressure in study, having trouble in interpersonal communication or even economic problems.

The above cases indicate the lack of regarding life in campus, which is actually the result of weak ideological and political education. Most of the students are suffering from mental deficiency and lack spiritual beliefs.

2.2 Lack of study motivation among the students in some of the universities

The education system is the main reason for this problem. After 12 years of national basic education, which is mainly based on paper examination with test-oriented education as the baton, the students have consumed all their interests and energy in study to get into colleges and universities. The regulation in current academy is easy to graduate but tough to entry for the students. The high school students manage to entry the university through intense competition in the entrance examination [3]. However, the testing method is much easier in the university, weakening the students' willingness to study hard. After experiencing the depressive study in high school, most of the students prefer loose pace in the university, which is the main reason of their study-weariness.

2.3 Emotional and interpersonal communication problems of college students

Some students fall in love early in the learning process, and the phenomenon of blind love is prominent in universities where love happens frequently. The students often get confused with emotional problems, especially like "why and how to be in love", because of the immaturity of dealing with emotion problems and wrong attitude to the difference of ideal and reality. Some small problem will bring about big problems when it occurs in life, thus causing suicide of some college students for love.

Conflict of sex orientation occurs in college students, and there are some gay and bisexual students, making up a special group in present university. Dissimilation of emotion orientation is often taboo in the current society in China, causing high stress of the special group.

In addition, interpersonal communication problems lead to conflict of thought of college students, and right consciousness and introversive character of some college students bring about more obstacles and stresses.

3 REASON OF DEFICIENCY IN IDEOLOGICAL AND POLITICAL EDUCATION OF COLLEGE STUDENTS

3.1 Open ideological environment of current ideological and political education

The external ideological environment that the college students are faced with at present is more complicated than that before. The current ideological environment is caused by collision and integration of pluralistic values, and the difference

and conflict of psychology, morals, politics and culture becomes more prominent. The globalization of current economy, the multi-polarization of the world political pattern and the pluralism of world and value outlook affect the college students. The western ideology including the behavior way of western young people is affecting the current college students, and the influence is within reach and everywhere. Besides, the development of internet, PC and intelligent mobile phone intensifies the permeation of the ideology, making a great challenge to the college education in ideology and politics.

The change of social environment, especially the adverse ideological values of external ideological environment is very bad for the college students, making the ideological and political education difficult for universities. Many thoughts collide with each other and Chinese society is in the transition period, causing confusion of many college students when they are faced with choice of life. The ideological and political education cannot give them a timely answer, making it unbelievable to the students.

3.2 Deficiency of ideological and political education itself

The current college thought in ideological and political education, including the way and content, is conservative and unsuitable to the college students at present. As a base of social development, Chinese socialist market economic system is in a process of supplement and adjustment. The collective value and individual benefit conflict with each other, causing the deficiency of ideological and political education in some colleges.

Although ideological and political education in colleges can be the spirit base of college students, it cannot satisfy the physical or spiritual requirement. The ideological and political education in some colleges only focuses on the teaching schedule, or the teachers are idle at teaching without considering the satisfaction of college students, thus making it unattractive and undeveloped.

The ideological and political education is considered as some empty preach by college students, and cannot help them find the spirit support or the feeling of time in fact, making it an empty shell.

4 IMPROVEMENT OF IDEOLOGICAL AND POLITICAL EDUCATION IN COLLEGES AT PRESENT

To improve ideological and political education in colleges, the attention to the students is important. Based on the deficiency of ideological and political education in colleges at present, the colleges should be adjusted and promote the students. So, it is important to reform the form and method of ideological and political education in colleges and supply the students with spirit.

4.1 To find the location of individual value in collective

The propaganda and education of value and life outlook to the college students should not only lay stress on the compromise and even sacrifice of individual value to collective value, but also the location of individual value in collective and find the common benefit of individual and collective value. Development of individual and society are unified, so ideological and political education in colleges should concern individual requirement and value based on value idea, thus developing the whole society at last.

Consequently, ideological and political education in colleges should motivate the potential of the students based on the satisfaction of individual value and the development of personality, and achieve individual and collective value together on basis of healthy social environment.

4.2 Improving teaching method of ideological and political education in colleges

The ideological and political education in colleges should be a democratic teaching idea, including the thought and idea of teachers and college students. In teaching process, the personality of the students should be respected and the students should have enough right to choose, thus building an equal relation between the teachers and students. The students should be guided to develop in a direction of self-management of morals, and learn the method of autoexcitation and self-discipline in ideological and political education in colleges.

The method and idea of teaching and the center of ideological and political education should be changed to make the students as the main part of teaching. The correct method of persuasion and listen should be used to keep two-way communication of the college students and teachers. The ideological problems of students should be found in communication, and the correct method used to solve corresponding problems. Different teaching method should be applied based on different college students, thus making the students beneficial to ideological and political education.

5 CONCLUSIONS

Although the ideological and political education in colleges is not a teaching of professional knowledge, only the college students with complete personality can realize the individual and collective value together in the process of social work. So, the teachers should pay attention to reform of teaching method and content to cultivate talents with strength.

REFERENCES

[1] Guan Shanshan, Countermeasure of Improving Attraction of Ideological and Political Education in Colleges, Journal of Luoyang Technology College, 2007(06): 79–82.
[2] Min Yongxin, Global View of Maxism and Whole Effectiveness of Ideological and Political Education in Colleges, Journal of Anqin Normal University (Social Science Edition), 2010(09): 1–6.
[3] Wang Youliang, Reform of Ideological and Political Education in Colleges, Heilongjiang Science and Technology Information, 2008(01): 162–167.

Future Information Engineering and Manufacturing Science – Zheng (Ed)
© 2015 Taylor & Francis Group, London, ISBN 978-1-138-02644-5

Protection measures of the interests of company's creditors

Weilin Huang
Management School, Jinan University, Guangzhou, Guangdong, China
Guangdong University of Finance, Guangzhou, Guangdong, China

ABSTRACT: With the transition of socialist market economy, more and more companies choose to issue company bonds to raise capital. Correspondingly, the effective protection of the creditor's interests becomes an important factor for market operation and the social stability. Currency, both the companies and their bonds-buyers are making all efforts to create a good mutual trust environment so as to protect the creditor's interests to the utmost extent. This study analyzed the necessity of the protection of creditor's interests, and then proposed several effective measures on that basis.

Keywords: Creditor's Interests; Company Law; Special Circumstances; Ex post facto Measures

1 INTRODUCTION

Compared with common creditors, the company creditors have a loan relationship with the company, which includes some obligations of both parties have to shoulder and some economic risks that may happen in the market. As capital-supporters for the improvement of companies, the interests of company creditors should be protected to the utmost extent. However, facts show that only contracts can not effectively help creditors to avoid risks, so people need to offer more reliable measures for the protection of creditors' interests from the angle of company and their creditors. Based on the important risks of creditors, some measures to protect their interests have attracted public attention. Different interests-protection demands deprived from different angles are significant, and offering necessary protection for the stability of social economy [1]. The emphasis and protection on the creditors' interests demonstrate the strategic desire of socialist market economy to expand the scale of market and capital, which is the developing trend of the function of national policies and market adjustment mechanism in bonds field.

2 NECESSITY OF THE PROTECTION OF CREDITORS' INTERESTS

The creditor is a technical term corresponding to a debtor. Most creditors offer their capitals to companies by investment or buying stocks. Large amount of capitals input to companies can help them expand scales and businesses, and enhancing the whole competitive powers. This kind of capital support certainly can not be gratuitous but constructed upon contracts aiming at exchanging bonus and service [2]. This integration and allocation of resources accompanied by interests is not only the demand of current fast-paced marketing and competition, but also the promotion of socialization production.

However, interest relation in this cooperation is very important and sensitive. Both the companies raising capitals by issuing bonds and corresponding investors shoulder respective risks. The creditors can help a company and they can also destroy a company. Many companies, especially those small-and-middle-sized companies with large amount of bonds, have suffered little or big bonds crisis when the capital changes of these companies were influenced by the public opinion. The mood fluctuation of creditors can make companies trapped in a dilemma. On the one hand, the running of companies can not be without the support of large amount of money; on the other hand, many creditors may exert pressure on companies. Under such a circumstance, many small-and-middle-sized companies are faced with restructuring, merger and even bankruptcy [3]. Therefore, the planning of the scale of bonds should be cautious and depended on the scale and running of different companies.

Compared with the companies' conservative attitude to their creditors, the creditors obviously have more autonomy. Most of them have enquired and learned the repayment period, debt service ratio and preferential clause by signing contracts; they also receive the necessary respect and service from debtors. However, in fact, creditors can hardly guarantee that the companies they choose have the good credit position, running abilities and professional anti-risk measures. Therefore, creditors can not wait for repayment after investment. On the contrary, they need to prevent risks at any time and make preparation for the protection of their interests. For example, when companies decide to adjust their production or management departments, the inappropriate management or failure integration can lead to the bankruptcy of companies. Then, creditors will be doomed to the same fate as the companies. Apart from normal debts termination, some companies adopt strategic measures to evade repayment of debts, which is an abominable measure to damage the creditors' interests. Thus it can be seen that the creditors' interests are exposed to many subjective or objective risks. The vulnerability of creditors' interests should attract attention from creditors, debtors and intermediary institutions to avoid the loss of creditors' interests. The happening of similar cases is alarming the creditors.

3 SEVERAL CONSTRUCTIVE MEASURES FOR THE PROTECTION OF CREDITORS' INTERESTS

3.1 *Strict observation and execution of new company law*

Section 1 of the Company Law of the People's Republic of China elaborates the necessity and importance of the protection of creditors' interests and emphasizes the respect and protection of their rights for those who have offered support to the construction of socialist market economy. This section satisfies the demand of future development of socialist market economy and can be the indispensable factor in maintaining the stability of market economy and guaranteeing the social stability. Code of Corporate Governance for Listed Companies in China issued in 2001 also emphasizes the role of protection of creditors' interests in reforming and upgrading company structure. The new company law enacted in 2006 encourages investment to promote the economic development. New provisions in this law attach more importance to the protection of creditors' interests, and aiming at the behaviors of companies.

The new company law further guarantees creditors' rights to know the register of company and its changes, financial situation, liquidation, and litigation of company. For example, people can know about the basic document and register of a company with the help of company registration authority. The relevant issues about company liquidation needs to be sent to creditors within a specified time limit and creditors' meeting needs

to be called when necessary to guarantee relevant issues are known by creditors. When a company has grave litigation, creditors should be notified in time and their rights to ask and know needs to be guaranteed. Besides, the disregard of corporate personality system, a system established by many years' of economic development and company practice is added in the new company law. When some companies are sued for misbehavior of their staff, they tend to transfer negative influence and relevant liabilities to creditors for the reason of independent legal personality, which takes advantage of flaws in law. Therefore, the disregard of corporate personality system is aimed at preventing the economic loss to creditors. The improvement of legal system can not only protect the legal rights and investment initiative of creditors, but also improve market rules of China, guaranteeing the fair and just economic development.

In addition, the improvement of company liquidation system also established a parclose for creditors' interests. The difference between creditors and shareholders is demonstrated in company liquidation. When a company conducts financial settlement, creditors have precedence over shareholders in satisfaction, which reflects their advantageous positions and is the respect for their generous support by offering large amount of capitals. A series of protection measures of new company law exempt creditors from suffering economic loss in many respects, which protects the stability and right to indemnification of creditors' capitals to the large extent and makes these creditors feel safe psychologically.

3.2 *Special protection for creditors' rights under special circumstances*

Once creditors choose to invest their capitals in a company and sign contracts with it, the rights and obligations of both parties are subject to certain restrictions. Creditors often have fewer obligations compared with companies, which have many restrictions on their business behaviors.

The Law of the People's Republic of China on Enterprise Bankruptcy adopted on August of 2006 introduced bankruptcy proceedings, which involves in shareholders, creditors and staff of companies. The law is aimed at protecting both individual rights and whole interests. The reorganization proceeding, a well-recognized transition mode in this law can effectively prevent the loss of social wealth and productivity. In this process, the plan of reorganization can be debated and considered by managers of the company and its creditors. Therefore, rights to speak of creditors can be guaranteed and their interests claims can be put up and protected by law.

On March 11 2013, CITIC, the entrusted manager of bonds called up meeting of creditors, but creditors of Chaori bonds refused to attend the meeting. The reason was that the Chaori bonds owned by CITIC had breached the contract for 6 days and many creditors thought entrusted manager had not fulfilled its duties in time, which caused loss to creditors' interests. In the process of having company bonds in hand, creditors are faced with the problems of debtor defaults and insufficient service of intermediary agencies. To solve these problems, creditors have to get the whole picture of company's credit status and condition of assets at early stage; meanwhile, they can exchange with professional intermediary agencies to reduce some risks at the beginning of investment.

3.3 Ex post facto measures of interests' loss

Although creditors reduced their risks of losing interests to the large extent when they decided to buy company bonds, these risks still exist in objective and subjective environment, which is a factor threatening creditors' interests. Creditors can have certain rights of control over their capitals when choosing company bonds. Once these capitals are invested in companies, they will lose rights of control over these capitals and their interests are subject to the performance of company management and operation. There are many reasons for the loss of creditors' interests, such as adjustment of company policy, the influence of market fluctuation, changes of economic situation, management level of companies and market competition.

The loss of creditors' interests includes the observation of code of conduct and full funding of shareholders as well as disregard of corporate personality. In 2005, Guangdong Juneng Economic Development Co. Ltd. and Guangzhou Brach of CCB reached an agreement called Contract for RMB Loan. By 2006, Juneng had repaid only about 3 million Yuan principal with interest (27 million Yuan principal with interest remain unpaid), and Guangzhou Brach of CCB failed to get back the remaining loans for several times. On May 26, Jingyi Company and Juneng signed an agreement on debts and assets reorganization. Three days later, they signed an agreement on share transfer and conducted large scale of share transfer. However, Jingyi did not pay corresponding money to Juneng although the Industrial and Commercial Bureau had finished the registration procedure. Guangzhou Brach of CCB thought they had not known the above information, which brought huge loss to CCB and further reduced Junneng's repayment ability. As a result, CCB launched a suit against Juneng for withdrawing

its behavior of share transfer. The court decided the behavior of share transfer to be invalidity and ordered Juneng to repay the remaining loans by stages. There are a lot of similar cases, but most creditors do not know how to get compensation. If creditors want to get timely and reasonable compensation in such cases, they need to learn the business situation of relevant companies in time by telephone inquiry or on-the-spot inquiry. Then they can make judgment and analysis on the decision-making or management of relevant companies, and they can also foresee the reorganization, merger and bankruptcy of relevant companies, which can better protect their rights and benefits.

In addition, using legal methods to sue the company for breach of contract is another way to protect creditors' interests. Once similar cases happen, creditors should launch suits quickly and accurately. They need to take relevant evidence to courts or corresponding departments for inquiry or suits. In litigant process, creditors should use inquiry link to analyze and determine the breach of contract and illegal behaviors of companies, asking for compensation for their losses and trying their best to eliminate the negative influence.

4 CONCLUSIONS

By analyzing previous cases between companies and their creditors, it can be concluded that protecting the rights and benefits of creditors can only be effective with the help of different kinds of measures. Apart from requiring companies to observe new company law and regulating their behaviors, creditors should be more active in protecting their rights instead of the mode of simple input and due receivables. This bilateral cooperation between companies and their creditors can effectively guarantee the full use of company bonds and further protect the interests of creditors. When the creditors' interests can be protected, the bonds investment can tend to be a virtuous cycle, which is the ideal cooperation between companies and their creditors.

REFERENCES

[1] Ma Mingming. *Bondholders Meeting System and Company Bonds Creditors Protection*. Journal of Hunan Economic Management College. 2010(3):34–39.
[2] Sun Yihong. *A Comparative Study on Legal System of Company bonds Creditor Group*. University of International Business and Economics. 2011(9):56–60.
[3] Liang Dong, Zhang Haole. *Legislative Thoughts on Creditor Protection System*. Finance and Accounting Monthly. 2011(1):70–75.

Future Information Engineering and Manufacturing Science – Zheng (Ed)
© 2015 Taylor & Francis Group, London, ISBN 978-1-138-02644-5

New media and cultivation of college students' core values

Xinbiao Zhang
Guangdong Food and Drug Vocational College, Guangzhou, Guangdong, China
South China Normal University, Guangzhou, Guangdong, China

Yingqi Chen
Guangdong Food and Drug Vocational College, Guangzhou, Guangdong, China

ABSTRACT: New media is not only an emerging media and a unique cultural form, but also a mechanism producing, expressing and disseminating ideology such as the socialist core values. In the process of cultivating college students' core values, characteristics of new media such as information production, expression and propagation should be deeply understood and exploited. Accordingly, college students' positive values will be fully guided by new media, thus cultivating and shaping a new generation firmly fulfilling the socialist core values in China.

Keywords: new media; college students; core values; cultivation; ideological and political education; ideology

1 INTRODUCTION

As a mental state of human-specific, sense of values means people's belief, faith and ideal system about fundamental values. On one hand, sense of values acts as value orientation and value pursuit, setting a certain goal of values; on the other hand, it appears as measure of value and evaluation criteria, resulting in a judgment on the object made by the subject. The 18th National Congress of the Communist Party of China first proposed the socialist core values, namely "promoting prosperity democracy, civilization and harmony, advocating freedom, equality, justice and rule of law, proposing patriotism, dedication, integrity and kindness", achieving a unity of value goal among nation, society and citizen [1]. As a mechanism producing, expressing and disseminating ideology, new media should play a more active and positive role in cultivating college students' core values, thus making students keep the socialist core values in mind and carry it out in practice.

2 CHARACTERISTICS OF NEW MEDIA

Impact and affect of mass media on society is increasingly important. In 1967, Marshall Mcluhan said "medium is message". Until now, the media has become "the fourth power" in society. Moreover, as to contemporary college students, "more than 90% of the external influence which leads to a basic understanding of the society, grasp of the rules of the game, and even the formation of outlook on life and values, is from the media" [2].

New Media is a kind of all-to-all propagation. With characteristics of mutual network and inter-activity, it breaks the boundaries of linear propagation in traditional media. Furthermore, quickly blooming into the dominant medium, it creates a new way of social life and behavior in information age, inevitably leading to a shock and change of society and culture.

2.1 *Limitless propagation in space-time*

In virtue of technology innovation, limit of information dissemination in time and space has been broken, and the width and depth of spreading extent also been enhanced. Hence, with the breaking of boundaries of traditional information transmission in means, platform and model, micro-blog and wechat have becoming the main ways of information transmission in network [3]. Besides, information of new media, both having a massive content and an instant velocity, is produced and released almost simultaneously.

2.2 *Real-time interactivity among audience*

In new media, the one-dimensional audience no longer exists. The audience is an information receiver, as well as a producer and disseminator. Moreover, audience's right to select and speak is growing in new media, and they make their voices heard through interacting with new media platforms.

2.3 *Superposition of effect in propagation*

New Media is a type of all-to-all propagation. During the process, individuals usually choose to publish and disseminate information according to their own ideas. Thereby, spread of individual-to-group and retransmit of individuals among audience will both result in superimposed of information effect in dissemination and circulation constantly and indefinitely, thus causing tremendous public controversy.

2.4 *Difficulty in selecting among audience*

As a way of transmitting information, new media has gradually penetrated into daily life such as shopping and dating, making information in every corner of society rapidly spread out in the first time, improving college students' desire for knowledge; meanwhile, based on technology convenience, new media makes the unorganized massive knowledge, information and values enter students' perspective in a more fragmented way through the terminal receiver, consequently influencing young college students' values. The deeper young students contact with the mass media, more difficult they choose and judge when facing knowledge and information.

3 INFLUENCE OF NEW MEDIA IN CULTIVATING COLLEGE STUDENTS' CORE VALUES

Owing to characteristics of wide coverage, strong impact, high trust and sound spreading mechanisms, the mass media is playing an increasingly important role in society today. It is not only a producer and communicator of knowledge and information, but also an expresser and transmitter. When it is changing people's way of life and thinking, it promotes the transformation of personality and formation of new values through advocating education, disseminating knowledge and delighting spirit at the same time.

3.1 *New media providing favorable conditions to foster core values*

Firstly, new media is conducive to enhancing college students' political identity. The proposition and formulation of core values disseminates rapidly through the new media, making the concept of political parties and the country be widely spread in the whole society, in turn achieving political mobilization to students and ordinary people. "Promoting prosperity democracy, civilization and harmony" is strengthening the consciousness of people and a nation, accordingly boosting college students' political identity to the party and the socialist cause.

Secondly, new media helps cultivate college students' civic responsibility. The interactivity of information spreading in mew media allows college students to change from audience to interlocutor. In the new media environment, values of "advocating freedom, equality, justice and rule of law" are spread, making college students have a richer understanding of national conditions and the community, further bringing them realize that concern and participation in politics is a kind of right, and also an inescapable responsibility.

Thirdly, new media is propitious to developing college students' consciousness of participating in social affairs. Having features of "close to reality, life, and the masses", new media builds an information transmitting and interacting platform in order to promote supervision of public opinion, thus providing convenient conditions to care for community and participate in public affairs for college students, eventually making them effectively participate in social practice and political training.

Finally, new media is in favor of formatting college students' healthy and positive personality. With the help of new media, college students can express their feelings and life demands freely and directly, and communicate with other participants with freedom. Through new media, values of "patriotism, dedication, integrity and kindness" can make students fully receive education of ideological mainstream. Meanwhile, students as a political body are manifested in the process.

3.2 *New media impacting on nurturing core values*

One is that information hegemony blurs college students' ideological tendencies. With the increasing maturity of new media technology, mass of information inside and outside is delivered intensively in limited and private time and space, especially western information hegemony which is in a hostile ideological state to China. Tied with cultural information, ideas and values, information is spread and transmitted, assaulting ideological position of developing countries, ultimately making young people blindly admire the west and become extremely easy to be influenced by western ideology.

The other is that information garbage blurs college students' value orientation. Government monopoly on information is broken by the Internet, especially by the birth and expansion of new media, and the one-dominator pattern of public opinion is eliminated quickly. Consequently, a mass of information is produced, duplicated and disseminated. On one hand, the space of free speech is broader than any other era in the past. "Due to underdevelopment of public domain, the Internet has almost become an only exit of ordinary citizens to resist information monopoly and make a sound." However, once the plausible, trade-offs and filtered values filling in new media such as the so called "chicken soup for the soul", "success" or "inspirational story" cross the border, they will not only become worthless information garbage but also blur people's values. On the other hand, corresponding to some college students' extreme psychology, some arguments, ideas and values which hide political purposes disseminate into the terminal through the point-to-point new media and are recognized. All of these above will lead to a deviation of core values as well as the ideas and norms it advocates, finally resulting in deviation values.

Lastly, vague boundary of information weakens college students' consciousness of nation and country. Due to the anonymous equality of subject and information popularization, the network power tends to be a flattening structure, an unbounded space and a direct participation. Accordingly, a virtual world of global integration created by new media blurs college students' concept of people and a nation in reality, making some college students believe that boundaries of country and nation disappear in network, so they may choose to get away from the real world. In addition, it is possible that global consciousness highlighted by new media reduces college students' consciousness of people and a nation, which may even dispel the collectivism and patriotism in China.

4 PRINCIPLES OF NEW MEDIA IN FOSTERING COLLEGE STUDENTS' CORE VALUES

4.1 *Principle of equal interaction*

In a equal, open and positive way, expression of values in mass media can inhibit and regulate those irrational and unhealthy values and interests; meanwhile, it actively guides those interests which are legitimate and healthy but may conflict with the fact, helping college students correctly understand and deal with conflicts between personal desires and reality. Hence, the new media should be made full use of to replace the previous one-dimensional model of values education, thus emphasizing dialogue and interaction between double subject or subjects, so as to creating a free scene of

communication: "No dialogue, no exchange; no communication, no real education".

4.2 *Principle of real-life situation*

Marx believed that "existence of people is the process of their real life", "social life is essentially practical". Only by practice will the purpose of fostering core values be fully achieved. Therefore, through new media, people should focus on life and identify college students' entry point in cognizing, selecting and practicing the core values, consequently making students experience, percept and rethink in a variety of real and simulated life situations through wechat, micro-blog and other new media. Thereby, college students' identification and recognition of people, society and country in values will be enhanced, finally internalizing identification and recognition of core values into their value goals and behavioral norms.

4.3 *Principle of recessive penetration*

Different from the dominant indoctrination education, cultivation of core values emphasize recessive penetration to college students on campus, social interaction and everyday life. Popularity and rapid development of micro-blog, wechat, SMS, mobile newspaper, campus BBS and other new media provides a wealth of carriers and platforms for the implicit penetration of core values. And so it does, all the new media mentioned above both reveal the contemporary college students' characteristics and emphasize the inter subjectivity of education, bringing students awareness and identification of core values imperceptibly, so as to achieving the purpose of practice.

5 APPROACHES OF NEW MEDIA IN NURTURING COLLEGE STUDENTS' CORE VALUES

By means of education and dissemination, the culture spread by correct values necessarily reflects socialism and the values it guides is inevitably the socialist core values. Therefore, as to cultivating college students' core values through new media, there are mainly three approaches, namely path of value in ideal and belief, path of cultural value and path of achievement by the subject.

5.1 *Path of ideal and belief: Cultivating value of practical ability of national consciousness*

In the process of information dissemination, signal dominant ideology should be strengthened by new media. Achievements of popularized ideological education should not only benefit college students, but also guide students to participate in social practice and fulfill the socialist core values. Besides, through interactively share and exchange images, pictures and articles, new media should make students understand national conditions and care the society; through organizing mini charity and competition, new media can attract college students to widely participate in public-spirited activities, cultivating their social responsibility, as well as spreading positive energy. Eventually, college students can achieve self-realization, social value and the Chinese dream.

5.2 Path of cultural value: Fostering capability of value awareness possessing consciousness of social responsibility

Through establishing and improving mechanisms of production, dissemination and supervision, new media should promote positive and healthy mainstream culture internally. Subject matter and technique of performance should reflect national construction requirement, so as to propagating mainstream values and digesting negative social impact brought by commercial cultural products, finally ensuring correct value orientation guided by mass media. In this way, college students can fully understand the sense of social responsibility and follow the law of value perception and judgment, thus fostering ability of value perception and judgment, further forming the ability of value consciousness. Namely, through self-awareness effort and self-discipline, individuals will turn the values which represent social progress trend into their value pursuit.

5.3 Path of achievement by the subject: Cultivating ability of selecting beyond healthy personality value

Besides uplifting self-consciousness and shaping youth idols, new media should also guide young people cross over individual interests and achieve interactive development between personal and social interests. Using modern media and ways of communication such as micro-blog, QQ group and wechat public platform, educators can contact with college students online and offline, listening and solving their practical problems encountered in learning, career planning, campus life,

workplace etiquette and training of professional ethics, ultimately enhancing their awareness of stakeholder.

All in all, new media is not only an emerging media and a unique cultural form, but also a mechanism producing, expressing and disseminating ideology such as the socialist core values. Undoubtedly, it will have a revolutionary impact on cultivating and building college students' core values. As to comprehensively shaping of college students, new media does not limit it in extrinsic behavior such as communication, sharing and learning, also it fundamentally remolds college students' character and spirit. Therefore, in the process of cultivating college students' core values, characteristics of new media such as information production, expression and propagation should necessarily be deeply understood and exploited, so as to fully guiding college students develop positive values, thus cultivating and shaping a new generation firmly fulfilling the socialist core values in China.

ACKNOWLEDGEMENTS

This work was funded by Party building research of universities in Guangdong, China in 2013 *Research of creating methods of Party building using network among vocational colleges*, project number (2013GZZZB13), as well as research of Guangdong Food and Drug Vocational College in 2013 *Research of education on college students' cultural quality—taking sino-foreign vocational colleges as a example*, project number (2013YR007).

REFERENCES

[1] Li Deshun. *Value Theory* (The 2nd edition), Beijing: China Renmin University Press, 2007.
[2] Wang Tao, Dai Jun. *Research of Changing Trajectory and Law on College Students' Values since Reform and Opening Up in 30 Years*. Higher Education Research, 2009 (10).
[3] Hu Yong. *Hubbub—Personal Expression and Public Discussion in Internet Era*. Guilin: Guangxi Normal University Press, 2008.
[4] Zhang Xinbiao. *Displacement of Value on Network Resentment—Analysis Based on Scheler Phenomenon*. Chengdu University of Technology (Social Science Edition), 2012 (1).

Future Information Engineering and Manufacturing Science – Zheng (Ed)
© 2015 Taylor & Francis Group, London, ISBN 978-1-138-02644-5

Analysis on measurement for creating English learning context

Shiqiao Lai
School of International Education, Guangxi University of Finance and Economics, Nanning, Guangxi, China

ABSTRACT: With the development of English teaching, English teaching mode based on various points of view has been proposed and widely concerned. At present, creation of context as an important factor contained in English learning, can greatly improve the quality of English teaching and learning in China. In the study, it is analyzed that an improved and expanded teaching condition, foreign teachers, exchange students and details of creating context can effectively reflect the importance of context in English teaching.

Keywords: English Language; Learning Context; Modern Education

1 INTRODUCTION

Since the 21st century, English education has been popular and more and more important in schools and educational institutions with the development of global village. The basic and high-quality English teaching should be realized. English teaching is constantly expanded on basis of the original definition to improve teaching quality in some schools with strong teaching force or famous English educational institutions. The base of language education is the scientific basic knowledge, and the soul is context [1]. Good context can be necessary for high-quality language learning, making the students start learning English in a high degree. Instead, it is useless to teach the students Chinese English and ignore context, thus causing a poor achievement and ability for the students.

2 IMPORTANCE OF CONTEXT IN ENGLISH LEARNING

Currently, English learning has been the subject with great concern and the update of the teaching method gives the English teachers certain pressure. With the expansion of the scale of global village, English, as an international language, is used by more and more people, and the requirement of English teaching constantly increases. English is used as the official language in a dozen more countries occupying 70 percent of wealth in the world. Not understanding English means disjunction with the world, and giving up English learning and use means falling behind other people. More and more parents demand the children to receive bilingual teaching at the beginning of the entrance

and form accurate and pure English pronunciation and use habit in the preschool stage of 6 or 7 years old [2]. So, the parents are willing to spend much time and energy supporting the children for English learning.

In English learning, the difficulties including English pronunciation, vocabulary, context, grammar, etc. cause Chinese language use habit, nonstandard pronunciation and Chinese English. Learning of a language contains not only listening, speaking, reading and writing, but also use, expression, understanding, etc. Pure written English cannot achieve the purpose of English learning and use. Context is an extensive definition covering many reasons, including non-linguistic factors like cultural background, concrete scene, etc. and linguistic factors like context, tone, etc. Mr. Wang Jianping considered that language environment is the context expressed by verbalism and the subjective factor not expressed by verbalism in language expression of people. It can comprehensively and accurately explain the factor of context involved in language learning. The 2014 national matriculation English test paper is analyzed by Teacher Wang Xiaowen of Lanzhou No. 1 middle school, and the total difficulty is not increased compared with that of last year. However, it lays more stress on understanding of the students to context, and reduces language rule. More test points are obtained by analysis of context, thus adding difficulty to the students. Some students with inflexible learning method can be easily to split hairs, thus affecting the understanding of the full text; some with flexible method and extensive knowledge can accurately understand the full text by analyzing context. The students taking examination can be distinguished in ability, thus making the point of future English teaching clear. In addition, context makes a great effect on English learning and use [3]. After situational context was proposed by Portland linguist Malinowski on the basis of concept of context, linguist field was divided into cultural context and situational context as a basis for language research. Context in English, including written narrative, preface, ending of oral language and the scene, character, position, identity, etc. of language use, can promote English learning and play a marginal role of improving ability of English use.

3 MEASUREMENT FOR CREATING ENGLISH LEARNING CONTEXT

At present, most language schools lay stress on a characteristic method for language memory and examination, and it cannot constantly improve ability of English use. So, there is much rising space for creating English learning context. Language reinforcement learning class or learning method direction is getting more and more. The context is improved slowly, thus avoiding limit of English teaching effect in time and space and making the students close to real English.

3.1 *Improvement of teaching condition*

Any teaching activity requires certain condition support in order to obtain a good result. English are linguistics with pluralistic teaching activities, and pure oral teaching cannot absolutely achieve an ideal effect. English context is different with other teaching factors, because it has a certain rigor in the layer of environment. Simple English environment is purely resource accumulation, and the students cannot purposely select English resource they are interested in, and accumulate the amount of English, thus making them unsatisfied with desire of learning. After a long time, the students will be confused because the disorderly English fragments cannot form arrangement, combination in order, ideal English context and thinking mode, making the improvement of English learning and use out of the question. So, English teaching resource is considered as the indispensable part for the improvement of English learning, and some measures are taken by some English educational institutions or schools. The primary adjustment is to improve the scientific English learning resource planning. The ordinary English learning resources, including English references, English movies, English and American TV series, etc., can be integrated by teachers and students. These data can be divided into part of standard speech, strengthening language habit and

popularizing culture context according to learning purpose based on fully understanding of the content and structure, so that the students can easily choose combined with their own learning scheme. Besides, they can be divided into British English and American English, or written English and oral English, thus providing the students with direction and maximizing optimization and use of the effective resources of many schools. In addition, plenty of English readers are introduced and bibliography expanded to professional fields such as applied science, social information, etc. to satisfy the English learning of high-level students in the aspect of non-mainstream. The expansion does not lose contact with the purpose of English learning, and makes the students more interested in English. Although they are possible to consider English as the medium of knowledge communication, the purpose of English learning can be achieved and the knowledge of English context strengthened.

3.2 *Function of foreign teachers and exchange students*

With the achievement of cultural communication of non-national boundaries, Chinese language education for foreigners and introduction of foreign teachers are more and more popular. Foreign teachers can speak English fluently, with new faces and strong region characters. When they enter the campus, English learning of students and teaching activities of teachers will be more energetic. However, the actual situation in China shows that a fraction of foreign teachers cannot drive the new wave. So, for some conservative schools, introduction of foreign teachers is still an important method of improving English learning of students.

The school should intensify the cooperation with famous international language schools and actively exchange the methods and results of English teaching of foreign teachers at present, thus making more opportunities for the students to contact with pure English. The students can communicate with foreign teachers in and after class, and the ability to adapt to English language environment and habit can be intensified in daily study and life. This ability can make the students form British or American mode of thinking, break through the limit of parent language and raise achievement of English.

In addition, exchange of students, as an effective method of helping students to improve English, can provide some students with better environment for English learning and form pure and mature English thought. It emerged during the period that the western scientific thoughts went to the east in modern China. Now, the excellent students of one country can be sent to another country to receive education for a period of time. They can learn foreign language, lifestyle and culture as much as possible, and adapt to pure English context, thus providing the maximum support for language learning. The schools can make all kinds of opportunity for exchange of students with the support and help of school and academic communication and specialists. Exchange of a great deal of students can not only make conditions to form better context for the students, but also make an effect of example of exchange students, thus optimizing context, making full use of the original foreign context and improving the comprehensive quality of the students in China.

3.3 *Grasp of details of context*

The context can be divided into wide and narrow sense. The context of wide sense contains not only English pronunciation, grammar, sentence and language habit that the students learn and contact in the process of English education, but also the time, occasion, atmosphere and tone by which the students try to express in English again. These factors can help realize the communication function of English by affecting psychological change, tone adjustment and semantic communication of the speakers to a large extent.

The context promotes the formation of communication because of five functions: matching, location, fill, generation and prediction. Grasp of the details of these five functions can effectively improve adaptability of the students in English context and promote the formation of flexible thinking method. Matching is the main function of English context, and different expressions can be derived from one semanteme in different contexts, expressed as "when you are in rome, you do like the romans do".

In official occasion of business English such as recruit and conference of company, etc, documented English is usually used. The word "conference" should be used instead of "meeting"; partial formal words like "descend", "decrease" and "decline" instead of "go down". The note in working environment can generate many expressions and records with different forms. Prediction is expressed as an example. The students cannot catch the meaning of every word or grammar in an examination or conversation, or completely understand the information because of differences of language habit and mode of thinking. At this time, they can judge according to emotion, likes and dislikes presented by language environment or atmosphere of full text, thus making the obtained information closer to the correct understanding and improving the English comprehensive ability of students.

In order to grasp the details of context, the students should have enough English knowledge storage, thus greatly improving English comprehensive ability. Some students cannot exactly understand the English paragraph in the examination, but they can give the correct answer by feeling. With the help of context, the students can estimate the composition of text, the meaning of paragraph and the emotion using the similar context and special vocabulary.

So, the context, as a part of English semanteme, plays an important role of foreshadowing in the use of oral English. The semantenme can be completely expressed by the look, body movement and interjection of the speakers and the time and occasion the conversation happens.

4 CONCLUSIONS

English learning laying stress on listening, speaking, reading and writing can only be adaptable to the written English examination in China. Perfect English learning demands the students to know the history of the English language, related English reading, film, television, language habit and culture. Only diversified English learning like this can show the value of English language in the actual use. Consequently, English education in China should focus on the improvement of language environment, including the teaching conditions and resources, and the grasp of details of related language environment in daily teaching, thus presenting an open state instead of single mode of teaching and making itself beneficial to students and English education.

REFERENCES

[1] Huang Weihe, Context Factor in Foreign Language Teaching, Journal of Hunan University of Commerce, 2010(10): 23–31.
[2] Yuan Cui, Difference of Meaning of Chinese and English Vocabulary and Cultural Introduction in Foreign Language Teaching, Journal of Anhui Agriculture University (Social Science Edition), 2010(7): 7–14.
[3] Wang Linchao, Relation and Method of Arousing Learning Motivation and Improving Learning Interest, Journal of HuBei Radio and TV University, 2011(8): 89–90.

Future Information Engineering and Manufacturing Science – Zheng (Ed)
© 2015 Taylor & Francis Group, London, ISBN 978-1-138-02644-5

Development strategies in aerospace cultural industry of Langfang in China

Jingang Yang & Juanwei Li
North China Institute of Aerospace Engineering, Langfang, Heibei, China

ABSTRACT: Aerospace cultural industry in Langfang has a solid foundation for its development. Aerospace culture, location advantages, economic and political policies have all contributed to its development. On this basis, current statuses of Langfang's early stage development in aerospace cultural industry are analyzed, problems are discussed, and development strategies—changing ideas to improve the strategic position of aerospace cultural industries—are proposed. Policy documents for development and plans of aerospace cultural industry should be improved. Reasonable plans for resources are needed to update the development model of aerospace cultural industry. Improvements of economic environment are required to realize multi-channel development of aerospace cultural industry. For aerospace cultural production, it is supposed to integrate its production order and improve its product property system. And finally, personnel training projects ought to be implemented to provide human resources for the development of aerospace cultural industry.

Keywords: Aerospace cultural industry; Location; Resource; Market; Policy

1   RELATIVE FACTORS IN THE DEVELOPMENT OF AEROSPACE CULTURAL INDUSTRY IN LANGFANG

1.1   *Location advantage of Langfang is the foundation of its development in aerospace cultural industry*

Location mainly refers to the position of an object or its connections with other objects in space. Determined by its characteristics, the development of aerospace cultural industry has special requirements for location [1]. Thus, Langfang has unique location advantages in its development of aerospace cultural industry.

First, from the perspective of geographical position, Langfang is located between two big cities—Beijing and Tianjin, relying on unique advantages of center cities, airports and harbors. Secondly, from the perspective of geographical environment, Langfang is rich of land and human resources. Environment there are beautiful and pleasant, and people there are honest and hospitable. Thirdly, from the perspective of consumption environment, Langfang has a consumer group consists of nearly 100 million people in Beijing city, Tianjing city and Henan province. Because of its industry and urban concentration, developed industry and commerce, huge market capacity and local residents' strong purchasing power, Langfang has become the golden area of north China.

1.2   *Strong aerospace industry of Langfang provides its aerospace cultural industry with rich resources*

Resource is one of the most important conditions for development in any industry. The unique position advantage of aerospace cultural industry in Langfang lies not only in its convenient transportation, but also in the resources that Beijing and Tianjin provided. Resource advantage of Langfang manifests itself in mainly three ways: first of all, Langfang neighbors Beijing Aerospace Headquarters to the north and Tianjin's rocket base to the south. Combines with local industries, including the production base of 529 factory, Huachuang Tianyuan Industrial Developing Co. Ltd., and the 6916 factory, which is closely related to the aerospace industry, Langfang has formed an aerospace industry group to provide this industry with sufficient meta resources. Besides, the North China Institute of Aerospace Engineering is located in Langfang. Its aerospace culture research base, aviation training center, party school, Qian Xuesen square, aerospace museum, etc., not only boost the development of aerospace culture, but also supply the development of aerospace cultural industry with rich human resources. Finally, International conventions in Langfang have provided the development of aerospace cultural industry with platforms for propaganda and exhibition [2].

1.3 Relevant economic and political policy guidance on the development of aerospace cultural industry

Political and economic policies are *"guiding towers"* for every industry. Therefore, the development of aerospace cultural industry should be carried out under relevant policies. Since the opening of the 5th Plenary Session of the 17th Central Committee, both central and local governments have increased their supports to cultural industries and put cultural development at an important position in regional development plans. Therefore, aerospace cultural industry in Langfang will have a rare opportunity for development. Especially since the opening of the 18th National Congress of the Communist Party of China, the idea—*"putting social benefits in the first place, adhering the unity of economic and social benefits, finally promoting a cultural prosperity and rapid development of cultural industry,"* has become one of the directions for development and structure optimization of aerospace cultural industry in Langfang.

In the *Cultural Industry Development Programme of Langfang City*, for the sake of making cultural industry become a pillar industry of Langfang, the development and prosperity plan for the next decade is established. As an important part of cultural industry, aerospace cultural industry is included in the programme, for it not only enriches programme's content, but also provides a fresh impetus for economic development [3]. In December 2012, Langfang set scientific and technological innovations as strategic supports of development, so as to promote *"higher"* and *"greener"* development in economy. Therefore, the *Guidance on Rapid Development of Innovative Economy* is developed and implemented, for the purpose of optimizing environment of science and technology innovations, making fully use of the science, technology and human resources in Beijing and Tianjin, improving the enthusiasm and initiatives for innovations and their industrialization. As derivatives of the most high-end scientific and technological industry, aerospace cultural industry not only has a publicity function in the development of aerospace industry, but it promotes the optimization and upgrading of industrial structure.

The analyses of key elements in Langfang' aerospace cultural industry have indicated some truths. First, the development of aerospace cultural industry needs a suitable location as its basis. Besides, powerful aerospace industry groups can be resources for industrial development. Finally, development of aerospace industry needs the guidance of related economic and political policies. Only by establishing a *"Trinity"* method can the basic elements of aerospace industry be formed.

2 CURRENT SITUATION AND PROBLEMS OF AEROSPACE CULTURAL INDUSTRY IN LANGFANG

2.1 Current situation of aerospace cultural industry in Langfang

Due to the unique advantages in location and resources, continuous efforts of local government and cooperations of enterprises and institutions, Langfang has made certain development in aerospace cultural industry. However, aerospace cultural industry accounts for only a small portion of the total development in cultural industry, while in terms of per capita availability and contribution to economic growth, the ratio is even lower. Overall, development of aerospace cultural industry is still in its infancy, and Langfang still has a long way to go before it forms the aerospace cultural industry chain. Specifically speaking, development of aerospace cultural industry is still in embryo, no complete industry chain has yet been formed. Besides, aerospace cultural industry has attained diversities in products, but the development of each product is uneven. Moreover, while the main body of aerospace cultural industry keeps growing, the degree of intensity is insufficient. Finally, though the export-oriented degree has improved, aerospace cultural industry is still less well-known when compared with other cultural industries.

2.2 Problems of aerospace cultural industry in Langfang

Problems of aerospace cultural industry in Langfang mainly manifest themselves in the following aspects.

First, the management system of aerospace cultural industry is imperfect. From the perspective of overall management system of China's cultural industry, problems such as *"administrative overlaps"*, *"integration of government administration with enterprise"* and *"multiple management"* can be found in different areas.

Secondly, policy guidances for aerospace cultural industry are missing. In general, policy documents for traditional cultural industries are more than that of aerospace cultural industry.

Thirdly, development modes of aerospace cultural industry are backward. Compared with traditional cultural industries, aerospace cultural industry is a field on the most advancing front. However, its development modes are relatively traditional. For example, corresponding souvenirs and toy models will subsequently appear on the market after the announcement of new spacecraft, but there is few products associated with other industries (such as tourism related simulated spacecraft). Thus the phenomenon that aerospace cultural products are low in technology content and lack of influence is generated.

Fourthly, the market expansion capability of aerospace cultural industry is weak. Langfang has great advantages in location and resources, providing a good condition for the development of aerospace cultural industry. However, fettered by the system of traditional cultural industry, aerospace cultural industry of Langfang is insufficient in market research and analysis, poor in the awareness of market expansion, and weak in marketing abilities.

Fifthly, the aerospace cultural industry market is disordered. There are various aerospace cultural products on the market, however, their intellectual property rights are lack of protection. Aerospace cultural industry is built on the development of aerospace industry. But in the industrialization process of aerospace culture, some unauthorized productions with distorted cultural intentions or misinterpreted scientific concepts cause not only misunderstanding of aerospace culture, but also a serious impact on the development of aerospace cultural industry.

Sixthly, aerospace cultural industry lacks talents. While having the North China Institute of Aerospace Engineering and other aerospace industry related enterprises and units, aerospace cultural industry in Langfang is still in short of professionals. Shortage of professionals that can really assume the developing task has become the "*bottleneck*" of aerospace cultural industry.

3 STRATEGIES AND SUGGESTIONS FOR DEVELOPMENT OF AEROSPACE CULTURAL INDUSTRY IN LANGFANG

According to its development situation and problems, measures can be taken in the following aspects to develop aerospace cultural industry in Langfang.

First, the role and strategic position of aerospace cultural industry need to be enhanced; its management system should be improved as soon as possible; concepts require to be converted. The specific measures can be summarized as "three abolishments and three establishments", which has three aspects of meaning. First, abolishing the concept that aerospace culture is a mere ideology, and establishing the awareness of economic value and industrial properties of aerospace culture; secondly, abolishing the opposing ideas between social and economic benefits, and establishing the market efficiency concept to meet social cultural needs; thirdly, abolishing the equivalence between aerospace development and aerospace cultural

development, and establishing the idea that aerospace cultural industry is irreplaceable. Thus, the "interrelation" phenomenon—development of aerospace culture and improvement of its management system are processing simultaneously—can be realized. Moreover, the competitiveness of the cultural industry is improved. And more importantly, the economic optimization of cultural industries is promoted.

Secondly, policies on the development and planning of aerospace cultural industry should be introduced as soon as possible, so as to guide the development of this industry. For instance, policies are required to be formulated, which covers the possession and usage of aerospace cultural resources, taxation, allocation of personnel, investment, financing, industrial structure, industrial organization, industrial technology and industrial markets. The formulation and perfection of these policies will directly affect the speed and level of the development of aerospace cultural industry.

Thirdly, by rational planning and integration of resources, the development model of aerospace cultural industry can update at a faster pace, and also, the technological content of aerospace cultural products can be improved. With a unified development plan and an integration of aerospace cultural resources, we should center on the development of aerospace industries and based on aerospace technologies. Then we can choose to build aerospace-themed projects such as a large-scale, intensive and diversified aerospace park. In this way, not only the integration of resources is benefited, but also the development model of aerospace cultural industry can update at a faster pace. Moreover, the technological content of aerospace cultural products can be improved.

Fourthly, multi-channel development of aerospace cultural industry can be achieved by improving the investment and financing environment. Development of aerospace cultural industry will benefit from many factors, such as increasing investment, improved financing environment, and also, multiple investors include governments, enterprises and private capitals. In addition, multi-agent investment will help to encourage public participation in the construction of aerospace cultural industry, and thus broaden its market.

Fifthly, besides the integration of aerospace cultural industries, we should unify aerospace cultural products' specifications, then standardize their market orders, and finally, improve their property right system. Compared with other cultural industries, ineffective regulations and imperfect policies have caused turmoils in the aerospace cultural production market. Improvements of the legal system should be addressed to eliminate such phenomenon. First, we should standardize the market order with legal norms and establish regulations accommodating to modern market economy. Secondly, management by laws should be strengthened, so as to ensure a benign development and standard production procedures of aerospace industry. Thirdly, property rights of aerospace cultural products should be protected by laws.

Sixthly, the implementation of talent cultivation project should speed up, so as to provide aerospace cultural industry with enough human resources. As an emerging industry, aerospace cultural industry has a very urgent demand for talents. Deng Xiaoping once said, "*Top talents should have chances to realize their abilities.*" Jiang Zemin also said, "*Talents are the most important resources of a country. In today's world, competitions for talents are fierce. One important reason for America's rapid development is that it collects large number of talents all over the word. With a political vision, we should set measures to cope with this issue. Moreover, the cultivation and utilization of human resources should be regarded as major events.*" As can be seen, talents can be formed by cultivation. Besides, personnel construction can provide sufficient development energies for related industries. Thus, the talent cultivation of aerospace cultural industry can be achieved in the following aspects. First, the personnel management system and incentive mechanism should be improved, so as to create a favorable social environment for personnel training and ability demonstration. Secondly, the specialty structure of professionals in this industry should be modified. Moreover, the structure level of professionals should be improved. In addition, actual combat training among different specialty structures and hierarchies should be conducted. Thirdly, municipal government and relevant departments of Langfang should plan to absorb professionals in multiple ways. The idea that "making fully use of possessed resources" should be realized. Fourthly, resources of education and aerospace-related enterprises should be integrated. To cultivate professionals, modes such as joint training of higher education and enterprises should be adopted. For instance, cooperated with the North China Institute of Aerospace Engineering, an Aerospace Technology Park in Gu'an has established a training base for students. This method can not only help strengthen the combat capability in personnel training, but also optimize the training mode of aerospace cultural industry.

4 CONCLUSIONS

Langfang has a very solid basis for developing aerospace culture for its unique position, convenient transportation and rich aerospace resources. Combined with attentions of national government, cooperations and guidances of local government have clearly pointed out the direction of aerospace development. Though possessed basic developing conditions and therefore reached some developments, there are still many problems in aerospace cultural industry. These problems can be solved in following ways: first, the strategic position of aerospace cultural industries should be improved; secondly, resources of aerospace cultural production should be integrated; thirdly, the personnel training method should be strengthened. In short, aerospace cultural industry in Langfang should be developed with rational exploitation of its location advantages. Thus we can enrich and promote the developments of aerospace cultural industry at the same time.

ACKNOWLEDGEMENTS

The work was funded by Langfang Municipal Science and Technology Bureau as a soft science project "*Development Modes of Aerospace Cultural Industry of Langfang in China*". Grant number: 2013023055, Phased Achievements.

REFERENCES

[1] LIN Xiu-mei, ZHANG Yi-li. The Analysis of Pulling Effect of Cultural Industry in Jilin Province, Journal of Northeast Normal University, 2014, 03, 112–117.
[2] CHANG Ling-chong. Path of Digital Communication of Tibetan Intangible Cultural, Heritage in the Age of Internet Journal of Minzu University of China, 2014, 03, 167–171.
[3] DUAN Jie, ZHANG Juan. On the Contribution of the Development of Shenzhen Cultural Creative Industry to Economic Growth Based on the Gray Prediction, China Population, Resources and Environment, 2014, S1, 457–460.

Future Information Engineering and Manufacturing Science – Zheng (Ed)
© 2015 Taylor & Francis Group, London, ISBN 978-1-138-02644-5

Study on new type of building construction technology based on BIM virtual technology

Haiying Peng
Shaanxi Vocational and Technical College, Xi'an, Shaanxi, China

ABSTRACT: The continuous development of building construction technology brings in high and new technology. Based on BIM in construction industry, the BIM virtual technology is aimed at changing the traditional pattern of building operations through technological superiority. By analyzing the concept, characteristics and application of BIM virtual technology, this study found that BIM virtual technology could reduce the construction cost to a large extent and enhance construction efficiency. However, there was some bottleneck for the application in some fields. The BIM virtual technology could be improved and comprehensively used by reasonably planning the market recourses.

Keywords: BIM Virtual Technology; Building Construction Technology; Study; New Type of Construction Technology

1 INTRODUCTION

The building construction industry has been rapidly developing over the recent years. On the premise of environment-orientation idea, it is the goal and trend for building industry to develop building construction with high and new technology so as to achieve high-efficiency and high-quality building construction. The BIM virtual technology is the technological reflection of building information modeling, which is featured with visualization and optimum design through the integration of building data [1]. For the building construction technology, the traditional construction technology inevitably can not meet the demand of modern technology access. Therefore, new type of building construction technology is desiderated to meet the market demand of building construction.

2 BIM VIRTUAL TECHNOLOGY

With the continuous development and application of modern building technology, some new type of building ideas have been used in the application of new construction technology combined with existing theoretical pattern. BIM virtual technology is a new popular type of building idea in current building industry. By adopting modern way of application, BIM virtual technology extends the development and application of building industry and building technology. Then, what is BIM virtual

technology? What advantages does this technology have in practical application?

2.1 Concept of BIM virtual technology

The BIM virtual technology is a technology based on building information modeling, which is similar to the simulation technology of 3D model. BIM is just the abbreviation of Building Information Modeling, which establishes model by integrating the information of building project and using these data as a basis. The BIM virtual technology often adopts vision technology of digitization and visualization, thus demonstrating the information of concrete building. In fact, The BIM virtual technology is more like a digitization tool, which uses model as the basis of data to achieve the sharing and transmitting of information [2]. For constructors, the BIM virtual technology can help them better command the understanding and handling of information, thus guaranteeing the enhancement of building efficiency and cost control. Therefore, the BIM virtual technology has a wide application in modern building construction.

2.2 Characteristics of BIM virtual technology

The BIM virtual technology is a necessity for the future development of building industry. Then, what are the characteristics of BIM virtual technology? What is the relationship between these characteristics and advantages of BIM virtual technology?

2.2.1 Visualization of BIM virtual technology

The visualization is the most advantageous characteristic of 3D simulation technology and BIM virtual technology has the characteristic of visualization. In building industry, although traditional architectural drawings can clearly demonstrate the construction process and blueprints, only professional technicians can understand these two-dimensional graphics [3]. However, the visualization of BIM virtual technology can get rid of this limitation and present a three-dimensional effects, which helps constructors understand the structure of buildings. Therefore, the construction efficiency is highly enhanced and the construction process is simplified.

2.2.2 Harmony of BIM virtual technology

In building construction, it is not rare to have problems of poor communication. In general, when problems arise, coordination comes. However, it often causes large amount of time to solve these problems, which can lead to duration delay and coordination difficulties. Hopefully, the coordination service in BIM virtual technology can solve these problems. As the BIM virtual technology is based on module data, the problems need to be coordinated can be found through data connection and coordination simulation. Then, coordination data will be produced to achieve the preventive disposal model of problems, which can narrow construction period to a large extent and guarantee the construction go on smoothly.

2.2.3 Imitativeness of BIM virtual technology

The BIM virtual technology is based on the information model of building data, which makes simulation operation possible. Differing from general simulation technology, the BIM virtual technology can achieve operability. The simulation of buildings is to conducting analog distribution of the dimension and data of buildings to achieve concrete effect through three-dimensional imaging technology. However, traditional simulation technology is an external imaging, which is hard to conduct simulation operation. On the contrary, the BIM virtual technology can achieve real simulation operation. By revising the data and setting up the model, technicians can conduct simulation analysis on potential problems of buildings. For example, in conducting energy conservation simulation, the way of data budget can be used to simulate. The advantage is that some problems happened in construction can be avoided, which can cause duration delay and uncontrolled cost.

2.2.4 Optimization of BIM virtual technology

Architecture optimization is the popular idea under the background of energy conservation and environment protection. Regardless of construction or building, energy conservation and environment protection need to be guaranteed. However, in practical construction, due to the design problems or environment problems, wastes and non-energy conservation are caused. The BIM virtual technology can solve these problems because it has the function of optimal design, which can conduct continuous optimal processing at the testing stage. For example, the BIM virtual technology can achieve optimal effect in project design and reasonable resource utilization. Then, the accidents and potential problems can be avoided in construction. Therefore, the optimal design of BIM virtual technology is aimed at making plans for construction at early stage so as to achieve the effect of modern construction technology.

3 NECESSITY OF BIM VIRTUAL TECHNOLOGY IN BUILDING CONSTRUCTION

Traditional building construction technology is an extensive pattern, which causes the waste of building resources and uncontrolled cost. In design stage of building construction, the limitation of design drawing can cause some accidents, or the unspecific on-the-spot investigation can lead to the duration delay of construction. Sometimes, the problems of consultation and coordination in construction can also lead to the duration delay. Therefore, there are a lot of problems in traditional building construction industry. The extensive resource waste and poor communication are the factors that lead to the uncontrolled cost, debasing the building quality. Therefore, in order to improve the above problems, high and new technology should be introduced.

Through the analysis on the advantages and characteristics of BIM virtual technology, it is necessary to apply this technology to building industry. The BIM virtual technology is especially effective in cost control and design. Using BIM virtual technology to conduct simulation and optimal design can enhance the quality of building construction and thus improve the regulations and standards of building construction.

4 STUDY ON NEW TYPE OF BUILDING TECHNOLOGY BASED ON BIM VIRTUAL TECHNOLOGY

Due to the current extensive pattern of building construction and the constraints on environment protection ideas, the building industry needs a reform. Based on the BIM virtual technology, a brand new pattern of building construction technology should be established so as to reduce the waste of building resources and lower the construction cost, achieving the modernization of building construction. Then, how to reform current building construction technology, and what is the new building construction technology based on BIM virtual technology?

4.1 Costing of building construction based on BIM virtual technology

Although the BIM virtual technology has a lot of advantages, there are still some problems in its application. The most important problem is the costing of BIM virtual technology. In practical building construction, the costing has some difficulties for the influence of some problems. These problems are listed below.

Firstly, the data size of costing is too large. Since every stage of building construction involves large amount of human resources, materials and machineries, the information is too large and complex, which brings difficulties to cost control. Therefore, it is common for construction projects to work to tight deadlines, leading to be busy enough with their own affairs. Therefore, the issue of optimal management and construction cost is put away temporarily.

Secondly, the poor coordination among departments causes the difficulties in conducting the costing. There are many posts in different departments of building construction, but the communication among different departments is rare, resulting in the improper costing. In addition, the costing is a comprehensive work, but the data collecting and exchanging can not be achieved in time, which also brings problems to the costing.

Thirdly, there are some difficulties in dividing the projects. Since the machinery management and staff distribution are related to some extent, the dividing work can not be single. Therefore, there may be impossible to costing when a project is divided into several parts, leading to overall dissimilarity.

However, the above problems can be solved with the help of BIM virtual technology.

Firstly, according to building construction, a cost database should be established based on BIM virtual technology and the visualization and classification management of data can be achieved. Through this management model, data costing and cost management can be clearly conducted.

Secondly, data related to cost should be timely input. Timely inputting these data can reduce the complexity of data management to some large extent. Besides, the computerized data can help reduce the cost of management and costing and improve the working efficiency.

4.2 Value demonstration of new building construction based on BIM virtual technology

The value of new building construction based on BIM virtual technology is clearly demonstrated. The effect of reform and management is prominent for traditional building construction technology. Then, how is the value of new building construction based on BIM virtual technology demonstrated?

Firstly, the visualization of imaging achieves the visualization management, improving the early design to the great extent. Through 3D simulation and visualization imaging, the design of building construction is more apparent and simplified, reducing the errors of design and construction.

Secondly, the construction cost is reduced to the great extent. Many operations can be accomplished through BIM virtual technology, thus reducing the loss caused by construction errors. Besides, BIM virtual technology has a good performance in optimal design. By optimizing the design alternatives, the BIM virtual technology has a great advantage in cost control.

Lastly, virtual operation improves the effectiveness of construction. With the help of BIM virtual technology, the test and evaluation that can not be done in practical operation can be achieved and compared with practical operation, which reduces the test cost and hidden dangers in construction.

Therefore, new building construction based on BIM virtual technology has prominent advantages.

5 CONCLUSIONS

The building industry is one of the great potential industries. However, with the continuous use and shortage of resources, how to reduce cost and improve building construction efficiency is an urgent problem in building industry. The application of high and new technology and the development of new building technology largely promote the development of building industry. Especially, the new building technology based on BIM virtual technology offers traditional building industry a new development pattern. In addition, in order to achieve the perfect connection between digital technology and building construction technology, the connection should be emphasized, guaranteeing that the development of technology and corresponding construction be related with the requirements of building industry. The advantage of BIM virtual technology is that it can achieve simulation of buildings and optimal operations through digitalization and optimal simulation, thus comprehensively improving the building cost and construction process. Therefore, since the new building construction technology based on BIM virtual technology is future developing trend, it is necessary to popularize this technology in building industry.

REFERENCES

[1] Wang Jianhong, Zhang Jinli, *On the Establishment of Prevention and Control System of Construction Quality*, Gansu Science and Technology, 2011(20):162–164.

[2] Yang Zhongcai, *How to Create Whole Process Quality Control System of Project Construction*, Development Guide to Building Materials, 2012(7):133–134.

[3] Jin Mingxun, *On the Establishment of Construction Quality Supervision System and Quality Control Methods*, Fortune World, 2010(22):20.

Future Information Engineering and Manufacturing Science – Zheng (Ed)
© 2015 Taylor & Francis Group, London, ISBN 978-1-138-02644-5

Application of aesthetic idea and spatial planning in landscape design

Changying Liu
Shaanxi University of Science and Technology, College of Art and Design, Xi'an, Shaanxi, China

ABSTRACT: The work studied the current application and practice of the landscape design from the views of aesthetic idea and spatial planning, with the specific application as the main consideration. According to the analysis, aesthetic idea and spatial planning are of practical guiding significance in landscape design. And the application mechanism in the process of design should be further specified. The innovation point in this work is analyzing the specific application of landscape design in multiple perspectives of western and Chinese traditional design and aesthetic idea.

Keywords: aesthetic idea; landscape design; spatial planning; garden design

1 INTRODUCTION

Urban landscape design covers the work in a lot of fields, such as road project, construction project, park design, square design, lighting technology, and forestation planning. During the rapid development of urban landscape, deficiencies exist in the aesthetic design including utilitarian purpose in ecological landscape, missing humanism aesthetics and natural aesthetics in urban landscape, leading to homoplasy in the lack of artistic aesthetic in urban landscape design [1]. Therefore, the design of urban landscape should be adjusted in spatial planning and aesthetic idea to meet people's demands of landscape design.

2 GENERATION AND APPLICATION OF THE BASIC LANDSCAPE DESIGN CONCEPT

The final product of landscape design is actually the service for the people in the corresponding area. As one of the important components in human settlements, landscape design does not only consider about the architectural design with physical function, but also the daily demands from the people who live in that area. Therefore, the design has to meet both the rational material needs and perceptual aesthetic needs, which is the modern artistic interpretation of the Chinese ancient saying *preferring eating without meat rather than living without bamboo.*

The modern landscape design started gradually in American since the 19th century. One of the classic designs is the New York Central Park. At that time, based on the original art of garden design, the American designers extracted the design art and skill of landscape for a new architectural subject, landscape architecture. As well as meeting the material needs from people in the corresponding area, the landscape design rebuilds the landscape in artistry and practicability [2]. Therefore, as a product of developing times, it indicates the improvement of people's living standard in the era of industrial civilization to alleviate people's anxiety caused by urban landscape. In current China, landscape design can solve the urban problems such as crowded residential condition, deteriorating habitation and heavy traffic.

3 APPLICATION OF AESTHETIC IDEA IN LANDSCAPE DESIGN

Concerning many psychological issues, the purpose of landscape design is meeting people's needs for ideal living environment to achieve a harmonious relation between human and habitation. Therefore, during landscape design, the relation among nature, society and human should be balanced at a harmony status, which is the process of pursuit in practice. In current China, living, residential and environmental problems are relatively serious. With the developing industrial civilization, problems of habitation pollution, heavy population and imbalanced ecological environment are the inevitable result of industrial development. Thus the priority of landscape design in China is alleviating the citizens' anxiety by offering peace and support with the aesthetic value. By utilizing the environment effectively, more people will be attracted to enjoy the beautiful scenery in their habitation. The landscape design can further inspire people's aesthetic needs of landscape construction. These needs will influence their inward consideration to improve their psychological demands.

3.1 Conception of landscape aesthetic: understanding people's psychological needs

The objective environment is actually the source of people's inner world. The mental consciousness in people's inner world is the reflection of their minds. With people's rational material needs fulfilled, superior needs will engage in perceptual and spiritual level. Therefore, the psychological needs of people's aesthetic become the premise of landscape design. The idea of *conception* was already proposed in the artistic practice of aesthetic in China's ancient time. Conception is the consideration of the whole artistic layout before the landscape design. People's aesthetic needs for landscape should always be the priority in any kind of design. Therefore, the perceptual function of landscape artwork as an aesthetic medium should be fully considered during the construction of landscape. The aesthetics contained in landscape is a sense of beauty combining illusion and reality. Only by combining the landscape designer's subjective feelings with the objective environment, the subject and object can be integrated. The moving effect of landscape design is not the geometric structure of the landscape but the original duty for it, which includes the designed function of this landscape, expected acceptance and inner feelings of the designer expressed by the landscape [3].

Therefore, the landscape design should be implied in the configuration. This implicit but revealing conception can express the common subject or prospect in people's inner world. Based on this conception, the design will be really moving to resonate with the viewers. Combining the objects in the nature with human's purposes to forge their emotion into the landscape, people can be further pleased to taste the precious life and the unique beauty of the world.

3.2 Conception construction of landscape aesthetics: further approach to psychological needs

The natural landscape is inimitable as none of the artifacts can match the extraordinary work of the nature. The essence of landscape design is a personal expression of natural emotion, accomplishing image building during imitating natural landscape to achieve human's implication in landscape. The artistic conception is an important concept in Chinese classic aesthetics, which can be found in many Chinese artworks indicating the trial and exploration of artistic conception. In landscape,

the artistic conception is the expression and observation of natural scenery, containing notable personal consciousness and aesthetics. The Chinese ancient litterateur Wang Guowei once said that, "All the words in landscape are emotional." The landscape without including human emotion is not worth watching. Therefore, the artistic conception actually guides the aesthetic formation of the landscape to reach people's psychological needs. By further approaching these needs the aesthetic realm of anatman will be achieved.

The artistic conception is based on people's basic needs and limited by all kinds of spatial and corporeal conditions. During landscape design, the effect of abundant philosophy and psychology are included under the finite objective conditions to create unique scenery with artistic expression. Building artistic conception in landscape aesthetics is not a simple accumulation but an interaction of personal emotion and scenery. Instead of specific setup of all the aesthetic concepts, landscape designer attracts the viewers' attention to the scenery out of the garden with the smart combination of specific scenery. From finite and tangible space to infinite and abstract space, the viewers' feelings will be highly satisfied by the combination of illusion and reality. Therefore, the construction of artistic conception in landscape aesthetics is building a broad and multi-dimensional aesthetic space, with both subjective and objective imagery included in the artistic conception of landscape. Emotion is a subjective factor and scenery is objective. Combining the unrestrained objective characteristic of the scenery with targeted subjective emotion requires strong aesthetic abilities in landscape design.

4 APPLICATION OF SPATIAL PLANNING CONCEPT IN LANDSCAPE DESIGN

The landscape design itself is a process of spatial structure design. Habitation is the major stage for human activities. No matter in public or private place, the landscape design has to coordinate with people's activity status. The spatial planning in landscape design can connect people from person to person, with nature as the platform and background to structure and humanize the function of various spaces. Landscape design is a style of art to avoid spatial homoplasy. And the spatial reconstruction can transform citizens' understanding and perception of the outdoor scenery to achieve the recognition and reformation of the whole landscape space.

4.1 Service function design of landscape space

The final purpose of the environment is to meet people's practical needs. Therefore, places such as the meeting place in residential quarters and the open area with multiple functions are always favored as outdoor living places. The design of open space is to achieve such function to meet people's practical needs and aesthetic requirements.

The open space in a specific human environment can fulfill its use value that is offering a relaxing and natural field for multiple activities. In such environment, people can release their stresses and break the obstacle in the society. The communication in new places will provide new experience and stimulation to them. In current China, high intensive residence takes the most part. People from different age groups, cultural degrees and occupations form a complicated and diversified residential atmosphere. As a behavior of human intervention, landscape design can improve the space function in space design to release people's stresses by adding stimulations in the environment.

4.2 *Ecological function design in spatial planning*

In current China, the intensive residence takes the major part. Combining the relevant points between domestic and foreign design, the landscape around residence should have ecological, aesthetic and practical function highlight the importance of these functions in landscape design. The ecological function of landscape design in spatial planning should not be ignored. Not only planting green plants but also integration planning concepts should be introduced to really achieve the ecological function in the limited space. The design of green area should meet the overall purpose of the residential quarter's spatial planning. The ecological function of landscape design should be integrated with the environment of the residential quarter and the spatial planning to achieve the best combination.

Therefore, the green landscape plant should be an important component in the ecological function of landscape design. For better ecological function in limited green landscape, breeds with reasonable price and trampling resistance should be introduced under daily maintenance. Residential ventilation and lighting can be optimized in spatial planning to achieve a pleasant environment. Trees on the road and ground should be utilized to improve the forestation in the important spots for a proper ecological landscape, where different design purposes should be identified to select the correct plants.

5 CONCLUSIONS

With the rapid development in China, the urbanization level keeps increasing in many regions, where the landscape is gradually developed. Currently, several important tourism cities have introduced the practice of urban landscape design, and many cities have already included landscape design into city planning. As a concept originally based on the natural scenery, landscape is the featured scenery in specific region. And landscape design is to characterize the scenery in the corresponding region based on the natural landscape with subjective initiative. The general process of landscape design is turning the idea into reality. From this point of view, landscape design is a landscape optimization with the help of architectural engineering, which requires experience to improve itself during practice.

REFERENCES

[1] Han Bingyue, Shen Shixian, *Landscape Design Based on Regional Characteristics*, Chinese Landscape Architecture, 2005(07): 61–67.
[2] Zhu Jie, *Inheritance and Development of Creative Graphics Design*, Hundred Schools in Arts, 2005(03): 138–139.
[3] Sunlei, Panyi, *Comparative Study on Metropolitan Park System in Boston and Green Lanes in Pearl River Delta—Taking Shenzhen as an Example*, Chinese Landscape Architecture, 2011(01): 17–21.

Future Information Engineering and Manufacturing Science – Zheng (Ed)
© 2015 Taylor & Francis Group, London, ISBN 978-1-138-02644-5

Marketization of the application of broadcast television

Yiping Gao
School of Literature, Liaocheng University, Liaocheng, Shandong, China

ABSTRACT: Analyzing the development of current radio and TV media during the marketization process of socialism, the work explored the development dilemma of broadcast and television from the aspects of new media and broadcast television's self-development, with the consideration of the marketization of the application of broadcast television from the aspects of brand operation, live broadcast and so forth. Through the analysis, it can be proved that if to really take the road to marketization, radio and TV products need to grasp the characteristics of the times and its own, and optimize its own operation mode, thus being able to form media competitiveness. The innovation of this work is to analyze the dilemma of the marketization of broadcast and television from a combination of internal and external causes.

Keywords: Broadcast Television; Marketization; Brand Operation

1 INTRODUCTION

Over five decades of development, China has more than 600 broadcasting stations and television stations as well as a multitude of radio and television programs at present. Meanwhile, China has also formed a relatively complete base strength in recording and editing radio and television programs, establishing a multi-level broadcast television network in the world's most populous nation. However, the marketization of Chinese broadcast television today is still not high enough. Nor is the development level of broadcast television to media function. Moreover, related broadcast television products have relatively fewer marketization varieties and a part of these products are in a relatively disorder competition state [1]. As a result, some problems gradually exposed under the background of socialist market economy so that the ways and methods to marketize the application of radio and television should be further thought and explored.

2 OUTLINE OF RADIO AND TV MEDIA'S DEVELOPMENT IN THE NEW AGE OF MARKETIZATION

Broadcast and television is a kind of transmission media which transmisses related images and voices within a certain range of area mainly through radio waves with different wavelengths. Therefore, it is a main mode of transmission for broadcast and television. Broadcast and television is generally divided into two aspects: one is to transmit the sound content purely, which is called *sound radio*; the other is to transmit the sound and images together, which belongs to *television broadcasting*. From the narrow point of view, broadcasting should not include effected signals transmitted by ultra short waves. Yet, the operation of broadcast and television is integrated in many places nowadays. Therefore, sound and image is of the same importance, which are an important carrier for radio and TV media to co-exist.

Broadcast and television is still playing a very important role in the contemporary age, which includes the following aspects: firstly, broadcast and television has the publicity function. It could take its own absolute advantage in mass media to transmiss the Party's related lines, guidelines and policies more timely as well as guide the current social ideology, forming an important propaganda function. Secondly, broadcast and television is equipped with education function [2]. It could be utilized to disseminate related scientific and cultural knowledge to the masses better, especially the natural science and cultural knowledge which a part of the masses have never contacted completely yet. It has very important effect on improving the scientific and cultural qualities of the masses in the whole society. What's more, current radio and television could still play a supervision role, which is to supervise some social behavior, including specific administrative behavior, economic activities and social activities and the like, by such a platform. Hence, it could enable to build and carry forward the social justice as well as suppress and rectify the illicit atmosphere.

3 DILEMMA OF MARKET-ORIENTED DEVELOPMENT CONFRONTED BY CURRENT RADIO AND TELEVISION MEDIA

3.1 Shock effect of new media on broadcast and television

The rapid development of new media has a relatively obvious shocking effect on traditional broadcast television and such media. The radio and television among traditional media has taken great changes in the whole survival state that it is faced with. Broadcast and television has also witnessed profound changes in original specific development environment and the whole internal frame, including a variety of network advertising methods. Meanwhile, related advertising modes of direct delivery all develop relying on new media. These are of very huge influence on traditional radio and television. For new development trends emerged in the media of the new world, radio and television will inevitably be confronted with very big challenges [3].

Under the development of new media, the amount of truly loyal viewers for radio and television is also rapidly decreasing, including a part of original young viewers. With the popularization of computer and development of Internet technology, the amount of time for young viewers to watch traditional TV and listen to the radio has constantly shortened, while the group size of Internet users in China is increasingly enlarging and the time they surf the Internet every week is unceasingly increasing, too. From this point, we can judge that the user group of radio and television products in traditional media will inevitably shrink to a certain extent, which kind of shrinking puts forward more and updated demands for the development of radio and television products.

3.2 Bottlenecks in the development process of radio and television products

China's reform and opening up has developed over thirty years. Undoubtedly, with the development of the times, the technology of radio and television itself has also developed greatly and broken through some important problems, such as camera technique. However, as far as the processing technology of radio and television products in China is concerned, the overall is in a relatively low-end level with many contradictions existed at present. Some important technology problems remained unresolved and broadcast television still exists a lot of practical development dilemmas from the perspective of the market.

The first aspect is that currently the masses of the people have a relatively strong demand for information while the products of radio and TV media are relatively in short. The radio and TV media products, as a transmission tool for information itself, provide a colorful daily life for the masses of the people. Yet, the broadcasting institutions of radio and TV media products in current China haven't increased the information of products yet, thus the product quality of current radio and TV products is not very high. Many radio and television programs contain relatively very little information, the quality of whose content still has a big space for improvement. In this way, the development of current radio and TV products is so greatly restricted that it is unable to satisfy the demand of the broad masses of the people for multi-structure cultural products.

The second aspect is that the development of current radio and television products has a relatively serious mixed identity. During the long process of development, the identity positioning embodied from China radio and television media is still not clear yet. Chinese TV and Broadcasting Administration and Bureau of Radio and Television have a very close relationship with each other. In other words, the producers of radio and TV products are always equipped with a mixed identity of both property and cause. Though entrepreneurial management method has been applied, such enterprises still adopt a traditional management method of public institutions. The original purpose of this method is to guarantee the public opinion and propaganda fields to serve for the current mainstream ideology as well as the broad masses of the people. However, in fact, TV and Broadcasting Administration, on account of such a dual identity, affect the market-oriented operation of radio and television and give birth to many disadvantages. Some TV and Broadcasting Administrations have a bad grasp of the degree of marketization while some others marketize excessively and abandon their own promotional role completely, which leads to a dilemma situation.

What's more, the present research and development team of radio and television products is short of more professional personnel. The contemporary economic development type is a kind of knowledge-based economic society. If any business wants to achieve real success, related talents are needed to provide support. The shortage of talents exists widely in the industry development of China radio and television. In particular, current radio and television industry has formed a certain public institution system, which demands more talents to bring out the original vitality of current radio and television in China.

4 THOUGHT ON THE WAYS TO MARKETIZE THE CURRENT RADIO AND TELEVISION

4.1 Media characteristics and key points that need to be mastered from the radio and television products in the current market-oriented age

The media industry in China have had significant development in the socialist market economy as well as gained relatively many achievements. However, all kinds of media at present gradually integrate with each other. China radio and television industry has suffered rapidly changing external environment. Therefore, the development of China broadcast and television need to be based on the new characteristics of media appeared in the present age first.

The first obvious characteristic of the current media is marketization. With China's reform and opening up gradually entering into deep-water district, traditional media also appears to be variously marketized. As a vital cultural force among traditional media, China radio and television industry puts more emphasis on the supply-and-demand dynamics of the whole market and related users' demand for this aspect under the socialist market condition. Its marketization has become an important way for the survival and development of current traditional media of various types.

Another important feature of media is that current media industry appears in an operation state of industrialization; the system between the investor and the issuer has become more clear and specific; equity of related media products has been distinguished more clearly. As a result, the whole country has formed a more united multimedia market. This feature gradually influences the management mode of blockette among the Chinese traditional media and forms an united management mechanism, integrating relevant media resources in an efficient way.

Under the development of current media, omnimedia has become an important media development tendency. Omnimedia is actually the integration of traditional print media, sound media and photo media. Relying on the electronics and computer technology, it could, to a larger extent, interpret the content tramsmissed in the media in multiple forms, showing a broader development space. Yet, omnimedia, being a young media method, has also gradually brought new challenges to traditional media such as radio and television.

4.2 Enhancing the influence of current live radio and TV broadcasting

Live broadcast, in fact, is an important form for radio and television to exert best its influence as a kind of media product. In this process, the radio and television media products should not only transfer the ideas organized and modified subsequently under the original traditional conditions, but also enhance more the live broadcast function of radio and television products themselves. Possessing more live radio resources itself, current radio and television becomes a real-time updated media product. Radio and TV media could live report more news, taking live broadcasting as a normalized product form. In this way, it could let people feel that radio and television media could always present the newest situation and keep updating its content. The masses could master the newest situation of some field by paying attention to the development status of radio and television products. Though the current Internet technology could realize the live broadcasting, the richness still cannot be compared with those of radio and television products. At present, all levels of radio and television are making innovations about the live broadcasting. Even though, there are still some technology problems about live broadcasting that need to be broken through, including delaying, making follow-up report arousing 24 hours continuous attention and how to make hot news and information attract the audience's attention in the live broadcasting process. All these are an important aspect that current radio and television practitioners should pay attention to.

4.3 Related programs and columns forming the brand management

At present, the demand of the broadcasting television's audience is gradually becoming refined, in which process the demand of the audience gradually become multi-leveled as well. Hence, radio and television need to process brand management of its own radio and television programs under the fierce market competition, forming some radio and TV media products with brand images. The radio and TV programmers need to show their own individualities more and form their own characteristics under the circumstance that the radio and TV programs in the whole market tend to be homogenized, thus forming their own brand personality by heterogeneity. Beside that, it is also very important to be good at designing all elements of the whole program. The contemporary age is an age of unabashed individuality, so the host broadcasting should be a brilliant and responsibility figure with individual character. In this way, it could arouse the attention of the masses of the people in the society combined with excellent and dynamic radio and television content. Finally, it could get more market share, becoming competitive brand products among the radio and television enterprises.

5 CONCLUSIONS

Broadcast television, in fact, is a very important kind of mass media, still occupying a large proportion in the current market. However, revolutionary development opportunities have emerged in the current communication field. For example, omnimedia, a new kind of media technology which grows up relying on electronics and computer technology, has an extensive and profound influence on broadcast television. Faced with such a development trend of omnimedia, how to further marketize the application of broadcast television and grasping the opportunities as well as challenges brought by the age should be explored and thought to a deep degree. Only by ultimately promoting the integration and development of new media and traditional media could the marketization of broadcast and television promote the significant development of the whole society's broadcast and television career. Current radio and TV media gradually tends to be marketized, which is unavoidable. Yet, marketization is a challenge as well as a very rare opportunity for radio and TV media. Current radio and television needs to truly implement its own branding strategy to promote the gradual marketization of radio and television from the aspects of talent structure, program forms and the like so that it could gain a space of its own in the fierce competition, enabling the radio and television to develop in a sustainable way.

REFERENCES

[1] Li Xiaoxia. *Research on the Television Media Industry*. Reformation & Strategy. 2005(09):67–69.
[2] Li Yanfeng. *Hubei Province Radio and TV Culture Industry Development Strategy Research*. Economic Tribune. 2005(16):10–11.
[3] Liu Hongyu & He Xiaoyi. *Seizing the Market Based on Local—Several Points of View on the Fierce Competition between Television Media*. Contemporary TV. 2003(09):72–73.

Future Information Engineering and Manufacturing Science – Zheng (Ed)
© *2015 Taylor & Francis Group, London, ISBN 978-1-138-02644-5*

Incentive mechanism of strategic innovation in human resource management

Dan Wu

Changsha Vocational and Technical College, Changsha, Hunan, China

ABSTRACT: At present, human resource management in enterprises is facing more-diversified factors, and HR is required to implement corresponding mechanism innovation in daily management. As an important factor impacting employees' benefits and enterprises' developing power, incentive mechanism should be well improved in order to ensure a good and sustainable development momentum for enterprises. Analysis indicates that enterprises should take humanistic concern as a fundamental starting point. Moreover, enterprises should make use of traditional competition mechanism and high-quality performance feedback management to achieve the best function of original incentive mechanism, ensuring scientific and effective human resource management.

Keywords: human resource management; innovation; incentive mechanism

1   INTRODUCTION

With more and more large-scale transfer and management of human resources, the optimization of management mode has become an important approach for enterprises to further explore their development space. Furthermore, employees' working status, mood and self-development demands are generating more and more obvious butterfly effects to the development of enterprises. After certain years' working in a company, each employee will inevitably experience a plateau. Besides, the continuous use of the same management mode of human resources will ultimately lead to the failure of its best effect [1]. Therefore, suitable and timely incentive and encouragement becomes the key to ensure the eternal youth and vitality for enterprises. In human resource management, the innovation and reform of incentive mechanism can meet this demand, and it will bring great improvement for both startup and grownup enterprises. However, such mechanism optimization requires active cooperation with employees and scientific HR plans. If the mechanism optimization is realized, it will greatly improve the environment and outlook of human resources for enterprises.

2   BACKGROUND OF HUMAN RESOURCE MANAGEMENT

As a flexible form of resources, human resource offers a variety of possibilities for its management, and the effect can be achieved and presented through different methods and means. In enterprises, the management effect of human resource can even be directly reflected through the performance and operation status of the company. Therefore, mechanism innovation of human resource management plays a pivotal role in enterprise management [2]. Human resource management mainly covers the following aspects: the strategy development of human resource, employee recruitment, employee training, performance management and salary management. In fact, all those management activities are ultimately the management of "people". What's more, those activities cannot ignore the emotion, efficiency and working conditions of individuals because they are directly related to the smooth and efficient operation of human resource system.

The concept of human resource management originated in late 1970s, and since then it has experienced different stages of baptisms and tests. Nowadays, enterprises pay more attention to achieve a more scientific and perfect management system. Modern society emphasizes the importance of talents. It is the most concerned and toughest issue for every HR to give maximum play of individual value. Nowadays, human resource management has far exceeded its original responsibilities. Modern human resource needs to satisfy employees' demands of career development and personal development. Besides, HR is devoted to achieve a scientific and standardized performance management [3]. At the same time, it tries to effectively and continuously realize the appreciation of human resources. Therefore, human resource management system also requires diversified and sustainable development.

Staff incentive mechanism can be fully used to create values for the whole management. More specifically, HR adopts incentive theories and related methods to satisfy or constrain employees' demands so as to affect their psychological state, thus influencing their daily work. The main purpose of this mechanism is to prompt employees to actively commit to their work and create profits for their departments or the entire enterprise. Professor William James from Harvard University pointed out that each employee only needed to play 20% to 30% of their abilities to maintain a stable existence in their work. This result indicates that HR faces great pressure to discover the real level and capability of each employee in order to achieve an efficient operation of the whole enterprise. However, at present most HR ignores the innovation and development of employee incentive mechanism. Besides, many so-called incentive mechanisms are just confined to the rules and regulations of work norms and reward-punishment systems. As a result, those so-called incentive mechanisms are totally different from the incentive systems of real senses, and it is impossible to achieve its actual incentive effect. Therefore, employees' enthusiasm to work cannot be motivated in the original system, and employees will not voluntarily improve their efficiency. As a result, the whole working group is lack of creativity since employees just follow old rule and adopt outdated methods. Such situation will bring dull working atmosphere to the whole group, impacting employees' efficiency and effectiveness and weakening the overall strength of the group. Therefore, incentive mechanism in human resource management needs suitable and timely innovation and reform to satisfy the needs of the whole group, thus realizing the sustainable development of enterprises.

3 INCENTIVE MECHANISM OF STRATEGIC INNOVATION IN HUMAN RESOURCE MANAGEMENT

3.1 *Humanistic concern as the most fundamental factor*

The most important aspect of human resource management is the management of people. On one hand, HR management refers to the optimization of personnel structure for the entire organization as well as the recruitment and appointment of department heads. On the other hand, it refers to providing psychological counseling for employees.

In both aspects, HR should pay attention to the working status and emotions of individuals. Therefore, enterprise managers should strive to establish a humanistic concern system. In addition to focusing on the structure and direction of human resource management, humanistic concern system needs to emphasize details and care about the working status of grass-roots employees. In fact, enterprises' humanistic concern for their employees is the basis and starting point of the whole human resources management. Moreover, it is the foundation for employees to "work hard" and be willing to give play to all their value under the management mechanism. Thus, enterprise managers should concern about employees' difficulties in both work and life. Besides, they should pay more attention to employees' development needs and career expectations. Otherwise, employees are likely to bear the attitude of demotivation, or they may quit and look for a new job.

Microsoft sets a good example in this aspect as it makes thousands of millionaires faithful to it. It is well known that Microsoft employees hold extremely high working enthusiasm and professionalism. Moreover, irresponsible employee behaviors such as resign and job-hopping rarely happen in Microsoft. The main reason for employees' loyalty to Microsoft is that the company cherishes and values talents very seriously. More specifically, Microsoft shows its humanistic concern to its employees through salaries and the concern for their life and work. Therefore, if employees have any new idea about their self-development, enterprises should provide the biggest technical support. Firstly, enterprises' effort helps employees to get improvement. In return, employees will pay a positive feedback to the company, directly contributing to improve the operation efficiency and economic benefits for the company. Therefore, HR needs to concern about employees' physical health and psychological emotions in the daily work and ensure them a high-standard living quality. Meanwhile, it is also important for HR to provide profound talent-cultivation plans and career-development space for employees. The developmental-talent management includes the expansion of talent-introducing channels, the enhancement of employee training and the emphasis of talent development direction. Those talent management plans can effectively generate a stronger motivation for the whole human resource system. Consequently, employees will make efforts to complete assigned tasks and strive to get further education and oversea study out of their demands on self-improvement. Such psychological attitude is conducive to form a positive working atmosphere and construct efficient human resource configuration. Therefore, the enterprise's humanistic concern for employees will gradually develop to its brand

and advantage, and employees will be able to provide typical quality samples for the optimal allocation of human resource.

3.2 *Competition mechanism as the main approach*

The incentive mechanism in human resource management actually plays the role of a booster in the whole management, and competition is the most important approach to achieve it. Competition refers to series of activities that individuals or organizations take to promote or fight against each other in order to achieve certain goals. Competition enables the activities of human society to achieve more significant effects, and make human realize self-awareness and self-criticism in society activities. "It is not the strongest of the species that survive, but the one most responsive to change." Competition, leading to the update of knowledge and skills and the improvement of work quality, has become a necessity for social development and even for a department or an enterprise. Once employees feel the pressure of competition, they will try hard to improve their working methods and efficiency, thus developing a new and efficient management mechanism. As a result, such virtuous cycle will ultimately provide the department or enterprise with continuous development power. Besides, employees can complete tasks in higher quality. Moreover, competition is good for setting up an entrepreneurial spirit and forming good competition consciousness among employees, providing enough positive energy for the whole management system.

McKinsey's "UP OR OUT" system gives a good interpretation of advanced competition mechanism in enterprises. On one hand, the company provides comprehensive training system for the staff in order to help them achieve certain professional skills and qualities, which is the "UP". On the other hand, the company implements the "knock-out" system for those employees who cannot meet the requirements and standards of enterprise managers, which is the "OUT". Such mechanism generates a talent pyramid within the enterprise and optimizes its human resource structure, providing a solid foundation for the operation of the entire enterprise. Besides, employees will form crisis consciousness. Employees with comprehensive qualities can skillfully "transfer the crisis into an opportunity" and win the competition through their abilities, avoiding being weeded out under the competition system. Thus, the competitions of employment, promotion and training opportunity enable talent selection and performance management to achieve its best function and value. This kind of modern working mechanism ensures the implementation of "use the capable ones, replace the average ones, and relief the incapable ones". Meanwhile, this mechanism is also the most ideal choice for business operation. Therefore, good competition mechanism can not only provide a relatively fair, just and open working environment for employees, but also enables employees to build the habit of entrepreneurial spirit. Meanwhile, the common cooperation mode in modern society presents new requirements for enterprises and employees in order to adapt to the fast working pace. It has become an approach utilized by most enterprises in human resource management. Moreover, it has become an important topic for further exploration of human resource management in the future.

3.3 *Emphasis on the importance of performance feedback*

Performance management is a symbolic section in human resource management. Generally, excellent system and rules can effectively guarantee the realization of employees' performance management and evaluation. Therefore, the performance management for employees in an enterprise is established on the basis of a mature system. In fact, many small-medium enterprises lack necessary foundation of performance management, resulting in inaccurate assessment and evaluation of employees' performances, efficiency, work hours and workload. Even worse, some enterprises are applying a disordered human-resource evaluation system; some unfair and unreasonable human-resource management approaches are utilized, putting human-resource management and its operation in potential crisis and danger. Therefore, large-scale enterprises and wise managers will make effort to establish standard and normative performance management system at the beginning of the management. Thus, it's convenient to implement macroscopic management of human resource.

A fair performance management mechanism is a reliable guarantee to enable employees to follow the instruction of the enterprise and work hard with assiduity. The basic performance feedback system of "more pay for more work, less pay for less work" requires further improvement in modern diversified enterprise management. Therefore, enterprises are likely to provide employees with more complete external conditions, more humane working environment and better credibility. Thus, employees can better satisfy enterprises' controlling and incentive management. Performance management applied in incentive mechanism assures employees to gradually achieve standardized and rationalized business in their daily work. Moreover, it is beneficial to gain employees' commitment and credit, enhancing their loyalty to the company. Thus employees are willing to start or continue to realize their value there. To achieve the function of performance feedback in enterprises, HR firstly needs to develop clear goals of management and ensure correct effects of other incentive mechanisms based on the main purpose of increasing yield and efficiency. Secondly, both regular and unscheduled performance feedback must be linked with employees' benefits, promotion and opportunities. Therefore, enough power for the operation of incentive mechanism and active cooperation from employees can be guaranteed for enterprises.

In addition, HR can adopt some advanced performance management modes that already have been proven to be practical through tests such as "morality-ability-diligence mode", "inspection-evaluation mode" and "joint-participation mode". All these modes well fit the basic purpose of incentive mechanism, realizing effective human resource management and scientific plan of the whole enterprise. Employees under the management of performance feedback can have two-way communication with enterprises. As a result, incentive mechanism can be further perfected in practice, and enterprises can achieve a better transparency of its operation and stronger cohesion among employees. Therefore, enterprises can acquire good system guarantee to effectively enhance its soft power.

4 CONCLUSIONS

Every employee in enterprises has certain personality and some common characteristics. Therefore, enterprises should make efforts to achieve a humanized management based on the understanding of these factors, greatly helping to establish a good mass base and organizational credit. This kind of "invisible" means enables enterprises to perfect their management mode and improve their operation efficiency. Human resource management is actually more about the elaboration of managing people. Fully understanding of employees' abilities, psychologies, emotions, future development and incorporating these factors into the allocation of resources helps to achieve the best role of incentive mechanism. Besides, HR should optimize the competition mode within departments or enterprises based on humanistic concern. Moreover, employees' performance management and feedback should also be emphasized. Consequently, management mode is equipped with the guidance of correct ideas and scientific approaches as the guarantee of success. Therefore, incentive mechanism in human resource manage will become the driving force for the continuous development of enterprises.

ACKNOWLEDGEMENT

This work was funded by a university scientific research project: Local Higher Vocational Colleges Building Incentive Mechanism Research in University-Enterprise Cooperation, Grant Number: 12C0988.

REFERENCES

[1] Guo Hongxia. Establishing and Applying Incentive Mechanism to Achieve Safe and Harmonious Development for Enterprises. Conference Proceedings of the 7th China Coal Economy Management BBS of Economic Management Professional Committee of China Coal Society. 2011: 187–191.

[2] Liu Linyu, Ran yanbo. Deficiency and Improvement of Incentive Mechanism in Human Resource Management in Chinese Public Sectors. Technology & Market. 2010(9): 89–100.

[3] Xiao Feng. Mechanism Innovation: System Guarantee of Talent Power. Chinese Talents. 2010(12): 23–34.

Future Information Engineering and Manufacturing Science – Zheng (Ed)
© 2015 Taylor & Francis Group, London, ISBN 978-1-138-02644-5

Regional studies on effects of the Ban for Free Plastic Bags

Ying Ji

Oxbridge College, Kunming University of Science and Technology, Kunming, Yunnan, China

ABSTRACT: In this work, the effects of the Ban for Free Plastic Bags in different social regions were analyzed to provide thoughts and analysis for its implementation. Analyses on the nature of the ban—environmental behavior and its implementation in farmers' markets and supermarkets—show the following results. The enforcement of the ban is good in supermarkets but inferior in farmers' markets. Because the present Ban for Free Plastic Bags is inadequate without any comprehensive alternatives. The innovation here is to conduct detailed analyses from different regions. The work also carried out multi-level analyses on phenomena based on the causes from subjective and objective bodies.

Keywords: the Ban for Free Plastic Bags; policy; region; plastic bag; market

1 OVERVIEW OF IMPLEMENTATION OF THE BAN FOR FREE PLASTIC BAGS

In 2008, the central government of China issued a notice about restricting the use of plastic bags in production and sales, which is called the Ban for Free Plastic Bags in daily life. Generally, this ban is to limit using common types of plastic bags—they're difficult to be degraded in a relatively natural environment. Massive landfill of plastic bags can cause soil compaction. And incineration may result in the release of chemical substances toxic to humans—benzene chemicals and dioxin. In recycling, cleaning plastic bags may lead to water pollution. What's more, producing the raw material of plastic bags, plastic, has a large consumption: threefold amount of oil will be consumed for every 1 ton of plastic. And oil resources are very precious [1].

Therefore, the Sate Council of China promulgated the Ban for Free Plastic Bags: the production, sale and use of plastic bags have been forbidden since June 2008. And all the markets such as stores, supermarkets and bazaars should obey the pay-to-use system of plastic bags—they can't provide free plastic bags. In addition, different types of plastic bags should be charged in all markets. And the sales can be conducted only if the price of a plastic bag is added up to the total price of a commodity. In fact, the central government has emphasized the prohibition of using plastic bags through repeated supervision, creating an atmosphere for forbidding such use.

Through the control of plastic bags' price, the Ban for Free Plastic Bags attempts to change the habits of the citizens and reduce their reliance on ultra-thin plastic bags. After the ban was promulgated, thick plastic bags, due to their degradablity, have become the substitutes for ultra-thin ones. Thus the Chinese government is trying to reduce the negative impact of white pollution. Ultimately, the fullest implementation of environmental protection in society can be realized via economic regulation [2].

2 RESEARCH BASIS OF EFFECTS OF THE BAN FOR FREE PLASTIC BAGS: BASIC RESEARCH ON ENVIRONMENTAL BEHAVIOR

In fact, the Ban for Free Plastic Bags is an environmental behavior. In a broader sense, environmental behaviors refer to specific behaviors influencing the quality of ecological environment. They can bring positive and negative impacts. In general, environmental behaviors tend to be positive according to the current literature. They have the subjective initiative while environment pollution usually doesn't. From this perspective, it's necessary to analyze the influence caused by the ban itself in order to truly understand the effect of its implementation. The analysis was launched based on the perspective of awareness and influence [3]. The ban's impacts on environment can be material influences coming from specific behaviors or effects on human motivation and awareness

In foreign countries, studies on impacts of environmental behaviors are relatively mature. They mainly investigate the effects generated by a specific behavior in the actual environment. Such research includes the recovery of related resources, energy consumption, education for environmental protection and formulation of environmental policy. However, the domestic researches of environmental behaviors are incomplete: some focus on environmental education, and some study public awareness of environmental protection. Thus the effects of the Ban for Free Plastic Bags should be studied in specifically divided regions rather than several fields of consciousness. A comprehensive study on the protection of environmental behaviors should start from means such as classification and organization.

3 EFFECTS OF THE BAN FOR FREE PLASTIC BAGS IN FARMERS' MARKETS

Ultra-thin plastic bags remain the mainstream in farmers' market of various provinces in China. The Ban for Free Plastic Bags has a modest influence on farmers' markets—ultra-thin plastic bags are still free there. The usage of ultra-thin plastic bags has decreased in farmers' markets when the ban was just issued. However, there is an increasing use of such bags in farmers' markets after half a year. Environment-friendly bags have gradually lost the general use value there.

3.1 Overview of effects of the Ban for Free Plastic Bags in farmers' markets

In current farmers' markets, most customers usually use a variety of plastic bags instead of nonwoven environment-friendly bags or shopping baskets. A few people use nonwoven bags to hold foods packed in ultra-thin plastic bags. This is, of course, not a direct use of environment-friendly bags. As a result, nonwoven shopping bags fail to promote the environment-friendly behaviors of more consumers.

Many traders in farmers' markets, however, are aware of the Ban for Free Plastic Bags. Interviews indicate that they took a wait-and-see attitude when the ban firstly appeared—that's why there's less use of plastic bags back then. A few traders nonetheless didn't stop using ultra-thin plastic bags based on the ban. Plus, the majority of customers have requested the use of plastic bags to carry food and exhibited strong resistance and resentment for paying the fees of plastic bags. Thus many traders have to adapt to market demand. Many customers in farmers' markets have pointed out that there is no distinguishing placement or physical isolation for food from various markets. Therefore, the primitive bacteria carried by one food, especially raw ingredients, can be transmitted to other foods. As an environmental behavior, the Ban for Free Plastic Bags

itself doesn't propose a more appropriate solution. In fact, the purpose of using plastic bags is to protect food hygiene. Although manufacturing plastic bags has a high expense, the cost is low after all. Normally, farmers' markets won't ask customers to pay for the fees of ultra-thin plastic bags. All in all, it's apparent that the Ban for Free Plastic Bags is just empty words on a page in farmers' markets.

3.2 Reason 1 for rebound of the Ban for Free Plastic Bags in farmers' markets: Imperfect policy

The Ban for Free Plastic Bags restricts the use of ultra-thin plastic bags in all the retail markets such as shopping malls, supermarkets and bazaars. Such restriction only limits the usage of plastic bags while other bags such as shutter-style ones are still available. What's more, farmers' markets are retail markets selling fresh food. Limited by the ban, fresh food such as tofu and diced retail meat can't be held in nonwoven environment-friendly bags. Consequently, farmers' markets witnessed a rebound of the ban since their nature is inapplicable to the ban. This is also the biggest difference between farmers' markets and supermarkets since supermarkets possess the conditions to implement the environment-friendly ban.

At present, nonwoven and paper bags are the major alternatives for plastic ones. However, paper bags have relatively high production cost, no antiseepage performance and poor heat insulation performance. In addition, the manufacture of paper bags requires a lot of wood, so such bags aren't very environmentally friendly. Its extensive use can even cause damages no less than ultra-thin plastic bags to environment. Moreover, nonwoven bags are expensive and can be easily polluted when used for holding foods. Thus it requires cleaning, consuming more water and time.

This shows that partial consumers purchased nonwoven bags when the ban first came out. Afterwards, they found out that using nonwoven bags brought about much inconvenience. Moreover, squeezing a variety of raw food materials into a nonwoven bag may cause cross-infection. People have to carry a number of different bags in order to avoid such situation, which is more inconvenient.

3.3 Reason 2 for rebound of the Ban for Free Plastic Bags in farmers' markets: Unclear responsibilities caused by multiple management

Currently, there's relatively loose and irregular management in many suburban farmers' markets, limiting the execution of the Ban for Free Plastic Bags. The managers of farmers' markets are reluctant to initiatively restrict traders' use of ultra-thin plastic bags. In their opinion, customers' free or disguised free use seems to have no large conflict with their interests. The ban yet states that the person who provides plastic bags will have to take a fine up to 10,000 RMB. However, the ban, as mere guiding regulations, isn't a codified law. Thus either the executor of the 10,000 RMB fine or the main responsibility is unclear. As a result, business departments, the National Development and Reform Commission, the Industrial and Commercial Bureau, the Environmental Protection Bureau and the Quality Supervision Bureau can intervene in the situation, which can easily lead to a multiple management. The ownership of the main and the first responsibility are indistinct. In the end, farmers' markets are located at the intersectional vacuum of the entire environmental behavior. In addition, the convenience of plastic bags, attracting more customers, can always acquire consumers' supports. The market managers didn't adamantly suppress the behaviors violating the ban since they're unwilling to overly depress the natural market demand

4 EFFECTS OF THE BAN FOR FREE PLASTIC BAGS IN SUPERMARKETS

Supermarkets, imported from abroad, are a new business mode. In addition to large-scale foreign supermarket chains, many domestic supermarkets have rapidly developed in recent years. Since the Ban for Free Plastic Bags was issued by the central government, most supermarket operators have implemented the pay-to-use plastic bags due to the nature of supermarkets. They're more likely to be the objects of evidence collection and attention. With the good enforcement of the policy, many consumers go to supermarkets with their own plastic bags or environmental sacks.

When going shopping in supermarkets, consumers usually take relatively small goods by hand. And supermarkets offer non-woven shopping bags while one-third of consumers will bring their own such bags. Meanwhile, a few customers will use the shutter-style bags of supermarkets to hold commodities, which indicates that economic adjustment is effective in supermarkets.

The staff in supermarkets has presented two views. View one: the use of ultra-thin plastic bags has increased in supermarkets. More consumers stay loyal to ultra-thin plastic bags after the ban was promulgated. View two: people use less

ultra-thin plastic bags in supermarkets than the former phase. At present, young people are the major group using plastic bags while older people and housewives usually prefer nonwoven bags.

Two conflicting views do exist. Although consumers have used less plastic bags since the central government introduced the ban, they haven't truly accepted the psychological gap from free to pay-to-use bags. Thus they often bring their own nonwoven bags instead. Young people nevertheless don't care the pay-to-use plastic bags due to the inexpensive price or stable income. Finally, the use of plastic bags gradually becomes a polarized situation.

5 DIVERSE EFFECTS OF THE BAN FOR FREE PLASTIC BAGS IN DIFFERENT REGIONS

The purpose of the Ban for Free Plastic Bags is to guide people's consumption behavior to gradually become environment-friendly through economic means. Although such environmental behavior has influenced markets to some extent, it hasn't flourished everywhere. Specifically, it's universal to use plastic bags in retail markets except supermarkets. And providing free plastic bags is even a rigid demand. Why can't the ban have a broader influence on society? Substantially, the reason is that the behavior can't achieve low-cost expenses. Many consumers have to pay significant costs for the environmental behavior or subconsciously think so. Thus they won't obey the requirement of the ban. What's more, the ban hasn't excellently solved the problems that some food isn't suitable for other package options. There isn't a perfect alternative for diced meat or other retail foods. Therefore, it's easy to have regional differences during the execution of the ban. On this aspect, plastic limitation can continue to be effective in supermarkets. On the other hand, alternatives should be considered for other types of markets in order to satisfy the consumption demands of different groups. Meanwhile, the efforts of environmental education and promotion are intensified to ultimately achieve the environmental purpose of the ban.

ACKNOWLEDGEMENT

The work was funded by the subject of Yunnan Education, Implementation and Restudy of "Plastic Limit"—A Case Study of Kunming, with Grant No.: 2011C010.

REFERENCES

[1] Zhou Juan, Zhou Zhijia. Study on Different Cases in Effects of the Ban for Free Plastic Bags. *Environment and Sustainable Development.* 2009(03) 4–7.

[2] Zhang Wenlei. Effects of Charging System for Plastic Bags in China. *Price: Theory & Practice.* 2009(04) 41–42.

[3] Dou Wei, Wang Yong. Analysis on the Ban for Free Plastic Bags and Effective Countermeasures. *Science and Management.* 2009(02) 43–46.

Future Information Engineering and Manufacturing Science – Zheng (Ed)
© 2015 Taylor & Francis Group, London, ISBN 978-1-138-02644-5

Expression of tragedy color in "Childhood sweetheart" by Higuchi Ichiyō

Fang Han
Ningxia University, Yinchuan, Ningxia, China

ABSTRACT: *"Childhood sweetheart"* is one of the main representative books written by Japanese female writer, Higuchi Ichiyō. Heavy realism color was represented in her works. Under the background of confliction between feudal society and western capitalism philosophy, the tragedy color of the whole story was created by the obscure feeling between the hero and heroine, Mei Lideng and Tengben Xinru. The main creation purpose is to show the misery destiny and humble social position of a lady who is under the lowest position. In this paper, the whole tragedy color expression is analyzed in the way that enhances the helpless of Tengben Xinru and humble position of Mei Dengli through the existence of A Zheng.

Keywords: Higuchi Ichiyō; *"Childhood sweetheart"*; tragedy color; emotional factor

1 INTRODUCTION

"Childhood sweetheart" is one of the main representative works written by Japanese female writer, Higuchi Ichiyō. The story was created in the period of Meiji restoration. Although new capitalism philosophy had already been delivered into Japan, the deep feudal thought of Japan conflicted obviously with new thought. In this situation, people who live in the lowest level had more bitter life, especially the female live in a very humble life. The writer had very deep experiences, and this is the reflection of leading role's life. Higuchi Ichiyō sought for help from one mountebank because of lack of money. And the mountebank requested to be her lover instead, promising to provide her all the found in return. This request was finally refused. This is the reason why writer has deep recognition of real life and expresses her feelings through works. There are many detailed romantic descriptions. The tragedy color of whole story was presented in circumstance, showing that love story was nothing and fragile when facing money and social position [1]. The traditional moral concept influences people in every social level, which creates background color of leading role's tragedy life.

2 EXISTENCE OF A ZHENG AGGRAVATES TRAGEDY COLOR OF MEI LIDENG'S AFFECTION

2.1 *Mei Lideng's feeling to Tengben Xinru—a mistake*

Mei Lideng, the heroine in *"Childhood sweetheart"*, a pretty who lives in a whorehouse has a fate to be a whore after growing up, while Tengben

Xinru has a fate to be a monk studying the Buddha after growing up. The leading role's love emotion is a mistake by the arrangement of fate and social difference. When Mei Lideng reaches developmental age, it means that the fate of becoming a whore is getting closer, while the Tengben Xinru's fate of being a monk is coming gradually [2]. The two of them fall in love with each other, and their emotional concept and motivation are very pure, however, the love is helpless because of their different social position and fate. The introduction of their social position and fate arrangement becomes the foundation of the tragedy of the whole story. The fate is the most helpless issue in *"Childhood sweetheart"*, and this is also the writer's reflection and projection about her helpless to the real society.

2.2 *Tragedy of Mei Lideng highlighted by a Zheng*

Mei Lideng is the leading role of the writer's work. Character description of Tengben Xinru and A Zheng serves as a foil to heroine. The character setting of A Zheng is a boy who has a cross on Mei Lideng. He loves her very much, staying surround her and always having endless talking topic with her. He likes to stay with her and decides to get married with her when he grows up [3].

Mei Lideng treats A Zheng friendly in the daily contact. Anyway, she knows that she doesn't like him, and her feeling to him is just friend although she always acts passionately to him. There was one scene of description in writer's work: A Zheng always says that he will get married with a pretty girl after growing up, so the Madame continuously asks him with whom he will get married, is she

A liu in the flower shop, Xi in the fruit shop or pretty Mei Lideng? Madame excludes one by one, finding that A Zheng moves backward to the corner with his flushed face when Madame mentioned Mei Lideng, while heroine's face stays normal and seems uninterested without flushed face.

"The people around see one lonely shadow went over the riverbank when darkness has fallen every day, catching a little arc-type lantern. They feel so chilly looking at his shadow. And what they saw is that A Zheng is collecting interest instead of his grandpa", described in the last chapter of "*Childhood sweetheart*". Mei Lideng's emotional attitude to A Zheng is expressed from the side position by describing the different thoughts of two of them when treating the same thing, which causes contrast when Mei Lideng see A Zheng in the later chapter to enhance the brutality of emotion. The tragedy was perfectly expressed by the existence of A Zheng and the emotional life of heroine.

3 HELPLESSNESS OF TENGBEN XINRU SERVERS AS A FOIL TO THE TRAGEDY OF MEI LIDENG

3.1 *Tengben Xinru restrained by the feudal ideological morality*

Tengben Xinru's father is the temple host, and he is ordained by fate after he grows up. Although he likes Mei Lideng, but he can't react against finalized fate for love by deviating from the traditional ideological education. "One rainy day, the hero broke his shoes while he walked by the whore house to deliver new cloth to his sister. The umbrella was blown away by wind while he bent down and fixed his shoes, with his cloth contaminated by rain. All his embarrassment was seen by heroine. And Mei Li intended to go for help, but she hid behind the door instead, watching him quietly with fast heartbeat. Her confidence was gone in front of him".

Mei Lideng's footstep can be clearly told by Tengben Xinru, anyway, he pretended to ignore anything today which makes Mei Lideng worry. When the heroine's mother shouted at her to let her come back home for shower, Mei Lideng responded loudly, trying to attract his notice while her face got flushed. She threw one silk cloth strip to him from the door, while Teng watched the strip stolidly and left without turning back. From that time on, Mei Lideng felt annoy. The strip was finally left behind. Tengben Xinru might has the same feeling about her, anyway, thinking about that he will leave from here to learn Buddha far way and never try to fight against the traditional moral spirit, he finally left the strip on the mud with his helplessness and Mei Lideng's sadness.

3.2 *Unwillingness for Tengben Xinru to give up*

Facing the reality and tradition, feeling is not the priority option. The fate has its arrangement, and that is why Tengben Xinru has to change his choice. There is no doubt that he will leave far way to study and let his feeling to Mei Lideng not happen. Tengben Xinru used his unique lonely method to say goodbye to her, and let him study with his empty mind, showing his feeling to her. There is one chapter in *"Childhood sweetheart"* described as: one cold morning, one daffodils was found near the door, with no idea of knowing who dropped it, Mei Lideng still places this flower into her flower vase, watching its loneliness and pride. It seems that her confidence was dropped by Tengben Xinru to the mud. She knows the daffodils language and its gesture, and she only wants to look at it quietly. She got a message that Tengben Xinru left her two days later after she picked the daffodils, wearing his Buddha cloth to learn Buddha. By the arrangement of the writer, the readers like to treat the flower as Tengben Xinru's goodbye gift to Mei Lideng or Tengben Xinru's feeling expression.

Mei Lideng's emotion has a refuge when she looks at the daffodils. She starts to believe that Tengben Xinru likes her although she never hears Tengben Xinru's expression. In the beginning, Tengben Xinru always ignores her, then he sends daffodils to her by stealth, which means giant progress in him and comforts Mei Lideng deeply. Self-comfort and perfect illusion of Mei Lideng finally didn't fight against the brutal reality that Tengben Xinru has already left. The description of the leading roles in *"Childhood sweetheart"* mostly focuses on the obscure conjecture. The hero who has expressed his feeling isn't accepted, while the heroine who hasn't expressed sends her massage. It is unknown that whether Mei Lideng has her own illusion love or that Tengben Xinru feels helpless about his feeling. Mei Lideng can't change any situation, only looking at the daffodils in the flower vase when she misses him. Their emotion has gone regardless of regret or confusion. Life has to move on, and no one want to stop to protect such visionary emotion, even looking at the daffodils is kind of luxury.

The helplessness of Tengben Xinru makes life of Mei Lideng more miserable. Choosing reality without hesitation in front of emotion and reality is not that only Tengben Xinru can do, but for the people who live in the deep down level and have very simple thought. They can only try to live in this world and to search for the additions. From the writer's emotional position, she expresses her sympathy to people who live deep down by her works.

4 HUMBLE SOCIAL POSITION OF MEI LIDENG

4.1 *Mei Lideng' destiny and fate arrangement*

Mei Lideng lived in the whorehouse when she was little, she was pretty, and it is destiny for her to be a whore, which is the same as the fact. Her sister was a very famous whore in an urban area of Tokyo. People also like Mei Lideng because of her sister, when she was 15. The destiny made Mei Lideng so much bitter and helplessness in both life and emotion. Living in the whorehouse, it is sure to be a whore when growing up from the aspect of living. €The social position is humble as a female, so she never tries to fight against it or try to change her fate. There is no equal right for a female who rely her life on male.

From the aspect of emotion, there is only loneliness and helplessness left for Mei Lideng when Tengben Xinru decided to leave, although she has feeling to him. Tengben Xinru can't continue his love together with Mei Lideng by the affection of traditional spirit and his negative activity of emotion. All she lost in the story is the result of society and feudal tradition. The fate of female is to become the addition of male in the male right society, despite that female may have different experiences. There is no reason of unfairness only because of the gender.

The end of Mei Lideng in the story is unclear with few description, the open-type end inspires people to guess the final fate of hers, trying to imagine their future life. Every misery experience is not described on purpose, implying the misery and helplessness of the whole society and the whole time. It is fate for the lost and damage of heroine in the story. All the unfair treatments of female in society seem not fair enough, however, every person has such experience, which shows that there is no such fairness at all. Life is ought to be like that. This makes people think that there is no need to be improved and to fight against. As a famous female representative, the writer understands the unfairness, and tries to encourage female who suffers, positioning herself as saver who was born to comfort bitter and disappointment of female. Many of her works are very good reflection of people's life in that period, possessing with literature value and precious historical value, contributing on the developing history of female social position in Japan.

5 CONCLUSIONS

The end of *"Childhood sweetheart"* is open style, impressing people with such tragedy color, and bringing lots of imaging space with the mood of sad. The whole article describes the helplessness of life with light sadness and attitude of no fighting. It is also the reflection of writer's bitter and her helplessness to the reality. The tragedy color of the whole story was analyzed through the existence of A Zheng, helplessness of Tengben Xinru and fate arrangement of Mei Lideng. The feelings of helplessness, bitter and sadness are well expressed by the straight or side description to the leading roles and other roles. The focusing point is the regret and sadness, which impress the readers greatly. The writer is good at misery description to her roles, the end of most her works are brutality and reality integrated with actual background, making people to rethink and recall.

REFERENCES

[1] Zhou Xiaoyu. *Separation of Fate-analysis on "Childhood sweetheart" by Higuchi Ichiyō*. Entrepreneur World. 2013(8):175–175.

[2] Li Hui. *Yamato of Lyric Poetry in teenage-literature features of "Childhood sweetheart" by Higuchi Ichiyō*. Journal of Language and Literature Studies, High Education Edition. 2012(6):21–22.

[3] Huang Ziyin. *Lost is Beauty—Feelings of "Childhood sweetheart" by Higuchi Ichiyō*. Journal of Language and Literature Studies, foundation education edition. 2011(8): 60–61.

Future Information Engineering and Manufacturing Science – Zheng (Ed)
© 2015 Taylor & Francis Group, London, ISBN 978-1-138-02644-5

Professional practice ability of young foreign language teachers in universities for nationalities

Lihua Yang

International Exchange College of Inner Mongolia University for the Nationalities, Tongliao, China

ABSTRACT: Professional practice ability of young foreign language teachers in universities for nationalities serves as one of the keys to make breakthroughs in foreign language teaching in universities for nationalities. This paper analyzed some defects in professional practice ability of young foreign language teachers in universities for nationalities through investigation. Additionally, this work offered several measures for increasing the ability. The shining point of this work lay in the foundation of particular purpose, content, and data for analysis of increasing professional practice ability of young foreign language teachers.

Keywords: Teachers in universities; foreign language teaching; practice ability

1 PURPOSE OF INVESTIGATION

At present, professional practice ability of foreign language teachers in universities for nationalities exposes inadequacy and receives different requirements from the universities for nationalities. Proceeding from current state, the investigation was conducted to produce reliable data. Based on the data, suggestions and measures for reform are specifically put forward in line with requirements of universities. Investigation mainly involves in teachers' profiles, including knowledge of foreign language, teaching concept, research capacity in teaching performance, and capacity for self-improvement and self-discovery [1]. Teaching evaluation for various teaching levels and teaching outcomes are also studied to get knowledge of current professional practice ability.

2 CONTENT OF INVESTIGATION

With development of educational reform, economy and the whole society in China, more requirements are attached on foreign language talents, making foreign language teaching more crucial in the entire education system. Universities owing strong national features and some teaching mode with national characteristics are much more likely to neglect foreign language teaching. As a result, professional practice ability of young foreign language teachers in universities for nationalities arouses overwhelming concerns fearing that a lack of professional practice ability will directly affect

the teaching effect. In order to further understand the current state of profession practice ability of young foreign language teachers in universities for nationalities, the investigation studied teachers' education, age, job title, knowledge, and practical performance as well as research achievements [2]. Besides, talk with the teachers conducted in the investigation is also conducive to the awareness of facts.

3 DATA IMPLICATION

Current configuration of young foreign language teachers in universities for nationalities shows that young teachers have grown into the pillar of foreign language teaching. Though proportion of young teachers is diversified in different universities, young foreign language teachers stably enlarge its percentage on the whole, resulting in the main power in the development of foreign language teaching. Data indicate that young foreign language teachers in universities for nationalities accounts for over 80% of the entire teachers, among which PhD is about 10% and master about 30% with the rest highly educated [3]. Therefore, it could be concluded that young teachers account for a large proportion playing a significant role in overall foreign language teaching in universities for nationalities. However, in terms of professional practice ability, some young teachers in universities for nationalities expose some defects, for example, about 30% lacks his/her own perception of professional practice ability. Some even neglect self-improvement while others who wrongly define and partially understand professional practice ability mix it up with issued paper and academic achievement. As a consequence, those teachers emphasize too much on professional practice ability.

4 INFLUENCING FACTORS OF PROFESSIONAL PRACTICE ABILITY OF YOUNG FOREIGN LANGUAGE TEACHERS IN UNIVERSITIES FOR NATIONALITIES

In universities for nationalities, foreign language course, naturally significant to foreign language major, is so important that other disciplines take it as compulsory course. Therefore, foreign language teaching plays a significant role in overall teaching in universities for nationalities. However, overall teaching shows some defects and disadvantages of professional practice ability of young foreign language teachers in universities for nationalities in China, which may largely limit cultivation of foreign language talents in the universities for nationalities.

4.1 *Restrictions from teaching concept*

Teaching concept, an important factor directly restricts foreign language teaching in universities for nationalities, has direct impact on foreign language teaching in private universities. In the eyes of universities for nationalities in China, there are two misunderstandings in foreign language teaching. Firstly, some teachers are still trapped in traditional teaching concept, especially the young teachers who take previous cases of outstanding teachers as models. Those teachers regard the outstanding teachers in their school time as examples and teach according to their learning experience. This teaching concept is so traditional that students may be adversely affected while young foreign language teachers are likely to over emphasize their own senses and judgments. Secondly, teaching mode with national characteristics creates some restrictions for teachers to link their teachings with students. Against this backdrop, no more improvement will students and young foreign language teachers get.

4.2 *Restrictions from teachers' professional quality*

In terms of professional practice ability of young foreign language teachers in universities for nationalities, their limited expertise directly affects their performance. In those universities, improvement of young foreign language teachers' professional practice ability is affected by teachers' limited foreign language level. Moreover, owing to a few class-hours, foreign language teaching cannot be reasonably conducted in non-foreign-language majors. In addition, in terms of the reins of teaching, young foreign language teachers in universities for nationalities are supposed to enhance the control over classes. If they are so dependent on original teaching plan that initiative of teaching is neglected when preparing and launching teaching, they get no chances to show their expertise. Besides, it is fundamentally not helpful with demonstration of their professional practice ability.

4.3 *Demands for more teaching passion*

Many young foreign language teachers emphasize too much on teaching outcome, resulting in much attention on research but neglect of teaching performance. Though this phenomenon is probably caused by teaching evaluation for foreign language teachers in universities for nationalities, their teaching attitude directly affects their professional practice ability. In some universities for nationalities, foreign language teachers hold a totally wrong opinion that articles equals their practice ability. As a result, professional practice ability is not likely to be perfected unless this perception and attitude are removed.

5 PRESSURE FACED BY FOREIGN LANGUAGE TEACHERS IN UNIVERSITIES FOR NATIONALITIES

As the education reform develops, new teaching objectives and tasks are proposed while more requirements are attached on teachers. In private universities for nationalities, multiple challenges—stimuli of teaching ability to a large extent—are faced by foreign language teachers while they expand professional quality. However, for a part of young teachers, these challenges mean kinds of pressure. Therefore, it is necessary and important to change the pressure into motivation in order to promote foreign language teachers' professional practice ability in private universities for nationalities.

5.1 *Teaching tasks as pressure*

Teaching task is the first pressure faced by teachers. According to current the number of foreign language teachers in private universities, some universities are still lacking teachers and teacher resources. In universities for nationalities, number of foreign language teachers is not proportional to their teaching tasks. In addition, teachers are overloaded with teaching tasks due to diversified national characteristics and foci of different majors in various universities for nationalities. Owing to enrollment expansion in universities for nationalities in China, foreign language teachers have to handle increasing teaching tasks, which essentially raise crucial challenges for them. In conclusion, for foreign language teachers in universities for nationalities, the overloaded teaching pressure cannot help improve their professional practice ability.

5.2 *Teaching objectives as pressure*

As with education reform, teaching objectives of school education also have changed. Changed objectives invisibly form some challenges for young teachers especially for the freshmen who need time to adapt to the change. Therefore, changed teaching objectives means a kind of pressure for foreign language teachers in universities for nationalities. Education reform also demands of higher education to input professionals and personalized talents with comprehensive skills into the society. In this context, teaching objectives are characterized by quality teaching. However, quality teaching invisibly forms other challenges and pressure for foreign language teachers who are accustomed to exam-oriented education. All changes in teaching objectives require teachers' timely adaptation for the reason that lack of knowledge, and improper adaptation and approaches will hinder improvement of practice ability.

5.3 *Teaching methods and evaluation as pressure*

Since every university has its own teaching environment, differences of teaching environment exist between universities for nationalities and other kind of universities. As with multimedia teaching and modernized foreign language teaching in most universities, foreign language teachers unexceptionally need to further follow this trend and reform teaching methods. Nevertheless, impropriety of reforming and mastering the teaching methods will certainly affects the teachers' teaching quality and demonstration of teaching ability. For foreign language teachers in universities for nationalities, another pressure is the imperfect teaching evaluation system. Therefore, more efforts are needed for breakthroughs both in theoretical knowledge and practice skills.

6 MEASURES FOR OPTIMIZATION OF PRACTICE ABILITY OF YOUNG FOREIGN LANGUAGE TEACHERS IN UNIVERSITIES FOR NATIONALITIES

As times and teaching environment constantly change, young foreign language teachers in universities for nationalities need to pay much more attention to the improvement of their professional practice ability to form a solid foundation for lifting the whole teaching level. The investigation concludes that inadequacy of professional practice ability of young foreign language teachers affected the teaching effect and blocked their self-development and progress. Proceeding from the situation, professional practice ability needs to be optimized, aiming to demonstrate the importance of the ability in multiple dimensions and improve it for young foreign language teachers in universities for nationalities.

6.1 *Training young foreign language teachers in universities for nationalities for good mental quality*

It is necessary for young foreign language teachers in universities for nationalities to be equipped with good mental quality. The reason is that young foreign language teachers in those universities are faced with both teaching pressure and relations with students. At present, teachers need to specially adjust their teaching mode according to mental and behavioral changes in students group and various characteristics and customs of students from different nationalities. In addition, young foreign language teachers sometimes are supposed to be mental health instructor and psychological guide to reduce students' mental problems and motivate them to learn. By training and developing good mental quality, young foreign language teachers in universities for nationalities will form a solid foundation to further improve their professional practice ability.

6.2 *Optimization of research ability*

No matter what disciplines teachers are engaged in, they are supposed to be equipped with research ability for it can reflect teachers' professional practice ability. As we all know, research activities provide teachers with platforms for combination of theory and practice. Therefore, for foreign language teachers in universities for nationalities, different students and various majors require them to conduct researches and study series of theories in order to meet different demands of teaching modes. However, young foreign language teachers' research ability is still inadequate, which does not mean they don't need it. Teachers' intention to keep pace with the development of society, universities and students also requires for good research ability. First of all, they need to have strong ability in researching teaching materials. Meanwhile, ability in researching teaching is another necessity for foreign language teachers in universities for nationalities. With the ability, teachers can select the best teaching methods and teaching modes to achieve good teaching effect with students by fully using modernized teaching platforms. In conclusion, it is smart to improve teachers' professional practice ability by promoting research ability.

6.3 *Reinforcement of expertise*

Relatively strong expertise serves as a basic quality for foreign language teachers in universities for nationalities. Only with strong foreign language expertise can teachers organize and perform teaching better than ever. Besides, teaching effect will be more apparent because of it. In a word, foreign language teachers ought to constantly keep self-improvement and reinforce their expertise on the basis of external help and self-expansion ability, by seizing every chance.

In conclusion, with development of education reform, young foreign language teachers in universities for nationalities are supposed to keep up with the trend. Learning the essence of the reform, they are about to innovate actively and grow in teaching performance, with an aim to motivate the teachers to improve professional practice ability.

ACKNOWLEDGEMENT

This work was funded by annual research project of education reform of State Ethnic Affairs Commission 2013—*Empirical Study on Cultivation Strategy of Professional Practice Ability of Young Foreign Language Teachers in Universities for Nationalities*, Grant Number: 13086.

REFERENCES

[1] Li Xiaohua, Current Situation and Strategies of Young College Teachers' Teaching Ability in Ethnic Minority Areas, *Journal of Inner Mongolia University for Nationalities (Social Science)*, 2013(4):102–104.
[2] Jiagn Shuping, Research and Developing Strategy on Current Situation of College Foreign Language Teachers Troop in Heibei Province, *Journal of Agricultural University of Heibei (Agriculture & Forestry Education)*, 2011(4):408–410.
[3] Zhang Yangbin, On Thinking of Improvement of Young College Teachers' Teaching Performance, *Hunan Social Sciences*, 2009(4):150–162.

Future Information Engineering and Manufacturing Science – Zheng (Ed)
© *2015 Taylor & Francis Group, London, ISBN 978-1-138-02644-5*

Theory and practice of violin basic-skill teaching

Gu Xu

College of Music and Dance, Xuchang University, Xuchang, Henan, China

ABSTRACT: Solid basic skills play vital importance in violin learning. However, students learning violin in conservatories in China are facing the problem of having poor basic skills of playing violin. Through analyzing of specific practicing methods based on the teaching theories of violin basic skills, the paper aimed to innovate practical training methods in order to improve students' basic skills of violin playing. After the practice of basic-skill training of violin teaching, the paper concluded that violin basic-skill training required students to study assiduously and perseveringly with scientific training methods. Therefore, only with students' perseverance and scientific methods they can improve their violin playing levels.

Keywords: violin; basic skills; pronunciation; teaching practice

1 INTRODUCTION

Violin creates beautiful music only with meticulous playing skills. In another words, only players equipped with skillful playing techniques can activate graceful music from violin. However, solid basic skills are the foundation of skillful techniques, leading to the palace of art without any shortcut. Therefore, the training of basic violin-playing skills must start from the enlightenment stage. If students didn't develop good habits during that stage, serious negative impact will happen to the future study of violin. Based on the analysis of violin basic-skill training methods, this paper introduced the teaching theory and experience—training the basic skills of playing violin—in order to improve the teaching level and innovate the traditional teaching ideas [1]. As the famous violin expert Galamian said, "nobody can learn violin just from books, and all the content about violin in books is just some basic principles; if students want to understand and flexibly grasp the specific content of those principles, they need lots of practices and exercises."

2 BASIC FEATURES OF VIOLIN BASIC-SKILL TRAINING

2.1 *Practicability*

Practicability is an important feature of violin basic-skill training. It requires to disassemble the entire movement into variety of small elements and then to master the skill through intensive training of these small elements. So students need to do some exercises about combined elements, and then repeat these basic skills as a warm-up exercise. Ultimately, these exercises should be developed into a habit in the future practice and learning, continuously improving students' playing skills. Besides, the practicability is also indicated in the daily training of violin. For example, a famous violin expert proposed: firstly, students could practice perfect fourth, perfect fifth and perfect eighth in the training of scale and pitch stage; then, practice of thirds and sevenths could be added in the training; finally, second and octave should be integrated into the practice. Such kind of training method has been proven to be timesaving and effective [2].

2.2 *Systematicness*

The training of violin's basic skills involves a complete system, obviously indicated in the violin teaching in Europe and the United States. In this system, the training is divided into many components including lots of details such as right and left hand practice, intonation and pronunciation practice. Left-hand practice includes the training of position shifting, vibrato, finger picking and other detail skills. The systematic training enables students to acquire the basic skills in a more comprehensive way and to play violin at their own will. At present, the relatively popular systematic-training mode includes the following sections of training: the first is the practice of standing posture and standardized violin holding method; the second is the basic-skill practice of right hand; the third is the basic-skill practice of left hand; the fourth is the practice of intonation; the fifth is the practice of scale, arpeggios and notasdoble; finally is the combined practice of all the above sections [3]. Thus, students can achieve skillful mastery through above comprehensive training of violin's basic skills.

2.3 *Persistence*

Basic skills cannot be achieved overnight. Instead, it is a long-term process of continuous study and repeated practice. The basic-skill practice of violin is included in the whole cycle of violin learning, and it cannot be interrupted during elementary school to university stage. However, the practice order and basic requirements of basic-skill training may change in different learning stages. Besides, the practice presents higher requirements to the changes and skills. Sassmannshaus teaching method, widely applied internationally at present, first divides the basic-skill training of violin into three stages: primary stage, intermediate stage and advanced stage. Meanwhile, it calls for students to study basic skills of different stages according to the differences of their learning abilities. For example, Sassmannshaus teaching method divides the basic-skill training of natural spiccato into primary stage, intermediate stage and advanced stage. In the primary stage, it requires students to flexibly master the method of pulling spiccato on empty strings. Following that, students need to practice on the basis of the combination of scales in intermediate stage. While in the advanced stage, students are required to practice music with spiccato.

The basic skills of some violin learners in China are relatively weak, so it is essential for them to keep learning. The practice of basic skills is a long-term sustainable process, and students should not give up while seeing no effects in short term. Therefore, students can make use of violin practicing-record method to stick to their practice in order to make the training sustainable. They can plan their practice schedule in a form and then fill in the actual practicing time after finishing the practice. Thus, students can be more focused on the practice, and it is also more likely to make this practice become a habit.

3 PRACTICE AND TRAINING METHODS OF VIOLIN BASIC-SKILL TEACHING

Basic-skill teaching of violin is a large-scale research topic, and mastering the specific training methods in practice is undoubtedly the most critical section. The following is a brief analysis of those important aspects.

3.1 *Practical training of scale and notasdoble*

3.1.1 *Scale practice*

Scale is undoubtedly the most important basic skill of violin. It plays several important roles in improving students' playing skills. Firstly, scale practice helps students to achieve a better intonation. Secondly, it enables students to be more familiar with the precise location of the fingerboard. Thirdly, it also practices the pronunciation. Finally, it helps students to master techniques of changing strength and bowings. In the basic-skill teaching of violin in China, the training of scale takes very important position, and many music schools include scale as an important part of exams. The combination of fast-way practice and slow-way practice is mainly adopted for scale training. Slow-way practice mainly trains students' abilities of intonation and position shifting, while fast-way practice mainly trains students' fingering techniques. In practice, the following methods can be applied for scale practice. First of all, students should accurately play the scales of perfect fourth, perfect fifth and perfect eighth, because those pure tones have a fixed standard—if students cannot pull the strings accurately, they cannot play it correctly. And this training belongs to basic-skill practice. Following that, students should integrate the seventh and the third tones into the practice, making the scales different. Finally, students should practice with the combination of the second and the sixth tones in order to understand and master the basic skills of scales in a short time. In addition, the practice of playing speed through continuously changing metronome speed is added in order to enhance students' proficiency.

3.1.2 *Notasdoble practice*

Notasdoble is very vital to train students' stability of holding the violin, and it is also beneficial to effectively improve students' ability of auditory identification. Notasdoble is one of the important basic skills of violin. Violin professor Zheng Shisheng believes that the earlier the notasdoble practice is, the better the students will achieve from it. Notasdoble is composed by third notasdoble, sixth notasdoble and eighth notasdoble. The training of third notasdoble is the most difficult. For this training, students can firstly use the thumb and middle finger to play the eighth scale on one string, and then use index finger and ring finger to finish this scale. After that, students can integrate these two kinds of playing methods, thus finishing the basic-skill training of third notasdoble. The training of sixth notasdoble is the practice of playing an eighth hexachord on one string. In this process, students can practice in turn with thumb and index finger at first, then with index and middle fingers, and finally with middle and ring fingers. Following all the above steps, students can finish the sixth notasdoble in a comprehensive way. As for the training of eighth notasdoble, students can firstly play octachord with octave on one string. Then, students should mainly play on the bass strings. At this moment, the training focuses on the strength of students' fingers on strings. So students can flexibly master the correct strength through continuous of such kind of practice, thus improving the flexibility of position shifting.

3.2 *Practical training methods of pronunciation*

Pronunciation is another important basic skill of violin playing, and it directly influences the timbre of violin. The pronunciation of violin mainly includes three aspects: pronunciation point, bow speed and bow pressure. According to the distance between the bridge and fingerboard of violin, there are five pronunciation points, namely the area close to the bridge, the area between the bridge and intermediate point, the intermediate-point area, the area between the intermediate point and fingerboard, the area close to the fingerboard. The shorter the distance from the high position is, the narrower the distance between pronunciation points is, thus creating different timbres. The key of pronunciation training is the practice of bow speed. The following methods can be used in the specific practice: students can practice with one to four different speeds in one bow in order to gradually grasp the technique of bow speed. The changes of violin timbre just simply manifest in the aspects bow speed, bow pressure and pronunciation point. Therefore, as for the basic-skill training of pronunciation, skillfully mastering the changes of the distances between fingerboard and bridge plus continuous practices of bow speed and bow pressure will surely ensure great improvement of the timbre of violin.

3.3 *Practicing methods of vibrato*

Vibrato is also an important basic skill of violin playing. In practical training, students can use an easy and practical method to practice vibrato, which is easy for students to master. Besides, this method will not result in the fatigue of hands muscles. The core of this method is to understand and grasp the breaking points of semiquaver, and then to play vibrato with reserved fingers. At the same time, other fingers also need to play vibrato. Although this is a small movement, it creates great difference if students do not play vibrato.

3.4 *Basic-skill training of left hand*

The basic-skill training of left hand mainly includes the following aspects. The first is the training of finger strength. If students use too much strength from fingers to pull strings, it is easy to cause muscle tension. Thus, the most appropriate strength is when the fingers can just press and hold down the strings, but students need continuous practices to master such strength. Meanwhile, the finger strength used on strings must be as small as possible, so that students can feel the elasticity of the strings. Besides, small strength on strings is also helpful for students to switch their fingers quickly, thus improving their performance skills. The second is the training of fingers to press strings. Fingers pressing strings are usually in a square shape and an expanded shape, both indicating the specific shape of fingers when pressing strings. These two shapes are used in the performance of semitone and whole tone, and it produces music with the movement of fingers rather than wrists. When playing semitone, only fingers need to move on strings but not wrists. The third is the training of reserved fingers. Reserved fingers are the key of left-hand basic skills. They are to reduce the redundant movements when playing violin, effectively improving the accuracy and flexibility of finger movements. Finally is the training of vibrato, belonging to the advanced part of left-hand basic skills. The vibrato can be produced through the inverse movements of lower fingers and higher fingers.

3.5 *Basic-skill training of right hand*

The basic-skill training of right hand mainly refers to the following aspects. The first is the basic-skill training of holding the bow. In practical training, students should continuously try to keep their fingers naturally falling down and slightly curved in order to hold the violin bow, thus to find the most relaxing way of holing the bow. Besides, it can guarantee the flexibility and slackness of the thumb. Meanwhile, the balance of pinky finger is also very important. According to the functions of fingers, a relatively fixed circle will be formed between the middle finger and thumb. The index finger will produce certain pressing strength, while the pinky finger will appropriately reduce the pressure for the finger at the end-side of the bow. In addition, students can adjust their index finger and ring finger to match the bow angle in order to complete the switch of different pronunciation points.

The second is the training of bowing skills. The key of bowing practice is to keep the bow straight. In the practice, students can try to make the bow and bridge form a parallel cross, thus producing the best resonance music. The key point in this process is to keep the straightness of the bow, and students can conduct segmented practice based on the bow as a separating point. The upper part and the lower part can be both divided into two sections, and students can compare each section to guarantee each bow section parallel with the bridge.

The third is the training of string shifting. The violin itself has seven planes: four monochords and three double strings. In the practice of string shifting, students need to pay attention to the stable level and to prevent the stress problem. At the same time, they must quickly switch the strings with wrists, thus to improve their proficiency of string shifting.

Finally is the training of playing slow bows. Slow-bow practice is mainly to improve players' control ability of violin bow. In the practice, students can play with semiquaver as a beat, and then play with 40 tones to 60 tones as a beat. Each bow has 16 beats. Meanwhile, the practice of appropriate strength should be added in the practice. Such training not only improves the timbre of the violin, but also improves player's control ability of the bow.

3.6 *Basic-skill training of intonation*

The basic skills of intonation are also very important. Students can practice with the speed of metronome 60. More specifically, students firstly should choose a musical phrase and then play a beat for each tone. After that, they stop for a beat and prepare to play the next tone. Through such training of each tone, students can feel the difference between fingers, effectively improve the intonation level. After continuous practice, students can gradually eliminate empty beat and increase the playing speed until they can play at a normal speed. Ultimately, the intonation will surely be greatly improved.

4 CONCLUSIONS

The level of violin playing is closely linked to solid basic skills, and students can only improve their basic skills of violin through constant practices with scientific and reasonable methods. The above six basic skills discussed in this paper is only the important parts among all other violin basic skills. Students need continuous practices in practical training in order to improve their basic skills of playing violin.

REFERENCES

[1] He Weidong. Musicality: Soul of Violin Playing. Northern Music. 2010(06): 102–103.
[2] Li Kaixiang, He Minnan. Basic-Skill Training of European and American Violin. Sichuan Drama. 2013(05): 87–90.
[3] Zhang Tianhong. Timbre Characteristics of Violin and Its Training Methods. Northern Music. 2009(03): 56–58.

Future Information Engineering and Manufacturing Science – Zheng (Ed)
© 2015 Taylor & Francis Group, London, ISBN 978-1-138-02644-5

Author index

Printed and bound by CPI Group (UK) Ltd, Croydon, CR0 4YY

18/10/2024

01776219-0007